Voice and Data
Communications Handbook

McGraw-Hill Computer Communications Series

In order to receive additional information on these or any other McGraw-Hill titles, in the United States please call 1-800-822-8158. In other countries, contact your local McGraw-Hill representative.

Voice and Data Communications Handbook

Bud Bates

Donald Gregory

McGraw-Hill, Inc.

New York San Francisco Washington, D.C. Auckland Bogotá
Caracas Lisbon London Madrid Mexico City Milan
Montreal New Delhi San Juan Singapore
Sydney Tokyo Toronto

McGraw-Hill

A Division of The **McGraw·Hill** Companies

Library of Congress Cataloging-in-Publication Data

Bates, Bud.
  Voice and data communications handbook / by Bud Bates, Donald
Gregory.
    p. cm.
  Includes index.
  ISBN 0-07-005147-X (h)
  1. Telecommunication—Handbooks, manuals, etc.   I. Gregory
Donald, (Donald W.) II. Title.
  TK5101.B317   1995
  384—dc20   95-22058
                                                                  CIP

1 2 3 4 5 6 7 8 9 BKP/BKP 9 0 0 9 8 7 6 5

ISBN 0-07-005147-X

The manuscript editor of this book was Anita Louise McCormick, the managing editor was Andrew Yoder, the supervising editor was Joanne Slike, the acquisitions editor was Jerry Papke, and the production supervisor was Katherine G. Brown. This book was set in ITC Century Light. It was composed in Blue Ridge Summit, Pa..

Printed and bound by Quebecor Book Press, Brattleboro, Vermont.

McGraw-Hill books are available at special quantity discounts to use as premiums and sales promotions, or for use in corporate training programs. For more information, please write to the Director of Special Sales, McGraw-Hill, 11 West 19th Street, New York, NY 10011. Or contact your local bookstore.

Product or brand names used in this book may be trade names or trademarks. Where we believe that there may be proprietary claims to such trade names or trademarks, the name has been used with an initial capital or it has been capitalized in the style used by the name claimant. Regardless of the capitalization used, all such names have been used in an editorial manner without any intent to convey endorsement of or other affiliation with the name claimant. Neither the author nor the publisher intends to express any judgment as to the validity or legal status of any such proprietary claims.

Contents

Introduction

Welcome to the world of telephony and telecommunications! You are about to embark on a descriptive and narrative overview of the telecommunications industry. This book is designed to help to clear the air for you. One of the major problems with this technology and industry is the use of jargon, or telephonese, which causes confusion and misunderstanding in the industry for users, purchasers, and vendors alike. Even professionals who have been in the industry for years have difficulty communicating in this business. The reason is simple: too many acronyms are used, many with multiple meanings. So, when talking to a voice telephony person, an acronym will mean one thing; to an engineer who has 20+ years experience, the same acronym will mean something completely different.

The rule, therefore, should be to disallow the use of telecommunications acronyms in any discussions you have with any vendor, carrier, or end user. If the propensity exists for them to use terms and acronyms, call a time-out. Have them explain all the alphabet soup they are spewing. You might be surprised to find that they can't explain the acronyms they are using. This will obviously cause you some concern, but fear not. They will ultimately get to the point. Further, because they won't be using all those buzz words, the communications flow should go smoother.

Now that the stage is set, let's get into a basic discussion of the telephony and telecommunications principles. No magic exists here, merely an understanding of what telecommunications is all about; the principles of a telephone; the line connections used; the forms of communications used; and an understanding of the telephone company networks. We intend to make it as simple as possible as we traverse the various techniques and terminology used throughout this book. Be aware, however, no matter how simple we attempt to make this information and how smoothly that we attempt to steer you through the roadguides outlined, this is a technical subject. Therefore, from time to time we might get a little "techie" blood in us. This is not to impress you, nor is it designed to confuse you; we just cannot think of a way to make it so rudimentary that it flows any better. At any rate, this book is designed to give you a fundamental understanding of the overall concepts used in the telecommunications arena, both voice and data.

The Format

The format of this book is to walk you right through the evolution of the industry as it progressed. Therefore, in Chapter 1, we attempt to discuss the evolution of the network as it pertains to the user. The invention of the basic telephone set was a monumental milestone because it marked the birth of the telephony world. From this invention, we know that a network of networks was designed. We approach the invention and the initial deployment of a network throughout the country based on the original Bell System. We look at the regulatory scene and the various legal issues that arose. As we all know, the monopoly was created to curtail competition with the universal access to a telephony network, but also as a wedge to recover the costs associated with the deployment of the capacities and services.

Chapters 2, 3, and 4 cover the basic characteristics of the voice evolution, from the telephone company and end-user perspective. We look at the basic characteristics of the human voice and how the characteristics formulated the way that the network was developed. This also implies that certain constraints were put in place to carry a plain old telephone call. Chapter 3 explains how a series of connections were laid out across the country, and ultimately across the world. This network evolution discusses the distinct dialing plans and how things have changed to accommodate phenomenal growth. Chapter 4 describes the way a voice call is handled through the basic telephone set. We also look at the way this set has changed over the years from a basic nondial telephone to the latest and greatest all-digital display phone with a built-in speakerphone. The set discussion also shows how the telephone set converts a sound wave into an electrical wave and prepares the electricity for transmission across the network.

Chapter 5 shows the changes that occurred in the network over the past 25 years, with the introduction of digital standards. The basic network was designed around an analog transport system, which served us well. However, as all things go, we outgrew the services of this analog capability. So it was only natural that the comparisons of capabilities and the benefits of digital systems would emerge. We hope that you will understand the differences between digital and analog after reading this chapter.

Chapters 6, 7, and 8 look at the players in the industry and the needs they will experience in attempting to serve a user. Chapter 6 deals with the different providers of service, the offerings they have and some of the financial considerations that make us want to use their services. Chapter 7 will clear the air in why things happen the way they do when we order a pair of wires, by comparing the differences and explaining what really is the difference between a line and a trunk. We further try to share some insight into why the old days of competition were so difficult, and why the newer competitive situation eliminates these roadblocks forever. Chapter 8 is a mathematical discussion of how we and the carriers need to constantly monitor the performance of our networks. Regardless of whose service is being considered, end users and carriers all must try to predict the volumes of traffic that will appear on the network at any single point in time. The ill-conceived past of trying to get too many calls on too few lines is gone. We now expect that at any time we should be able to access this network and get to any other end point in a moment, on demand. What happens when all does not go according to plan? What are the risks that we have too few or too many circuits? How do we compensate for traffic loads that are variable, and where can we get the data to conduct our analyses? All of these points should become crystal clear after this discussion.

Chapters 9, 10, and 11 compare the capabilities and features of the equipment that can be plugged into the telephone network through some form of connecting arrangement. Chapter 9 looks at the larger equipment market called the PBX, which is used by organizations around the world. This section will explain what the PBX really is and how it is geared to function. From there we also look at an alternative to ownership of the big iron, by exposing you to Centrex (a service offering by the players in the telephone companies), as a means of giving you PBX features and capabilities without the cost of building your own. Chapter 10 steps down to the lower end equipment and how it works for the smaller organization, the branch office of the larger organizations, or anyone else who has a need for multiple sets, but wants to compare costs and conveniences. Chapter 11 covers the add-on equipment to our telephone networking equipment. The voice processing, such as voice mail, automatic call distribution, and automatic attendants, all constitute ways of serving customers without labor-intensive human resources. However, we caution you of the risks associated with these techniques because many organizations have readily made a mess of implementing these aspects of communications. They have turned their telecommunications systems into their worst public relations representative to their customers.

Chapter 12 views a subject that will normally cause even the strong of heart to shudder: the use of an analog telephone line to transmit our mission-critical data. How can we make digital data look like a voice call? What are the tools to get the information in a useable and understandable form? This section gets quite lengthy and from time to time, a bit technical. However, you need to understand this concept. The world of today and tomorrow will demand the use of data transmission techniques. We must understand how to keep pace with the demands of the data world across a dial-up telephone voice network. If not, all things will grind to a halt. So, take this section slowly and a piece at a time; it will all come together at the end of this section.

Chapter 13 compares the use of a digital network instead of an analog network. The use of T1 and T3 services is escalating rapidly, bringing users rewards that they could not believe were possible. This technique gets us back into the all-digital transport system, and makes the data more reliable.

In an all-digital world, voice is data and so is digital data. We have no choice, so we must learn how it works and embrace this technology before we get left behind. We attempt to make this discussion simple, but again some parts are a bit complicated. Read this one in earnest; it is your future.

We had a lot of fun as we tried to think of the best way to present Chapter 14 to you. This section gets a little comical in the discussion of the standards that we use in just about all aspects of the industry. Designed around the data world, there are a series of services and protocols. These are all based on standards, or de facto standards. We compare the standards as they stack up against the granddaddy of them all, the OSI model. We think that you'll really enjoy the way that we present this very complex model, and how we can try to make all things simple if we take it one step at a time. Other operating standards exist in the data world, most predominant is the SNA world of IBM, the DNA world from DEC, and the TCP/IP world from the makers of the Internet. So, draw your own conclusions and see if this makes more sense after you get through this chapter.

Chapter 15 deals with the evolution of a packet network using an international standard known as X.25. We are not talking about some sci-fi robot or formula. This is a very well-documented standard for breaking large problems (our data) into smaller problems for reliable delivery across the network. It served us well and will be around for awhile, so learn how this can also be taken into account. Our comparison and dialogue in this section is sure to amuse you, but will also give you a very simple analogy on how X.25 works for you!

Chapters 16, 17, and 18 all deal with another way of moving data across a communications system, but on a localized basis. In the early 1980s a technique called a local area network emerged from the rubble of our data communications systems. It was designed to make the data communications between and among computers spread around our buildings user friendly and easily accessible. So we consider the differences between what a LAN is and is not. Then we compare in a chapter each the way an Ethernet and a Token Ring works. You will find this triad most informative, but possibly a bit tricky in the discussions. We discuss the limitations, strengths, distances, and speeds of each of the topologies used. Also, the way to get the best from your data dollars is covered in these sections.

Chapters 19, 20, and 21 look at taking the LAN out of the building and making it more of a public rather than a localized service. We look at the ways of creating a CAN, a MAN, or a WAN connectivity solution. Each of these will be ways of creating wider area coverage and higher speed connections. In the section on the frame relay we describe how X.25 has evolved into a faster wide area network (WAN) connection solution. In the SMDS discussion, we look at the metropolitan area network (MAN) solution to get the data across the network fast and efficient. In the fiber optical connection using a higher-speed token-passing ring discussion of FDDI, we see the campus area network emerging (CAN). These are all addressed in some detail, so they will require some added time to understand the concept and the reasons for using them. Allow some extra time when viewing these sections.

Chapter 22 takes a comparative look at the terms used on our data networks. The baseband versus broadband discussion should really clear this discussion up for you. Too often we hear the words bantered around, but no one understands just what they mean. We will attempt to show you how and why each of the choices of the types of cable, the capacities of the coaxial and the multiplexing schemes will deliver high-speed communications to the desktop on a single platform.

Chapters 23 and 24 deal with some of the latest and greatest "hot buttons" in the industry today. The first is the beginning of the emergence of ISDN, which does not mean "innovations subscribers don't need" but means that we are about to roll out several techniques to dial up a lot of voice and data on an all-digital network. The switched services of a digital network will make your current analog modems look like the turtle and ISDN will represent the hare. Only this time, the hare wins. Yes it was slow catching up, but it finally is ready for prime time. Services are appearing every day. But, we do explain what to watch for and what doesn't quite fit yet. The ATM discussion is an emerging fast packet system that makes X.25 and Frame Relay look weak and ATM is packet switching on steroids. The ATM discussion does not refer to the automated teller machines that are so prolific in the banking industry. But if you read this section and don't scream "I want it now," you missed something. But you'll see!

After all these discussions of the wired network solutions, we then take a different approach with Chapters 25, 26, and 27. This is a discussion of connectivity without wires. We look at cellular communications and personal communications for today and the future. The telephony and the data side of wireless does not equal that of the fiber world, but there is some movement here. Enjoy learning how the wired and wireless future will share the same tracks on the trail toward the turn of the century. We also look at wireless radio in the form of microwave (not the ovens) and satellite communications. These approaches are merely to compare the connectivity solutions against the old reliable twisted pairs of wire. Then of course, we added a section on the use of light beams without the fiber, called infrared transmission. This section details how close our communications needs can be met without the use of wires, and also shows the strengths and weaknesses of this transport system.

The last four chapters are all add-on items that we felt the average user may have some interest, but would not be foaming at the bit to get specific details on. The video conferencing section describes how this telecommunications technique has come a long way in terms of standards, availability and cost advantages. This is discussed in Chapter 28. Then with the costs of any communications method, we included a section of rudimentary cost justification techniques that will help to describe how to sell a system or network concept to management. This covers the basics of financial justifications in Chapter 29. A separate chapter on the use of facsimile or fax machines describes the evolution and the way things are going. This is a form of e-mail that most people have gotten very accustomed to. So we compare how the systems transmit and receive their information, as well as the evolution of the fax cards used in our PCs and notebook computers today all covered in Chapter 30. As the final chapter unfolds in Chapter 31, we discuss the capabilities of our cabling systems from copper twisted wires, coaxial wires, and fiber optics. The capacities and composition of these wiring systems is outlined to give the reader an understanding of what to look for and what to steer clear of.

We hope that we have intrigued you to read on. This is not a novel, it cannot be read from cover to cover. So allot some time each day and take a chapter or a triad of chapters together to gain an appreciation for the overall world of telecommunications. There is no reason why this book cannot give you the tools necessary to deal with the novice or pro alike. The acronyms and glossary included in the back of the book should help you with a quick reference to deal with the issues at hand. Take some time to familiarize yourself with the ideas of the book, use the examples and analogies. Enjoy the stories and heed their message. Let's have some fun!

The History of Telecommunications

In 1876, an inventor named Alexander Graham Bell was awarded a patent for one of the most significant devices in our lives, the telephone set. Mr. Bell had spent years trying to develop a method of communicating with his wife. Mrs. Bell was deaf, so Alexander was looking for a means of establishing a way to convert sound into some other form of communications, so that his wife could understand him. While she was in a hospital for the deaf in Boston, Massachusetts, he was busy at work trying to solve his basic problem of communications.

Having some experience with the way the telegraph worked, where coded messages could be sent across a cable, Alexander Bell decided to mimic this means of communications. Using the basic philosophy that communications could be converted from sound into electricity, he would be able to speak into a communications device, which in turn would convert the human sound wave into electrical energy. This electrical energy could then be used to produce a coded message similar to a telegraph message. This all sounded good, but led to long bouts with its development. Many frustrations were experienced by Bell and his assistant, Dr. Watson. No matter how hard they tried, the success of the transmission of voice into an electrical current seemed to allude these two inventors.

Then one day, fate struck. While working in the lab by himself (Dr. Watson was there, but had left for a moment) Dr. Bell spilled some acid on his worktable. This acid acted as a catalyst to produce a battery effect. Without realizing what had actually occurred at the time, Dr. Bell called out for Dr. Watson. His call of "Dr. Watson, come here I need you!" was activated across the experimental device they had set up for the communicator. His voice carried across a wire to a second room, where Dr. Watson was working. Hearing the call for help, Watson ran to the aid of his partner.

They discovered that if a battery is applied across the electrical circuit (the wires) while the user speaks, the sound wave produced by the human voice could be carried across this same pair of wires to a receiving end set up to accept this electrical current and convert the electricity back into sound. Hence, on that fateful day, the birth of a new industry occurred. The telephone set was invented!

Bell and Watson were elated. They were the inventors of what we have grown dependent on, the basic voice communications telephone set. This was in light of a similar activity taking place elsewhere.

Who Really Invented the Telephone?

There was a race from the outset between Dr. Bell and a German inventor named Dr. Elisha Gray, who had used a similar concept to Bell's, yet was too late in completing the necessary documentation to do anything about it. Some say that Gray had actually invented the telephone before Bell but was too late in filing at the U.S. patent office. Mr. Gray filed suit against Bell for stealing his idea, but lost.

This all led to the birth of the telephony and telecommunications industry, which has proliferated to the point that we can't live without it. We have made the world smaller, and attached devices to the telephony network, such as computers, fax machines, video conferencing equipment, and even interfaces to LANs. All of this has led to the approach of users that they can't live, or do their jobs, without the use of a telephone.

Since Bell originally invented the telephone many events have taken place, which we'll examine in closer detail to give you an appreciation of the industry in general. Incidentally, a theory exists that even Thomas A. Edison was the inventor of the telephone. With so many theories about the invention of such a remarkable piece of equipment, it's hard to believe that it proliferated as much as it did.

Evolution of the Telecommunications Industry

Historically, some of the important dates in the evolution of the telecommunications industry are shown in Table 1.1.

TABLE 1.1   Summary of Early Events in Telecommunications Industry

Year	Event
1872	Western Electric formed, Bell worked for the western electric company while developing the telephone.
1876	Bell granted a patent for his invention (March 7, 1876).
1877	Bell offered to sell his invention to Western Union Telegraph for $100,000. Western Union rejected the offer.
1877	The Bell Telephone Company was formed.
1877–1878	Western Union formed a subsidiary company called the American Speaking Telephone Company to market their own telephone set (speaker developed by Tom Edison, receiver developed by Elisha Gray) the first competitor to Bell System.
1878	New England Telephone Company was formed as one of the first competitors to the Bell Telephone Company.
1878	Bell Telephone Company sued the American Speaking Telephone Company for patent infringement.
1879	Bell Telephone Company and New England Telephone Company merged and formed a new name called National Bell Telephone Co.
1879	Bell and Western Union case settled. Western Union agreed to stay out of the telephone business, Bell agreed to stay out of the telegraph business in areas that Western Union operated.
1880	American Bell Telephone Company was new name for the Bell System.
1882	American Bell entered into an agreement with Western Electric. WECO to be the sole manufacturer of Bell equipment.
1884	First long distance line installed between Boston and New York.
1885	AT&T was formed as a subsidiary of American Bell to provide long distance and telephone service for communities around the country and the world.
1887	Interstate Commerce Commission formed to regulate interstate carriers.
1893	Bell's first patent expired, opening the door for competition without patent infringements.
1899	AT&T acquired the assets of American Bell.
1910	AT&T aggressively fights competition by acquiring controlling interest in Western Union.
1910	Mann-Elkins Act added to the Interstate Commerce Act to regulate the activities of telecommunications industry.
1913	Department of Justice considering antitrust against Bell System, a commitment made by Woodrow Wilson to break up private monopolies.
1913–1914	AT&T agrees to divest its Western Union stock and stop buying up independent telephone companies in a commitment by AT&T VP Kingsbury, in return for dropping antitrust action.

TABLE 1.1   Summary of Early Events in Telecommunications Industry (Continued)

Year	Event
1918	President Wilson placed the telephone and telegraph systems under control of the Post Office (until 1919).
1921	Graham-Willis Act established the telephone company as a natural monopoly.
1934	Federal Communications Commission created to regulate interstate, maritime and international communications. Congress also established "universal service" as its goal. The FCC investigated the Bell System operations and DOJ began to formulate a major antitrust action against Bell. However, WWII postponed action because it considered Bell critical to national defense.
1935	Public Utilities Commissions were formed to regulate intrastate communications and rate setting. Also, the PUCs were instrumental in formulating revenue sharing for calls using more than one carrier.

The First Telephone Companies Formed

Once the telephone was invented, and the proliferation started throughout the country, a group of telephone companies began to form independently to allow for interconnectivity. This technology started out with many saying that the telephone was a frivolous toy and would only be used by the affluent. How wrong they were!

The first competitive telephone company to form was the American Speaking Telephone Company. It was funded by Western Union Telegraph Company and used the pieces developed by both Gray and Edison. Edison had developed a transmitter that improved upon the performance of the set, so this transmitter was used. The receiver was an invention of Elisha Gray. The combined transmitter and receiver (Edison/Gray) set was an enhancement over the Bell invention. Edison was also used to assist in the distribution of the wiring to local business and residential users. Edison's knowledge of electricity and the use of power distribution systems in the form of grids were applied to the telephone business. Of course, the Bell Telephone Company was actively challenging the American Speaking Telephone Company's position on the basis that the company's use of telephony were infringements on Mr. Bell's patents.

The New England Telephone Company followed the development after the other two companies mentioned previously, in building both telephones and network services. This was in direct competition with the Bell Telephone Company. A problem existed with this whole scenario: any user who wished to connect with a local telephone company, of which there were now three, could not communicate with another neighbor or business connected to the other companies. Interconnectivity did not exist. What this meant is that if I needed to speak to three different businesses that used the services of the three operating companies, I would need three separate telephone lines and three different telephone sets on my desk. This was ludicrous to conceive, but true.

The Regulatory Scene

The industry was rampant with suits and counter-suits during the early stages of deploying the networks. Bell was active in trying to displace the "fake" companies on the basis of the patent infringements. Many of the cases had a bearing on the future of the industry, in particular with the formation of the Telephone Companies as monopolies. Some of the key events are listed in Table 1.1.

This was quite a turn of events; AT&T (as we know it) started out as a subsidiary and ultimately became the dominant force in the market. All of this happened in the first 23 years of the industry.

While this was all happening, many smaller companies started up, either building telephones or networks. By the 1920s nearly 9000 independent telephone companies were in operation. This number shook out a little because AT&T would not let the independents interconnect to the AT&T network. As a result, many were financially strapped and went bankrupt. AT&T began acquiring these failing operations, causing the number of independents to diminish. About 1400 independent operating companies are still in existence. AT&T had the single largest network in place and began to undercut prices, buying out competitors in an attempt to drive the smaller companies out of business. This left gaps in coverage for customers served by the independents, who could not connect to the major cities where AT&T had the market.

AT&T began to cave in to public pressure. A statement from an AT&T Vice President (Kingsbury), outlined new positions to the attorney general. Kingsbury agreed to:

- Relinquish its holdings in Western Union.
- Stop the practice of acquiring the independents, unless authorized by the Interstate Commerce Commission (ICC.)
- Allow the independents to interconnect to the AT&T network.

Still another major event in the industry was the "Graham-Willis Act" that established the Telephone System as a natural monopoly. For years AT&T had dominated the industry, providing all telephone equipment and service through its normal distribution channels, the telephone companies. The network was the sacred cow, with only AT&T products and services allowed for interconnection to the networks. The philosophy was that any other product might harm the network, thereby keeping the monopoly over the products and services to be installed.

"Hush-a-phone"

In 1947, the FCC permitted the connection of customer-owned and provided recorders, through special protective arrangements. These devices were used to record conversations, but were required to sound a beep when recording took place.

In the late 1940s, other manufactured devices could be attached to the telephone lines, so long as they used a protection device. This was an appeasement to the telephone companies who stated that the connection of a foreign piece of equipment on the network would disrupt the network, decrease its efficiency and jeopardize national security. The carriers, who controlled the network and the revenues achieved through this total control, did not like these devices.

In the 1950s, a device known as the "Hush-a-Phone," an acoustically coupled device designed to eliminate noise and increase privacy, was tested before the FCC, whereby the FCC agreed with AT&T's position to prevent "foreign" devices from being connected to the telephone network. This decision was later overturned by an Appeals Court. Because the acoustic coupler was a nonelectrical device, it should pose no threat to the integrity of the network. The U.S. Naval Department was one of the users who fought for the use of the Hush-a-Phone. After losing the battle in court, AT&T suggested that if a recording device was allowed on the network, then a beep should be sounded on the line periodically (every 15 seconds). This beep would alert the user at the far end that the conversation was being recorded. The Bell system, therefore, suggested that if a beep is introduced to the line, then an electrical input was required. They fought for the right to install the equipment on the line that provided the beep. This was made possible, because the Hush-a-Phone decision was on the basis of a nonelectrical input. The Bell system was then able to charge the Naval department for the equipment that introduced the beep. All was settled.

In 1956, the Department of Justice filed an antitrust suit against the Bell System that had been postponed because of WWII. In general, the suit was aimed at getting AT&T to divest itself of Western Electric. This was finally settled in what was known as the 1956 Consent Decree. The result was that AT&T could retain ownership of Western Electric if it only produced products for the Bell-operated companies. This decree went on to deny access to the Bell System from offering commercial data processing services, but limited them to providing telecommunications services under regulation. This still left the Bell System as the regulated monopoly with no competition for equipment and services in their operating areas.

The Introduction of Competition

In 1968, however, the first true case was tested by a company called Carter Electronics of Texas. Carter made a device to interconnect a mobile radio to the telephone network via an acoustic coupler. The FCC ruling was paramount, in that they allowed direct electrical connection of devices to the network, so long as they used a protective coupler arrangement. The decision had monumental impact because prior to the Carterphone decision, manufacturers had a very limited outlet for their products. They typically sold their products to the independent operating telephone companies. Now the flood gates were opened, and these manufacturers had a whole new world of opportunity to market their products. Even the foreign manufacturers sped into the equipment market, offering end-user products that had previously been totally controlled by the Bell System. The European and Japanese marketeers were quick into the PBX and terminal equipment business. However, the use of a protective device was still an area of frustration with these equipment vendors. The Bell-rented protective arrangements were a recurring cost that made the purchase of other equipment less attractive. The manufacturers continued to complain about this. These complaints resulted in the institution of an FCC registration program for all products that could be attached to the telephone network in 1975. As long as a manufacturer could pass the requirements of the FCC registration (Part 68), then it could be attached to the network without the use of these protective devices. The stage was set after the Carterphone decision, with many companies competing to build connecting devices. This began what was called the *interconnect business*.

Still, additional activity took place. In the 1960s, a small microwave carrier company (MCI) began constructing a microwave network between Chicago and St. Louis. MCI's initial intent was to provide alternate private line services in a high-volume corridor. MCI took its interconnection request to the courts and won the ability to interconnect its network with the telephone company network. MCI further began to offer switched services which was immediately referred to the FCC. The legal actions nearly put MCI into bankruptcy, but they prevailed and began the "other common carrier" industry. One more nail in the monopolistic coffin for AT&T.

The Divestiture Agreement

In 1982, the most monumental decision to be made since the inception of the monopoly in the telecommunications business was reached. In light of all the computer inquiries and consent decrees that were taking place, the Department of Justice was still after AT&T, Western Electric, and Bell Labs. This was an action that had started in early 1974. It was aimed at the complete breakup of the Bell System. In early 1981, the suit finally made it into the courts, but was surprisingly brought to a halt on January 8,1982 when an agreement was reached to drop the suit and submit a modification of the 1956 Consent Decree to the courts.

In order to achieve the goals targeted, AT&T would be relieved of the limitations keeping it from other unregulated markets. However, in return for this relief, the courts demanded the breakup of the Bell System. Because the Bell system operated in a monopoly and enjoyed the noncompetitive environment, it was to remain under a regulated status. The Modified Final Judgment (MFJ) basically considers the provisioning of local dial tone services as a monopoly and everything else shall be a competitive operating environment. Therefore, the local dial tone business provided by the local exchange carriers would remain regulated and AT&T could pursue and operate in the competitive long distance, equipment manufacture, and sale markets. What was once built as a single entity, owned and operated as an integrated system, was now about to be broken into multiple pieces. The 23 Bell Operating Companies would be divested from the organization of AT&T (which retained its Western Electric Manufacturing, Bell Labs, and Long Lines Divisions). Some of the salient points of the MFJ include the following:

- AT&T had to transfer a sufficient number of personnel, assets, and access to technical information to the Bell Operating Companies (BOCs) or whatever new organization that was owned by the BOCs so that they could provide exchange services or access to exchange services without any ties to AT&T.

- All long distance services, links, personnel, and other facilities had to be relinquished by the BOCs and turned over to AT&T.

- All existing licensing agreements and contracts between AT&T, or its subsidiaries, and the BOCs had to be terminated.

- Equal access for all interexchange carriers into the BOCs switching systems must be provided. This breaks down the special privilege that AT&T enjoyed over its competition. The equal access for all carriers was to be provided within two years, with the only exceptions being electromechanical systems where it was cost prohibitive, or with companies serving less than 10,000 users.

- After the breakup, the BOCs could provide but not manufacture equipment.

- After the divestiture the BOCs could produce and distribute directories to subscribers.

- Carrying calls between and among offices was defined as local exchange services and interexchange services. The local exchange carriers (LECs) would hand off any call of an interexchange basis to an interexchange carrier (IEC). To define the boundaries of who carries the call and who shares in the revenue, a number of local access and transport areas (LATAs) were created. The calls inside a LATA were the responsibility of the LEC and calls that left the LATA were the responsibility of the IEC.

- Joint ownership or participation in the network between AT&T and the BOCs is prohibited. Although this was a common practice before the breakup, it was not supported after the breakup.

The provisions of the MFJ were strictly aimed at the Bell System. This was a point of confusion for many people because they thought that the whole network would fall apart, which it did not. The areas of most confusion arose when the MFJ was decided upon in early 1982 with an implementation date of January 1, 1984. Fear, uncertainty, and doubt reigned in the industry.

The business users were prepared for the breakup, but not much happened in the residential area. Although the BOCs and AT&T tried to communicate the message to end users, they were highly unsuccessful. Ten years after the breakup, many residential users still did not know that there was a difference between the suppliers. This is a sad epitaph for the industry, that a communications company was unable to communicate successfully with its customers. The small business and residential users were so accustomed to the "cradle to grave" services offered by their BOC that they did not grasp the significance of this breakup. No longer would one call do it all. Separate companies provided dial tone services and long distance services.

The divestiture of the Bell system made for advances in the industry that might not have occurred for some time to come. With the introduction of competition and the ability of other suppliers to enter new markets, the evolution of the network and service offerings was dramatically escalated. Digital networks were already used in the Bell system for years, but with the breakup, the proliferation of new digital architectures was rolled out to customers at a rapid pace. Further, when divestiture took place, the costs associated with local dial tone increased by as much as 50–60% on a monthly recurring basis. However, long distance services and usage charges decreased by 60% or better. This might sound like a break-even analogy, but the ratio of local to long distance service is an 20:80 rule. Twenty percent of a customer's actual monthly bill is for the fixed recurring charge, whereas 80% of the bill is for the variable usage-sensitive costs associated with long distance. From this, you could suggest that we would gladly accept a 50–60% increase on 20% of our bill, if the remaining 80% of our bill is reduced by 60%.

To add to the confusion, whenever a customer called for a service that did not fall into the scope of the provider called, the standard answer given was that the "judge" didn't allow the vendor to provide such services. This reference to Judge Harold Greene of the Federal District Court who presided over the suit and the MFJ rulings, was a standard escape clause used by LECs and IECs alike. Unfortunately, this was an agreement reached by the court and AT&T rather than an actual mandate. Thus, the burden of setting the rules in the telecommunications industry was placed on the shoulders of the courts.

Tariffs

Another escape clause used on a regular basis is that "the tariffs don't allow this service," which is a clear misnomer. The BOCs and the AT&T organization write tariffs and submit them to their respective judicature, whether it is the FCC for interstate traffic for AT&T, or the public utilities commission for the BOCs. An easy way to avoid offering a service or delaying a customer request is to suggest that the tariff doesn't allow this. However, who writes the tariff? If a service or something out of the ordinary is not covered in the tariff, then the logical solution is to write a new one. This is very rare that a new tariff will be written.

What is a tariff? A tariff is a description of the service, offering an appropriate rate that will be charged for that service, and the rules under which the service is provided. It is the basic agreement of terms for a service between the customer and the provider that must be submitted to the regulators for approval before any changes in service can be offered. In the U.S., there are 50 different regulators that rule on the tariffs, one for each state. This leaves room for plenty of disparity between offerings and rates. Furthermore, the FCC governs the rates and services for the long distance suppliers, primarily AT&T though. This makes it very difficult to understand what the offerings are and what pricing strategies are in force. A user typically does not understand nor has the patience to read through a tariff. Therefore, we are at the mercy of relegating the overseer function to the public utilities commission and the FCC. We can only hope that they are performing a reasonable job of ensuring that the rates and offerings meet the needs of all concerned parties in a fair and equitable manner.

In the rest of the world, such elaborate rate settings and guidelines differ, depending on the country. Many of the telephone companies are under direct control of the local government. The telephone service then is relegated to the post office agencies; these are called Post Telephone and Telegraph organizations. The name stands for the services provided: postal, telephony and telegraphy. Privatization is now starting to be enacted in several parts of the world, where the telephone services are being removed from government control and passed on to private organizations. It stands that these organizations will face stiff competition for the services, much the same that has happened here in the United States.

Chapter

2

Voice Characteristics

The telephone set, as it was designed, is used to transmit sound to the receiving party. The basic telephone set that was designed by Alexander Bell carried his voice between rooms from his set to the one where Watson was working. So it began. If the telephone network and the telephone set were designed to carry voice signals from one point to another, then the characteristics of the voice signal must be known. A great amount of time and effort went into studying the actual variables associated with sound as the vocal chords create it. From these studies, one can derive the fundamental concepts of a telecommunications network. This implies that the telephone set is designed to carry only voice. In reality, this is correct because all of the effort was to carry voice in its truest possible form from the sender to receiver. The word *telecommunications* involves this very basic concept. A definition is in order here because we will be covering the characteristics of the telephone networks and its ability to carry the human voice.

Telecommunications: The transmission of information, in the form of voice, data, video or images, across a distance over a medium, from a sender to a receiver, in a usable and understandable manner.

The definition is complicated, but can be simplified in such a way as to break the two pieces apart and state them differently.

Tele: From the Greek root meaning from afar. This implies that the information will be carried over some distance, from as little as a few inches to as much as thousands of miles.

Communicate: From the Greek root meaning to make common. The implication here is that if we are to send any form of information across the medium, then it must be understandable or else it is wasted energy.

The Medium

A *medium* is any form of transmission capacity that is used to carry signals. In the telecommunications world, this can be in the form of copper wires, coaxial cables, optical fiber (glass), or air. There are other forms of media; for example, this book is using paper as the medium to get the information to the reader. But we are dealing with a telephony concept, so we will stick to these types of analogies. Depending on the medium used, the distances information is transmitted can vary greatly. However, strides have been made since the late 1870s when the telephone set was first invented. Now we can transmit any information over thousands of miles as simply as we do over inches. The information takes on a form of electrical energy to recreate the sound waves that we generate with our voices, in their representative form, so that the distant end will be able to understand the message.

Sound

Sound is nothing more than the banging of air molecules together at a rapid rate. As we generate sound with our vocal chords, we are banging these air waves together to produce intelligible information that can be used and understood by others. This is in the form of air pressure changes. Hold your hand in front of your mouth as you speak. Feel the air pressure hitting your hand? This is the effect of sound as we generate the changes in air pressure. Our vocal chords, therefore, are moving back and forth and banging the molecules of air together.

There is an old question, "If a tree falls in the forest, and no one is there to hear it, does it create sound?" The answer must be an obvious yes. Although no one is there to hear the information, the air pressure changes from the tree falling to the ground—regardless of where it falls—must still be producing sound. If we put a tape recorder in the forest and the tree falls, it is most likely that the sound will be captured on the recorder. Even though no one is there, the recorder still captures the noises created from the air changes.

The telephone converts these sound waves to their analog equivalent in the form of electrical pulses, which then are carried across the telephone network. We can, therefore, assume that in order to communicate across these wires, we must have some form of sound. The sound will be converted into electricity. The electricity is then sent across the telephone wires.

Sound is the banging together of air molecules at a rapid pace. This is called *compression* and *rarefaction*. Regardless, the human voice produces sound at a constantly changing set of frequencies (speed) and amplitudes (loudness). The human voice changes these variables of frequency and amplitude in cycles per second. The vocal chords compress the air molecules at a rate of between 100 and 5000 times per second. To recreate the sound in good faith, the sound waves (or air pressure changes) are converted from sound into electricity. This is what the telephone set is doing for us. For a more detailed explanation of how this happens, refer to Chapter 4.

As the electrical equivalent of sound is created, compare the sound wave to that of an electrical wave. Electricity is typically generated in an analog form by rotating the electromagnetic energy around a center point. As the energy is on the rise, it increases the amplitude to a peak level in decibels, then begins to fall. Because the wave is concurrent, it will have both a positive and a negative side of the electrical field. Therefore, as the signal decreases from the peak of the positive energy, it moves back toward the zero line. As with a magnet there are two poles, the positive and negative sides. Therefore, as the wave gets to the zero line, it will continue to fall to the negative side of the voltage line until it hits some peak at the bottom side of the energy field. From there, it will in turn rise back up to the zero line (value) again. This, in effect, constitutes a 360° cycle around the base line, or one complete rotation. This rotation is called a *sinusoidal wave*. This sinusoidal wave (Figure 2.1) is the analogous recreation of the human sound wave in its electrical form. This form of one wave is called one *hertz*, named after the gentleman who discovered this concept. Hertz is normally abbreviated as Hz.

The human sound wave will have a constantly changing variable energy in both signal strength (the amplitude) and the number of rotations around the baseline over a period of time (the frequency). Voice creates these

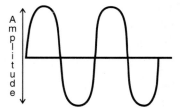

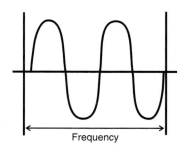

Figure 2.1 The human sound wave produces both frequency and amplitude changes.

waves from 100 to 5000 times per second, as stated. Therefore, the voice is characteristic of creating 5000 cycles of information per second. This is abbreviated into 5 kilohertz. Kilo is represented as thousands, represented in kHz.

The human ear is responsive to variations of frequency and amplitude at rates from 25 to 22,000 hertz. This means that the ear can receive and discern all of the information contained in the human voice. This is important in telephony.

Differences exist in human responsiveness to sound. For example, older humans can discern sounds ranging up to 7 to 8 kHz, as a result of abuses and deterioration of the eardrums. A person who has fired a weapon (rifle or handgun) without proper protection of their ears will have damaged the upper and lower frequency responses. Whereas, a younger person who has not had the chance to damage the ears, will be able to receive and discern sound waves in the 16- to 18-kHz range. Although when we see these youngsters carrying those "boomboxes" on their shoulder, with the box blaring away and sitting directly next to the youngster's ear, we can only imagine how the ear will respond in a matter of time. They are inadvertently destroying the frequency responses that the ear will be able to selectively respond to.

The telephone company realized that the majority of usable information in a human conversation will fit in a 3-kHz range; therefore, it provides a pair of telephone wires, typically made of copper, that will carry all of the usable information on it. The telephone networks were built to carry speech, and the most commonly carried signal on the network is in the electrical equivalent of speech (voice). The telephone transmitter converts the acoustic signal (sound wave) that is generated in the human speaker's larynx into electrical waves.

Actually, the analog waves can be represented in frequency and analog changes over a broad spectrum (or band) from approximately 30 Hz to about 10 kHz. However, most of the usable and understandable energy falls in the spectrum of 200 to 3500 Hz. If we subtract the differences between the high end and the low end, the spectrum is 3300 Hz wide, or 3.3 kHz. It is not necessary to recreate all of the speech waveforms precisely to get an acceptable transmission of human speech across the telephone network. This is because the ear is not really that sensitive to very fine distinctions in the frequency changes, and the human brain can make up for any variations in the speech form by interpretation. Of course, if something does not come across the wire clearly enough, the human brain will intervene and cause the mouth to say "What?" This in turn will cause the transmitting end to regenerate the signal over again by repeating themselves.

Because the cost of transmitting the signal across a telephone network is directly proportional to the amount of energy that must be carried, the telephone company uses a "bandpass filter" on the circuit. Commercially acceptable and usable information is transmitted in what is called a *band-limited channel*. This means that all of the usable information is allowed to pass onto the circuit, but the extraneous information that does not add significantly to the conversation is filtered off. If the extra energy is put onto the wire and carried from end to end, the costs will go up at an equal rate; this is wasteful.

What Is Bandwidth?

One of the toughest concepts for anyone to understand is the discussion of bandwidth. To the novice, it becomes even more perplexing when bantered about by telephone company personnel, engineers, and others. However, it is the basis of most of what we do in the telecommunications and telephony world.

Think of bandwidth as a water pipe or a garden hose. The greater the size of the pipe, the larger volume of water that will flow through the pipe. The smaller the pipe, the smaller the flow of water. Now, in telecommunications terms, think of bandwidth as a communications pipe. The bigger the pipe, the more information that will flow through it. The smaller the pipe, the less information that will be carried through the pipe. Assume, for example, that we have a lawn and we need to water it regularly to keep it green and moist. If we have an average-sized lawn, the job can be done with the standard ⅝" garden hose, the type that can be bought in any hardware store. When we turn on the spigot attached to our regular water pipes, a sufficient flow of water comes out of the hose. This is a regulated flow with several control mechanisms in place:

- The water pipes are perhaps ¾" in diameter, allowing a certain amount of flow through them to begin with.
- The garden hose is slightly smaller, so as the flow from the larger pipe to the smaller diameter garden hose occurs, the flow is restricted, but pressure allows the flow to be constant.
- The spigot also can be used to turn the water on full bore, or constrained to a specific flow that is more to our liking.

This is a band-limited pipe that uses several constraints to control the flow of water through the pipe. Enough is allowed through to do the job, but any more would probably cause flooding and over-saturation in certain areas. Therefore, the limitations meet our needs without being wasteful.

Now, let's assume that we also want to water another area. Let's assume that we want to handle the watering of the Super Dome if real grass was on the ground. This football field is much larger than the average home lawn, so if we use the tools that we have in the home watering scheme, things will be tougher. Imagine trying to water this lawn with the average ⅝" garden hose! It would take forever. Not only would it take forever to get from one end to the other, by the time we got to the opposite end, we would have to start the job all over again. The first end will be parched dry, due to the length of time it took to get from one end to the other.

This is ineffective. So to do the job, we will have to purchase a garden hose that is 6" in diameter. Now we hook this 6" hose to a spigot that is also 6", and connected to a water pipe of at least 6–8" in diameter. The flow of water through this pipe will be significantly greater than that of the ⅝" garden hose. Thus, the job can be done in a reasonable amount of time.

Bandwidth is similar to this garden hose analogy as shown in Figure 2.2. It is the range of frequencies that can be carried across a given transmission channel. If more information is sent, more bandwidth is necessary. A typical telephone channel (line) is provided by the telephone company that will carry 3 kHz.

Therefore, this is a 3-kHz channel, which has 3 kHz of bandwidth. This is fine because all the usable information of a voice-grade conversation is contained in this amount of bandwidth. This was all covered in previous paragraphs. In actuality, the telephone companies break the available electromagnetic spectrum into slices, each about 4 kHz wide. Then these 4-kHz slices (called *channels*) are limited with bandpass filters, as shown in Figure 2.3. Consider the spigot analogy, turning the water on faster or slower via the spigot valve. The result is that we receive 3 kHz of the available 4-kHz slices.

Other forms of bandwidth requirements exist. For comparative purposes, we can see the differences of what capacities are used in various forms of bandwidth allocations as shown in Table 2.1.

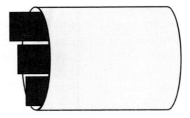

Figure 2.2 Bandwidth can be compared to a pipe or hose.

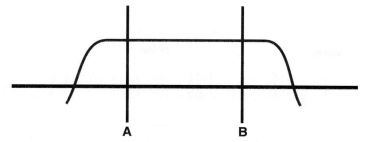

Figure 2.3 Four kHz of bandwidth.

**TABLE 2.1 Summary of Channel Capacities
to Carry Different Forms of Information**

Service	Bandwidth allotted
Voice channel	3 kHz
Hi-fidelity music	15 kHz
CD stereo player	22 kHz
FM radio station	200 kHz
TV channel on CATV	6 MHz
TV channel actually used	4.5 kHz

MHz is a new term here. It represents millions of frequency changes per second. You can imagine the amount of information carried in a TV signal, when sound, motion, and voice are all on the same channel. Think of how a video signal on a TV station would look if the channel was restricted to carrying only 3 kHz of information. By the time some moving picture was created on the set, the viewer would have lost interest.

The frequency spectrum of a TV channel as shown in Table 2.1 shows that a CATV channel is actually allocated 6 MHz of capacity. Yet, the amount actually used approximates 4.5 MHz. The difference is the band limitation on the channel, much the same as the voice channel. In this case, the amount of flow is in the millions of cycles per second (a big pipe is needed here). The band-limited channel uses bandpass filters so that there is a guard band between the TV channels. In other words, as you watch channel 3 on your TV set, the filters are placed on the line to prevent frequencies and information from channels 2 and 4 from overflowing onto channel 3 (Figure 2.4). These guard bands are placed on every channel, thereby restricting the 6-MHz channel to only 4.5 MHz of usable information.

The same concept holds true for just about all of the channel capacities used in the telecommunications industry. The available bandwidth is a func-

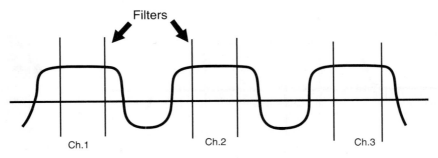

Figure 2.4 The bandpass filter separates the channels.

tion of need and cost. You get what you need, but any more will be too expensive.

Voices

The human sound, or voice, therefore, is a consolidation of the waves of electrical energy carried across the given channel capacity. Humans generate a combination of amplitude and frequency changes in a continuing flow. If the changes are held constant, then the conversation becomes monotone, highly unacceptable for the average conversation. Instead, if everything was held constant, the recipient of the information would be lulled to sleep. We use the variations in our voice to reflect the deviations in emotion, accentuation or articulation and emphasis on certain points. Every voice generates a different pattern of amplitude and frequency changes. This is what separates us from the masses.

It is interesting to note that the female voice pattern typically generates more amplitude changes than frequency. This accounts for the higher pitch in the female voice. Conversely, the male vocal pattern generates more frequency changes than amplitude shifts. This accounts for the low pitch, more grainy conversations of the male. These are averages because every human is different and the norm can be deviated from at any time. Suffice it to say, when a person is highly emotional, there are definite pattern shifts from their normal voice patterns. The pitch might go up by varying degrees showing the results of stress or anger.

The telephone links are installed to handle this widely varying pattern of shifts. However, from time to time, the voice pattern might exceed the differences in the allotted bandwidth. Then the frequencies in the voice go above or below the normal range. Therefore, the bandpass filters start to filter off the excesses. What can result is a pattern of tinny or flattened speech conversations. This can be also heard if someone uses the letters "F" and "S" where the frequency ranges on these two sounds might exceed the bandpass ranges, they get flattened.

Other Services

Because this telephone network was built to carry the analog equivalent of the human speech patterns, the other services that we wish to communicate, such as data, facsimile, images, and video must be applied to the same constraints of voice calls. The network allows any telephone set to contact any other telephone set across a 3-kHz bandwidth. If you want to communicate anything else, such as data, the data speed must be accommodated on this same 3-kHz bandwidth. This invokes a limitation on the speed of our data communications channel capacities. Because a modem can transmit signals quicker, they must be constrained into the size of the pipe. A video conference transmission will also reflect the nonreal time motion because of the channel capacities on a dial-up basis.

If you need to move more information across the channels, we will have a couple of choices. One option is to dedicate a high-speed line between the two or more end points we need to communicate with. However, this approach will eliminate the possibility of "any to any" connection and require forethought well in advance of communicating with the end points. Leased lines between the points might be underutilized, and could be more expensive.

Another option is to have the bandpass filters moved out to (for example) 8 kHz between the end points. Unfortunately though, this is similar in the constraints. We would have to know every location that we would be communicating with, ameliorating the benefits of the any to any connection. Secondly, there is no guarantee that the switching systems will route the call over the same path on a switched dial-up connection every time. This would force us back into a 3-kHz channel along the route selected if it is different every time. This obviates the benefits of the wider channel capacity.

3

The Telephone Network

A *network* is a series of interconnections that when all tied together form a cohesive and ubiquitous connectivity arrangement. Whew! That sounds ominous, but to make this a little simpler, look at the components of what constitutes the telephone network.

Generally, a network is a series of interconnection points. The telephone companies over the years have been developing the connections throughout the country and the world, so that a level of cost-effective service can be provided to its customers. In order to build out this connectivity, the telephone companies install wires to the customer's door, whether business or residential. The spot where these wires terminate is called the *demarcation point*. The position of the demarcation point depends on the legal issues involved. In the beginning days of the telephone network, the telephone companies owned everything, so they ran the wires to an interface point, then connected their telephone sets to the wires at the customer's end. This was the cradle to grave service that allowed them to proliferate the connections throughout their entire operating area.

New regulation in the United States, in effect since the divestiture agreement discussed in Chapter 1, changed this connection to a point at the entrance to the customer's building. From there, the customers hook up their own equipment, items they purchase from a myriad of other sources. In the rest of the world, where full divestiture has not yet taken place, the telephone companies (or PTTs) still own the equipment. Other areas of the world have a hybrid system, where customers might or might not own their equipment. The combinations of this arrangement are almost limitless, depending on the degree of privatization and deregulation. However, the one issue that is common in most of the world to date, the local provider owns the wires from the outside world to the entrance of the customer's building. This is called the *local loop*.

A Topology of Connections Is Used

At the local loop, the topological layout of the wires has traditionally been a single wire pair or multiple pairs of wires strung to the customer's location. This has always been an issue of money. Just how many pairs of wires are needed for the connection of a single line set to the network? The answer is obvious. But for other types of service, such as digital circuits and connections, the answer is two. Depending on the customer, the number of wires run to the location has been contingent on the need vs. the cost. As a result, the use of a single or dual pair of wires has been the norm. More recently, the local providers have been installing a four-pair (eight wires) connection to the customer location. This is because the customer (both business and residential) has begun to use voice lines, separate fax lines and still separate data communications hookups. Each of these require a two-wire interface, so the need for multiple pairs has grown. It is far less expensive to install multiple pairs the first time than to install a single pair of wires every time the customer asks for a new service. So, the topology is a dedicated local connection of one or more pairs from the telephone provider to the customer location. This is also called a *star configuration*. The telephone company connection to the customer is from a centralized point called a *central office* (CO). Using a star configuration, all wires home back to a centralized point, the CO.

Once the hundreds or thousands of wire pairs get to the CO, things change. The provider at that point might be using a different topology. They can use a star configuration to a hierarchy of other locations in the network layout, or they can use a ring. The ring is becoming a far more prevalent method of connection for the locals. Although we might also show the ring as a triangle, it is still a functional and logical ring. This star/ring combination constitutes the bulk of the networking topologies today.

At the local telephone company's (or PTTs) office, the wires are terminated in what is called a *wire center*. The wire center is nothing more than a very large extension of the customer's hookup. Thousands of customers come together in this centralized point. From the wire center, a series of spokes are run out to the customer direction, or to other central offices, higher level offices in the hierarchy, or wherever they need to go. The wire center is also called a *frame*, where all the wires are connected to the frame.

At this frame, a series of cross-connections are made. These will either be to other wires that go to other locations or to a switching center where the telephone company's central computer (in older offices this can be an electromechanical system) resides. This is called the *switch*. Most of the equipment today is a stored-program common-controlled computer system that just happens to process cross connections for telephone calls. Remember one fundamental fact: the telephone network was designed to carry analogous electrical signals across a pair of wires to recreate a voice conversation at both ends. This network was built to carry voice. Only recently have we been transmitting other forms of communication, such as facsimile, data, and video.

The switch makes routing decisions based on some parameter, such as the digits dialed by the customer. As these decisions are being made very quickly, a cross connection is made in logic. This means that the switch sets up a logical connection to another set of wires. The connection can be back to the frame where the wires serve a neighboring pair of wires connected to our next-door neighbor, or to another connection that links another central office. The possibilities are only limited by the physical arrangements in the office itself.

Between and among the offices built by the carriers (the local and the long distance providers) is a set of connections usually laid out in a ring, but it can also be a star configuration. These are called the *facilities* that carry traffic.

Throughout this network, more or less connections are installed, depending on the anticipated calling patterns of the user population. Sometimes there are many connections among many offices. Other times, it can be simple and single connections.

Tied all together then is a series of local links to the customer locations, through a central office where switching and routing decisions are made, then on out to a myriad of other connections from telephone companies, long distance suppliers, and other providers. This is the basis of the telephone network.

The Local Loop

Our interface to the telephone company network is the single-line telephone set covered in Chapter 4. It stands to reason that we need to connect this set to the telephone company central office (CO). The pair of twisted wires running from the telephone company's CO is called the *local loop*. Each subscriber, or customer, is delivered at least one pair of wires per telephone line.

There are exceptions to this rule in rural areas where the telephone company might share multiple users on a single pair of wires. This is called a *party line* and is again a financial decision. If the number of users demanding telephone service exceeds the number of pairs available, they might well offer the service on a party or shared set of wires.

The phone company distributes its outside plant, or distribution, to the customer by running large bundles of twisted pairs toward the customer location (Figure 3.1). This is done using feeders, which are composed of 50 to over 3000 pairs of wires.

The feeders are run to splice points or breakout points called *manholes* or *handholes*. At this point, the splicing of two reels of cable will take place, assuming that the cable on a single reel was not sufficient. A lateral distribution can also take place here. *Lateral distribution* is the breakout of a number of pairs to run in a different direction. The lateral distribution or branch feeder is then strung to various customer locations. The end of the pair to the final customer location is called the *customer pair* or *station drop*.

It is in this outside plant, from the CO to the customer location, that 90% of all problems will occur. This is not to imply that the telco is doing a lousy job of delivering service to the customer. These cables are exposed more to cable cuts because of construction (commonly called *backhoe fade*), flooding at the splice locations, rodent damage, and many other risks.

In the analog dialup telephone network, each pair of the local loop is designed to carry a single telephone call to service voice conversations. This is a proven technology that works for the most part, and continues to get better as the technologies advance. The cables can be delivered via a telephone pole, buried conduit, or direct buried cables. Either way, the service is one that we are familiar with and feel comfortable with.

What has just been described is the connection at the local portion of the network. From there the local connectivity must be extended out to other

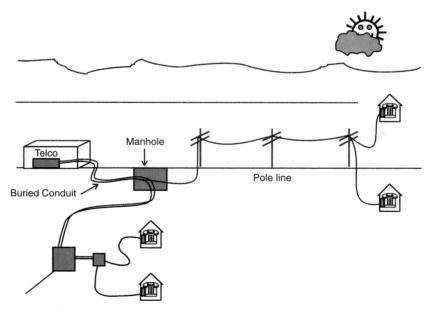

Figure 3.1 The local loop.

locations in and around a metropolitan area, or across the country. The connections to other types of offices are then required.

The Network Hierarchy (Pre-1984)

Prior to 1984, the network was owned by AT&T and its local Bell operating telephone companies (BOCs). It evolved through a series of interconnections based on volumes of calls and growth. A layered hierarchy of office connections was designed around a five-level architecture. Each of these layers was designed around the concept of call completion. The offices are connected together with wires of various types, called *trunks*. These trunks can be twisted pairs of wire, coaxial cables (like the CATV wire), radio (such as microwave), or fiberoptics. The trunks vary in their capacities, but generally high-usage trunks are used to connect between offices. Figure 3.2 shows the hierarchy prior to the divestiture of AT&T, with the five levels evident.

The class-5 office is the *local exchange,* or end office. It delivers dial tone to the customer. The end office is the closest connection to the end user. This can also have the name of a branch office. Think of a tree. The end of the branches is where all the activity takes place, and the customers are the leaves hanging off the branches. Calls between exchanges in a geographical area are connected by direct trunks between two end offices. These are called *inter-office trunks*. There are over 19,000 end offices in the U.S. alone that provide basic dial tone services.

The class-4 office is the *toll center*. A call going between two end offices that are not connected together will be routed to the class 4. The toll center is also used as the connection to the long-distance network, calls where added costs are incurred when a connection is made. This toll center could also be called a tandem office, meaning that calls would have to pass through (or tandem through) this location to get somewhere else on the network. A basic arrangement of a tandem switching system is shown in Figure 3.3. The tandem office usually does not provide dial tone services to the end user. However, this is a variable where a single office might provide various functions. The tandem office can also be just a toll-connecting arrangement that is a pass through from various class-5 offices to the toll centers. Again, this is a variable depending on the arrangements made by the telephone providers. The ratio of toll centers that serve local long distance is approximately 9 to 1. Prior to divestiture, there were approximately 940 toll centers.

The class-3 office is the *primary center*. Calls destined within the same state area would be passed from the local toll office to the primary center for completion. These locations are served with high-usage trunks that are used strictly for passing calls through it from one toll center to another. The primary centers would never serve dial tones to an end user. The number of primary centers prior to divestiture were approximately 170, spread across

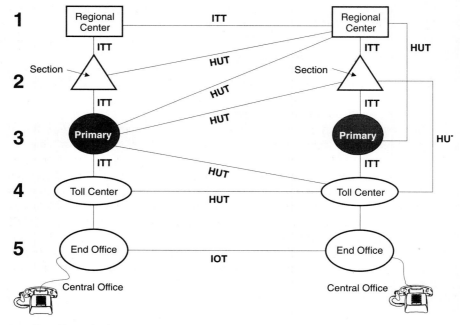

Figure 3.2 Network hierarchy pre-1984.

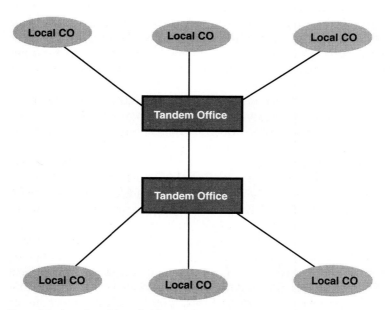

Figure 3.3 A tandem (class 4) office arrangement.

the country among the various operating telephone companies (both Bell and independent operating companies).

The class-2 office is the *sectional center*. A sectional center was typically the main state switching system used for interstate toll connections designated for the processing of long-distance calls from section to section. There were approximately 50+ sectional centers before the divestiture of the Bell System. These offices did not serve any end users, but would serve between primary centers in the country.

The class-1 office is the *regional center*. Ten regional centers existed across the country. Each center was tasked with the final set-up of calls on a region-by-region basis. However, the regional centers constitute one of the most sophisticated computer systems in the world. The regional centers continually updated each other regarding the status of every circuit in the network. These centers are required to reroute traffic around trouble spots (e.g., failed equipment or circuits) and to keep each other informed at all times. As mentioned, this was all prior to the divestiture of the local Bell operating companies by AT&T. The number of regional centers has changed to seven in the AT&T network, and might be consolidated into four mega centers in the future.

The Network Hierarchy (Post 1984)

After 1984 the network took a dramatic turn, with the separation of the Bell Operating Companies (BOCs) from AT&T. Many users screamed that things would fall apart, with service being affected. None of this came true, however. This doesn't mean that there was not a lot of confusion, there certainly was. However, things just kept humming along for the most part and calls were completed through this series of interconnecting points called the network.

The hierarchy of the network shown in Figure 3.4 introduced a new set of terms and connections. The BOCs were classified the same as independent telephone companies. They are all called *local exchange carriers (LECs)*. The seven spin-off companies formed as a result of the divestiture became Regional Bell Holding Companies (RBHC), which had regulated arms called the *Regional Bell Operating Companies (RBOC)*. Each RBOC had the Bell Operating Companies in its geographical area. Additionally, the RBHCs also had an unregulated side of the business, where new ventures could be entered into (such as equipment sales, finance, real estate, etc.).

Equal access, or the ability of every interexchange carrier (IEC or IXC) to connect to the Bell operating companies for long-distance service became a reality. Equal access was designed to allow the same access to other long-distance competitors that AT&T had always enjoyed prior to divestiture. Prior to divestiture, a customer attempting to use an alternative long-distance supplier would have to dial a 7- or 10-digit telephone number to

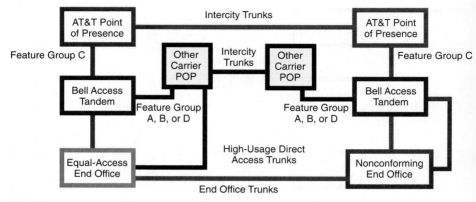

Figure 3.4 The network after divestiture.

get to this supplier's switch. Then, upon completion of this connection, a computer would answer the call and place a tone on the line. From there, the caller would have to enter a 7- to 11-digit authorization code. This code identified the caller by telephone number, caller name and address, and the billing arrangements. Only after the computer (switch) verified this information would it then send dial tone to the caller's ear. The caller would then have to dial the 10-digit telephone number of the requested party. This could involve very lengthy and frustrating call set-up times—especially when the called number was busy.

Users chose not to use these alternative carriers because of the time, the number of digits required, and the frustration of busy call attempts. That is, unless the organization forced the user to dial across the carriers' networks. The reason for all of the digits was simple. The telephone company did not pass on the caller information to the alternate carrier (MCI, Sprint, ITT, etc.) that they did to AT&T. Thus, the choice of many callers was AT&T because it was simpler. Now the caller information is passed on in an equal basis, hence the access is equal.

However, some of the independents and BOCs who haven't yet upgraded their offices do not connect to these IECs. These are called *nonconforming end offices*. The *point of presence (POP)* is the point where the IEC and the LEC are interconnected. This usually refers to AT&T, MCI, Sprint, and other carrier's offices. This name is now starting to change to the *point of interface* (POI) as opposed to POP. The main reason is that a POP implies that a switching system is at the location, which is not always true. The point of interface is where the two parties, Bell or independent, and the carrier connect. This can be a closet in a basement or hotel, connected to a dedicated trunk to another part of the country. All of this should be transparent to the end user, however.

The Public-Switched Network

The US public-switched network is the largest and the best in the world. Over the years, the network has proliferated to even the most remote locations around the country. The United States primary call-carrying capacity is through the public-switched network. Because this is the environment that AT&T and the Bell Operating Companies built, we still refer to it as the *Bell System*. However, as we've already seen, significant changes have taken place to change that environment.

The public network allows access to the end office, connects through the long-distance network and delivers to the end point. This makes the cycle complete. Many companies use the switched network exclusively; others have created variations depending on need, finances, and size. The goal of the network hierarchies is to complete the call in the least amount of time and over the shortest possible route. The network is dynamic enough, however, to pass the call along longer routes through the hierarchy to complete the call in the first attempt wherever possible.

The North American Numbering Plan

The network numbering plan was designed to allow for the quick and discreet connection to any phone in the country. The North American numbering plan, as it is called, works on a series of 10 numbers as shown in Table 3.1.

The Area Code

Note that there have been some changes in this numbering plan. When it originally was formulated, the telephone numbers were divided into three sets of sequences. The first was a three-digit area code [or numbering plan assignment (NPA)]. This started with a digit from 2 to 9 in the first slot of the sequence. In the second slot, the number was set as a 0 or a 1. In the third slot, it could be any digit from 1 to 0 (0 being the 10 digit). The reasons behind this sequence were very clever. For example, the first digit did not use a 0 or 1 because these digits were used for access to operators (0) or operator services

TABLE 3.1 The North American Numbering Plan as It Evolved

Timing	Area code	Central office code	Station subscriber number
Original	N 0/1 X	N N X	X X X X
Pre-1995	N 0/1 X	N X X	X X X X
Post-1995	N X X	N X X	X X X X

such as credit card calling, etc. The 1 was used to send a significant digit to the local switching office indicating that the call was long-distance; therefore, the switch could immediately start setting up a toll connection to the toll center or tandem office. Thus, the exclusion of 0 and 1 in the first digit of the area code facilitated the quicker call set-up. In the second slot the digit was only a 1 or 0. This was used by the switching office equipment in a screening mode. As soon as the system sampled the second digit and saw a 0 or 1, it knew that this three-digit sequence was an area code. The third slot was any digit, so it didn't matter, it was just processed normally. Back in the early 1960s, we recognized that we were running out of the area codes, given that there were only 160 available ($8 \times 2 \times 10 = 160$). In reality, only 152 were allowed for use by the various states because certain ones were allotted for special services (the area codes with an N11 were always reserved, e.g., 211, 311, 411, . . . etc.). Very close administration of the area code assignments held this until 1995, when the inevitable had to occur. From the outset, the use of the entire numbering plan was limited, but it held up for over one hundred years.

The Exchange Code

Following the area code is another three-digit sequence, called the *exchange code*. This is a central office designator that lists the possible number of central office codes that can be used in each area code. The exchange code originally was set up in the sequence NNX, meaning that the first and second numbers used the digits 2 through 9, for the same reasons stated in this area code. 1 and 0 were reserved for operator and long-distance access and the 0/1 exclusion in the exchange code prevented this three digit number from being confused with an area code. In the third number slot of the exchange code, any digit could be used. Clearly the greatest limitation in the exchange codes was that we would run out of exchange codes first. With NNX we have 640 possible exchange numbers to use ($8 \times 8 \times 10 = 640$), but these were used very quickly.

In the late 1960s, we began using an exchange code that changed the sequence to NXX or expanding the number of exchange codes in each of the area codes to 800. This added some relief to the numbering plan, but when using the NXX sequence the need arose for a forced 1 in advance of the 10-digit telephone number so that the call screening and number interpretation in a switch wouldn't get confused. This met with some resistance, but ultimately customers got used to the idea. The first two locations to use this revised numbering plan were Los Angeles and New York City back around early 1971.

The Subscriber Extension

The last sequence in the numbering plan is the subscriber extension number. This is a four-digit sequence that can use any digit in all of the slots, allotting 10,000 customer telephone numbers in each of the exchange codes. Because the four-digit sequence can be composed of any numbers, the intent is to give every subscriber their own unique telephone number. This hasn't changed as yet. However, the possibility that we can still run out of numbers always exists. Therefore, a couple of ideas have been bandied around: add two, three, or four more digits to the end of the telephone numbering plan or add one or two more digits to the area code or exchange code.

In either method, users will be asked to dial more digits, an idea that is never popular, but might become a necessity. However, this is far more complicated than just adding a few more digits here and there. The whole world will be impacted by any such decision and the length of time to implement such a global change will be extraordinary.

Private Networks

Many companies, depending on their size and need, create or build their own networks. These networks are usually justified on the basis of cost, availability of lines and facilities, and special need. Often, a mix of technologies are used, such as private microwave, satellite communications, fiberoptics, and infrared transmission. The initial and the ongoing costs of private ownership can be quite expensive (Figure 3.5).

Because the costs for a private network must be borne by the end user, oftentimes the telecommunications department managers are under pressure to cross subsidize the costs in other ways. Many companies with a private network have been subjected to criticisms from end users that the costs are higher than the public-switched network costs. Individual sites or

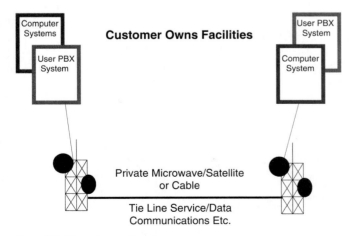

Figure 3.5 The private network.

departments, therefore, begin reducing their own use of the private facilities. This increases the burden on remaining users, who then foot the bill for lightly used networks.

Still others, who have built their own networks, have stated that they are amassing huge savings. When placed in a position of defending their networks, they can produce statistics of the savings reaped on a company-wide basis. If, however, they have a problem with underutilization, they can sell off some of their excess capacity.

This gives these organizations a little breathing room. Collections, maintenance, and administration costs tend to increase when reselling takes place. There is no single best answer here; each organization must look at its needs, availability, costs, and services before deciding which network to use.

Hybrid Networks

Some companies have to wrestle with the decision of whether to use a private or a public-switched network. This decision is not an easy one because oftentimes the numbers do not play out well. The return on their investment just will not be there.

As a result, these organizations use a mix of services based on both private and public networks (Figure 3.6). The high-end usage services, heavy volumes between two or more major company locations, will be connected via private facilities; the lower volume locations will utilize the switched network. This usually works out better financially for the organization, because the costs can be fully justified on a location-by-location basis. Pressure to install private line facilities comes from the integration of voice, data, video,

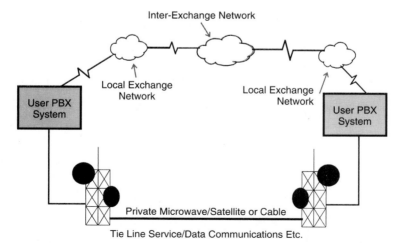

Figure 3.6 The hybrid network.

graphics, and facsimile transmissions. Only by combining these services across common circuitry will many organizations realize a true savings.

Local Access and Transport Areas (LATAs)

Local access and transport area is a term introduced with the divestiture agreement in 1984. One of the problems facing the court system was dealing with the long-distance versus local calling areas. It was a revenue sharing concern more than anything else. AT&T forced the issue by demanding that some form of revenue sharing be put in place so that a single LEC would not have the option of handling all calls and cutting out the IECs.

To solve this problem, the agreement was reached that stated the LECs would still maintain a monopoly on local dial tone and local calling. They would be restricted from carrying long-distance traffic, which would fall under the domain of the IECs. However, many states cover very large geographical areas, and calling from one end to another would be considered long distance. The country was broken down into 195 separate bounded areas for local calling (some local tolls are allowed) based on the density of population in the area. This again was a money thing. The interconnection between two LATAs would be done by the IECs.

A whole new set of acronyms emerged as a result of the divestiture agreement; LATA is only one of them. To complicate things even more, there are four types of calling using the LATA concept:

- Intrastate-IntraLATA belongs to the LECs
- Intrastate-InterLATA belongs to the IECs
- Interstate-InterLATA belongs to the IECs
- Interstate-IntraLATA can be either/or, but was originally given to the LEC

The result of this is a very confused public, and sometimes very confusing tariffs. For example, a call from Philadelphia to Los Angeles can be carried on the IEC network for as little as $0.10 per minute, yet a call from Philadelphia to Wilmington can cost as much as $0.40 per minute. The difference here is that the 3000-mile call is regulated on the long-distance basis and FCC jurisdiction. However, the 29-mile call between Philadelphia and Delaware is regulated by BOC tariff and the PUCs. These anomalies will change and are doing so now, but this is the confusion that can crop up when various players are trying to protect their interests.

Wiring Connections: Hooking Things Up

The telco uses a variety of connections to bring the service to the customer locations. The typical connection is the two-wire service that we keep talking about. This two-wire interface to the network is terminated in a demarcation point as required by law. The DEMARC is the point of least penetration into the customer's premises, typically within 12" of where the telco cable comes up into the building. Normally, telco terminates in a block; this can be the standard modular block for a single-line telephone. If the customer has multiple lines, telco will terminate in a 66 block, or an RJ21X. These are fancy names for their termination points.

The typical modular connector uses an RJ11C for telephones connected to a 2-pair interface (not to be confused with the two wires), or an RJ45X as a 4-pair interface for both voice and data. Another version of connector for digital service is an 8-conductor (4-pair), called the *RJ48X*.

When a telco brings in a digital circuit, it will terminate the 4-wire circuit into a newer RJ68 or a smart jack. There is no major mystique in any of these connectors. The number is strictly a uniform service code so that they can keep it all straight. However, when ordering a circuit, the telco will ask you how you want it terminated. The rule of thumb, in a multiline environment, is to use the RJ21X (which is a 66 block with an amphenol connector on it). Sounds complex, doesn't it? A single line will terminate in an RJ-11C or RJ-12.

Types of Communications

The next area is the directional nature of your communications channel. Three basic forms of communications channels can be selected. They are:

One way (simplex)

This is a service that is one way and only one way. You can use it to either transmit or to receive. This is not a common channel for telephony (voice) because there are very rare occasions when we speak and everyone else only listens. Feedback is one of the capabilities we pride in our communications, which would be eliminated in a one-way conversation. Broadcast television is an example of simplex communications.

Two-way alternating (half duplex)

This is the normal channel we use in a conversation. We speak to a listener, then we listen while someone else speaks. The telephone conversations we engage in are normally half duplex. Although the line is capable of handling a transmission in each direction, the human brain can't deal well with simultaneous transmit and receive.

Two-way simultaneous (duplex) or full duplex

This is used in data communications where a device can be sending to a computer and receiving from the computer at the same time. The direction of the information can be at differing speeds, such as 1200 bits/s toward the computer from the keyboard (humans cannot type much faster than that) and 9600 bits/s from the computer.

Equipment

Equipment in the telephony and telecommunications business is as varied and as complex as you can get. The mix of goods and services is as large as your imagination. However, the standard types are the ones that constitute the end points on the network. The computerization of our equipment over the years has led to significant variations yet improvements. The devices that hook up to the network are covered in various other chapters, but to summarize some of these connections as they pertain to being a part of the network we see the following pieces:

- The single line set
- The multi-line set (called a *key set*)
- The PBX (private branch exchange)
- Modems (data communications device)
- Multiplexers (get more users on a single line)
- ACD (automatic call distributors)
- VMS (voice mail systems)
- AA (automated attendant)
- Radio systems
- Cellular telephones
- Facsimile machines
- Centrex (central office exchange)

This is a sampling of the types of equipment and services you will encounter in dealing with vendors in this business.

4

The Telephone Set

Most people take the telephone set for granted. How can this one device that we have known and used since we were old enough to stand up be worth any form of mention? Why should we even care? This is not an uncommon concern. Since the early days of the telephone network installation, users have accepted this device as the norm. There was no real concern. We had the device, and all we had to do was pick it up and use it. It always worked!

For the most part this is true. However, from time to time these devices do break, and when they do, we feel lost. We do not have the foggiest idea of what went wrong or what to do to fix it. Yet, the mystique is not just a carryover from the good old days when the telephone set was reliable and equivalent to a "boat anchor." It seems that we were always told, prior to divestiture, that we were not allowed to even take this set apart. Because we did not own it, we were not allowed to tinker with it. The poor user who might have taken the set apart and not known how to reassemble the set would rue the day that he or she had to call the local telephone company and admit committing the worst of all possible telephone crimes. Users would conjure up a vision of being put away for life, or worse yet: losing our privileges to use the telephone network. Telco would turn off our dial tone, take the set away and leave us without a means of communicating. How awful!

This scenario might sound far fetched for many novices in this industry, but surely the old timers around can remember the stories well. The authors (not to imply that they are old timers, that's a matter of mind) have heard of many similar cases. One story goes like this:

Back several years, a farmer in St. Louis, MO wanted to know just what was inside the telephone set. So, one night he decided to check it out for himself. Family members, realizing his plans, recognized that this was a major problem. They knew from the local telco that you were not allowed to open the cover of the set—to do so would violate all warranty and utility commission rules. They couldn't bear the thought of having no telephone in their household. They begged and pleaded with the farmer to let it go, not to fall prey to the mystique of the set.

Unfortunately, the farmer would not be daunted by fear, uncertainty, and doubt. The decision was made. As a means of acquiescing to his family, he agreed to do this after dark, when no one could see from outdoors. That night the farmer covered all of the windows of the farmhouse with blankets and proceeded to quench his thirst for knowledge about the telephone set. As he opened the set, he found a nest of wires running from hither to there. There was no magic to be found here. Not to be outdone he quickly screwed the set back together, then proceeded to unscrew the mouthpiece from the handset. Alas, when he untwisted the mouthpiece, it came off the handset. When the set came apart, a group a small black particles fell out of the mouthpiece and rolled around on the floor. Before the farmer was able to catch these little black pellets, they rolled down between the slots on the floorboards. Back then the floorboards were installed with spacers in them, leaving large cracks between the boards. Well as one might imagine, these pieces of whatever they were fell through the floorboards and were lost forever. Now the farmer was in a spot. What could they have been? Why were they inside the telephone set and why were they loose? Regardless of the answer, another thought loomed up in the farmer's mind. The dreaded thought of having to call the local telco and ask them to send someone out to fix the set wrought fear in the deepest crevices of the farmer's mind. How would he explain that he was merely inquisitive? What would they do? Would he lose his telephone privileges? Would they take away his dial tone and telephone set? For how long?

Being a quick study, the farmer recognized that the pellets, whatever they were, resembled something he had seen before. Aha! He remembered that this stuff looked just like gunpowder. The problem was solved. He would merely re-pack the mouthpiece of his set with the granules of gunpowder and be done with it. So he did . . . We heard that his next phone call was a real blast!

All kidding aside, this scenario is probably unfounded. But certain issues were brought up here to make us stop and think. When the original telephone network was installed under the monopoly and controlled by AT&T and its Bell Operating Companies (the independents were not as severe in their treatment), we were left with the fear that if we tinkered with the telephone set and broke it, we would lose our phone privileges. Can you imagine that? Of course since the breakup of the Bell System, things have changed considerably. The set now belongs to the end user. We can do anything we want with it or to it. If we break it, we buy a new one. Simple?

Despite this frivolous thought process there must be some significance to just what this set is all about.

The Function of the Telephone Set

Chapter 2 covered the characteristics of voice. We stated that the voice can create sound wave changes ranging from 100 to 5000 times per second, and that it was a constantly changing variable of amplitudes and frequencies. From this understanding of how a human voice works, we then must understand that the network was built to carry voice. How then does the voice get put onto a telephone company network and carried to the other end of the line. The sound waves must travel down the one pair of wires that the telephone company has delivered. The challenge is to take sound then, and convert it into something that the telephone network deals with: electricity. Yes, electricity is what runs across copper wires. The telephone company delivers a pair of copper wires to our door then attaches a telephone set to those wires. So, the function of the telephone set is to convert sound into electricity. It will therefore act as a change agent (Figure 4.1).

Conversely, when a voice is sent down the wires as electricity, the process must be reversed at the receiving end. Therefore, the telephone set must also function as a converter from electricity back into sound. The change process must work in two ways. The "from" and "to" functions are what the set must accomplish. Figure 4.2 is a representation of the reverse process.

But wait, there's more. On the top of the telephone set, there are two little buttons that are spring loaded. They must have a function too. The *switchhook*, as it is called, is a mechanism to originate a call and to finalize a call. When you press the "springy-dingy" buttons on the top of the set, the dial tone goes away. If you lift the handset off the base of the telephone, you

Figure 4.1 The telephone set converts sound into electricity.

Figure 4.2 The set also converts electricity back into sound.

hear the dial tone. So, these things "request" and "relinquish" the dial tone from the telephone company.

By now, you will notice that the telephone is a very complex piece of equipment. The invention of Alexander G. Bell has come a long way from its founding. But functionally, it still does the same thing. The first telephone sets had no dial pads, they required that all calls be handled by a telephone company operator. In order to make a call, you would have to flash the operator and request the called party. The next evolution was a crank phone, you lifted the handset and turned a mechanical crank which sent a ringing voltage down to the telephone company operator (Figure 4.3). The operator still had to place the call, but the crank was a means of signaling the operator when you wanted service. Other variations kept appearing as time went on. This was inefficient, so the manual call placement was eventually replaced with the rotary telephone set. Later came the Touch Tone set, which was designed to improve the efficiency and speed the process. Still later, the process involved the digital telset that created a whole new wave of interfaces to the network. Figure 4.4 shows a group of variations on the telephone set. The evolution has added features and functions to the set, but the primary function is still the same after these 118+ years. So, let's back up a little and look at the pieces of a telephone set, at least generically. The telephone set consists of multiple pieces. Each of the pieces serves at least one function and has a role. Without one or more of the pieces, communications might not take place. They are all intertwined to work together so that we have constant service at any time.

Figure 4.3 The crank phone allowed the user to generate a signal to the operator by turning the crank.

Figure 4.4 A potpourri of telephones has evolved.

The Pieces

The telephone set is made up of the following major components:

The base

The base is the component that all of the components are either housed inside or attached to (Figure 4.5). It is typically a plastic or space-age healed plastic that will withstand many impacts, such as dropping it from a table, desk, etc. This case protects the inner workings and other components from exposure to the most dangerous part of a telephone network: humans. We tend to spill things on the set. We also drop it or knock it over. When we are frustrated on the phone, we slam it down or beat it on the desk. We take out our anger and frustrations with other humans on this one device that is always supposed to work. Without the casing to protect it, we would never have service. This casing also protects the insides from dust and other elements that are free flowing in the air.

The handset

The handset (Figure 4.6) houses two separate components, the transmitter and the receiver functions. The handset is designed to ergonomically fit the distances from the mouth to the ear. The transmitter is located at the bottom portion of the handset, so it is positioned in front of the mouth when we hold the handset to our face. The receiver is located at the top of the handset so that it rests against our outer ear when we hold the handset to our face. The first handsets only consisted of an ear mechanism, the transmitter function was a stationary mount on the face of the base (Figure 4.7).

Figure 4.5 The base of the set houses a far more complex system.

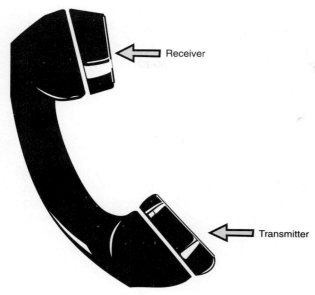

Figure 4.6 The handset houses the transmitter and receiver.

Figure 4.7 Older models had a stationary mouthpiece, but a movable earpiece.

Now to the inner pieces of the handset. Figure 4.8 shows the handset with the transmitter and receiver functions. Note that the wires also connect the pieces. There are two wires running from the ear piece and two wires running from the mouthpiece. Remember that the telephone is going to convert sound into electricity. The wires will carry the electricity created at the mouthpiece to the telephone set. The second set of wires will carry electric-

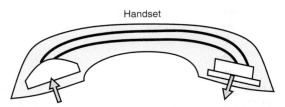

Handset

Figure 4.8 The transmitter and receiver are connected with two wires each inside the handset.

ity from the telephone set to the ear piece. So, there is a four-wire circuit in the handset, two wires for the transmit and two wires for the receive. Each side of the function has its own two-wire electrical circuit at this point. Modern-day functions of the following parts are still the same. However, newer solid-state componentry have replaced the pieces, but they functionally do the same things as the pieces described in the following paragraphs.

The transmitter

As mentioned in the discussion about the farmer wanting to see the inside of the telephone set, the transmitter houses some very sophisticated componentry. The first portion of the transmitter houses a diaphragm just under the outer casing of the mouthpiece. The diaphragm is a very sensitive membrane that will vibrate with air pressure changes. When we speak into the mouthpiece of the telephone set, we are really creating air pressure changes with our vocal cords. This sound pressure causing air pressure changes will move at the same frequency of the changes that we produce by vibrating our vocal chords (larynx). The pressure changes cause the membrane on the diaphragm to move back and forth at the same frequency (Figure 4.9). Behind the diaphragm are loosely packed carbon particles (the black pellets the farmer dropped). When the diaphragm moves back and forth, it causes the carbon particles to move back and forth. This produces an electrical charge that is occurring at the same frequency and amplitude of the sound pressure changes we create with our voice. The result is an analog sine wave that is directly (mostly) proportional to the frequency and amplitude of our voice, or the electrical equivalent of our sound. This electrical energy will be carried away from the transmitter across the wires that exist behind the mouthpiece.

The receiver

The receiver function is shown in Figure 4.10. When voice or other telecommunications in the form of electrical energy are being transmitted to us from the far end (the other party on the phone call), the network carries this electricity to the telephone set. The wires behind the receiver are run up to an electromagnet that is mounted inside the receiver portion of the handset. The electromagnet is attached to a diaphragm similar to that in the mouthpiece. The electricity comes through the wires to the magnet, and the magnet will vibrate at the same frequency of the received energy. As the magnet vibrates, it causes the membrane on the diaphragm to move back and forth. This back-and-forth motion causes air pressure changes (vibrations) that are at the same frequency of the transmitted energy. As the air pressure changes occur, they cause the air waves to bounce off the inner eardrum, which is what produces sound to our ears. See Figure 4.11 for the eardrum comparison.

The transmitter/receiver functions complete the conversion of sound to electricity and electricity back to sound. We often refer to the handset functions as the *E&M* which many people refer to as the "ear and mouth" communications process. The E represents the direction of flow to the ear, or to the receiver and the M represents the direction of energy from the mouth or the transmitter. In actuality, E&M means something different, but that is for a later discussion, possibly the sequel to this book.

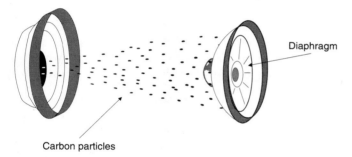

Figure 4.9 The transmitter had a diaphragm and loosely packed carbon particles that create the electricity.

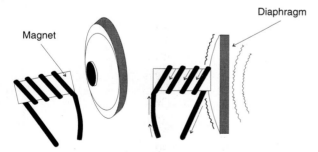

Figure 4.10 The receiver uses an electromagnet to change the electricity back into air pressure.

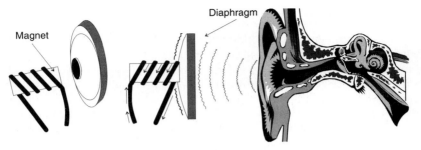

Figure 4.11 The eardrum receives air pressure, changes it, and converts air pressure into sound.

The connector or the handset cord

Coming from the handset and proceeding to the base of the telset is a curly-coiled wire. It doesn't have to be curly-coiled, that is to make it easier to manage. However, inside the curly-coiled wire are four wires. These are the two from the transmitter and receiver respectively. The modern sets use a connector that is called an *RJ11C*. This is the four-wire connector that has a little spring-loaded clip on it to plug and unplug the connector into the female receptacle. The cord is double-ended with RJ11C, male connectors. This is also a matter of convenience because the cords and the connectors are the most likely pieces to wear out, it makes sense to use a plug-and-play module. Older versions of this cord used a screw-down version (called a *spade lug*) but this required that the telephone set be opened up to get at the screw mounts. Remember this was an area that we were not allowed into.

A funny thing happens along this connector cord though. When it gets to the base of the telephone set, it gets plugged into the female version of the RJ11. But inside this jack, a hybrid arrangement takes place. One of the transmit and one of the receive wires are connected to each other (Figure 4.12). The other two wires that continue into the set are covered in the next section. The important part of this is to understand why the transmit wire gets connected to the receive wire. This creates a loopback arrangement of sorts. The transmitted sound is looped right back to our ear. We can hear everything we say in the conversation because the energy is transmitted back to the ear piece. You are probably asking yourself why this is done. The process produces what is called *side tone*. This side tone is a means of letting us know that an electrical connection across the network still exists. When you speak into a telephone during a conversation, and you hear yourself, you recognize that there is still a connection to the distant end. When you speak into the set, and you hear nothing or the sound is dead, you recognize that there is no one out there. In other words, you have been cut off. This is part of your basic training in telephone usage. Although no one ever explained this to you, it was a learned process through experience.

The inside of the telephone

Once the wires are run from the handset to the base of the telephone, two wires go into the set. Actually, these wires meet up with a whole network of cross connects (connections from place to place) inside the set. But, the two wires are what we are concerned with. As shown in Figure 4.13, one of the wires, the transmit wire, is connected inside the set to the dial pad (whether rotary or tone dial). This is used as an addressing mechanism. The telephone network deals with the use of a decimal numbering system.

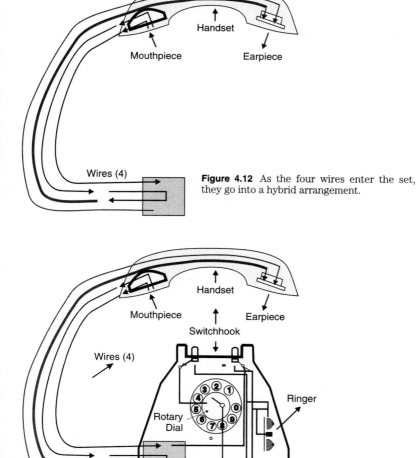

Figure 4.12 As the four wires enter the set, they go into a hybrid arrangement.

Figure 4.13 The transmitter wire is connected to the dial pad and the switchhook. The receive wire connects to the ringer.

This is in the form of a sequence of digits that we dial to the called party. To produce these digits the dial pad is used to send the information out the telephone company for delivery to the called party. This wire connected to the dial pad is also cross connected to the switchhook (the springy-dingies) where the spring-loaded buttons exist. The same wire is used to connect to another jack (RJ11, RJ12, or other), which is connected to the outside world.

The second wire that came from the handset cord is connected through the network of wires to a ringer. The ringer can be in the form of a bell or a turkey warbler (electronic ringer), but the function is what is important here. The ringer is a receive wire function. It is used to alert you that someone wishes to speak to you. If you did not have a ringer, every now and then, you'd have to pick up the handset and listen to see if someone is on the line. This would be inconvenient, not to mention how the caller would feel if it was undetermined if you would ever pick up the handset. They would be in limbo forever. So, the ringer is a crucial component of the set and the process.

The switchhook

Already mentioned, but equally important, is the switchhook function. This is the portion of the set that actually gets the dial tone from the telephone company or when we are through with the call, gives the dial tone back. In reality, what is happening is that an electrical wiring circuit is physically connected from the telephone company office to our telephone set. This circuit gives us access to the services offered by the telephone company, mainly the dial tone and the ability to make and receive calls. Under normal conditions, we do not have a dial tone at our set. This is a service that we must request. To request the service, we lift the handset off the switchhook (called *going off hook*). When we do this, the little buttons pop up. At the same time, they "make" the circuit by adding a resistance across the wires. The telephone company equipment is constantly monitoring all of its connections to end users. When the central office computer (now they are computers) sees a change in the resistance across the wires, it recognizes this as a request for dial tone. If dial tone is available, it will be provided immediately. The time it takes us to go "off hook" and bring the handset up to our ear, the dial tone is there. Actually, it gets there much faster than that if everything is functioning properly. As we use the line and make calls, we will use the other components in the telephone set, but the central office computer will monitor the line (wires) for the time when we are finished with our call. When we hang up a call, we place the handset back into the cradle. This pushes the spring-loaded buttons back down, breaking the circuit. This causes the central office computer to see the change on the wires and recognize that we are all done. Therefore, the central office computer takes the dial tone away from us and uses it for other users who require it. Then, the process starts all over again.

The dial pad

It really does not matter whether you have a newer set or older one. The dial pad uses the set to perform the addressing service to the network. When we have our dial tone, we then start a sequence of dialed digits giving the central office an indication with whom we wish to speak. As mentioned, the first telephone sets did not have a dial pad. Everything was manually performed through the telephone company operators. If you needed to speak to someone, you got the operator by going off hook. The off-hook process created the "make" on the circuit that lit a light or created a visual indication to the telephone company operator. The operator would answer and converse with you. At this point, you would pass on the name or number of the requested party and the operator would provide the connection. It was quite simple, but as things got busier, the human involvement became an issue. You could be on the line waiting for an operator forever, if they were busy at the central office.

Hence, the first attempt to automate the process was to give us a rotary telephone set (Figure 4.14). Because this was the first attempt at automating the process, it was well received. However, over time it became a nuisance. The rotary set is an electromechanical device. As you dialed the number, you would crank the dial in a clockwise direction. When releasing the dial, the counterclockwise return to normal position created a series of electrical interrupts or pulses across the line to the telephone company office. The CO would interpret these pulses, and make the necessary routing decisions. The main problem which led to frustrations was that statistically to make a call and speak to the intended party, four attempts were necessary. Either the network was busy; the called party was busy; the phone at the receiving end went unanswered; or some other event got in the way.

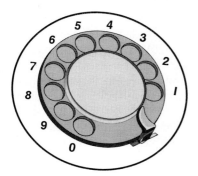

Figure 4.14 The rotary dial pad.

Thus, users felt frustration with the use of this set; especially because it took so long to dial a telephone number.

The pulses are generated at 10 per second (10 PPS) which also created long call set-up times because of the slowness of the dial pad. Additionally, because this is electromechanical, the system was prone to failures. The rotary set might stick, miss-dial the digits, or other such influences would wreak havoc on the user. Users began to complain, and the manufacturers had to do something about it. The primary manufacturer at the time was Western Electric, a division of AT&T. Why even bother to mention a set that is so old? At least 35% of the U.S. telephones are still using pulse dialing. There are many reasons for this, but some of the major ones are:

- *Cost* Many people only use the phone infrequently. The cost of touch tone dialing is typically $1.00 more per month. Therefore, these people won't pay the price.

- *Technology* Many independent LECs have old technology and have not yet planned to upgrade their equipment. Therefore, the service is only rotary.

- *Gimmees* A lot of folks who now own their own phone have subscribed to some magazines and received free phones as a reward for signing up. These phones might have push buttons on the set. However, when you use it, the conversion from tone to pulse occurs inside the telephone set. One need only hear this once to recognize the distinctive click click in the ear to know this is a rotary set.

The second edition of the dial pad then became the Touch Tone phone. This changed the look and feel of the dialing process. Instead of using the rotary set to electromechanically produce the dialed digits, AT&T's Western Electric Division created a tone dialing sequence. Using a group of tones that fall into the spectrum of the voice frequency, the tone dialer has a set of buttons that will send out frequency-based tones instead of electrical impulses. The touch-tone telephone set was created to speed up dialing and call processing; reduce user frustration; increase productivity and allow for additional services through the dial pad. The touch tones are actually a dual-tone multifrequency (DTMF) set. Each of the numbers represented on the dial pad have both an X and Y matrix or crosspoint. This was a decision that Western Electric made to create a number of tone generators that could be used at the set, and tone receivers to be used at the Central Office. Western Electric did not want to have too many of these tone generators/receivers, because they are expensive. Therefore, by using a dual-tone arrangement, they could get away with using only half of the generators needed to represent the same number of digits. See Figure 4.15 for the tone dial pad and note the frequencies as they appear on the chart. When you press a number, two separate tones are generated and

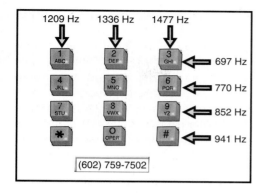

Figure 4.15 The Touch Tone phone.

sent to the CO simultaneously. Each number on the dial pad has its own distinct set of dual tones, so the system recognizes each as a discrete number. This allows for quicker dialing, due to the fact that the tones are generated at 23 pulses per second (23 PPS); much quicker than the rotary set.

As the tone dial was introduced, a 12-button pad was created, adding two extra numbers to the telephone. The older rotary set only had 10 numbers (decimal). These new buttons allowed for new features and capabilities to be introduced in the telephone world, using the pound (#) and star (*) keys as delimiters when dialing into computers, etc. The extra keys allow for feature activation in Centrex and PBX-type systems (for example: to put a call on hold, dial #4, to retrieve the call back *4). These extras also squared off the dial pad to make it easier to form.

Tone dialing also helps to get the call through the network quicker because the tones pulse out at a much faster rate than that of rotary pulsing. This improves network performance because the network processes the call set-up and tear down faster, creates happier users because they can dial their numbers much quicker rather than the tedium of a rotary dial sequence, and promotes an overall satisfaction with the network.

One example of improvements is with the call set-up time. In the rotary call, a call set-up was 43 seconds; whereas with tone dialing the call can be set up in approximately 6 to 10 seconds. Other improvements, such as improved signaling systems, have reduced this set-up time further, but the tones were the first big thrust in gaining time back from the network. Is this important? Sure it is! Think about the hundreds of millions of calls placed in any given month. If you had a system that took 43 seconds of call set-up (which the older system not only took that long, but tied up a line across the network for the same set-up period) and you could reduce it to 10 seconds, you have reclaimed 33 seconds of line utilization (which was not billable time) per call, for other users. This is big bucks no matter how you look at it. There is a paradox in this situation, however.

Think about this for just a second: you as a user of the telephone network have a rotary dial set. You want to place a call from Boston to San Francisco (approximately 3000 miles). You place the call and tie up 3000 miles of wire for as much as 43 seconds. This is nonbillable by the carrier. The clock only starts when the call gets answered. So, now the carriers offer you the ability to speed up the process by using a tone dialer. You are obviously going to be happy with this arrangement because you can dial so much faster. The carrier stands to save 33 seconds (remember back then the cost per minute was $0.65) and get the call billing going that much faster. For this savings of approximately $0.30 to $0.35 per call that you make (which all rolls back into the carriers' profits), you have the privilege of paying $0.75 to $1.00 per month more on your phone bill for the convenience of tone dialing. Isn't this a wonderful industry?

The ringer

Already discussed, the ringer serves the purpose of alerting the called party that someone wishes to converse. The ringer has no mystique other than it must be "told" via an electrical current to ring. The current from the central equipment out at the telephone company office is sent to a ringing generator that will produce the current across the line. A -48-Vdc current used from a battery in the office produces a 120-Vac output across the telephone line to the customer location. This is as simple as it needs to be. Occasionally, differences appear in the way these things are laid out or the voltages that are applied to the line, but we are using norms in our examples.

Tip and ring

Tip and ring is a phrase that telco installers or repairers use all of the time. They love to talk about tip and ring. This is a description of the two wires that are connected to the telephone set from the outside world. You will recall that the four wires that left the handset went through a hybrid arrangement as they came into the telephone set. At this entrance point, only two wires were carried into the set to connect to the signaling systems inside the phone. The two wires, one from the transmitter and one from the receiver, now exit the set and proceed toward the telephone company connection. They call the transmit wire the "tip" and the receive wire the "ring." This nomenclature is easier described as the transmit and receive wire, but telephone company personnel held the words over from a bygone era. *Tip and ring* comes from the use of the old operator cord boards. When the operator tested the line, the tip of an RCA jack was used. If a call was already in motion on a particular line, the operator would hear static on the line when "tipping." If the line was not in use, the operator would not get any static from the tipping; therefore, the operator would insert the jack on a cord board into the designated extension port on the board. This insertion would then cause a ringing voltage to be generated across the wires and your phone to ring. This stuff has all but disappeared here in the U.S., but in several other parts of the world, cord boards still exist. That's the main reason for the mention of where this all came from.

Newer Sets

The telephone is a remarkable set that evolved over time. Conceptually, nothing has really changed, but the pieces continue to move on. Today's digital telephone sets provide far more than just a plain old analog telephone call. The older set functionally provided the interface to the network provided by the various carriers. This interface, with all of the electrical and mechanical components, got us to where we are today. However, progress continues. Users wanted more features and functions from the devices placed on the desk or wall in the office and residences. Consequently, the migration to solid-state, microprocessor-controlled telephone sets became the wave of the future. The newer sets are digital. That means that they perform the same functions with less voltage, very discrete events are pre-programmed into the sets, and other characteristics of digital transmission will add to the quality of the conversation. Digital transmission is covered in later chapters (Chapter 5 discusses the differences between analog and digital, and Chapter 13 discusses the T1 digital transport system).

The telephone set was a clear target to roll digital technology into. The newer sets are varied by the number of manufacturers. Anyone can now manufacture a telephone set, or system for that matter, so long as they abide by certain registration techniques and parameters. A newer set is shown in Figure 4.16. On this set, several added features that enhance the call placement and process are incorporated into the telephone set rather than being served by an external device. For example, some of the features that are now becoming the norm are:

Speed dialing

A user can preprogram from 1 to 100 numbers into the memory of the set. By using a feature button or a pre-programmed button on the set, the telephone will dial a much longer string of digits (usually up to 33 digits long) and connect you with the requested party.

Call hold

A hold function will place the handset in the mute mode, both the transmitter and the receiver function are disengaged for a period of time. Using this hold feature, a visual indicator will alert you that you have done some-

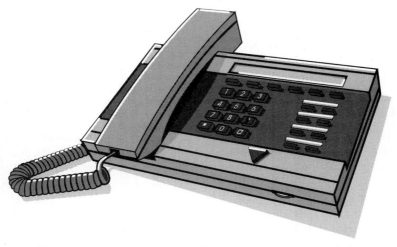

Figure 4.16 Newer digital sets are feature-rich, but still functionally perform the same things.

thing. In the case of hold, a flashing light will let you know that the call is still there but on what is called *hard hold*. This serves as a mechanical function by keeping the link active, but disengaging the set until the line button is pressed again to reconnect the parties together.

Call transfer

Assuming this feature is available at the central office or PBX equipment within your area, a call can be transferred to another line. This function is available in many central offices today. To activate it, you must press the hook switch down and draw what is called *recall dial tone*. Then you must dial the third-party number (the transferred to) and wait for the line to ring. It's a good idea to wait for the new called party to answer and announce that the call is being transferred, then hang up. However, many people don't provide this courtesy. A problem exists with activating this feature; it requires that you press the hook switch down for a short period (usually 100 to 200 milliseconds) to draw the recall dial tone. Most people fall into one of two categories: "fat fingers," where they hold the hook switch down for 1 to 2 seconds and cut off the caller; or "skinny fingers," where they press and release the buttons too fast and never get the recall dial tone. Consequently, a preprogrammed feature button on this newer set will perform these functions for the exact amount of time and prevent the problems associated with using transfer.

Conference call

This is similar to the transfer feature, but in this case, the third party called will be added onto the call, rather than transferring the call to this third party. The conference feature allows the three parties to converse without the need to worry about cutting each other off.

Redial last number

Using this feature, the set has its own internal memory, so it can automatically dial the last number called—even if it is not a speed number. This is great when you encounter a busy tone on the network. The problem with last number redial comes into play when the call requires a string of digits that are beyond the normal 11 digits for a phone call. Pauses are required in some cases, but this feature will send the 11 digits, plus the credit card number and any other digits dialed in the last sequence. This can cause some confusion, but it is a time saver. Redial can also be a preprogrammed thing where some sets upon encountering a busy tone will redial up to 99 times in a preset time period (over the next 60 to 90 minutes as an example). Not all sets will do this. The degree of sophistication is directly proportional to the dollars spent for the set.

Built-in speakerphone

Many sets have the ability to turn on a speakerphone for hands-free two-way communications. This can be simplex, which means that one must be careful not to overtalk the other end. If overtalk occurs, then the conversation becomes choppy because the set is bouncing between the transmitter and receiver, cutting off words or syllables. The other option which some sets will have is a duplex speakerphone. This means that simultaneous talk and receive can occur without cutting in and out. This duplex circuitry has separate paths for the transmitter and receiver.

Hands-free dialing

Similar to speakerphone service, but not the same, hands-free dialing allows just that. Without picking up the handset to the face, a user can go off-hook in the electronics by pressing the line button or an "on" button, then dial. This frees the users' hands for other tasks while the connection is being made. However, once the called party answers the phone, the caller must pick up the handset and converse. The speaker is only a receive speaker not a two-way. Many people get confused over this feature, and cause a lot of grief or hangups when they attempt to use it improperly.

Displays

Where features are available in the central office, a display set can be used to show the incoming caller identification. This is not ubiquitous and legal everywhere, but it is coming. Where the incoming caller display is not available or used, the display can still be used to display the number that the caller dials. Further, when the call is dialed, a timer can be used to keep the chronology of the length of call, etc. This can be used for client billing and chargeback or for reminders of what time it is; in case you have a meeting.

5

Analog versus Digital Transmission

Introduction

In the data and voice communications field, one of the most common points of confusion are the words "analog" and "digital." Most people recognize that the term *digital* refers to the expression of information in terms of ones and zeros. But few can easily make the mental connection between this fact and the real-world requirements of moving voice signals expressed in a "ones" and "zeros" format. In dealing with the term *analog*, everyone merely refers to voice communications. This chapter will help you to make and understand the connection between these two terms. Not that you will become an electrical engineer, but you will be able to converse with the best of them! If there is a single element that must be understood, it is that analog and digital are means used to move information across any medium. Some media typically deal with analog, others deal with digital, and the rest deal with both. Thus, this understanding will hopefully help you to understand the reasons why the different transmission systems are used.

Analog Transmission Systems

As mentioned earlier, "the network" was originally developed solely to provide voice communications services. The communications circuits AT&T built through the Bell Systems and their own communications capabilities used strictly analog technology. Yes, it changed into a digital world, but it was built around analog communications for voice. But what does this mean?

In an analog communications system, the initial signal (in this case, the spoken word) is directly translated into an electrical signal. Chapter 4 describes how the telephone set accomplishes this task in more detail. The characteristics of an analog signal deal with two constantly changing variables: the amplitude and the frequency of the signal. The strength (amplitude) of the electrical signal varies with the loudness of the voice, and the frequency of the electrical signal varies with the pitch or tone of the voice. Both variables (amplitude and frequency) change proportionately with the original sound waves.

If the signal would be monitored with an oscilloscope, the displayed pattern would visibly and recognizably change as the sound changed. This is shown in Figure 5.1, where the signal characteristics can be visibly shown on an oscilloscope. For a more graphic example, think of the visual displays in some discos, where the music is fed into a light display that varies as the music changes. In an analog communications system, instead of light, the electrical signal changes. Figure 5.2 shows an analogy to this concept, where a music beat is shown. The lights of a disc jockey's equipment will change and flash with the beat of the music. In this case, the lights are changing based on the electrical characteristics of the signal. What is actually happening? In telecommunications terms, you could say that the actual sound is produced by banging air waves together. This banging of air waves is really the movement of the air molecules. Technically this is called *compression* and *rarefaction*. But, who wants to get technical?

The human voice is an interesting phenomenon. As we speak, we generate sound. We actually bang the air waves together quite a few times in a relatively short period of time. This was discussed in Chapter 3, but to stick with this thought, we create an analogous sinusoidal wave. This wave is then converted into its electrical equivalent. Quite simply, if you were to take the sounds created by the voice, and modify them based on the pitch and the strength of the sound, you could produce this sinusoidal wave (Figure 5.3). The wave is what would happen electrically if we were to use a magnetic field to change the sound into electricity. Working around a base line of zero electrical voltage, we then would have a 360° rotation of electri-

Figure 5.1 Using an oscilloscope, you can see the characteristics of an analog signal.

Figure 5.2 Comparing the analog wave to how the beat of the music changes lights in a DJ's equipment.

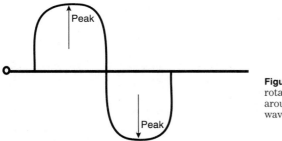

Figure 5.3 The sinusoidal wave rotates 360° of electrical current around a magnetic force. This is one wave or revolution, called a hertz.

cal current. This 360° wave around the zero line is called a *hertz*, named after the electrical engineer who documented this concept. The wave starts at the zero line and rises as the amount of energy increases. This will ultimately peak at some point and begin its descent. The electrical energy will then proceed from the height on a downward slope to the zero line. From here, it will continue in its slope below the zero line to a point below the zero line (or the negative side of the wave). Continuing to a peak voltage on this side of the line, the energy will hit its peak value then begin an upward slope back up to the zero line. The wave has made a complete 360° cycle. One complete cycle constitutes one hertz.

The human vocal chords can bang the air waves together between 100 and 5000 times per second. Simply stated then, the voice is producing up to

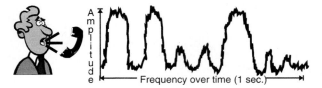

Figure 5.4 The human voice produces usable information in a range of 300 to 3300 Hz per second.

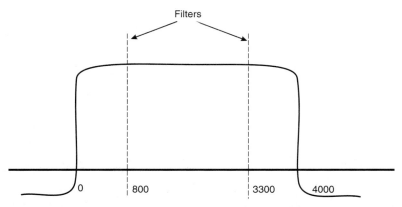

Figure 5.5 The telephone companies band limited their 4-kHz channels to a usable 3 kHz with bandpass filters.

5000 cycles per second. When this is converted into its electrical equivalent, there are up to 5000 hertz. We do not like to state such big numbers, so we abbreviate this to 5 kilo (meaning thousand) hertz—5 kHz, for short. From this electrical equivalent, we now have an analogous "or look alike" to our sound waves. This is a constantly changing variable of electrical energy. Both the amplitude and the frequency will change from 100 to 5000 times per second. This is something that the telephone companies learned to deal with from the beginning of our communications industry. Over time though, they found that the human voice will generate from 300 to 3300 cycle changes per second or 3 kHz of electrical cycles as a norm (Figure 5.4). As with any communications channel capacity, the telephone company did not want or need to give users any more than they needed to carry a voice conversation. Over the years, as the network expanded, the telephone companies limited the bandwidth of a telephone call to that of a 3-kHz channel capacity. This was a money issue. Given a limited amount of bandwidth, how can we allow a human conversation to take place across the network and produce reasonable representations of the original voice? And how can we do it cheaply? The range of frequencies that are generated are from 300 to 3300 Hz (Figure 5.5). The telephone company therefore will limit our use of the channel to just that amount. In radio frequency spectrum, the telephone companies divided all of their capacities into 4-kHz slices. On each of these 4-kHz slices (wires or radio channels), they install frequency bandpass filters at 300 and at 3300 Hz. Anything that falls in the middle of this allocation of RF spectrum will be allowed to pass. Anything that falls outside of this range will be filtered out (thrown away). This is what they call a *band-limited channel*. Because the voice can go as high as 5 kHz, there will be cases where we hit the higher range of frequencies with our conversation (such as words with the SSSS sound and the FFF sound), and it will be flattened out through the filters. This might sound unreasonable, but what it really produces is a little fuzziness on the line. The human ear and the telephone equipment is not sensitive enough to recognize any significant problems. Once the electrical equivalent of the sound is created (through the telephone set or another device), the electricity is sent down the wires.

When a call is proceeding down the wires, resistance in the wires to the electrical signal immediately begin to diminish its amplitude. The signal gets weaker and weaker. This weakening of the energy will eventually lead to the total absorption of the electrical energy, or what will be the loss of the signal beyond recognition. This is called *attenuation* of the electrical signal. The signal can only travel so far before it runs out of strength and disappears. An analogy to this is like a relay race runner who is trained to run around a quarter mile track as fast as he or she can. By the time the runner gets to the end of the quarter mile run, all strength and energy is expended. The runner is out of steam (Figure 5.6). If the runner did not have someone to pass the baton off to, but had to go around the track again, the second loop would take forever. That is if the runner made it around at all before passing out and falling flat on the ground.

Figure 5.6 The runner around the track is "out of energy," so he must hand the baton to a new runner.

In order to keep the signal moving along the wires, amplifiers are used to boost the signal strength (Figure 5.7). These amplifiers are usually spaced about 15,000 to 18,000 feet apart. Typically, we only need one amplifier between the customer location and the telephone company central office; two, at most, are needed. The central office is usually located close to the customer location, or within 5 to 7 miles on average. Only in remote locations will the central office be farther away.

A second activity takes place on these wires concurrent with the transmission. Noise (introduced line loss, frayed wires, lightning, electrical inductance, etc.) is ever present in the form of white noise or hiss. It also begins to cause a deterioration of the signal. Noise is always on the line, but its increased presence from cabling problems causes the degradation of the signal. Unfortunately, as the signal proceeds down the line the noise and the signal begin to intertwine. Amplifiers cannot discern the noise from the actual conversation. Thus, the amplifier not only boosts the signal, it also boosts the noise. This creates a stronger, but a noisier signal, which leads to some extremely noisy circuits (Figure 5.8). The results of amplification are cumulative over distance. The more amplifiers that must be used, the worse the overall signal gets. Many of the longer-distance circuits were constantly being amplified. This again is not to imply that the telco doesn't want to do any better. It is a function of the equipment, the electrical characteristics, and the finances combined.

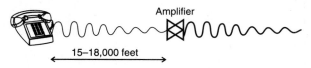

Figure 5.7 Amplifiers are used to boost the signal strength up along the wires.

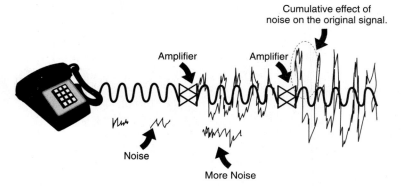

Figure 5.8 The amplifier is amplifying both the telephone signal and the line noise.

It is interesting to note here that some of the latest advances in digital signal processing (DSP) could now be used to "clean up" somewhat an amplified voice transmission if analog technology were still in wide use. But a better approach to producing clean signals was developed many years ago.

Analog-to-Digital Conversion

Because the analog version of the network was a problem for the network suppliers and the telcos alike, both migrated to a digital form of communicating. In order to convert the voice conversation from analog to digital, a device called an analog-to-digital (A/D) converter employs a sampling technique.

Sampling refers to the process of measuring (see how we are creeping up on quantifying a signal?) representative portions of a signal over time. We make the assumption that chronologically adjacent portions will differ only slightly. If the samples are taken frequently enough, and played back faithfully at the other end, the ear will not be able to differentiate the playback from the original. A (nondigital) sampling technique is used in movies and other video applications. When a movie is made, there is no truly continuous record of the images; instead, a series of still images, sampling the reality at 30 samples (or "frames") per second, is recorded and later presented to the viewer. Normally, the viewer cannot distinguish between playback of the samples and the real thing.

As mentioned earlier, the bandwidth of the audio signal we wish to transmit is 3000 hertz (3300 minus 300). Based on the Nyquist theorem (which states that one should sample at a rate at least twice the maximum frequency of the line), the minimum sampling rate would be 6600 hertz (2 times the 3300 Hz). In fact, a somewhat higher rate of 8000 hertz (samples per second), is used. This is used to address the higher range of frequencies on a conversation, such as those SSS and FFF sounds that were filtered out in the analog world (Figure 5.9).

Each sample measures the amplitude level of the voice signal at a particular point in time. One sample comprises eight *bits,* where a bit represents a one or a zero. An eight-bit character or *byte* can represent any decimal number from 0 to 255 (00000000 is zero, 00000001 is one, 00000010 is two, 00000011 is three, 0000100 is four, 0000101 is five, and so on up to 11111111 which equals 255). Therefore, there are a total of 256 possible levels, sufficient enough to recreate the analog signal in good faith at the receiving end. More sample values would produce a higher-quality replication, but the ear is not sensitive enough to discern the differences.

Eight thousand samples per second, where each sample requires eight bits, generates a digital stream of data at the rate of 64,000 bits per second. We know this as the digital signal 0 (DS0), the digitized equivalent of one voice channel. The bits are each in the form of a *square wave,* as contrasted to the familiar sinusoidal wave that is typically seen on an oscilloscope.

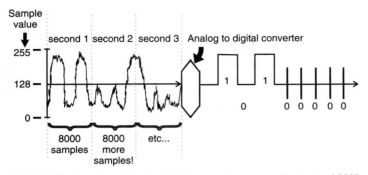

Figure 5.9 During the analog-to-digital conversion, a sampling rate of 8000 hertz is used.

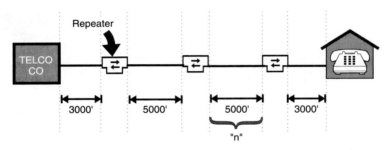

Figure 5.10 Repeaters are stationed at approximately each mile of line between the telephone company and the end user.

The square wave travels down the same pair of wires we are accustomed to, but in order to handle the digital signal, the amplifiers are removed. Additional equipment (such as loading coils) are also removed. In total, the entire circuit is reengineered from end to end. In place of the removed equipment a digital regenerator (or regenerative repeater) is used. The strength of the digital signal is based on a 3-volt pulse of a very short duration; therefore, the repeaters must be placed closer together, typically every mile (Figure 5.10). This requires more equipment, which is an expense to the telcos and network suppliers alike. But when the samples are played back at the far end, customers receive the transmitted signal in a form indistinguishable from that sent from the original A/D converter at the sending location.

Digital Signaling

As mentioned earlier, digital signals consist of ones and zeros. But even this statement can be a bit confusing; how does one insert a "one" or a "zero" onto a wire? Unlike the varying levels used in analog transmission, signaling on a digital circuit is very simple; two different voltage levels (e.g., +3 volts) are sent to represent ones, while no voltage represents a zero. This is called a *unipolar signal*. The receiving equipment measures the voltage level once each bit time.

A *bit time* is the length of time that it takes to transmit one bit. For example, at 9600 bits per second, one bit time equals ¹⁄₉₆₀₀ of a second (about 104 microseconds). The measurement is taken in the middle of the bit time, when the likelihood of getting a precise level is maximized (Figure 5.11).

There are actually two types of digital signaling. The previous description presents the type of signals used for local (that is, non-telco based) digital signaling. A slightly more sophisticated scheme, called either *alternate mark inversion (AMI)* or *bipolar signaling*, is used for wide-area digital transmissions. Chapter 13 includes a detailed discussion of AMI.

Because voice is inherently analog (i.e., it varies continuously, rather than occurring in a discrete, numerically related pattern), an analog-to-digital (A/D) conversion must be performed to a digital format if you want to transmit voice digitally. Later paragraphs cover how to change an analog signal into a digital format. But first, we must answer the important question, "Why bother to convert to digital?" The answer is quality.

Analog signal paths must use amplifiers if the paths cover any significant distance. Because amplifiers retain and even strengthen noise in a conversation, the more amplifiers used in a transmission path, the noisier will be the background. So, when you use an analog circuit to call your grandfather in the old country to wish him a happy birthday, you will be most fortunate if either of you can understand the other over the background noise. To compensate for this problem of a noisy overseas (or long distance) call, we use an age-old philosophy. Yell!

By yelling into the phone to overcome the noise, we do little to help the situation. When we raise our voices to new heights, we increase the amplitude. This makes the signal very strong (stronger than what the telephone companies expect). The telephone companies have, however, installed

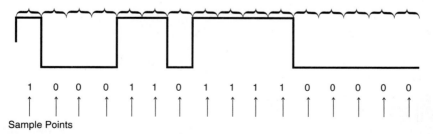

Sample Points

Figure 5.11 The measurement is taken in the middle of the bit time when the likelihood of getting a precise level is maximized.

equipment on the line that will hold the height somewhat to a level. These *pads* as they are called sense that the amplitude is much higher than normal, so they introduce loss to compensate and bring the amplitude down. This just serves to compound the problem, because the loss in decibels (dB) will steal away the amplitude, but do nothing for the noise. So, we have gained nothing (Figure 5.12).

The digital equivalent of an amplifier is a *repeater.* We compare this to what we commonly refer to as a "schizophrenic cookie monster." Why that? Well, let us explain:

As the digital pulse is placed onto the wires, only the 1s (+3 volts) need to be of any concern. But as the pulses are placed onto the wire, it immediately begins to propagate down the wires. The wire will immediately begin to act on this pulse by diminishing its strength. The result will be a constantly decaying pulse of electricity (Figure 5.13). As this decaying process begins immediately, the likelihood of the pulse falling into obscurity is very high. So this "cookie monster" is placed on the link at approximately every 5000 to 6000 feet. As the pulse declines in strength, it must be kept at sufficient strength to be recognized as a pulse and not the result of noise. This distance is used to keep the signal above what is called the *pulse detection threshold level.* If the strength of the pulse falls below this detection threshold, it will be mistaken for noise and ignored.

When the pulse arrives at the repeater, the repeater will act like the cookie monster. It will literally absorb the signal—eat it up and make it dis-

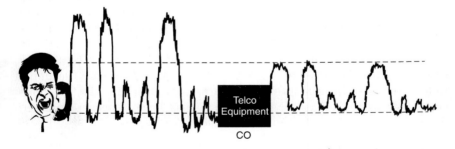

Figure 5.12 By yelling into the set, the amplitude is increased, but equipment at the central office "pads" the signal to a norm. Therefore, nothing is gained by yelling.

Figure 5.13 The pulses of electricity constantly decay as they travel down the line.

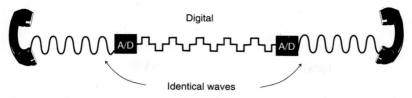

Figure 5.14 If digital is used, it is possible to deliver an identical signal to the receiver as what was inserted at the transmitter.

appear. The cookie monster has done its job. However, the other side of this cookie monster is schizophrenic. Therefore, it will basically say, "there was a pulse there, so I'll be a sport and put a brand new pulse on the line in the original's place." Thus, a brand new 3-volt pulse will be introduced. Thus, the signal begins to propagate immediately down the next leg of the circuit. And the process starts all over again.

But where an amplifier simply strengthens the input signal, a repeater detects the 1s and 0s of the original signal and retransmits them out the other side as good as new. In particular, any noise introduced into the signal between the source and the repeater is completely ignored as long as it stays below a certain tolerance level. Amplitude decreases will still occur between repeaters, and noise will be introduced, but so long as a repeater can recognize the original pulses (ones), those are the only parts of the received signal that will be used to build the new output signal. Thus, if a communications path is built with digital technology, it is entirely possible to deliver to the ultimate receiver a signal that is identical to that sent from the first digital transmitter in the path (Figure 5.14). Because the pulse is short duration and a discreet value of energy, rather than a constantly varying voltage level, more repeaters are required on the link. Telcos must use three to four times as much equipment on the same distance of wire with digital communications at the local loop.

Note the last part of the previous statement ". . . from the first digital transmitter in the path." Because voice is analog, there will be a conversion at some point. From the voice source to that first conversion point, noise can be introduced that will remain in the signal. Obviously, the closer to the voice source that the A/D conversion takes place, the better.

If you have a conversation with someone in another state or another country, and that conversation is as clear as though the person to whom you are speaking were right in front of you, you can be sure that most if not all of the technology involved is digital, rather than analog. Such quality results in happier customers, cleaner transmission, improved data communications, and happier carriers. This is why all the hype exists in the industry on the benefits of digital transmission.

There is another benefit to using digital, rather than analog transmission, although it directly benefits the service providers more than their customers: digital technology facilitates combining and manipulation of signals, providing a carrier more flexibility in how it moves the signal from source to destination.

Digital Data in an Analog World

We have been focusing on voice transmission, a communications technology that starts and ends with an analog signal—a voice. But what about data communications? Because it starts out as digital, doesn't that simplify things? The answer is no, not so you'd notice.

Until the last few decades, digital service was not available to users. Since then, an entire technology has come into being that provides methods for carrying digital signals across the analog voice network. The critical component in this technology is the *modem*, a contraction of the terms *modu*lator and *demo*dulator. In a nutshell, a modem's function is to provide digital-to-analog conversion (and of course the reverse) for a digital device that needs to communicate with a remote digital device across the analog telephone network. Figure 5.15 shows this multiple conversion process.

Because in fact telephone companies' backbones are now mostly digital (with the local loop signals being digitized at the CO), a communications path using modems includes not one, nor even two, but rather a minimum of four digital-to-analog or analog-to-digital conversions—and that's only one way!

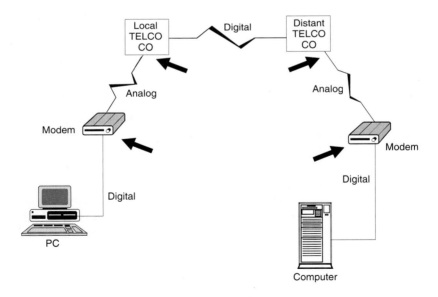

Figure 5.15 Four Analog-to-Digital or Digital-to-Analog Conversions!

Modems and other data communications equipment are addressed in more detail in Chapter 12.

Most voice and data transmission systems now use digital technology, with a few key exceptions. An exception that has particular impact is the telephone local loop. The local loop (sometimes called the "last mile" or other less printable terms) has been held back because of the vast installed base of analog telephones and amplifiers. However, even this last bastion of analog transmission is under assault.

A digital technology and telco service offering called integrated services for digital networks (ISDN) is slowly spreading. ISDN can provide digital voice transmission right up to and including the telephone. Chapter 23 covers ISDN in greater detail.

Digital transmission is the way the world is going. While in the past the cost of digital devices exceeded analog equivalents, the price curve for the analog devices has long since flattened—whereas the cost of digital components has steadily dropped, and continues to do so. Until the local loop is fully "digitized," we will operate in a world of mixed analog and digital components. But except for that local loop, expect virtually every other communications medium to employ digital technology.

6

Carriers

Local Exchange Carriers

The local exchange carrier (LEC) is the telephone company serving your area, and providing you with dial-tone services. The LEC came into existence after the breakup of the Bell System from the AT&T network. The modified final judgment decree specified that the local telephone company, or dial-tone provider, would be kept separate after divestiture. Prior to 1984, the carriers were called the Bell Operating Companies (BOC), and this is how they still are referred to. The names of the telephone companies attached to the older Bell System have changed a little, but some have tried to keep their identity as much as possible. Examples of this identity are the names still used in the industry today like: Illinois Bell Telephone Company, Bell of Pennsylvania, and so on. However, the telephone companies were fighting to keep the Bell logo at the same time the parent organization (AT&T) was also trying to keep certain identities. The local carriers won out in this battle and managed to keep the logo and the Bell name. Surprisingly enough, since the divestiture of the AT&T and Bell Systems in 1984, people still do not separate the two entities. For the public, this is the toughest part of the whole scenario. They still think of the system as one big entity, rather than organizations that were separated as a result of the court decisions.

In early 1994, 10 years after the breakup occurred, there are still users who do not recognize that the change ever occurred. This, of course, leads to a lot of confusion because the problems that occur on a daily basis are always blamed on the telephone companies, regardless of where the problem actually resides. Additionally, the independent telephone companies, of which there are hundreds (1400+), who were always referred to as *independents* are local exchange carriers. These organizations, although not part of the Bell System, still provide the basic dial tone for many communities. Some of the larger of these independents include GTE Telephone (and the inclusion of their buyout of CONTEL Telephone Systems), Commonwealth Telephone, Standard Telephone Company, Centel, etc. The numbers and the areas served by the various independents varies, but the functionality is the same. They all provide local dial tone service as their primary charter. Now they are all considered LECs because the term fits better as a function of the service they provide.

The LECs provide more than just the dial tone. Since the divestiture, when they were required by law to separate the customer premises equipment part of their service offering, the LECs have a clearly defined demarcation point where the telephone company service stops and the customer takes over. This demarcation point, also referred to as the *DEMARC*, is the point where the rate and tariff process starts and stops. Figure 6.1 shows the termination of a LEC circuit in a demarcation point. In this particular reference, the demarcation point is a network interface unit (NIU), which is a smart version of the RJ-11 jack. To keep some semblance of how this all evolved, remember that the telephone companies provided "soup to nuts" in the early stages of the industry. They provided:

- The dial tone
- The local loop of wire connecting to their equipment and the customer location
- The telephone set to interface to the network
- The interface to the local and long distance portion of the networks and the access to the long distance network
- The billing and collection functions for all services
- The installation of all related pieces and components to allow the end user to access their network
- The maintenance and repair functions

As this was a single interface, the customer's only responsibility was to make a single call that did it all. The use of the network led to additional billings, so the only actual responsibility of the customer or user is to pay the bill. This all changed in the 1984 divestiture of the network and telephone company interfaces. Consequently, the customer now must decide who to talk to regarding services and equipment needs.

Some of the primary services that the local exchange providers or LECs provide follow. This is a representative sampling of their services, not an all-inclusive or exhaustive list.

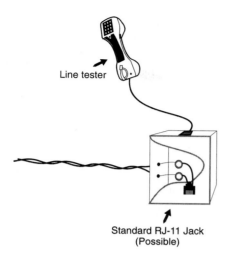

Line tester

Figure 6.1 The LEC terminates the two wires at the NIU (DEMARC) where the LEC stops, and the customer takes over. In this case the LEC may provide a hard-wired lug connector or an RJ-11 jack. The added RJ-11 jack on top of the NIU is for testing.

Standard RJ-11 Jack
(Possible)

Local Dial Tone (Single Line and Party Line) Service

This is the primary bread and butter service and the main service for which the telephone companies are chartered to provide. The local dial tone can be in the form of a single-line service, that being one registered user per telephone pair of wires, or party line service where multiple registered users share a single pair of wires. The use of party line service was always a financial consideration, particularly in the rural areas of the country. Many telephone suppliers did not have a financial justification to install multitudes of wires outside of the major downtown areas. Therefore, as the rural users requested dial tone, the cost of running the wires out to their locations was exorbitant. Consequently, the telephone companies allowed for the connection to multiple customers (two to four) on a single cable pair, with a distinctive telephone number and ringing tone for each user. This should not be confused with a multiplexing scheme. Only one of the users could use the dial tone line at a time. Others on the same physical pair of wires had to wait for the line to be available before they could either make or receive a call. If a local community was on party line service and the user "A" (Figure 6.2) is on the phone, the wires are busy. When user "B" wants to make a call, the process requires that "B" pick up the handset and listen to ensure that the line is free and dial tone is heard. Because "A" is already on the line, "B" will be able to hear the entire conversation on the wires, but will not be able to make a call until "A" gets off the line. Additionally, if "C" is expecting a call, as long as "A" is on the line an incoming call cannot make it to "C" because the line is busy. This required the cooperation of the users sharing the service and a form of mutual respect. Because any one of these users can listen to the conversation, it was a good neighbor policy that said

Figure 6.2 With "party line" service, multiple users share the same pair of wires. This arrangement was used in many rural areas when the LEC did not have enough pairs to serve all subscribers.

the user upon hearing a conversation on the line would hang up and not eavesdrop on the other party. Also, a form of cooperation or scheduling was informally used when "B" wants to make the call, he/she will tell "A" who is already on the line. Therefore, "A" will hurry the call to its conclusion so that "B" can have the use of the line.

This all worked well in rural America where neighbors were more dependent on each other and were far more cooperative in their communities. However, in the cities where this was also an offering at a reduced rate, the system really could become a major problem. Suppose that "A" had a teenage child who liked to talk on the line for hours. "B" and "C" would continually be blocked out from the line. Or, as they would continue to pick up the phone and hearing a teen on the line would ask the teen to get off the line so that they could make a call. This could lead to significant hard feelings among the neighbors and several confrontations. Thus, the use of party line service was moved away from as much as possible. The rural parts of the country still have this service because of limited facilities in their area, and no other option exists. This is a variable depending on the local provider services and the geographical location.

Centrex Service

Centrex (central office exchange service) is a technique the LECs offer as their flagship service for business users. What Centrex is all about is the subject of several discussions in the industry. When the Bell system originally was united with the AT&T organization, the rental of telephone service included the private branch exchanges (PBX). Because this was a rental service, or leased over longer periods of time, the Bell system always had the edge over competition. After all, the supplier of the telephone dial tone also provided the interface equipment at the customer's location. Therefore, nothing was necessary for the customer to do but use the service and pay the bill. However, as competition was gaining ground over the existing installed base of equipment as provided by the Bell system, the telephone companies realized that some of the reasoning behind the move to competitive products was a limiting factor of the products they had to offer. Many customers were locked into long-term equipment agreements and the AT&T systems were not advancing as fast as the competitive products. This was the result of trying to keep pace with the market and at the same time, protect the embedded base of equipment without mass replacements that would be expensive and difficult to cope with. While this was ongoing, the central offices were moving ahead of the PBX marketplace in that they were replacing the old electromechanical systems with the newer electronic switching systems (ESS). Rather than replace both ends, the customer and the central office equipment, the logical selection of a central office-based telephone system was offered. The customer therefore could rent a partition of the central office both hardware and software to act as the PBX surrogate. Further, the Centrex offering allowed the customer to get away from the hardware system roller coaster. Hardware changes can be very expensive, particularly with the constant upgrades that were being introduced by the vendors.

The use of Centrex service was therefore a stepping stone for the Bell operating companies and the independent telephone operating companies, but once the break-up of the system occurred, the Centrex service became their flagship product to serve the large business customer. As it evolved, the offering became far more viable to the smaller customer with as few as two or three lines. The Centrex offering provides the features and functions of a PBX without the heavy investment requirements.

Business Service (Direct Inward Dial and Direct Outward Dial Lines and Trunks)

Beyond the basic dial tone services for business and residence customers, the LECs offer the rental of in- and outward dial services for the business user. The use of *direct inward dial* services allows the customer or outside caller to dial directly into a user's telephone via an extension number. The call gets passed into the telephone system (PBX) from the central office switching system and redirected to the called party's extension telephone. Just as the caller can dial directly into the called party's extension on the telephone system, an added capability exists on the telephone systems to allow the extension user to dial directly outside the system, called *direct outward dial*. This provides the users to dial an access code (number 9 as an example) that tells the telephone system to find a line or trunk to the outside world and return dial tone. Upon receipt of this second dial tone, the station user then dials a local 7-digit or a 10-digit telephone number. No operator or manual assistance is required via this access method. Hence the term *direct outward dialing*. More in-depth explanations of DID and DOD services are covered in greater detail in Chapter 8.

Residential Service

Clearly, the LECs provide both business and residential services. The delivery of residential dial tone is also one of the primary objectives of the LECs. These services include the single line and the party-line services. However, the LECs are always looking for newer sources of revenue. Therefore, they offer various features and functions that are included in the residential package. The first and foremost that the residential user might have as a choice is the standard flat rate residential service. In some cases, the LECs are also offering a measured usage service at a reduced rate. Flat-rate service can cost between $10 to $18 per month, depending on the local regulatory rates that are approved. To help reduce the high monthly costs for the nontypical user (a user who does not make very many calls per month) the LECs offer the residential user the ability to rent a reduced rate dial tone line for $8.50 to $10.00 per month, plus a usage charge for every call that is made (a message unit might be $0.05 to $0.07 per call) in excess of some predetermined number of calls. The third option is to use a further reduction of the monthly charge to a rate of $4.50 to $6.00 per month, plus a message usage charge for every call. Regardless of the selection made, the customer typically tries to match to specific individual needs. Although this is attractive, there are some cases where the telephone company will offer the reduced-rate service compared to the flat rate, but once a measured rate is selected the customer might not be able to go back to the flat rate. This would be a critical service decision if the caller has a variable usage pattern because the usage might be higher than the flat-rate service (Table 6.1). In this table, the cost for service is compared at various usage levels.

This table assumes that the cost of the monthly service for flat rate at $15.00, measured with first 50 calls free at $10.00, and measured with no free messages at $6.50. Each call above the allowance is set at $0.07. The rates for these services will vary from state to state because the tariffs are all different. However, the rates for the per message is a new variable because the cost per call is now becoming distance and time sensitive. A call, for example, might be $0.07 for the first five minutes, then an added $0.07 for each three minutes. Other variables exist and the combinations are innumerable.

TABLE 6.1 A Comparison of the Various Residential Service
Offerings with Measured vs. Flat Rate Service

Description	Flat rate @$15.00/mo.	Measured rate at $0.07/call	
		50 calls	No calls
10 calls per month	$15.00	$10.00	$ 6.50
50 calls per month	$15.00	$10.00	$10.00
100 calls per month	$15.00	$13.50	$17.00
200 calls per month	$15.00	$20.50	$20.50

Local Calling Services (DDD)

Direct distance dialing is nothing more than the ability to place a call at any distance within the LEC's area of service. This includes the basic residential and business offerings at a monthly rate for the line and the usage. This is not unlike a regular business line other than the fact that the business user can access the dial tone and dial wherever he or she wishes without the assistance of an operator. The acronym or abbreviation DDD is characteristic of the changes that have occurred over time. In the late 1960s, an operator was required to make some calls on a long-distance or international basis. This part of the evolution of the network was a must because the manual intervention was expensive and labor intensive. In the earlier days of the telephone networks, this was not a major problem, but as the usage picked up and more dependency on the telecommunications networks was beginning to grow, the carriers realized that this access was a must.

Pay Phone Service

Clearly evident anywhere in the LEC's service area will be the existence of pay phones (coin telephones). The LEC offers the access to their network and access to the long-distance network via either a public telephone service, a semipublic telephone service or a private telephone service. Each of these is handled differently based on the access method chosen. The public pay phone is obvious when traveling down the streets of any city or town. These are the phone booths that are positioned in full view for use and access by anyone who wishes to make a call. The pay phone or coin-operated phone is designed to allow access to the network for the masses. If a user does not have a phone in their home or office, then the public phone is available for outgoing calls to the network on a pay as you go basis. There is no monthly rental charge for this service because it is shared by the general public. However, the cost per call or per minute will be more expensive due to the fact that the coin-operated phone typically needs the operator assistance or automated assistance to complete a call. This is the most expensive type of call available on a dial-up basis. Along with this dial-up service, the user can use a credit card to complete a call through the carrier of choice, or an 800 toll-free call can be made without the need to deposit a coin. The dial tone accessing the LEC's network is present, so a call can be initiated without the use of a coin.

The semipublic telephone service allows an organization or business to rent the set and line through the LEC, and place it within their offices or lobby of their building, but also allows them to deny access to the general public. The rental will normally be based on a flat rate per month, less a portion of the income that is generated by the pay phone. Normally, there is a minimum amount of revenue guaranteed to the LEC by the renter of the pay phone. If, however, the minimum is not met, the renter must pay the difference. The variables of the pricing arrangement are a complex area of the local tariffs.

The final type of pay phone service is the use of a private coin telephone service. In this particular scenario, the pay phone can be rented from the LEC or purchased from any third party by the customer. The customer also rents the access line to the LEC's network for a fee (flat rates are still available in some areas) then selects the long-distance carrier to carry any calls from this phone. The customer can allow selected users access to this phone, which is normally located well within the confines of the customer's location. This might be in the cafeteria or other lobby location. Therefore, anyone off the street cannot just walk into the building to use the phone because it is not in a public place. From this phone, the customer can select the method and the price to be charged to the user on a cost-per-call or a cost-per-minute. The LEC carries the local call and keeps track of the minutes of usage, for billing and collection purposes. From this arrangement, the LEC might get a percentage of the revenue for the administration and handling of the service. Or, another option is that the LEC merely bills the customer for the basic service at the tariffed rate, and all revenues go the owner of the pay phone. This has become a competitive tool for customers since the divestiture of the telephone system.

Private Lines and Tie Lines

These are leased services rented from the LEC on a flat-rate basis. A private line would indicate the service is dedicated to the user rather than accessible to others. This is a contrast to the public-switched telephone network where many users compete for and share the lines and trunks from the LEC. The private line is typically a flat-rate service connecting two locations of an organization. In the case of a tie line (terminal interface equipment), the LEC will install a private line connection between two customer locations located in the same local access and transport area (LATA). The customer can use the tie line to link two telephone systems together (PBXs). The connection can be via a two-wire with E&M signaling (which actually uses four wires) or a four-wire with E&M signaling (which uses six wires) circuit. The variable nature is caused by the customer equipment being used. Clearly, the four-wire E&M circuit will provide a higher quality service because twice the bandwidth is available. Using six wires allows two wires for transmit, two wires for receive, and two wires for signaling outside the bandwidth of the circuit. After drawing internal dial tone from the PBX internal extension, a user then dials some selected dial code (i.e., dial 7 for tie lines). Once the dial code is entered, the PBX will draw a second dial tone from the remote PBX on the other end of the tie line. From this second dial tone, the end user then dials a three or four digit extension number to reach the called party at the other end. This access method allows for the flat-rate connection between two company locations. There is no switching or routing of the call across the network, it always takes the same path on the dedicated circuit. The LEC only has to provide the dedicated pair of wires from location "A" through the wire center at the local central office and then right out the back end to another central office or directly to a cable pair to the second location. This is shown in Figure 6.3 for the connection between the two customer locations in different central offices. In Figure 6.4, the connection is the same, but is served by a single central office. In both cases, the central office only provides the service through the wire center rather than routing the connection through the switching center. The end user gains the access to the remote site without having to dial

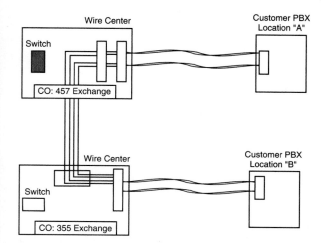

Figure 6.3 A tie line linking two customer locations in different COs in the LEC area. The wire centers are connected with physical connections rather than access switched services.

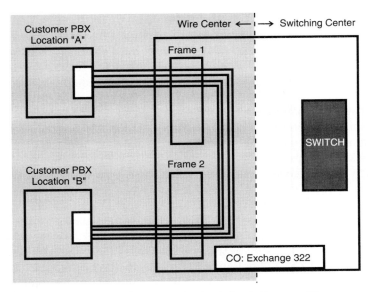

Figure 6.4 Two locations linked together by tie line in the same CO.

an outside line (9th level is typical in a PBX), and then dialing a 7-digit direct inward dial extension number. This reduces the number of digits that the user dials and eliminates the additional usage-sensitive costs of message units for local calls. Depending on the number of interoffice calls made between these two sites, the savings realized can be significant. A flat-rate service might prove to be advantageous over usage sensitive rates. The LEC charges for the service on a mileage basis for the dedicated pair of wires between the two locations.

Foreign Exchange Service

Foreign exchange service (FX) is a leased-line service that allows a customer to draw dial tone from a remote central office in the LEC's service area. The primary use of the foreign exchange is to use a local telephone number for customers to call. The local 7-digit telephone number is connected to a leased line that goes to central office in a remote location, and is then passed along to the subscriber's location. This will eliminate a long-distance call from a customer to a supplier, allowing the appearance of a local presence. The user pays the LEC for a local dial-tone line at the foreign central office and the mileage charges for the dedicated private line to the user's location. An example of this service is a customer located in Wilmington, Delaware calling a supplier in Philadelphia, Pennsylvania. Normally, this is a long-distance intralata call between the two cities served by the LEC. The distance between these two cities is roughly 29 miles. At the current tariff rates, this call costs $0.40 per minute. Because the customer calls the supplier on a regular basis to place an order, it is advantageous from the supplier's perspective to allow the customer to call free of charge. Therefore, the supplier rents a foreign exchange service from the LEC. Normally the customer calls (215) 555-1111, but using the FX service the customer now calls a local 7-digit telephone number (302) 777-1234. By calling this local number, the customer does not incur any local long-distance charges and will not hesitate to call the supplier on a regular basis. From the supplier's perspective, the leased line from Wilmington to Philadelphia will cost in the vicinity of $200.00 per month. Given the cost for the toll call at $0.40 per minute the break-even point is 500 minutes. Because the customer makes calls that can last up to 10 minutes each, then 50 calls justify the circuit. Taking this to the next logical step, the supplier who is paying the cost of the monthly circuit charge can also use the FX when calling back to the customer. The service can be set up in a one-way in, one-way out or a two-way service. Therefore, the return calls will help to justify the expense between the two sites and reduce the overall cost per minute for the usage fees. Although there are other ways to handle this same service, other than the use of FX, this is an example only. Keep in mind that the service in this context is offered by the LEC and is a local toll replacement. If the capability is required between locations in other parts of the country, the LEC will only provide the last mile on each end, while the interexchange carriers provide the interstate, interlata portion of the circuit. Table 6.2 shows a summary of the one- or two-way traffic, compared to the local calling costs. This is calculated at a standard call of 10 minutes in duration. Other variations of the pricing arrangements can be calculated based on the LEC's tariffs in the specific areas around the country. Each variation carries different pricing configurations and significantly different cost justifications.

Table 6.2 reflects a total number of 10-minute calls made between the two locations and only two parties. If the number of customers increases, the total number of calls and minutes used on a single line will be limited. There is only one line between the two central offices and the supplier location. Therefore, only one conversation can be accommodated at a time. The numbers of calls and minutes will be limited, as the total usage on a single circuit will not be equal to eight hours per day. This is because of the downtime when one caller hangs up and the other initiates a call. The circuit will only provide a certain percentage of availability to both ends. Typically, a single circuit will only result in 3 to 3½ hours of total talk time availability. Therefore, if the number of calls exceeds those reflected, another FX line might be required. One can see, however, that the pricing differences and savings opportunities are substantial. The use of FX lines are a mainstream for the LEC in providing cost-effective solutions for businesses.

TABLE 6.2 A Summary of Possible Savings with FX Service

Number of calls per month at 10 minutes per call	Tariffed toll calls	FX service fixed rate	Difference
50	$ 200.00	$200.00	$0
75	$ 300.00	$200.00	($ 100.00)
100	$ 400.00	$200.00	($ 200.00)
150	$ 600.00	$200.00	($ 400.00)
200	$ 800.00	$200.00	($ 600.00)
300	$1,200.00	$200.00	($1,000.00)
400	$1,600.00	$200.00	($1,400.00)

WATS (Intrastate, Intralata)

Wide area telecommunications services (WATS) is a special tariff that the Bell System created back in the 1970s. The original purpose of a WATS line allowed large customers to access the local and long distance network at a reduced rate. In effect, it is a volume discount offering that AT&T (the Bell System) created for the very large user or for the smaller business who had specific volumes of calls to geographical areas of the country. The way the WATS tariff originally worked varied depending on the volumes of calls that the customer had. This included the following types of service:

Bands

The country, exclusive of the state the customer is located in, was originally broken down into concentric circles or areas of coverage depending on the user location. In Figure 6.5, the original banding is shown for the original five bands for the east coast, using Massachusetts as a reference point. In each case, the band selected by the customer allowed for calling into states by area code contained in the band. In the example of the east coast, the band 1 line selected allowed calls into the eight New England states. Any call made into the area codes covered by this line were carried without any restrictions. However, if a caller attempted to place a call into a state that was not part of this band, they were given a reorder (fast busy) tone, meaning that service to that area is denied.

If the need to call a greater area of the country existed, then the next higher band was offered. The band 2 line would allow expanded calling coverage to the next concentric circle of states. Included in the band 2 service area is the area covered in band 1. Thus, as the concentric circles got larger, the coverage became inclusive. Using a band 5 WATS line allowed calls anywhere in the contiguous United States. The costs for the lines varied by banding, with one bearing the least cost and five bearing the highest cost. Companies using these services had to select the zones that they planned to call into. The burden was on the user to configure their lines based on the communities of interest of their calling patterns.

Obviously, the selection of a mix of bands 1 through 5 was the preferred choice. A company did not want users to arbitrarily call a band 1 state (or area code) on a band 5 line because of the difference in cost for the service. The carrier, in this case AT&T, created this configuration to appease the large customers without offering the same discounted rates to the smaller user. The FCC and the local public utilities commissions were very sensitive to the differentiation of pricing among groups of users. Therefore, AT&T worked around a technique of offering WATS without the standard call detail that was associated with the standard long-distance direct-distance dialing service. The customer was responsible to capture the details of the calls for allocation purposes through extended call detail recording capabilities on the PBX or Centrex service and the customer paid a flat rate for the service whether the line was used or not. Thus a difference in the service justified the difference in prices.

Full-time WATS

A full-time service offering with WATS allowed the customer to pay a flat rate per line regardless of the number of calls or hours used. Actually, the line was not truly a flat rate; it allowed up to 240 hours of usage per month (8 hours per day at 30 days per month). Upon reaching the 240 hours of usage, a premium was applied, but at an extremely low rate per additional

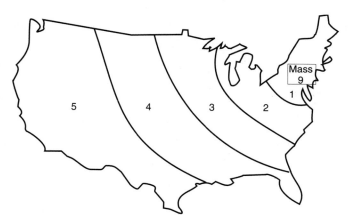

Figure 6.5 Original WATS bands consisted of five contiguous "bands" emanating from the originating point. Band 1 lines cannot be used to call areas beyond its contiguous band.

hour. The flat-rate costs were dependent upon the band selected. In Table 6.3, the original pricing is shown for the rates from the New England area. These differed depending on the location of the user.

Measured rate WATS

Measured rate service became the option for either the smaller user who had certain volumes, or the larger user who had a mix of flat and measured rate lines. The measured rate allowed a customer to use the first 10 hours at a fee. Each additional hour of usage bore an hourly rate at a higher cost, but still less than direct dial rates. Many customers used the measured rate service as an entry level into WATS service. Upon reaching a sustained volume, at some threshold of a break-even point, the customer then converted to flat-rate services. This was more complex to manage and monitor on a regular basis. However, the typical break-even point ranged in the 80 hours of usage. Once again, the onus rested with the customer to monitor and reconfigure the mix of services based on sustained use.

LEC intrastate WATS

Where AT&T offered the interstate calling services, the local operating telephone companies offered the in-state service. Because the LEC is constrained to offer services in their LATA only, they therefore have to limit the distances covered by their WATS line capability. The LECs offer long distance intralata WATS at a reduced rate over the toll call rates. Much like the IEC version of WATS, when this was first introduced, the intrastate service had to be separated from the interstate service. The reason for this separation is obvious, because the FCC controls the rates for the interstate interlata offerings, whereas the local PUCs control the LATA and state rates. To offer some volume discount to their large business customers, the LECs had to come up with a discount plan through a tariff filing that would be ap-

TABLE 6.3 Original WATS Flat Rate Pricing for 240 Hours Usage per Month

WATS band	Flat rate monthly cost
1	$ 900
2	$1,100
3	$1,265
4	$1,440
5	$1,675

proved by the local PUC. WATS on an intrastate basis was the answer. The use of WATS on an intralata intrastate basis is founded on the basic line charge per month (typically $37.50 per month) plus usage charges based on hours of usage. The use of the line increases and therefore decreases the cost per hour. Many organizations have been able to reduce their costs per minute on WATS to a reasonable $0.10 to $0.15 per minute as opposed to the direct dial rates of $0.20 to $0.25 per minute, or more. Clearly, the benefits of volume discounts are the savings that can be achieved.

One Plus WATS

Once the costs for WATS service became a commodity, the carriers had to come up with a new offering to appease the smaller business. To do this, the One Plus WATS was offered. Using an access line to the long-distance network, within the LEC's domain, the user can place long distance calls on a single access line to any point in the LATA. Therefore, separate lines were not required for the customer to use a discounted plan. As a matter of fact, the LEC can provide the same service to route calls that are local alongside of the calls that are toll. Consequently, the business user can use the service even though the telephone system might not accommodate added lines for special purpose use. Discounts can be had through a billing arrangement rather than through special circuits. This makes the service readily available to the masses and saves the customers a significant amount of time and money over their current mode of operation.

800 Service

Introduced over 25 years ago, the use of a toll-free arrangement became the service to vie for. While businesses were reaping the benefits of reduced outbound calls through the WATS tariffs, the other side of the business dealt with the incoming traffic. As users recognized, the cost of telecommunications expenses was escalating dramatically. Customers who were located geographically at a long distance from the supplier had some hesitation to call long distance to place orders. Further, the supplier's sales force out on the road selling goods and services were forced to run around with pockets full of change or a telephone credit card. The credit card was kept as a perk for some, only senior executives were issued a credit card to place calls back to the office from the road. Consequently, many users of the telephone network set up a work-around procedure to overcome the cost of long-distance calling back to the office. A customer or an employee of the organization would dial an operator assisted call, collect person-to-person. This was the most expensive cost to make a call. However, they would also ask for themselves or some fictitious-named individual. This would be used as an indicator that the called party answering the incoming collect call should refuse the call by saying that the individual was not there. The operator would then suggest that when the asked for party returned, call "the caller" back at this number. The called organization would then hang up and grab an out WATS line and return the call immediately. This saved the user a good amount of money on the telephone bill because the call was sent out at a reduced rate. Further, the incoming collect call based on a person-to-person calling rate would not be billed to the recipient of the call because the call was literally not accepted. To overcome this use of the network for the incoming fictitious numbers, the carriers had to come up with a solution. They could not bill these calls, even though they tied up their network and human resources in an attempt to complete them. This was also used by residential customers that had children in schools or who were traveling, whereby the student could call home asking for himself—the signal that they arrived all right.

Thus, the 800 or toll-free calling service was introduced. This is nothing more than a reverse-billing WATS service where the callers can enjoy the benefit of calling at any time and not have to worry about the cost of the call. The owner of the 800 WATS service had to worry about the volume of calls and the origin of the calling party. The LECs offer in WATS as part of the potpourri of services for the intrastate intralata areas, to allow the reverse charges of the calls. This is a tariffed item similar to the outward service where a monthly access fee is charged for the line plus usage. When the 800 service was first introduced, the same flat- and measured-rate services were available. As the commodity business of WATS caught on, the same evolution took hold that a customer could use the same access line for incoming regular calls and for incoming 800 calls. The limitation is that only one call per line can be active at a time. If the customer calling a supplier uses the line for an 800 service call, then others trying to dial direct into the called party will encounter a busy tone. With the possibility of having an 800 number, the supplier can accomplish the receipt of a call without inconveniencing their customers through an expensive call to the location. This will allow the customer to call toll free to the supplier regardless of the distance, feeling no remorse about the long-distance call whatsoever. 800 area code numbers are rapidly depleting, so the carriers will be offering "888" as an additional toll-free numbering scheme. This might confuse the general public

Directory Services

The LEC also offers a directory assistance for a fee. The typical arrangement offered by the LEC is to look up called parties' numbers through an operator-assisted call. Once the number is obtained, a voice response unit takes over, playing the number out to the interested party. The typical fee for this type of directory assistance is from $0.35 to $0.50 per call. Some companies offer an initial three free operator assists for directory assistance. Newer versions of this service are emerging, where the operator will still key in the name of the requested number then the voice response unit offers to complete the call upon delivery of the message. The fee for this connection and automatic dialing is from $0.35 to $0.85, depending on the location. These are newer methods of offering assistance and at the same time, generating new revenue streams.

Although it might sound excessive or impractical to have a calling party agree to this arrangement, one need only think about trying to get a telephone number from a phone booth while on the road. First, the pen always runs out of ink at the time when the number is being announced or second, the rain starts to pour while trying to write on a piece of paper. Either scenario is both frustrating and normal. Try to remember a number in your head compounds the problem all the more; therefore, this added service is not as far fetched as it might sound.

OPX (Off-Premises Extensions)

Another form of leased-line service offered by the LEC is the use of off-premises extensions. Let's assume that an organization is in a large office building. This user has plenty of growth from a telephone system (ala a PBX) perspective, but not as much square footage to sustain growth of human resources. If, in fact, the organization runs out of space, it becomes necessary to move people off-site for either short-term periods while reconstruction efforts can be accommodated, or for longer-term periods that could be months or years. In either case, moving people off-site becomes a problem with managing the telecommunications flow. A new phone system might be required as a purchase at the off-site location. Along with the new phone system, new lines will be brought into the building for the branch or department that is relocated. This means that customers might have to remember two different numbers when calling the organization. This can be inconvenient for the customer, and any such inconvenience could lead to an unhappy customer. By all stretch of the imaginations, telecommunications should be friendly, not inconvenient.

Above and beyond the problems caused by this arrangement for customers, one also must look at what it might do to the internal communications flow of the organization. When dialing between offices in the same primary building, one merely dials a 4- or 5-digit telephone extension number. However, when calling this remote group after getting internal dial tone, the user must dial 9, followed by a 7-digit telephone number associated with the newer building. At this point, the call might just go to a receptionist who screens all incoming calls, because direct access to an individual is probably impractical. The receptionist will then either "buzz" the called party on an intercom or page through the building for the called party to pick up line one (or two, three, etc.) Frustrations in the delay of this process might build. The process can also fail through constant cut-off conditions or through extended hold times while waiting for the intended party to come to the phone. Therefore, morale can be affected within the organization. Another issue is that the remote group will have to make long distance and local calls from this new system which might be more expensive. This gives these folks the feeling of being orphaned children as far as the organization is concerned.

To overcome this problem, the use of off-premises extensions might fit the need better than the separate system as described above. This means that every user being relocated will maintain their extension number from the primary telephone system, but that the LEC will run (or install) private lines between the two locations. Every extension will have a two-wire interface to the telecommunications system and a single-line set at a remote building. The LEC merely provides the wires from the initial location to a demarcation point at the remote location (Figure 6.6). The LEC bills this on a per-station mileage base per month, plus a one-time installation fee. Mileage is handled differently by each provider, so the pricing is deliberately avoided in this discussion. The customer then plugs into the outside lines from the LEC with a piece of equipment (i.e., a single-line set). The LEC runs the wires through the wire center at the central office(s) involved, as with other private line services. The dial tone at the end unit is drawn from the customer's switch (PBX), not from the local central office. All features and functions available to the primary location, local and long-distance services, and extension users are all accessible from this OPX. The only difference is that the buildings are separated and that the single-line set might not give these remote users identical capabilities as available in the office. This is an analog telephone set for the

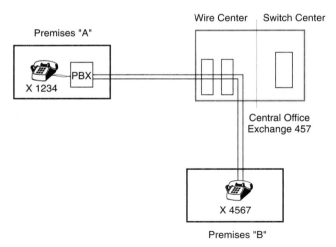

Figure 6.6 Using off-premises extension, the LEC provides a 2-wire facility (leased line) between the two locations through the wire center. Dial tone at location B comes from the PBX in location A.

most part because such digital display sets and feature buttons are not available on a single-line analog desk set. However, the users are still treated as part of the organization and it is transparent to the callers that this separation is in effect. That is the primary goal of the OPX, and it does work. Many opinions are brought on this issue where some would suggest that the OPX is a better solution, while others suggest that separate systems work better. This is a matter of personal preferences and should be analyzed in the actual circumstance.

Access to Interexchange Carriers (Equal Access)

The LECs also provide the access to the long-distance suppliers on an equal access basis since 1984. Prior to 1984, the LEC provided access to the interexchange carriers based on the interconnection arrangements that they have set up. Typically, this was through the AT&T network because that was the way things were done back then. But when equal access was mandated, the LECs had to allow connections to any carrier of the customer's choosing. This is done on a user selection, not as a fee base service. Any long-distance carrier who is marketing in a specific area shall be granted access to the customers on an equal basis as was provided to AT&T. The IECs will be charged for the access fees that they build into their rate structure and are passed along to the user. The access is based on the type of connection available and the distances separating the carrier's point of presence (POP) and the LEC's central offices involved. The customer is given a choice of the primary interexchange carrier (PIC) for each line installed. The LEC then sets a database on how to switch or hand off calls destined to the IEC from each customer. The connections are shown in Figure 6.7. The LEC might hand the call up to the access tandem switch (class 4 office) or optionally have connections directly to the IEC in an equal access end-office (class 5 office).

Inter-Exchange (IEC/IXC) Carriers

The *inter-exchange carrier (IEC or IXC)* is the long-distance carrier. In general terms, this carrier connects to the LEC with circuits from its point of presence (POP) to the central office at the local telephone company. Most of the major players in this arena are IECs, such as Sprint, MCI, AT&T, Cable & Wireless, etc.

The IECs can carry either long-distance switched and/or private line service. The major players provide both. Some of the IECs, however, primarily provide private line service, depending on their charter. An example would be Williams Telecommunications (WILTEL). Wiltel is currently in negotiations with another long-distance supplier known as *LDDS*. If and when this merger takes place, the combined Wiltel/LDDS organization stands to be-

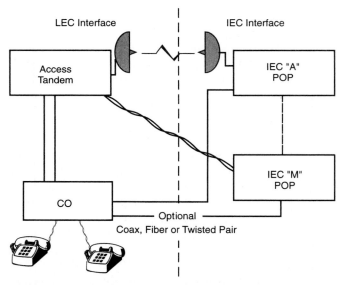

Figure 6.7 The LEC hands calls off to the IEC depending on the customer's selection of which IEC will carry the call. Various forms of media
can be used to interconnect the LEC and IEC offices.

come the number two supplier in the industry. There are some benefits to
providing switched services which at this point is a commodity service.

Typical services provided by the IECs include interstate and interlata capabilities as follows:

Switched long distance (DDD)

Called direct-distance dialing, the IECs provide for the end user to pick up
a phone and dial anywhere in the continental United States. So long as the
end user has selected the IEC as the primary interexchange carrier of
choice and the end user is served by an office that has equal access, the
user merely has to dial one (1) plus the 7- or 10-digit telephone number desired. The carrier will then use the capability of the switched public network to route the call to the appropriate end point. Many of the carriers
have the ability to let the user access their network through the one plus
network access or through a leased line into the carrier's point of presence
(POP). The POP is either the access point into a switching system that will
route the call to the appropriate state, or it can be a closet in which the carrier has terminated equipment on a leased high capacity circuit.

To better explain this concept, let's assume that a long-distance provider
wishes to set up an operation in downtown Philadelphia. However, the clos-

est points of access into the carrier's long-distance facilities (the microwave radio systems or the fiberoptic cabling system owned or leased by this particular carrier) are either in Washington, DC or New York City. To begin service in the Philadelphia area, the carrier could rent space, buy a switching system and run cables or radio systems out into the area (Figure 6.8). A second option is to rent space and put some equipment (such as a multiplexer or a rack of dedicated wires) in a closet or a hotel room. As the carrier begins to add customers, the LEC connections from the customer location to the IEC location will be on a two-wire or four-wire local loop. Rather than terminating into a switching system, the LEC's wires will be terminated onto a frame or backboard in the rental space defined above. The IEC will then take the wires from the LEC demarcation point and cross connect them to a different pair of wires running from Philadelphia to Washington or New York (Figure 6.9), where the carrier has a closet approach. At this point, the IEC will then bring the customer's connection into the switching system located within the boundaries of their network connections.

Although this is not a major problem, the IECs using this system are carrying all the traffic from a single point to maximize their usage of the circuits emanating from their offices. However, if things go wrong, the testing and troubleshooting can be prolonged because the connections are run across multiple extra miles of wires to get to the switch, adding to prolonged delays

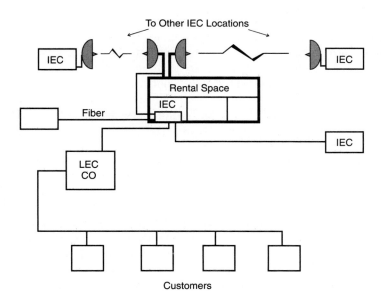

Figure 6.8 The IEC rents space and buys a switching system and connects to other POPs located in the area. This is a true switching center.

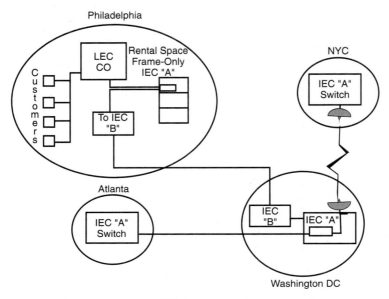

Figure 6.9 To open a new area, IEC A may rent closet space only to provide a physical interface to the LEC wires. From there, the IEC can lease lines from a competitor or install their own lines to the city where a switch is located.

and possible complications with the integrity and quality of the circuit. Many of the IECs start out using this closet approach until they increase the volumes in the location, then they build a switch in the area when the volume has sufficiently justified the expense. Others keep the arrangement set-up this way to maximize the total throughput at their primary locations and never plan any additional sites, yet they still market in the areas where they do not really have a switch. They do advertise that they have a POP in the area, which gets to the crux of the definition. Their point of presence is different than other points of presence. One should at least understand just what the differences are and ask the carrier how they handle the interconnection.

Credit Card Service

All of the IECs (and the LECs for that matter) offer the ability to use their network services through the introduction of a credit card or calling card. This is a matter of convenience for their customer base. When a customer is traveling, the need exists to use hotel/motel phones, pay phones and generally any other service that allows access to the long-distance networks. Rather than have a customer run around the country side with a pocket full of quarters ($0.25 pieces), that would be a very difficult thing to do, and discourage the use of the network, the credit card allows the caller to access the network at will and use the service indiscriminately. This service carries a surcharge with it. For every call placed on the carrier's network, an initial surcharge of between $0.50 and $1.00 per call is tacked onto the initial minute of the call. Given that the average length of calls on the network are five minutes long, this adds a $0.20 per minute penalty on the call. If a caller does average the five plus minutes, the cost of the call becomes close to the direct distance dial rate. The benefits and discounts of using a single carrier help to offset this surcharge.

When the calls initially had to be handled by an operator assisted arrangement, the surcharge was introduced to offset the cost of the personnel needed to process these credit card calls. However, this is a misnomer today because the automated process is now used where a customer dials the zero (0) plus the ten-digit telephone number and waits for a tone (or bong). After receiving the tone, the customer then continues by dialing the credit card number to be charged. From there, the call gets processed across the network and completed. For billing purposes, this is straightforward. But, no operator ever came in on the call, since the automation took care of the entire call. Why then do the carriers continue to charge the operator assisted rates? Adding some insult to injury, if the need arises to need an operator to complete the call for whatever reason (not on a Touch Tone phone, the dial pad is shunted out after the zero plus 10 digits are dialed, etc.), then the carrier bills a premium for the operator to assist in the completion of the call. This sounds like a double whammy in terms of the billing mechanism, and technically it is. The carriers will have the user believe that the more expensive services of using a human in the call completion has been reduced when using the automated process. One never knows. As a management function, it is very difficult to know because the calls being placed are all being made by others while on the road. The end user is primarily interested in ease of call completion, so they do not worry about the cost issue. Moreover, these end users will not be as intent to report when problems occur on the line. Many of the carriers will offer a credit for calls that are poor quality or for cut-off situations where the end user has to dial the called party a second time, and incur the credit card setup cost for a second time. However, the users are either unaware that this is an option, or they could care less because this is just another inconvenient step. One could possibly save a significant amount of money by educating the user and by following up on the billing at the end of each month.

WATS Service

Just as with the LECs, the IECs all have some form of WATS service as a volume discounted offering. The more a customer uses, the less expensive the cost per minute will be. Everything is a commodity these days. To use the WATS services of many of the carriers these days, customers have several options (Table 6.4).

TABLE 6.4 A Summary of the WATS Variations and Pricing Arrangements

WATS service	Initial cost	Variable costs
Dedicated lines	$37.50	Based on usage by bands. The usage follows a rate structure: First 10 hours Next 25 hours All over 40 hours
Dial One WATS		Cost per minute based on distance and time of day.
Reach Out America	$10.00	Cost per minute based on distance and time of day. Minimal discounts applied.
Virtual WATS (SDN/VPN)	Access fee into VPN or SDN	Special billing based on distance, time of day and volumes used.
Special Tariff 12 WATS service	Access fees at the T1 or T3 rates	Special pricing based on long-term contracts of three to five years. Greatest discounts apply for WATS.
Pro WATS	$5.00	Plus cost per minute on distance and time of day.
Multi-location WATS	$2,500.00	Flat rate to consolidate bills plus largest discounts possible on WATS.

In each of the cases shown in the table, the options are many, indicating that the carriers are concerned with offering some form of discounted service to users, depending on the volume of calls and the distances that will be called. In every case, the user still benefits over the regular long-distance (DDD) rates, but the variables are really becoming commodity items again. When looking at the options customers can easily get confused, yet with all of the changes going on in this industry, you cannot assume that the carrier's representative will be able or willing to point out added options or discounts. The best way to consider using this type of service is to consider no more than an annual contract period or a month-to-month contract with the carrier. Therefore, the user will not be locked into a pricing structure that prohibits on-the-fly changes and adjustments as the usage changes. The carriers will, however, discourage this option stating that the customer stands to benefit from longer-term contracts based on reductions offered over and above the standard offerings. Keep this straight and in perspective. The value of the end user is critical to the IEC as a result of the fact that equal access has been administered and implemented because there is little or no price sensitivity on the cost per minute. The differences between carriers can be as little as $0.002 (stated as two mills of a cent), whereas if the carrier sincerely wants the business the differences can be far more significant. Let's assume that an IEC can deliver a cost per minute to a customer for long-distance calls at $0.20. What is the carrier's actual cost, as opposed to their rates charged? Interestingly, the carriers (at least the major players) can deliver a call at their cost for $0.03 per minute. Now in order to generate profits, they might mark this up to approximately $0.06 per minute (representing a 100% gross profit).

So, how can they sell service to the end user at $0.20? Well, first they have to pay access fees to the LECs to gain the access on the LEC's wires to the customer's premises. The typical access fees that are charged on a call-by-call basis is $0.07 per minute per end. Looking at the total number, the cost per minute is shown in Table 6.5. This table covers a call from San Jose, California to Philadelphia, Pennsylvania.

Obviously, the numbers reflected in this table are for the casual user of a WATS service, not the larger users who have thousands of minutes to hundreds of thousands of minutes of usage. However, to overcome these costs and to add some degree of new attractiveness to the end-user community, the IECs are now offering bundled services on T1 access lines. This works out that the IEC or the customer rents the T1 line as a dedicated link. The cost of the leased line is fixed, doing away with the usage sensitivity of the calls. At a cost of between $350 to $400 per month, the cost per minute is reduced. In Table 6.6, the costs of the calls are shown. In this scenario, the comparison of using 24 individual WATS lines vs. the cost of using a T1 are compared.

TABLE 6.5 The Total Cost Picture Charged at $0.20 Per Minute to the End User

Description	Fees
Local access charges at the LEC end in San Jose	$0.07
Interexchange carrier cost	$0.03
Interexchange carrier profits	$0.03
LEC access fees in Philadelphia	$0.07
Total	$0.20

TABLE 6.6 Comparing Dedicated WATS (Analog) Lines to a T1 Service

Description	Using dedicated WATS lines	Using a T1 in lieu of WATS lines
Line rental	24*$37.50=$900.00	$400.00
Cost per minute on access	$0.07*24=$16.80	-0-
Cost per minute for WATS	$0.15*24=$3.60	$3.60
Cost of 100 hours usage	[60*100*(3.60+16.80)]+900.00	[60*100*3.60]+400.00
Total	$123,300.00	$22,000.00

800 and 900 Service Offerings

As mentioned, the 800 service is a derivative of WATS line service. When initially introduced in the AT&T network 25 years ago, the 800 service was called *In-WATS*. This was a mechanism to replace the inward collect calling services. As the name changed, the service was made more flexible. A call coming into the 800 number is billed at a usage-sensitive rate based on the band that the call comes in from. Further, as a call is routed and delivered into an 800 service, the customer can designate that the network deliver the call to different locations or to specific agents in a call group based on where the call is originated. This inbound service can be called 800 service, 800 Megacom service, etc. Further, small business services (such as Readyline or 800 Starter line) are available. In each of these cases, the service is priced according to the needs of the customer. For example, the 800 service will be used by a medium-sized customer who selects dedicated lines for the inbound service. The 800 Megacom service will be used by larger customers who use a T1 line to bundle 24 In-WATS lines together. The Readyline service would be used by the small branch office that does not have line space on their existing telephone system, so they route the calls into an existing incoming line that is shared with a regular incoming dial tone line. Lastly, the Starter line would be used in a very small office or a home office environment where the customer uses the least volumes routed to a specific telephone number shared with a business or residential line. The prices vary with the choice selected, but the flexibility is what is being highlighted here. Table 6.7 shows the different uses as orders of magnitude pricing, rather than specific pricing arrangements.

900 services are pay-per-use offerings that have drawn mixed emotions and reviews. When first introduced, the 900 offerings were provided as a means of pay-per-call providers, such as the lottery services, dial a porn, dial a joke, etc., services. They could be accessed by all callers willing to pay for the service charge associated with the call. The owner of the 900 service offering establishes an account with the IEC. From there the owner then establishes a rate for the service ($2.00 per call, $25.00 per call, $0.50 per minute, or any other derivative of this pricing scheme). The IEC assigns a

TABLE 6.7 A Summary of Various In WATS Offerings

Service offering	Average pricing
IN WATS 800 service lines	$37.50 per line per month plus usage at $0.15+ per minute
800 Megacom service	$400.00 plus usage at $0.10–0.12 per minute
Readyline service	$20.00 per month plus $0.20 per minute
Starter line service	$6.00 per month plus $0.26 per minute

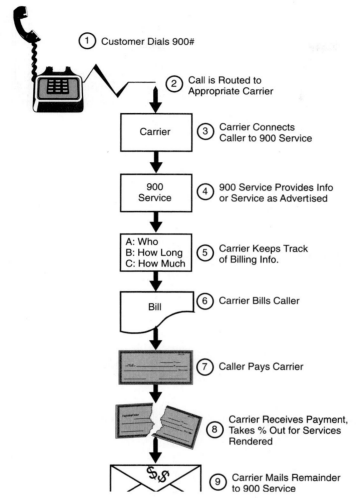

Figure 6.10 The flow of a 900 call process.

900 area code number that the owner then advertises. The next step is that the IEC establishes a rate to be charged to the owner. Now the customers start to call the 900 number, based on the inducement to call, regardless of the offering. As the caller dials the number, the call is delivered to the advertised location of the owner. This is then billed to the caller based on the advertised rate on their telephone bill. The IEC bills the call, collects the revenue, holds out a piece of the revenue based on the agreed to price (10% or more) and after collecting the revenue sends a check for the difference to the customer (owner). See Figure 6.10 for the flow of this operation. This service has created many millionaires in the industry.

However, to overcome these premium services, many organizations disallow the access to 900 area code numbers. They use the features of the telephone company to block this area code from access by their individual users, or they use the features of their internal telephone systems to deny access to the area codes. But, unfortunately, one of the changes that has occurred in the industry are new uses of this 900 service. For example, Novell, a manufacturer of network operating systems software for LANs, has introduced a 900 service for customer support and maintenance. This, of course, means that the telecommunications managers in business must allow some calls, but deny others. Confusion will reign in this environment.

International (IDDD) Access

The IECs allow for gateway access into the international direct distance dialing through their service offerings. The call is carried across the IEC's network, then delivered to the international carrier at a midpoint. This is an arrangement where the IEC and the international carrier share the revenue. IDDD is a service that will allow a user to pick up a phone anywhere and access an international country code dialing sequence, followed by a city code then the variable-length telephone number of the called party. The IEC does the billing based on a cost-per-minute, collects the revenue for international call then does a revenue-sharing arrangement with the international carrier. This makes it simple for the user of the service; by receiving one bill, the call can be logged and appropriately charged without having to pull together different pieces of a call from various bills.

Foreign Exchange Service (FX)

From an IEC perspective, the foreign exchange service that spans across LATA or state boundaries falls into their domain. A foreign exchange line from Dallas to New York (or any other two points under the IEC domain) is provided as a leased line with dial tone coming from the foreign central office. The IEC can bill this at a monthly rate for the mileage points between the two cities. They are also responsible for the end-to-end connection if the customer so desires. Many IECs and LECs will allow for the long-haul portion of the circuit to be the responsibility of the IEC and the billing for the local dial-tone service in the remote city becomes a billing responsibility of the LEC. This is less desirable because the customer must then assemble various portions of bills to gather all of the facts and costs of an FX line. Primarily, the service will be billed at a flat rate, however originating and terminating minutes might bear a cost from the LEC at the end of the circuit. One should verify all costs associated with the line as a means of cost justification. The use of a foreign exchange line between the two points mentioned might have applications other than the obvious benefit to a customer. Initially, the calls made by a customer to a supplier's location can be made on a foreign exchange. This allows the customer to place a local call that is then connected to a long-distance line 1000 miles away, avoiding the toll call for the customer. However, another application of the FX is for a Dallas-based company that is tentative about opening an office in New York. Rather than renting an office, buying or renting furniture, hiring a staff, and trying to break into a new market, this company can take a special group of agents to deal with future New York customers. The customer can call a local 7-digit telephone number, implied with this is the local presence of the supplier. Some companies feel more comfortable dealing with organizations that are local. Using the FX, the supplier can present the image of local presence. Therefore, for a minor amount of investment, the supplier can then begin marketing in the New York area. If the volume of calls is sufficient to warrant a staff in the New York area, then the Dallas-based organization can staff an office. However, if the volume and acceptance level do not warrant a full-time staff, the supplier can still service the customer base in New York without the investment in real estate and office furnishings. This is an effective use of telecommunications as a strategic corporate resource.

Off-Premises Extensions (OPX)

When a IEC's customer has many locations around the country, or around many other LATAs, the IEC can provide private line service in the form of an off-premises extension. This has already been discussed in the LEC environment, so the extent of the IEC's involvement is the extension of this private line across boundaries that relegate the service offering to the IEC domain. Once again, the private line draws dial tone services from inside the customer's owned and operated PBX or Centrex offering. The IEC merely provides a quality circuit to any point outside the originating LATA at a flat-rate fee. Figure 6.11 is a representation of the use of an OPX in two different states. The cost is dependent on the mileage associated with the circuit. There is no magic in this offering; the IECs provide a lot of connections with this service.

Operator/Directory Assistance

Just as with the LECs, IECs also provide operator services to assist callers with the completion of a call. A fee is charged for the intervention of the operator in call completion. More and more services are being displaced from the human intervention to the use of interactive voice response and recognition services (IVR). More than one carrier offers the ability to get operator assistance and directory assistance through the initial contact with a human, but then reverts immediately to IVR system. As more of these services

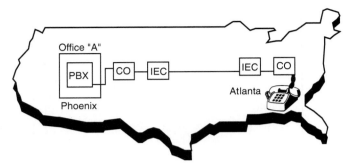

Figure 6.11 IECs can provide OPX between two different states on a mileage (month-to-month) basis. This is a standard offering from the IECs.

are provided, the use of the IVR will provide the number, offer to complete the call automatically for an added fee, and if a busy tone is encountered, allow the caller to automatically leave a voice message on the network to be redialed on a regular basis until the message gets through. This is also a fee-added service. Others are now using voice response so that the callers on the IEC network can use a voice-activated capability to the 10 most frequently called numbers in a pre-established database. The Foncard2 offering allows a user to pre-establish calls to "home," "office," "lawyer," etc. By merely saying the word into the IVR, the call can automatically be dialed for the caller. This is a convenience, but not a showstopper.

Remote Call Forwarding (RCF)

Remote call forwarding, as the name implies, is the ability to forward calls dialed into a number from the original location to a remote location across the IEC network. For a fee, the customer rents a dial-tone line from the LEC, then establishes an automatic call-forwarding arrangement to a remote site in another city outside the LATA or in another state. The customer then pays for the long-distance call from the original site to the forwarded site (typically about $0.25 per minute) based on distance and time of day. The customer portrays a similar situation to the FX in that the caller is dialing a local 7-digit telephone number, but the call is routed to a remote site in another city (Figure 6.12). This is particularly useful in lieu of a private line service, where the private line can be an expensive proposition. If the calling volume is low, the use of call forwarding on a call-by-call basis might be less expensive. Thus, the option is straightforward based on the outlay of money each month. If the volume stays low, then the RCF remains in place. However, if the volume of calls increases and the price increases proportionately, then a private line FX or other service can be used.

Another option is the use of the 800 service, but this automatically implies that the called party is far away or long distance—even if this isn't the case.

Value-Added Carriers

Other carriers emerged into the network business, called *value-added carriers* or *value-added network suppliers (VAN)*. Many of these initially were data communications suppliers. They provided services, such as dial-up data communications, packet switching, and other miscellaneous services. Initially, most were based on circuits leased from AT&T. The primary VANs were Telenet (acquired by Sprint) and Tymnet (acquired by British Telecom). As long as they provided additional services (or value added) and they could be reasonably priced, a niche existed for them.

However, many of these terms have slipped into the background because every carrier is offering some form of IEC, and VAN service as part of their portfolio of products today. Even the LECs are building out these services on an Intra-LATA basis. So the distinction is becoming very fuzzy. The terms are still used, so be aware.

Newer offerings, such as service bureaus that offer voice messaging services, store and forward message switching, voice to text/text to voice, voice recognition services, and other variations are also considered VANs.

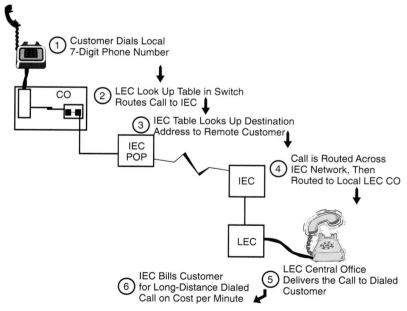

Figure 6.12 The remote call forwarding process.

Alternate Operator Services (AOS)

Alternate operator services (AOS) began when divestiture and deregulation became a reality. Pay phone services, which were normally part of operator-assisted calling, became a whole new battlefield. Anyone who wished to become an AOS was free to do so. You could buy a pay phone, rent a local dial tone line from the LEC, select the IEC who would carry the call, and be in business. The IEC would bill you for the long-distance calls made on your line. However, the caller would be billing the call on a credit card, so you could charge any rate you wished. For example, a long-distance call at $0.30/min. could be billed to the credit card at $2.00/min. Although this sounds unrealistic, this scenario was a reality. Users were unaware of the rates they were being charged because there was no human operator involved. They found out the rate when the bill appeared at the end of the month. The AOS folks got deeper into the long-distance business by making arrangements with hotels and motels to carry their long-distance traffic from guests, and offering commissions to the hotel/motel. The rates were even higher by now; the hotel was charging a surcharge for every credit call you made in your room ($0.50 to $0.75 per call), and getting a commission at the same time.

Although the customer was perplexed, options were limited. Many of the AOS and the hotels blocked the customer from the 10XXX access. You were stuck with the carrier and the rates. Your only option was to find a different phone that allowed access to the IECs directly or through the 10XXX route. The FCC has finally stepped in and demanded that the AOS carriers allow the caller a choice. The AOS carriers have begun to adjust their rates and practices to get the heat off them. Watch your phone bills and compare rates; opportunities exist to create large savings by educating your users and frequent travelers.

Aggregators

Still a newer phenomenon in the industry is the existence of the aggregator. *Aggregation* is a spin-off of AT&T's tariffs for large companies. AT&T offered larger discounts to organizations with multiple locations. The smaller locations of these big firms were billed for their long distance and WATS usage as single entities. In an effort to keep the customer and be competitive against the competition, AT&T began offering a multilocation WATS tariff. All bills were consolidated and billed at the greater volumes, regardless of the number of locations and the volume of calls per site. There is a fixed fee for this service and certain guarantees of volume. But to the large company, these conditions were easy to meet. What then of the small- to medium-sized company? They had no benefit, even if multiple locations existed. The volumes, guarantees, and fixed fees were so severe that the smaller companies had no way to use this service.

Enter the aggregator. These started out as consultants and entrepreneurs who saw an opportunity. If they could pay the fixed fee, then recruit small- and mid-sized companies to use aggregation rates, the smaller organization could save some money, and the aggregator could make a profit. The carrier typically charges the aggregator an annual fee of $30,000. The aggregator in turn recruits hundreds of smaller businesses to sign up. The aggregator then calls the IEC and tells it to bill the customers usage for all sites at the reduced rate.

Some aggregators even convinced their new customers that they should sign long-term agreements (1-3-5 years) with them, and possibly even split the savings (50/50).

Competitive Access Providers (CAPs)

The local loop has been protected as a monopoly since early in this century. The LEC was the only carrier allowed to deliver dial tone. Regardless of the cost, service, or quality of the lines, users had no options, except to bypass the LEC with their own equipment. This might include microwave, copper, or satellite communications. However, bypass only served to get access to the IEC or to another customer premise. The telephone company still provided dial tone and local-switched services. However, a group of entrepreneurs began to lay fiberoptics in the ground, connecting multitenant buildings in major cities. The vendors began to offer all-digital, fiberoptic bypass solutions to the customers. Bell's turf was being invaded. They offer reduced rates, volume discounts, and quick delivery times for the service. These vendors are beginning to offer some attractive options to the end user. They might also offer some relief to the LECs, even though they are perceived as the enemy today.

The LECs might gain more from the presence of these carriers/vendors than they believe might be possible.

Resellers

Resellers have been around for a long time, although they were considered OCCs at one time. The reseller usually owns systems, or switches, which are designed to route calls at the least possible cost. They rent/lease their lines from a variety of carriers (IEC) in various configurations. For example, they will look at volumes of calls to specific cities and rent FX, tie lines, WATS services, etc. They then resell the long-distance service to the end user. The user can be business or residential. The volume of calls they carry gives them the ability to offer switched services to their customers at reduced rates over the other IECs. This is like a combination of aggregation, IEC, OCC, and VAN all in one.

The reseller's networks are becoming quite sophisticated. Using the least-expensive services from a variety of carriers, they have built very elaborate dial plans, alternate routes, etc. They do render bills to the customer and provide a mix of custom billing arrangements including reports, location, account codes, etc. Yet, they make money and the customer saves at the same time. The larger resellers are also connected to the LECs via the equal access capability, thereby making it easy for the customer to use the service.

At one time, many of the resellers were considered "fly by night" operations. Today, this industry has shaken out, and the reseller is now a respected part of the network.

Newer opportunities include services, such as switchless resellers. A reseller of long-distance services can negotiate volume pricing deals with the "big 4" suppliers. Rather than owning equipment, the switchless reseller merely resells the long-distance services and has the actual carrier handle the call processing. Call detail can be passed on to the reseller so the reseller can bill the user.

Chapter

7

Lines Versus Trunks

Introduction

We present in this chapter a number of service offerings available from various carriers. Note that with few exceptions, the offerings in this chapter are based on a set of identical technologies. They differ only in exactly where and how the connections are made, how they are billed, and what information is transmitted over the circuit in addition to the voice signal itself. Underlying all of these offerings are the concepts of "line" and "trunk." We will use the term *circuit* to refer to a general point-to-point voice-grade physical connection. We will also see that there might be no physical difference between a line and a trunk because a two- or four-wire facility might be used for either. The real difference is the functional use of the wires that is applied to this linkage. The easiest way to define the two follows.

The term *line* unfortunately refers to more than one type of circuit. In most cases, it includes a connection configured to support a normal voice calling load generated by one individual (typically about 10 minutes per hour for a business user). But in the case of a PBX (see Chapter 9), the term *line* usually corresponds to one connection from the PBX to a desktop. In the case of Centrex, a line is normally one physical connection from the customer site to the CO. With a key system, a line corresponds to one telephone number—but it might also be referred to as a *trunk* (Figure 7.1).

The term *trunk* normally refers to a circuit configured to support the calling loads generated by a group of users—possibly many thousands of users. Thus, a general-use circuit from a PBX to a CO would usually be described—and billed—as a trunk (but see DID lines). Connections between

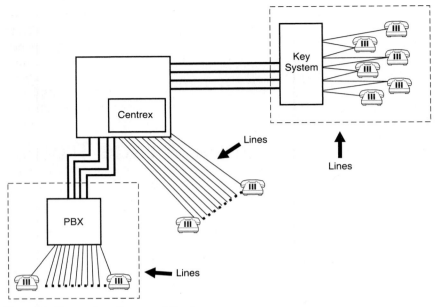

Figure 7.1 Places where lines are defined.

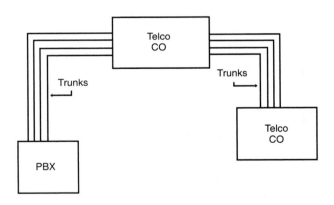

Figure 7.2 Trunks are used between intelligent switching systems.

COs or higher in the network hierarchy would also be referred to as *trunks*. But note that these trunks are (or at least can be) physically identical to lines. Why then the different terminology? This is shown in Figure 7.2 with reference to a trunk.

The ability of any given switching system such as a CO or a PBX to establish connections is limited. For example, although a PBX might be able to support 200 connections or *ports*, it might only be able to actually provide 80 paths at any one time. In such a case, if 80 people connected to 80 other people (some of them possibly off site), that would account for 160 of the ports; if any of the remaining 40 telephones or ports attempted to be serviced, it would fail. That is, a user could pick up the telephone and not receive a dial tone. Some systems are configured so that no such failures can happen. In the previous example, if only 160 physical connections were made to the PBX, then it could provide simultaneous service to all of them. Such a configuration is described as *blocking* (Figure 7.3), whereas Figure 7.4 represents a nonblocking environment.

Normally, a PBX's connections to the CO are configured so that a much higher utilization than 10 minutes per hour is achieved on those ports; a pri-

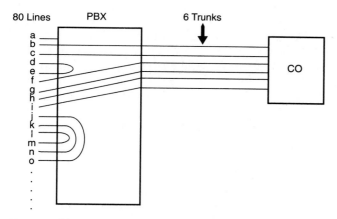

Figure 7.3 Blocking in the PBX arena.

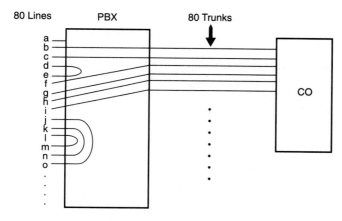

Figure 7.4 Nonblocking represents a 1-to-1 ratio.

mary benefit of a PBX is the ability to buy fewer telco connections than one has telephones. The CO must be configured so that it can provide connection services to such trunks at this higher utilization rate, thus using more of the CO's overall switching and connection capacity (COs are not normally configured as nonblocking switches). So, the telco will naturally bill a PBX trunk at a higher rate than a single business line—even though the PBX trunk might be physically identical to that single line.

As a final comparison of the definition of line versus trunk, then would be as follows:

A line is an end point from a central switching service, such as a CO or a PABX. The line is represented as the end point on the pair of wires regardless of where the intelligence resides. A line carries one single conversation at a time on the physical channel capacity. It is a billable location for the telephone companies.

A trunk on the other hand connects between two intelligent switching systems. The trunk might be a single circuit carrying a single call at a time, or it might be a bundled service that is multiplexed and carries multiple simultaneous conversations. The difference is that a trunk will be used for switching and routing decisions from the switching offices (CO or PABX). The trunk is continually used rather than occasionally used. It is a billable address that can have additional sub addressing capabilities behind it. In a telephone company world, it is the connection between and among other offices in the hierarchy discussed in Chapter 3. In the private user (customer) world, it might be a single connection to the intelligent PABX from the CO. These distinctions offer some variations in billing and utilization.

With this in mind, here are some common configurations.

DID

DID refers to direct inward dialing. From a caller's point of view, this service is in place if the caller can dial a 10-digit number from the outside, and reach a specific individual without operator (live or automated) intervention. Thus, Centrex normally inherently supports this capability without any additional configuration—everyone already has their own telephone number. A true key system (where telephone numbers are normally shared) can only do it if any given telephone number has only a single appearance.

But DID is usually referred to in the context of a PBX. It is a specific PBX feature that must be enabled and configured, with elements set up both within the PBX and also with the telco. Consider as an example a new site intended to support 1100 employees, each with his or her own telephone connected to a PBX.

The first step in arranging DID is to reserve the telephone numbers for all those employees. Let's say that the main company telephone number is

555-1234. The Telecommunications manager will request a block of DID numbers from the telco, probably about 2000. The telco might say, "Your DID numbers are 555-2200 through 555-4199." Notice that while there is a good chance the block will have the same exchange as the main number, it probably will not include (and one would not want it to include) the main number. The company will pay for these numbers on a monthly basis, but nowhere near as much as would be required for actual telephone lines. So far, the only thing arranged is the reservation of the block of numbers themselves. These numbers will not be given out by the telco to anyone else. The Telecommunications manager will assign each employee one of the numbers in the DID block.

Next, the Telecommunications manager must determine how many trunks (or DID lines) in the trunk group will be required to support the calls from outside to the company's employees. These are inbound only, and are in addition to the normal in-out or inbound trunks that serve the main operator, so they must be engineered to a very low level of blocking indeed (see Chapter 8 for more information on Traffic Engineering). With DID, the telco passes on to the customer PBX the responsibility of handling answer supervision (e.g., busy signals). The DID link is shown in Figure 7.5.

So, if an external customer calls Jane at extension 2313, the customer will dial 555-2313. The telco CO will seize the next available trunk in the DID group (if no trunk is available, the caller will receive a busy signal), and signal along it that there is a call for extension 2313. At that point, if extension 2313 is busy, the PBX must deal with it; the CO is merely passing along the signals. Possible PBX actions include forwarding to a message center, generating a busy signal, or forwarding the call to a specified alternate extension.

DID is most often used to reduce or eliminate the manpower required for a central answering position. The more calls that customers can place directly, the fewer must be answered by the company operator. On the other hand, some companies prefer to have all incoming calls answered by some-

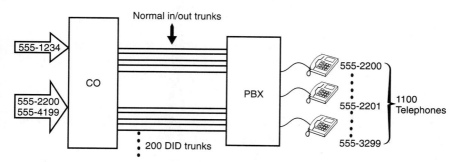

Figure 7.5 DID trunks link the PBX and CO together.

one trained in how that company wants its telephones to be answered (e.g., "Thank you for calling Kay's deli! How can I help you" vs. "Hello?"). It would generally be a mistake for a Telecommunications manager to make decisions regarding whether DID is to be implemented without consulting with company management.

DOD

DOD refers to direct outward dialing. If an employee can dial and reach an outside number without internal operator intervention, then the company has implemented DOD. In the past, when less-sophisticated telephone systems were available, it was not uncommon for a company to route all of its outbound calls through an internal operator. The operator's responsibility was both to screen calls ("no, you may not call Australia from that telephone") and to route the calls over the appropriate facilities (e.g., the right WATS line—see below). With the advent of modern PBXs and Centrex, such limitations can be programmed, if desired, on a telephone by telephone, or even user-by-user basis, eliminating the requirement to involve an operator in outbound calls. *DOD* is a term not often used these days because few companies consider not providing it.

FX

FX, not to be confused with FAX, refers to a foreign exchange circuit. In this case, "foreign" refers to a CO other than one's own local CO, not to a location outside the country.

Consider the case of an airline that wishes to locate all of its reservations clerks in Atlanta. It cannot expect all of its customers to pay long-distance charges to make reservations. What are its alternatives? One possibility is a group of 800 circuits. Indeed, it will probably have a large number of those, but 800 trunks cover large areas (and are priced accordingly). What about service for customers calling from large, high-density metropolitan centers, such as Chicago? Perhaps a more focused service might be more cost-effective. The FX grouping is shown in Figure 7.6.

Think of an FX line (or trunk) as two-thirds of a dedicated point-to-point (or "tie") connection. It starts at the customer's location, connects to the local CO, and extends from there to another foreign CO anywhere in the country. There is a fixed monthly charge for all that mileage; but there are no usage-sensitive charges for these miles. At the foreign CO, it is "open." It has a telephone number associated with that foreign CO. Calls made to that number ring at the customer's location. Calls made from the customer's location over the FX line emanate from the foreign CO, incurring only local charges for the call from the foreign CO to the called location.

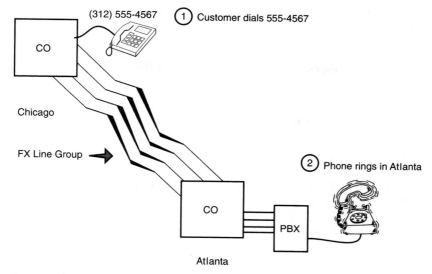

Figure 7.6 A foreign exchange (FX) connection between two major cities.

FX lines are often used by companies to provide a local number that customers can call in cities where those companies do not in fact have offices. In the airline's case, it could arrange a group of FX lines from its Atlanta offices to a Chicago CO. All of the lines could share one Chicago local telephone number. People from anywhere could call the number, but normally only Chicagoans would, because it would appear only in their telephone book—and it would be a local call only for them. If the airline wished to allow it, service representatives could also place calls from Atlanta to Chicago over the FX lines. The calls would be billed as though they were placed from within Chicago. Perhaps calls notifying customers of changed flight information might be placed this way.

OPX

OPX refers to off premises extension. An OPX line permits a telephone not at a company's location to function to all intents and purposes as though it is located at the company's location. This capability becomes particularly interesting with the recent increase in telecommuting. Suppose an employee plans to work at home. One of the problems to overcome in such a case is the isolation such a worker might experience. Providing the employee a telephone that looks like an internal line at the company might help to reduce the problem. Others calling the line within the company will dial an internal extension, which will ring at the employee's home; if the employee wishes to make a long-distance call, he or she usually just dials 9 and the rest of number just as though they were at a desk at the company's location. An OPX link is shown in Figure 7.7.

As with an FX line, an OPX connects from the company's location to the local CO, then continues via whatever intervening COs are necessary until it terminates directly on a telephone at another location. A key difference from an FX, however, is that on the PBX an OPX is connected and configured as a telephone rather than a trunk. This results in a limitation on the type of service provided: normally, only an analog telephone can be used at the end of an OPX because the digital signaling between a PBX and its proprietary telephones will probably not successfully make it through the various analog and digital circuits that make up the OPX. This limitation is not normally a show-stopper; rather, it just imposes on the telecommunications manager the need to configure the PBX to support a certain number of analog telephones as well as the digital telephones that might be used in-house.

Tie Lines

We discussed in Chapter 3 the general categories of dial-up and private leased or dedicated point-to-point circuits. In the voice communications industry, *tie line* (also sometimes called a *tie trunk*) refers to a private point-to-point circuit used to connect two voice facilities. For example, a dedicated link between customer PBXs at two different locations would be referred to as a *tie line* as shown in Figure 7.8. Other examples of tie lines might include a link between a PBX and a Centrex system as shown in Figure 7.9, or one between two Centrex systems. In all of these cases, one

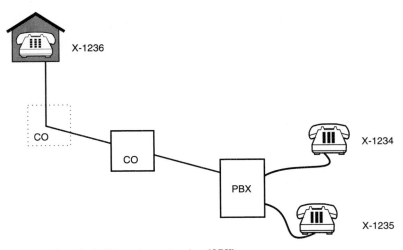

Figure 7.7 A typical off-premises extension (OPX).

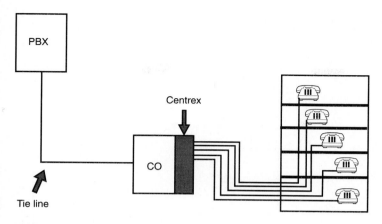

Figure 7.8 A tie line connecting two PBXs over distances (states).

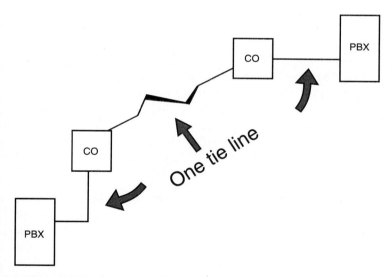

Figure 7.9 A tie line between a Centrex and a PBX arrangement.

would be equally correct to refer to the circuits as private or leased lines. On the other hand, if one of the connected systems is not a voice system, the term *tie line* would not normally be used. *TIE* stands for terminal interface equipment (Figure 7.10).

WATS

WATS is an abbreviation for wide area telephone service. WATS lines come in two flavors: in-WATS and out-WATS. Another name for in-WATS is 800

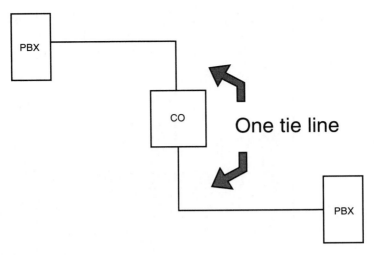

Figure 7.10 A tie line connecting two PBXs locally.

service. When most people refer to a WATS line, they mean an out-WATS facility. Both services are merely billing arrangements for reduced billing of long-distance calls based on a fixed monthly fee and discounts for larger calling volumes. 800 service also has the characteristic of reversing the charges to the called party.

Historically, WATS lines have been separate facilities (physically identical to local PBX trunks or private lines). Their geographic coverage was also banded; thus, one might have had a WATS line that only reached adjacent states (band 1), or all of the "lower 48" United States (band 5), or some intermediate variation. For out-WATS, either one's PBX had to be smart enough to recognize the dialed area and choose the correct outgoing facility, or users had to dial special codes to select the right WATS line. Any given band included all closer bands (but, of course, billed the calls at the higher rate for the wider band), and this caused a certain amount of difficulty in either configuration or training. Because different physical facilities went to different regions, this also resulted in a traffic engineering nightmare. This complication was all the more unreasonable because WATS calls (both in and out) are handled identically to all non-WATS calls; WATS is really only a bulk billing arrangement for calls that would otherwise be considered direct distance dialing (DDD or toll calls).

WATS service has never been free, although some of the older tariffs did have points where all calls above a certain (rather large) volume were free. Those tariffs are long gone; all calls now cost on a per-minute basis. The only variable is the per-minute charge, which does decrease as the calling volume increases.

One significant improvement is that WATS-type volume discount billing can now be set up on existing trunks; no longer is it required to have separate facilities into the local CO to have such an arrangement.

Private

Any circuit leased from a carrier from a point on one customer premises to another point on a customer's premises (even the same premises) can be described as a private line or circuit. Of course, if an organization builds its own facilities (e.g., a microwave link across a metropolitan area), these facilities would also be described as private circuits. In either case, the alternative is normally a dial-up link.

Many factors go into the decision as to whether to set up a private facility (of either type). Some reasons why a company might set up a private link include:

- Private analog circuits can be tuned for higher performance (both in terms of speed and reliability) than can dial-up facilities.
- Many types of digital facilities are only available on a private basis.
- Management and troubleshooting of private facilities can be more tightly controlled than in a dial-up environment.
- High call or data volumes would generate higher charges on the public dial network than on a nonusage-sensitive private network.

Reasons to go with the public dial network include:

- Low call or data volumes cannot justify a leased link.
- Unwillingness or inability to coordinate and manage a private network (do you really want to be your own telephone company?).
- A large number of small locations which would be uneconomic to connect with private links.

With few exceptions, there is not a "right" decision on this issue; rather it is based on a set of trade-offs involving economic, management, and performance considerations. What might make the most sense today might be obviously uneconomic tomorrow.

A classic example of this kind of change is the decision as to whether to build a private voice communications network of tie-line-connected PBXs (called a *tandem network*—this is unrelated to Tandem computers). Many large companies built such networks for sound economic and functional reasons between the early 1970s and mid 1980s. But the emergence of virtual networks, such as AT&T's Software Defined Network (SDN) made many of these private networks uneconomic by comparison. They continued to function, but many companies have retired them because, in most cases, the virtual networks are far less costly while delivering most of the characteristics that justified the construction of the original private networks.

Comments on Line and Trunk Networking

Some of us might remember the early days of competition in the long-distance arena. Remember the way we had to connect to the alternate long-distance suppliers, like Sprint and MCI. The sequence was covered in Chapter 3, but it is worth restating here to show the reasons why things occurred the way they did. This scenario will clear up how the networks all came together after divestiture and how things have improved with the use of trunks instead of lines. The play-out of a call followed this sequence.

A customer might sign up with a long-distance supplier other than AT&T prior to 1984. As a new customer of the competitors, several discounts were offered over the long-distance tariffed rates from AT&T. In order to use the service, the long-distance supplier would issue the customer an 800 number to call the competitor's network, or a special number. This special number was a 7-digit telephone number that could be a regular local number in the area, a 950-XXXX number, or a foreign exchange telephone number from a major metropolitan area. The choices were based on the density of the carrier's services in the customer metropolitan or geographical area.

The customer would issue this telephone number to all internal users. Along with the 7- or 10-digit telephone number into the carrier's network (we'll use MCI from this point on for simplicity), another 10-digit number, called an *authorization number* was issued. This might be a unique number for every individual in the organization. Or, it might be a global number used by the entire organization.

The caller (end user) now wants to make a long distance call from his or her office. So the sequence begins like this:

- Pick up the phone and get dial tone, then dial "9" for an outside line.
- Dial 1-800-Cal-lMCI or in numeric (1-800-225-5624), for example.
- Wait for a connection. As the call proceeds a ring tone will be heard, then the MCI system would answer and provide a "computer tone," which sounds much like a steady high-pitched tone.
- Upon getting the computer tone, dial 1234567890, or whatever 10-digit authorization number was assigned to the organization. Wait for computer to confirm this number.

- After the computer acknowledges and verifies that the 10-digit authorization code is valid, it will return a dial tone to you.
- Now dial the 10-digit telephone number of the party you wish to speak with: (602) 555-0121.
- Wait for the call to proceed and ring. Hope and pray that the call gets answered, is not busy, and is clear enough to hold a conversation on. Otherwise, start all over at point one.

The users would obviously become very frustrated with this procedure. This is especially true if the called parties were busy or if they needed to make multiple calls, such as a telemarketing group. The need to dial 32 or 33 digits just to get a call through was frustrating enough. But, if they did not follow the company guidelines and dialed AT&T directly, they only had to dial 12 digits. This is significant, particularly when there were thousands of calls being made per month. The accumulated waste of time might have cost the organization more in productivity losses than the savings they were trying to achieve with MCI's service.

So, why did MCI require all of these digits? The answer is simple, they had no choice. When competition first began, AT&T was the owner of the Bell System. To preclude the competitive threat, they controlled how the network was set up. MCI had to rent telephone lines from the local Bell Telephone Company. At the central office, this line was connected from the CO to the MCI computer. The call was a completed call the minute the computer answered the incoming request. MCI did not get any of the information that is passed along from CO to CO, or CO to LD supplier because they were on the wrong side of the switching system. They were on the line side, not the trunk side. AT&T was on the trunk side of the switch, so all of the caller ID information was passed along from switch to switch, thus they did not require all the extra digits.

When divestiture took place and the Bell System was broken apart from the AT&T network, then equal access was allowed. Prior to that, AT&T controlled the network and made sure that equal access would not be a reality, or they priced the equal-access connectivity so high that no vendor could afford it.

Now that all things and carriers are equal in the eyes of the MFJ, the carriers (such as MCI, Sprint, LDDS, etc.) can all be connected to the trunk side of the system. Now caller ID information, called *automatic identification of outward dialed (AIOD)* or *automatic number identification (ANI)* and many other names, is passed on to any carrier that is connected to the local or toll switches on the trunk side of the network. Calls are passed from intelligent switch to intelligent switch, routed through the network to an end point before a termination takes place. The world is a better place for this. MCI and its peers are now all able to offer the same limited dial sequence that AT&T has always enjoyed. Further, now they get even better access to the systems and are offered services called *feature groups*, allowing for flat-rate billing, call screening, and multiplexed services on high-speed trunks. This makes them as attractive as using any of the long-distance services that they were competing with in the past.

8

Traffic Engineering

The art of conducting true traffic engineering studies has all but died. In the early days of telecommunications systems, the telco planning and capacity engineers spent days and weeks designing their switching and trunking systems to provide optimal performance. Since the inception of computer technology, programs have been written that can perform all of the calculations and the various iterations necessary to fine-tune a network. What used to take the telco engineers weeks now can be accomplished in minutes of processing time. Beyond the telco engineers, telecommunications managers and designers were equally concerned and attempted to model their networks for the best accessibility, greatest utilization of their lines and trunks, at the most reasonable price. From the discussion in earlier chapters, options for usage sensitive services exist. Using the best mix of WATS lines, for example, could reap savings in the past that were exceptional. In all actuality, this is not an engineering case, but a mathematical computation of all probable events that can take place in a given time frame. There is no engineering taking place, but a modeling of the traffic needs of the organization and the access lines necessary to support these traffic needs with a given level of service. It is the calculation of a random number of arrival events taking place over a specified period of time. In the telecommunications arena, this is the number of possible first attempts to get an outside line, or for an incoming line to ring into an organization on the first try. Because there are a finite number of time events, but an infinite number of possible attempts that can be achieved in a time period, we attempt to define the best possible level of service within a specified period of time. For the installation of lines and trunks, the goal is to serve the maximum number of callers at a single point in time (usually a 1-hour period). Because the arrivals of calls are random, the design must equate to an hour's worth of possible arrivals, that being 3600. This number is obtained by multiplying the number of 1-second increments in a minute times the number of minutes in an hour (60 sec. × 60 min.). From this possible random one-second arrival event, the concept of serving the worst-case scenario in a one-hour period (the busy hour) is calculated to arrive at the number of lines or trunks needed to serve the heaviest load in a one-hour period. For the rest of the day, there will be too many facilities, but for the busy hour, there will be an adequate number to deliver the level of service that has been decided upon.

An example of this would be similar to the traffic designers of the road system. These people are chartered with the responsibility of creating road systems that will allow the maximum amount of cars and trucks to traverse the roads in a city, town, state, etc. Because a good deal of money will be invested in the building of the road system, the engineers are requested to allow for the most amount of vehicles to travel the road system within a reasonable cost estimate. From there, the engineers create a model for the traffic based on the hourly distribution of the vehicles on the road. They will conduct counts at random times, clicking off the number of vehicles passing an area within a specified period of time. Then they allow for a certain amount of new traffic, or growth as traffic patterns might change. Armed with this information, they design a road system with the appropriate number of lanes to support the steady flow of traffic at a constant rate of speed (whatever is allowed by law in the area). Knowing these pieces of the usage patterns on a daily and hourly basis, the engineers then design their roads to maximize the total throughput from end-to-end on the road that will be built. Reality starts to set in when we perceive the actual daily conditions on our current roads. In the early morning and late afternoon rush hours, traffic gets snarled and grinds to a halt for extended periods of time. The engineers have accomplished their mission: they have maximized the traffic on a road system; on every square inch of road there is a car. Unfortunately, during rush hour, one critical ingredient gets forgotten. These cars were supposed to move across the road system at a constant rate of speed and exit the road within a certain time frame. However, as we all know, they are all parked on the highway system; no one is moving, so congestion exists. Taking this one step beyond the rush hours though, the engineers designed a road system that allows users to get on the road and move at a steady rate of speed from an entrance point to an exit point. The level of service during nonrush hours is much better. Actually, the system is over-designed for these off-hours periods, somewhat inefficient use of the roads. But, you cannot add and delete lanes of highway for selected portions of the day, they are fixed in concrete or asphalt forever.

This is exactly the same scenario that a telephone traffic study must follow. During the average hour of the day, all calls should be processed into or out of the organization, with limited delays. The amount of incoming and outgoing lines are crucial to ensure that the communications process works efficiently. However, the number of access lines are going to be fixed all day long because the copper or other medium brought into the building is not something that can be taken away or added on an hourly basis. For certain periods of the day, more calls might be required. This is called the *busy hour*. The telco or telecommunications analyst typically tries to design the number of lines to support the worst hour of the day. Yet for the rest of the day, too many lines are available. The more lines that are sitting around at a fixed cost, the more expensive the organization's costs are. This is based on a mathematical probability and must look at the dynamics of the calling patterns of the users or customers. Consider though what the risks of a poorly designed network are:

- If the customer attempts to dial into an organization to place an order and the lines are all busy, then that customer might hang up and place the order with a competitor.

- If the outgoing traffic within an organization is severely blocked due to busy conditions, then the users might not be able to get mission-critical information dispensed as needed.

When busy conditions arise, users might find alternative ways of getting their job done. They might rent or lease their own lines, bypass the cost-effective facilities provided, or lastly place the calls via credit card or third-party arrangements through an operator assisted call. In any case, all of the options will be far more costly to the organization.

Therefore, the designer of the services must consider the levels of service to be provided. This requires a thorough understanding of the traffic to be handled. From this perspective, a designer will have to consider the information that is available for study. This plays out in the paragraphs that follow. Information gathering is essential.

Where Can the Information Be Obtained?

Analyze the traffic that is coming in and going out of the organization. Access to this information might be a difficult portion of the desired study. To obtain the information there are several points that can be used:

- The telephone bills for all incoming 800-service calls are good indicators of the total volume. But, this bill will only show the completed calls, not the ones that encounter a busy tone when trying to call into the organization.

- The telephone bills for all outgoing calls from an organization. These might be difficult statistics to obtain, depending on how the billing systems work. For example, if a group of departments receive their own bills and process them for payment, the information might be difficult to get. Or as another situation arises, the billing cycle might be different for different types of lines and trunks. In many cases, the direct-dialed billing for the main telephone listed numbers are received on the first of the month, whereas the long-distance portions of the bill might be on a cycle of the 10th of the month. This requires an inventory of all the line and trunk billings that are processed each month. Working with either the providers or the internal accounting department might well get this all started in the right direction.

The call processing time must be considered when analyzing any requirement for access lines. Often, the novice telecommunications analyst ignores the amount of set-up time for calls and, therefore, could miss a sizable chunk of time to be factored into a calculation. This figure can be obtained by dialing several calls and concluding the average amount of time necessary until the call is completed to its destination point. Remember that the telco and long-distance carriers will bill on the duration of a call from the time it is answered until it is terminated, with some rounding of up to six seconds possible. Therefore, the added time a circuit is held up for call set-up is important in calculating the total usage time because the line or trunk is inaccessible to others for this time.

Many of the telephone systems, call distribution systems, etc., have management and statistical reports available. These might not be turned on or reports might not be generated on a regular basis, but they can be obtained when necessary.

Telco and long-distance suppliers alike can gather statistics on what is called a *busy study* and a *peg count*. The busy study can usually be conducted for a week, reflecting all attempts into the incoming line group of an organization, whether completed or not. Further, this study can indicate the hourly distribution of the incoming calls so that some hourly sensitivities are afforded in the design study.

The wrong mix of telephone lines will be frustrating for customers and users alike. Yet, the overkill situation might be too expensive. The charter of the telecommunications department within an organization has always been "to provide the proper mix of goods and services to accomplish the calling needs of the organization, increase productivity, and reduce the bottom-line expenses to the organization." All too often, we have made the mistake of trying to get to a bottom-line figure, forgetting the other portion of the mission statement at all costs.

Accomplishing the Mission

Armed with the calling information, the next step then is to look at the role being accomplished. What will the design effort accomplish? You can create a design to improve customer service, reduce costs, or improve user connectivity. Another option is one of just attempting to configure a new setup. Regardless of the choice, the basic elements are the same. Upon successfully garnering a mission, there are several steps that can be taken to perform the design. This is the traffic engineering function at its best. The steps that should be taken are highlighted in Table 8.1. These are the basic steps that one would follow for just about any design of a configuration. Each of these is covered in greater detail later.

TABLE 8.1   The Steps in Accomplishing the Traffic Engineering Function

Basic steps	Action
1. Determine the traffic demand that will have to be satisfied.	Gather the data of incoming calls, peg counts, busy studies and other sources of information.
2. Convert the information from a monthly basis to a distribution table based on daily and hourly statistics.	From the monthly information, plot the data into an hourly distribution, or a daily distribution so that the design can be used.
3. Determine the appropriate traffic engineering tools to be used. The tools will include Poisson, Erlang, or Extended Erlang.	Choosing a tool that will assist in creating a rudimentary design, given the historical data can be assimilated into a design tool.
4. Apply the information into the modeling tool.	Using the tools and tables, apply the data to the tool, to obtain the desired format.
5. Select the grade of service to be delivered.	By establishing the grade of service, the percentage of calls blocked vs. service within a specified period can be predicted.
6. Extract the information from the tools and the tables, depending on the method chosen.	Actually once the data is submitted and the grade of service is established, extracting the information is fairly simple.
7. Apply the information back to a monthly distribution.	From the hourly distribution, the monthly traffic that will be handled, as opposed to the blocked and overflow traffic can be assessed.
8. Try various iterations of grade of service.	Since the data is known, the iterations will allow for usage sensitivities and variable conditions that can be accommodated.
9. Select the right configuration based on the available information.	Using the information gathered in the iterations then select the design that best fits the need. From this design, costs can be estimated and equipment needs will be established.

Using the Information

From these steps, the information can now be set up for the calculation process. This will entail all of these steps to calculate the actual number of circuits needed to serve a need. There is no mystique here, just a formal mathematical computation of the data against a given set of rules.

Gather the Data

As already stated, the ability to gather the data will be contingent upon our ability to assimilate the various bills, reports, and studies that are available from the local telco and from the in-house hardware systems. From these reports, the amount of traffic to be served will be easier to extract. Differences will exist from the outgoing and the incoming traffic. In all cases, it might be preferential to give a better grade of service to incoming customer calls than to outgoing administrative calls. This all depends on the individual organization. Assuming that all of the data is gathered, let's play out an example of an outgoing calling pattern for an organization. This example will be based on a typical bill where the average hours of usage are 500 per month. Note that this is the aggregated amount of calling in the one-month period. The data are then accumulated and estimated averages are placed on the total number of calls. For the sake of this example, the average length of call is estimated at 4 minutes. The total hours are then divided by 4 minutes per call to come up with the total number of calls.

$$\frac{\text{Total hours} * 60 \text{ minutes per hour}}{4 \text{ minutes per call}} = \frac{500 * 60}{4} = \frac{30000}{4}$$

$$= 7500 \text{ calls per month} \quad (8.1)$$

Added to the total calls per month, a certain amount of extra time must be added. The 500 hours figure reflects what the telco will bill for the call time, not for processing time. As a matter of simplifying this, the average time allotted per call for set-up and tear down, which is time-consumed on the circuit, but not billed, will be 30 seconds per call. Where the 30 seconds is derived includes a tone-dialed call from beginning until the call is answered and the billing begins. This is broken down into the following components in Table 8.2.

Thus the added call processing time must be accumulated with the hours of usage for the month. Therefore, these two numbers are added together. If this is omitted, the engineer might design a system with a certain blockage level (grade of service) and not be able to deliver that level of service. Look at the amount of time added to the hours of usage.

TABLE 8.2 Added Time to Call Processing

Action	Time allotted
1. Dialing an eleven digit telephone number, from a touch-tone phone	8 seconds
2. Delay in telephone system (PBX) network processing	6 seconds
3. Delay in network processing and regenerating digits	7 seconds
4. Ringing, assuming three rings (1 sec ring, two second pause)	9 seconds
Total time allowed	30

Total hours usage + (total calls * added processing time)

= Total time

= 500 hours + (7500 * 30 seconds/60 seconds)

= 500 hours + (62.5 hours) = Total time of 562.5 hours   (8.2)

Note that if the extra 62.5 hours were not added into the total number, the design could be off by a factor of 12% in the overall traffic. This is a common mistake that people make when designing their networks.

Convert the Data to Daily or Hourly Usage

Although the telephone billing information is possibly a source of the information, the calling patterns on a daily or hourly basis can become quite tedious. However, if a design is to be effective, the information must be converted into a usable and understandable format. The indication of traffic occurrences will assist in determining just how much arrives within a given point or arrival time. The best way to achieve a result is to look at an hourly distribution. From the telephone bills, the total hourly distribution will likely vary from day to day and hour by hour. Consequently, scanning of the paper bills into a computer database, or receiving the information on some electronic medium (i.e., diskette, tape, CD, etc.), will allow a telecommunications manager or designer to read the information into a PC or computer system. From this read, a sorting of dates and hours can be achieved. This will give the best results. However, when the information is a combination of paper or other form of data, the choice is to draw some averages of the calling patterns.

Typically, a pattern of distribution will exist that can be used in assimilating the information into a calculable form. The call distribution for many organizations is shown in Table 8.3. This table reflects the hourly distribution averages for a typical example, it is not to be construed as applicable to any specific organization. Further, this distribution will be based on the services offered by the organization. Call centers might well experience an average hourly distribution, whereas a manufacturer might see a dual-mode distribution curve (Figure 8.1). The example used in Table 8.3 is geared toward a manufacturing environment, with no traffic being carried during the lunch hour. This is usually a fair assumption.

The distribution would then be spread across a dual-mode distribution curve. This attempts to define the number of circuits required to satisfy the distribution of calls in the worst hour, as already stated.

These assumptions are made on the basis of various inputs and can be modified. Given that the information is now placed in an hourly percentage, the next step is to convert the data into hours of traffic in the hourly distribution. From there, the process can begin to take place.

Choose the Appropriate Tool

Several tools exist to model the data. The primary ones include Poisson Distribution, Erlang B, and Extended Erlang B. The best way to describe the tools is the way the iterations take place and why they do so. In Poisson distribution, a mathematical distribution of the loads placed on the number

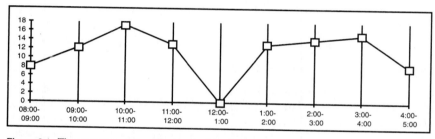

Figure 8.1 The average usage reflects a dual mode pattern.

TABLE 8.3 Hourly Distribution of Calls for a Manufacturing Organization

Hourly distribution	% of calls per hour
08:00–09:00	8
09:00–10:00	13
10:00–11:00	17
11:00–12:00	12
1:00–2:00	12
2:00–3:00	16
3:00–4:00	15
4:00–5:00	7
Total	100%

of circuits will not be calculated. Rather, the numerical values assigned with the overall distribution as set in the tables are merely placed into a grade of service level and a number of circuits is selected. The load is assumed to be even over the number of circuits chosen. This is OK for some calculations, but if additional distribution is required, then the other tools will be used. For a comparative look at these tools, first look at how the Poisson distribution will work. Remember that the monthly traffic is 500 hours. From a daily perspective, some added criteria is needed. The added parameters are shown in the following list:

1. The number of business days in the month is 22
2. The average day is 8 hours long
3. The hourly distribution is shown in Table 8.3
4. The average length of call is 4 minutes
5. The call processing time is 30 seconds

Poisson Distribution

From all of these numbers then, a set of parameters can be established. One added factor is that the Poisson distribution model uses a slightly different calculation. Remember in the earlier section of this chapter that the statement was to consider the random arrivals of calls (incoming or outgoing) based on a 1-second arrival rate. Therefore, the number of possible arrival rates in an hourly distribution is 3600 (60 minutes × 60 seconds). The Poisson distribution takes into account all 3600 possible arrival rates and calculates these on the number or circuits necessary to support this traffic load. The telco engineers use a term called *Centum* (or hundred) call seconds (CCS). Rather than divide and multiply the hourly traffic by 3600, the theory is that if we divide the 3600 by 100, we arrive at a traffic load of 36 CCS per hour of traffic. Therefore, to use the Poisson distribution, all traffic is converted to CCS prior to submitting it into the model. Once the traffic is converted into CCS, the model will determine the number of CCS that can be carried within a specific design.

Therefore, if you use these assumptions, the traffic load can be converted and then inserted. The resultant traffic load is therefore summarized in the following list. The traffic is based on a worst-case scenario in the busy hour. A grade of service level will be assigned to determine the number of circuits necessary.

1. Total traffic hours carried   500
2. Added traffic for call set-up   62.5
3. Total hours divided by 22 business days (562.5/22)   25.57 daily hours

4. Daily hours multiplied by busy hour percentage $(25.57 * 0.17)$ 4.3469 busy hour traffic

5. Busy hour traffic multiplied by 36 CCS $(4.3469 * 36)$ 156.48 CCS

6. Grade of service desired   P.05

Determine the Grade of Service Desired

As already mentioned, the grade of service desired is the worst-case statement for the busiest hour of the day. The GOS is used to determine the number or percentage of calls that will experience a busy tone on the first attempt during the busy hour. A grade of service of P.05 (probability of 5%) means that 5 of 100 callers might encounter a busy on the first attempt. The service actually experienced might be much better than the P.05, but the worst case is 5 of 100. Using this information and the average busy hour load of 156.48 CCS that must be carried, the Poisson distribution model can be applied. A sample Poisson distribution table is shown in Table 8.4. This table is read in the following way:

Read across to the grade of service desired (P.05). At the column headed P.05, read down until the calculated load value is found. In many cases, the actual value of the load will not be even; therefore, round the number in the table up to the next value on the table. From the load number, read left until the column labeled number of circuits, where the numerical number is the circuits that will be required to carry the traffic. In this case, the number of circuits required to carry this traffic is 9.

Using the information in the highlighted area, reading across left to right then down, lastly back across from right to left, the answer is 9. This means that the number of circuits necessary to support the busy hour traffic is shown. However, during nonbusy hours, less circuits are used, which should deliver a far better grade of service. Rather than go into the details of the actual grades of service for each of the other hours, these are shown in Table 8.5. These grades of service are extrapolated from the information in the percent of calls in each of the hours, and the number of CCS loads for each of these loads.

To conduct some sensitivities on this design, the number of CCSs can be kept constant and the grade of service can be either increased or decreased, with the results reflecting more or less circuit requirements necessary to support this new service level. You can play this out in various sensitivities, which is why the excerpt from the Poisson distribution table includes P.01–0.05, 0.07 and 0.10. Bear in mind that the sensitivities still include the differences of the expected hourly load. Worst case, this is the level of service. You might well expect that better grades can be achieved because this is not a perfect science and the rounding effects of the calculations can become a factor. Further, in arriving at the hourly percentages

**TABLE 8.4   An Excerpt of a Poisson
Distribution Table**

Number circuits needed	Grade of service					
	P.01	P.02	P.03	P.05	P.07	P.10
1	0.4	0.7	1.1	1.8	2.6	3.8
2	5.4	7.7	9.6	12.7	15.4	19.2
3	15.7	20.3	23.8	29.4	33.9	39.7
4	29.6	36.5	41.5	49.1	55.2	62.8
5	46.0	55.0	61.4	70.9	78.3	87.6
6	64.3	75.2	82.8	94	103	113
7	83.9	96.6	105	118	128	140
8	105.0	119	129	143	154	168
9	126.0	142	153	169	181	196
10	149	166	178	195	208	224
11	172	191	204	222	236	253
12	195	216	230	249	264	282
13	220	241	256	277	292	311
14	244	267	283	305	321	341
15	269	293	310	333	350	371
16	295	320	337	361	379	401
17	320	347	365	390	409	431

**TABLE 8.5   Sensitivity of the Same Data with
Differing Grades of Service Reflect the Additional
or Less Trunks Needed**

Grade of service	Number of trunks @ 156CCS	Difference
P.01	11	+2
P.02	10	+1
P.03	9–10	0, +1
P.05	9	0
P.07	8–9	0, −1
P.10	8	−1

of the monthly calls, some estimates were used. If the organization has periods of fluctuation during certain months of the year, or days of the month, then these calculations could be off significantly. But the data supporting the way the calculations were performed did not reflect these monthly deviations. A true monthly detailed distribution would be more precise, but this would take much longer to analyze the actual call details.

Erlang Distribution

Mathematical computations of the traffic offered to a group of circuits can take a different form, that uses different assumptions. One such variation is the use of Erlang B techniques as opposed to the Poisson distribution. Erlang B was developed in the early 1900s by K. Erlang of the Copenhagen Telephone Company. The assumption used is that if a caller encounters a denial or busy tone when attempting to place a call, the caller will be routed to a more expensive circuit (such as the long-distance network called *DDD*) or will give up trying to make the call. This would be devastating in a call center operation if callers experienced a busy tone, then hung up and called the competition. However, the assumptions were that some alternative facility would be available. Erlang B was more effective in calculating call completion and attempts on an outward dialed basis, where callers could be routed through an automatic route selection process to a more expensive link. The network would not accommodate a queuing arrangement in the assumptions modeled with Erlang B computations.

Other models were created using the process of modeling after Erlang assumptions, these introduced various options in the design of a circuit group. A summary of these variations is shown in Table 8.6, which compares the techniques and the basis for using the variation. The Erlang table comparisons assume that a queuing technique can be introduced and accepted.

For the purposes of describing a queuing technique, you can imagine a busy branch office of the local bank. During periods of heavy load, the bank might have two or three tellers available to service its customers. However, on paydays in many major corporations, the lunch hour becomes a very congested period for the banks. Even though they have three tellers there to handle their customers, there will be more customers at any one time than there are tellers. Thus, the bank installs the old cattle car line arrangement. Using a series of weaves or redirection cords, the bank attempts to get the customers to line up one behind the other. This allows the bank employees to serve customers on a first-come, first served basis. Because the arrival of the customers is serial, as they form (or queue) in the line, they should get serviced in the order in which they arrived. Of course, the length of time it will take to get serviced depends on the arrival rate and what position the customer holds in the queue. During busy conditions, the bank manager seeing the queue

TABLE 8.6 Variations of the Erlang Modeling Techniques

Model	Handling	Choice considerations
Erlang B	No queuing	Use a route advancement technique automatically routing the traffic to the next available route, even if more expensive to service the call.
Extended Erlang B	No queuing	If all circuits are busy the caller must hang up and try again later. No advancement of routes are used.
Erlang C	Queuing allowed	Calls are queued by a computer or an operator and wait until the next available circuit becomes available.
Equivalent Queue Extended Erlang B	Queuing allowed	Calls are queued for a specific time; if a circuit is still not available the system may then route the call to a more expensive route or give the caller a choice to try again later.

building, might decide to activate more employees into the servicing of the queue. This might require that other employees with different functions be pulled into the customer processing. Supervisors and managers of other departments might also be activated until the queue is handled. All of these arrangements work on the basis of creating additional resources for the handling of the traffic. However, if the customers were to arrive during this rush hour, and the bank had taken a position that only one customer at a time was allowed in the bank, the picture would change dramatically. Instead of being serviced, the customers would be told to go to a different branch office for service. This would infuriate their customers and could prove far more costly. It would only take one occasion like this and the customers would find a new bank to handle their accounts. So, goes the Erlang technique, you can be queued for the same group of facilities, or you can be sent out across the network to a more expensive solution.

Each of the Erlang theories is a better gauge of the blockage and grade of service than the Poisson distribution theory listed earlier. Thus, they have been used in the modeling techniques most recently since computers have been used in the modeling of traffic systems. An Erlang is equivalent to one hour's traffic in a one-hour period, or stated against the Poisson model, an Erlang is 36 CCS.

Using the same data as applied to the Poisson distribution model, a computation can be made with the Erlang method. Therefore, the data to be used in the Erlang B formula is 500 hours of usage, with an average length of call set at 4.5 minutes (this is the 4-minute average plus the 30-second overhead). The application of the formula actually works out as shown. This formula is used to calculate the number of circuits required to deliver a specified grade of service.

$$E(C, T) = \frac{\dfrac{T^c}{C!}}{\displaystyle\sum_{i=o}^{C} \dfrac{T^i}{i!}} \tag{8.3}$$

where T = Traffic attempts, C = Circuits available

To make this a little simpler, the following shortcut can be applied using tables that have already been created, then applying a simpler format to the data and deriving the number of circuits from the table shown below. To create the scenario, look to the following formula:

$$\text{Erlangs} = \frac{N \times A}{3600} \tag{8.4}$$

where N = number of calls handled in the busy hour, and A = the average length of calls in seconds

As stated, the Erlang result from the formula is then applied across a table in the same way as the Poisson distribution model to arrive at the appropriate number of circuits needed. Use Table 8.7.

You can see from these tables that some sensitivities exist, and that the difference between the number of circuits needed and the various blockages that could be experienced will produce differing results. An organization that is in the mode of trying various iterations can get between a P.01 and P.05 with the possibility of minimal additions or deletions to the calculations. However, this is not just about circuits needed. The obvious deduction that can be drawn revolves around the difference to "customer service" and the ensuing numbers of devices required to deliver that grade of service. What also goes with this is the possibility of staffing a call center operation. The number and cost of circuits are variables that are not necessarily that important anymore. Because the cost of renting additional circuits has all but become a commodity pricing arrangement, the number of circuits are irrelevant, other than the degree of blockage allowed. The cost of a WATS line, or an 800 service these days are based on the monthly charge of $37.50 worst case. Thereafter, the usage cost is based on an average of hours used per circuit. However, as shown in Chapters 6 and 7, these costs are really based on the averages base on the total number of hours used divided by the number of circuits, then applied to certain rate steps (0 to 15 hours, 15 to 40 hours, and over 40 hours). With this in mind, assuming that each circuit will carry at least 40 hours of traffic, each added hour is a marginal increment cost-wise.

TABLE 8.7 Erlang B Tables

Grade of service # circuits	P.01	P.02	P.03	P.05	P.10
1					
2					
3					
4	0.87	1.09		1.52	22.05
5	1.36	1.66		2.23	2.87
6	1.91	2.28		2.96	3.76
7	2.50	2.94		3.74	4.66
8	3.13	3.62		4.54	5.60
9	3.78	4.34		5.37	6.54
10	4.46	5.08		6.22	7.50
11	5.16	5.84		7.08	8.49
12	5.88	6.62		7.95	9.47
13	6.61	7.40		8.83	10.46
14	7.35	8.21		9.72	11.47
15	8.11	9.02		10.64	12.48
16	8.87	9.83		11.54	13.51
17	9.65	10.66		12.46	14.52
18	10.44	11.49		13.37	15.55
19	11.23	12.33		14.31	16.58

But the degree of frustration delivered to the caller will be the actual deciding factor. Now, applying this degree of frustration and the number of other ancillary devices is where the calculation of traffic demand will be used.

For example, if using the tables and drawing the conclusion that a 5% (P.05) blockage level will be delivered to the first-attempt callers and the telecommunications manager decides to see what the impact of a 1% blockage will require, the calculation will take on new meanings. This can be extended to the numbers of:

- Ports required on the automatic call distribution system
- Agents required to service the calls
- Terminal devices for the agent answering positions
- Floor space and desk space for the agent to conduct business

- Benefits associated with hiring the agents
- Payroll costs associated with each agent
- Personal computers or mainframe terminal devices for customer lookup and inquiry
- Wiring

The costs associated with each of these devices, and the human resource costs, can amount to a significant number. The marginal cost to add one added agent could be as much as $100,000 per annum (benefits and payroll taken into account) plus the capital costs associated with the equipment purchase and maintenance of the devices. This size number is not to be taken lightly, so a true analysis will be required to prove in the benefits. The issues are not always related to costs, but a cost to benefit ratio could be applied that will either justify or prove that a better grade of service is warranted.

If the decision to improve levels from a P.05 to a P.01 grade of service is considered, then the additional costs can be compared to the expected increases in sales, revenues, or service benefits. For general discussion, many organizations use a 3:1 cost/benefit ratio before selecting a new system or a change to an existing one. If an organization meets or exceeds the 3:1 ratio the decision is usually easily justified. The greater the ratio, the easier the sell to management. A representative cost/benefit ratio equation is:

$$Cost\ benefit\ ratio = \frac{Benefit}{Total\ cost}$$

$$= \frac{450,000}{100,000} \qquad (8.5)$$

$$= 4.5$$

However, if the cost/benefit is below the 3:1 ratio, then one will have to struggle to gain management's acceptance, or have some alternate plans ready in case the funding cannot be accommodated. The example used in the figure is for a new design that will cost a total of $100,000. However, sales is expecting to create $450,000 of new revenue from the installation of this system. Bear in mind this is on the basis of cost to benefit. If one was to create a profitability index on this installation, then the picture could change dramatically. What is the true benefit? What other choice would the organization have to implement a new program to generate $450,000 of new sales? If the differing projects pointed to a cost of implementing some other technique (marketing campaign, promotions, etc.), then the cost comparisons could be used. For example, if sales would have to spend $600,000 in marketing campaigns and promotions, then the difference would be between the two choices ($100,000 vs. $600,000 yields a 6:1 benefit cost ratio). This allows a business decision to be made based on the facts rather than strictly some "guesstimate" or emotional facts.

This is what traffic engineering is really all about. When this can be fully justified, then a business decision can be supported and defended with senior management who is chartered with the custodianship of the organization's spending and the maximization of shareholder wealth. Therefore, the selling of communications lines and equipment is not for the telecommunications department, but really in support of the organization. Consequently, this is to support the call center operation and they should have an input into the budgeting process followed by the benefits expected from the installation of the new configuration or equipment. The use of traffic engineering is a means to get to an end, not the end in itself. Once this is clearly understood, the telecommunications department can proceed from there.

9

Equipment:
Private Branch Exchanges

Private Branch Exchange (PBX)

A *private branch exchange (PBX)* is a large organization's typical telephone system. In this environment, an organization, served by a central office dial tone from the local exchange company, might need the capacity of high-volume calling and handling services. Clearly, a single-line telephone set with a dial-tone line for each user will work. But, it will only just work! It will not satisfy the needs of the organization.

Additionally, it will be expensive. Assume that a dial-tone line costs $20.00 per month. If the organization has a multitude of users, the cost per month will be significant. Table 9.1 highlights some of the typical costs associated with basic dial tone for various numbers of employees. These numbers are representative only, but they should get our point across. The table reflects the basic monthly and the annualized cost of renting a dial-tone line from the local carrier.

You can clearly see from these numbers that the use of a basic dial-tone service can get quite expensive. As a matter of fact, many organizations now say that their telecommunications cost is the number two expense items in their corporate expense register, only second by personnel costs. This is both good and bad. It is good that organizations are depending on telecommunications more, as opposed to a more expensive alternative (such as travel, personnel, and other sales and marketing costs). Pound for pound, telecommunications expenses still produce a greater return on every dollar spent.

TABLE 9.1 A Summary of Costs for Dial-Tone Lines for the Organization

Number of users	Monthly cost @$20.00	Annualized cost
100	$ 2,000.00	$ 24,000.00
500	$ 10,000.00	$ 120,000.00
1,000	$ 20,000.00	$ 240,000.00
2,500	$ 50,000.00	$ 600,000.00
10,000	$200,000.00	$2,400,000.00

TABLE 9.2 A Summary of Single-Line Set Equipment Costs

Number of users	Cost of equipment
100	$ 6,000.00
500	$ 30,000.00
1,000	$ 60,000.00
2,500	$150,000.00
10,000	$600,000.00

But back to the point. The costs can be staggering to a financial or senior managerial person in an organization. But the dial-tone line costs listed above only give the user the dial-tone access. This is a full-time dedicated access line for two-way service for every single user. If you add just a single-line telephone set for each of these users, then there is some capital costs associated with the ownership of these lines. Table 9.2 compares a single line set for every user, at a base price of $60 per single-line telephone set. These are, again, basic assumptions for the purchase of these sets, one could do better.

Again, you can see that the equipment costs can mount quickly. But what is wrong with this picture? Well for starters, the single-line set limits what the user can do with the basic dial-tone service. Also, the single-line set does not allow for intercommunication between the users within the organization, unless they tie up their dial-tone line as follows:

- Grab the dial tone by going off hook
- When dial tone is received, dial the digits (seven) of the desired internal party
- When the ring is generated and the party answers, hold a conversation

But this ties up two complete outside lines for the two parties to converse. If a customer tries to call either of these two parties, the customer will get a busy tone. That is unless the call "hunts" to some other number. If the call does hunt, then a third outside line is occupied for the duration of taking a message at the rollover line. Customers can be denied access, and get frustrated. All of this while two parties could be talking to each other in the next office (Figure 9.1). Note that however long the wires run back to the central office where the dial tone is provided, the call uses twice that to get the two conversationalists together. Clearly, this is not an optimized use of the telecommunications services.

It should be obvious from the discussion above that larger organizations require the larger capacity and capability of a private branch exchange (PBX). These systems have names that come in many flavors such as the Private Automated Branch Exchange (PABX), Computerized Branch Exchange (CBX), Digital Branch Exchange (DBX), Integrated Branch Exchange (IBX), and Nippon Electric Automated Exchange (NEAX).

These names basically mean the same thing. They are just a different vendor acronym used to differentiate their specific products. The generic term PBX is a private (customer owned and operated) branch exchange (like a central office it switches and routes calls internally or externally and provides dial tone to the internal users). The PBX marketplace is inundated with acronyms and features. However, they all do similar things: They

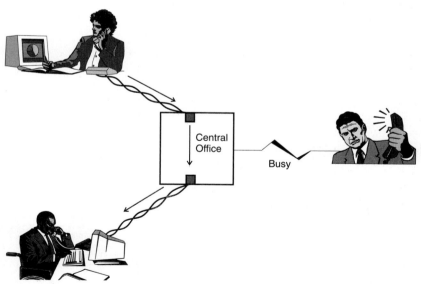

Figure 9.1 The single line is busy when A talks to B, therefore when the customer calls in, a busy tone is received.

process primarily voice calls for the organization. These devices are computer systems that just happen to do voice. Now they do other things, such as provide data communications and data access.

On average, the all-digital PBX will cost approximately $750 to $1000 per station. A station is the end-user device, and includes the cost of all associated hardware to support that telephone set. Included in this generic price is the card inside the computer that provides the dial tone, the logic, and a portion of the common equipment that serves many users, the telephone set, the wiring, and the installation.

The components of the PBX are shown in Figure 9.2.

- The central processor unit, or CPU, is the computer inside the system; the brains.

- Memory, any computer needs some amount of memory.

- Stations, or telephone sets, which are also called lines.

- Trunks, which are the telco CO trunks terminated into a PBX.

- A network that switches inside the system.

- Cabinets that house all the components.

- Information transfer or bus that carries the information to and from the computer.

- Console or switchboard where the operator controls the flow of incoming calls, etc.

- Common logic, power cards, etc.

- Battery back-up.

- The wiring infrastructure.

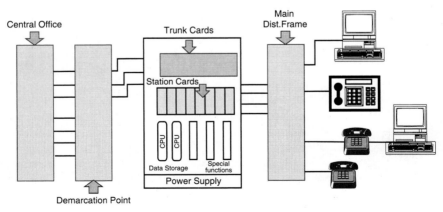

Figure 9.2 The components of a PBX.

The PBX is a stored-program, common-controlled device. As a telephone system, it is a resource-sharing system that provides the ability to access dial tone and outside trunks for the end user. This stored-program controlled system today is an all-digital architecture. In older versions, the PBX could be an analog system. Newer systems, however, are all digital. It would not make sense to produce an older technology for a modern-day telephone system.

Analog Systems

What is meant here is that the analog system used analog components to handle the call set-up and tear-down for the entire system. A voice call is introduced into the system in much the same way for a business or residential user's input is introduced to the telephone company network. As the user generates a call, the telephone handset is picked up from the cradle. At this point, an I/O (input/output) request signal is sent to the main architecture of the PBX, usually a computer. Once the signal is sent to the common control, the system then returns dial tone. The user then dials the digits for the party desired. This dialing sequence is done in-band on the wires in the talk path of the caller. The digits either rotary (pulse) or tone (DTMF) are sent down the wires to the telephone system.

From there, the telephone system kicks in and generates a request through the architecture to a trunk card. The trunk card serves as the interface to the central office to request outside dial tone. The PBX, upon receiving dial tone at the trunk card interface, then regenerates the pulses or the tones across the line to the central office. The CO will process these digits in the same manner that it processes individual line requests from a residential user. This is the easiest way to process the information from the telephone company's perspective.

Digital PBX

All newer systems are basically digital. As a computer architecture, the system processes the information in its digital format. The use of a digital coder/decoder (codec) in the telephone set will convert the analog voice conversation into a digital format. The digital signals are then carried down the wires to the PBX heart (the CPU) for processing. If a call must go outside to the world, the PBX will have to determine the best route to process the call onto. In the case where the call will be traversing the telephone company's central office links on an analog circuit, the PBX must format the information for the outside link. In this case, a digital-to-analog conversion will take place. Even if the call is to traverse a digital link to the world, the PBX might have to go through a digital-to-digital conversion. This is because the digital signal at the PBX interface is a unipolar signal, whereas the signal to the telephone company is a bi-polar signal (Figure 9.3).

The list of vendors selling and supporting PBX systems is quite lengthy. These are offered by the manufacturers directly to the customer, or through a distributor. The options are many. The two largest suppliers of systems in the U.S. are AT&T and Northern Telecomm Inc. This ranking is based on number of systems sold, rather than a qualification of "best," although you might establish that the quantity sold is a reflection of some qualitative measure. Table 9.3 shows the top players in the U.S., based on sales volumes. It is interesting to note that the top two command better than 50% of the U.S. market.

The PBX market has been plagued recently with soft sales. This is a function of the recession, the rightsizing and downsizing of corporate America, and the overall unsettled market from a technological standpoint. End users are uncertain of what to buy and when on the market curve they should buy. Therefore, the vendors have had to resort to major markdowns and often throw in several other goodies. The buyer's market prevails in the PBX industry. As a result significant discounts can be achieved if you work with the vendor and understand the product being offered. Many of the

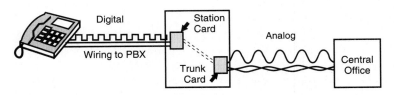

Figure 9.3 The digital signal from the PBX is converted to analog to the telco.

TABLE 9.3 A Summary of the Big Players in the U.S. PBX Market

Top players in the U.S. marketplace
AT&T
Northern Telecomm (NORTEL)
Rolm/Siemens
NEC
GTE
Intecom
Fujitsu
Hitachi
Mitel

TABLE 9.4 A Summary of Costs for a 1000 Line Digital PBX

Item	Price
Cost of hardware, software, training, all telephone sets and interfaces with installation of the hardware	$ 350,000.00
Cost of wiring and installation for the building infrastructure	$ 350,000.00
Markup and profit	$ 300,000.00
Total	$1,000,000.00

vendors will also compete severely with their distributors. Remember that this is a buyer's market. In Table 9.4 is a summary of how the costs would look for the acquisition of a digital PBX, the basic telephone system for the organization. This table reflects three important pieces in the billing arrangements. It would not be unethical to see how the vendors price out their systems against this model. In the table we will use an average price-per-port of $1000. The costs associated with a 1000-user system would, therefore be as appears in table above.

Another item of note is the third line item, that being profit. We always want our vendors to survive for another day, no two ways about that. However, we do not want to pay a 30% total markup on a system for profit. In actuality, the margin is 37%, and we will see why later. This is unheard of. So, the discounts that might be passed along from the vendor might well be from the profit picture. Suppose that the vendor offers a discount of 20% off the top of the price. The total price is $1,000,000 and the discount is 20%, so you can expect to pay $800,000. That should make you feel pretty good, to get a $200,000 discount off the top of your system. But, wait! What if the vendor came back and said that the total discount is only $70,000.00. Where did we go wrong? Well the issue is where the numbers are being calculated. The vendor discounted the 20% from the top of the system cost ($350,000 × 0.2 = $70,000). Now you are paying around $930,000.00 total for the system, installed. That is not exactly what you thought you were getting a discount on! The vendor will explain that the cost of the wiring cannot be discounted because they use a subcontractor and have to pay this third party for the installation. True, but the vendor also marks up the cost of the wiring and installation. That $350,000 fee to install and wire the system is probably only $280,000 to $300,000 from the subcontractor. So, the manufacturer or distributor is getting a piece of the pie for the installation too!

Yes, this is true. Regardless of how we slice and dice the numbers, this is still a very lucrative sale for the vendor. With a $50,000 to $70,000 markup on the wiring (which is conservative), a $300,000 profit margin and the remaining cost of the system ($280,000), you can imagine just how much the vendor is making on this system, can't you? Well now look at the margins based on this new evidence (Table 9.5).

Can you see anything wrong with this picture? Even though the vendor has given a 20% discount to you, and you feel so special for negotiating such a difficult deal for the vendor, and a great one for the organization, the overall margin of profit that the vendor has achieved is still 37%. This still leaves a lot of room for negotiation before the deal is done. If you consider that there is still room to cut the cost in the profit margin, the profits on the sub-contracted piece of the wiring and the overall system cost, then the dealing has only begun. In many cases, the ability to sub-contract the wiring (for example) might produce more productive and competitive results. This is the case where many organizations will act as the general contractor for the overall telephone system, then contract for the wiring separately from the telephones. An example of the wiring costs might look like the numbers shown in Table 9.6, where a separate contract is issued for the installation of a four-pair cable installed at 1000 user locations, the horizontal wiring between the telephone closets and the main distribution frame, and any ancillary cabling needed to implement the system.

Keep in mind that these figures are generic, and will require separate bids from various installation companies. If, however, you compare this figure now, and recognize that the wiring contractor has already built in the

TABLE 9.5   A Summary of New Profit Margins with a 20% Discount on the System Cost

Item	Original cost	New cost	Profit	% Margin
PBX system	$ 350,000.00	$280,000.00		
Wiring and installation	$ 350,000.00	$350,000.00	$ 70,000.00*	20%
Margin and profit	$ 300,000.00	$300,000.00	$300,000.00	30%
Total	$1,000,000.00	$930,000.00	$370,000.00	37%

 * Excludes the wiring contractor's profit margin for the installation and wiring

TABLE 9.6   A Summary of Wiring Costs

Cost per user location	Extended price
Cost of wiring a 1000 user system @$250–280	$250,000–280,000
Cost for PBX manufacturer @$350	$350,000
Difference	$70,000–100,000

TABLE 9.7 A Summary of New Cost Structure and Margins

Item	Cost	Margins
PBX	$280,000	
Markup	$300,000	
Sub-total	$580,000	115%
Wiring (as a separate contract and separate margins)	$280,000	15%
Total	$860,000	

TABLE 9.8 Final Configuration of System Pricing

Item	Original pricing	Revised pricing	Difference
PBX	$ 350,000	$280,000	($ 70,000)
PBX markup (@30% of contract)	$ 300,000	$ 84,000	($216,000)
Wiring	$ 350,000	$280,000	($ 70,000)
Totals	$1,000,000	$644,000	($356,000)
Percentage			35.6%

necessary profit margins to make money on the installation, then the PBX price now has a different perspective. The margin for the hardware, installation, and warrantee on the PBX is now subject to serious negotiation (Table 9.7).

As you can now imagine, the cost for the telephone system is $280,000 with a profit margin of $300,000 (over 100% markup). No vendor will ever approach this structure, these are comparative pricing scenarios. However, if you consider that a 30% markup is what the vendor is entitled to, the following summary gives us a whole new structure to deal from. The intent is not to jeopardize the stability and profitability of the supplier, but to maximize the comfort between the two parties. This case will obviously consume a lot of time and effort. But, the overall results are significant (Table 9.8).

Clearly the price has changed significantly! The system is now being considered at approximately $644 per user now instead of $1000. This accounts for a $356,000 discount overall. This is the way you can look at using the system pricing, rather than just accepting standard pricing. The pricing can vary quite a bit from the original proposal.

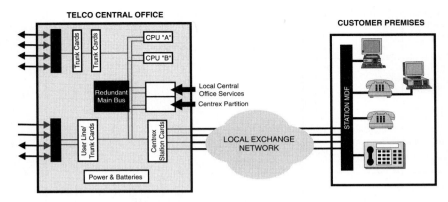

Figure 9.4   Central office Centrex components.

Central Office Centrex (CENTREX)

An alternative to the purchase of a telephone system is called *Centrex*. Centrex is a service offering from the local exchange carrier (LEC). It stands for central exchange, or PBX services provided from the central office (exchange). The LECs, either the Bell Telephone Companies or the independent telephone companies, all have a service offering. The Centrex service is a partition inside the CO which provides telephone service on a private basis to a business (Figure 9.4).

Centrex is usually rented on a line-by-line basis, month to month. Some companies have long-term agreements to hold down costs, but charging a monthly rate is more prevalent. Costs per line can range from $20 to $25 a month, depending on the company and the serving telco. The user must still buy the station equipment (telephone sets). However, with very large organizations, a long-term contract can be negotiated with the provider. This will result in costs that are in the range of $10 to $12 per month for a Centrex line. The result is a monthly savings of $10 to $15 per month per line. With a 1000-line system, this can be a significant savings. An example is the 1000-line system that we were using in the PBX comparison. Here, a system can cost approximately $12,000 per month on a long-term agreement. The $12,000 is for the dial tone and features, anything else is an extra.

The components of a Centrex are as follows:

- The central office, which is the serving office
- The CPU or computer
- Station cards
- Switching network

- Line cards
- Memory
- Common logic and power cards
- Bus to carry information to and from the CPU

Centrex Service

The primary suppliers, as already stated are the LECs. They provide the service as part of the central office function. The most commonly used systems in the North American COs are Northern Telecomm and AT&T, although NEC, Ericsson, and others might be used.

The list of Centrex providers includes:

- Bell Operating Companies (The seven regional Bell Operating Companies and their 23 operating telephone companies are the largest sellers of Centrex)
- Independent Operating Companies (such as CONTEL GTE, CENTEL, COMMONWEALTH, etc.)
- Resellers of CENTREX service (Many arrangements have been made with resellers around the country)
- Bell authorized agents (authorized agents of the LEC, but not employees of the company)
- Consultants (Special resellers who have vested interest in their customers' behalf, and a vested interest in the commissions they might receive from the LEC for the additions to the Centrex)

Peripheral Devices

The list of peripheral devices for the key systems, PBX, and Centrex markets is virtually unlimited. The devices range from items as simple as an external bell to a very sophisticated management system. The prices are too numerous to list here (and change too frequently!), but there is still a lot of negotiating room in any component you might need.

Some of the devices that might appear in the picture are:

- Automatic call distribution
- Voice mail
- Automated attendant
- Call detail recording
- Modem pools

- Multiplexers
- Head sets
- Display sets (telephones)
- Paging systems
- Least cost call routing
- Network management systems
- Design tools
- Answering machines

10

Key Telephone Systems

Another form of equipment is used in a smaller environment. Where the user office is small, or a branch office is involved, the equipment used is called a *key telephone system*. This is another form of resource-sharing device, used to reduce the number of outside telephone lines and provide access to many end users.

The equipment selection depends on the size of the organization and the needs for communications connectivity. Many small branch offices, small businesses, home office businesses and other forms of organizations require a limited amount of telephone sets and can get by with a key telephone system. This type of system usually comes in a packaged amount of telephone sets to outside lines. Note the reference to lines in this example, where the telephone company sees the system as an end-user connection, much the same as a single-line telephone set. Therefore the customer gets by with the use of lines instead of trunks. Although they functionally do the same things, lines are typically less expensive monthly than are trunks. The noun-nomenclature for a key system is normally outlined in the number of lines and the number of sets; such as 1648 equates to 16 outside telephone lines, and 48 telephone sets comprise the system. These are maximums on the systems, less amounts can be used where appropriate.

A key telephone system is comprised of the following pieces:

- The key service unit, or KSU is the heart of the telephone system.
- Line cards are the interfaces to the telephone company lines. They effectively provide the "off hook" and "on hook" signals to the central office in lieu of the individual telephone sets.

- Station cards are the cards that control the intelligence and interface to the end users a key system. The station card interfaces with the end user and the line cards for the access control of the system.

- Intercom cards are not always required. Some systems use an intercommunication card for internal connectivity, whereas others have this built in to their functionality on the backplane of the system.

- Telephone sets are variable-user interfaces, whether they are single line or digital multiline sets. The intelligence that a user is allowed to access within the system is controlled through the telephone set.

- Power supplies and logic cards.

- The wiring infrastructure that is used to connect the sets to the KSU central processor unit, or CPU, the brains of the system.

These pieces are shown in a graphical representation in Figure 10.1. Most key telephone systems today are computer-stored, programmed-controlled systems, similar to the PBX discussed in Chapter 9. As a matter of fact, it is becoming far more difficult to differentiate the PBX from the key system. Key systems can be as large as 200+ ports, whereas PBXs now come with as few as 50+ ports. Functionally, these two types of systems provide the same services and features. The differences are so subtle, it is becoming almost a moot point to call these systems by differing names. These systems are

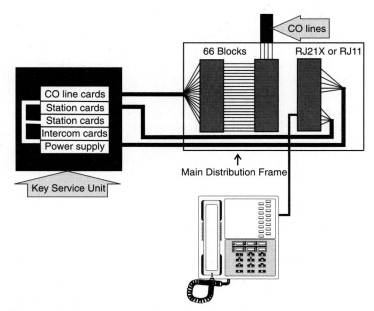

Figure 10.1 The key telephone system components.

used extensively throughout the industry, so they are fairly perfected in terms of connections being simple, are feature-rich, and have significant penetration in the business community.

The key systems are used by larger organizations too. In departments where a grouping of persons require connectivity to each other, or where a consolidation of features are required behind a Centrex or PBX, the key system offers some solutions. Although many organizations try to emulate the key system in a PBX, with multiline sets, it is often more prudent to install the key service unit. Many dual systems provide the access to the features of the PBX but the clustering of workers of a key system.

Why Key Systems?

It is not uncommon to hear business users question the need for a key system. The obvious choice is to go to the telephone company and rent a single line for each user in the organization, branch office, sales office or home-based business. However, the same arguments that were discussed in the opening of the PBX chapter hold equally true for the key system justification. One would be frivolous to rent a single line for every user in a sales or service office. The intent of the sales office is to have an office location for the sales force to schedule work and drop into from time to time for meetings and report writing. However, the sales force might be road based, with primary emphasis on visiting their customers and serving the customer's needs. From this analysis, there should not be a one-to-one relationship of employees to outside lines. This, of course assumes that the sales force are out calling on customers. If the charter is to telemarket, the entire scenario changes dramatically. When the office force does drop in at varying intervals, it is unlikely that all of them will be around at once. Therefore, a three-to-one ratio can usually be achieved, with three telephone sets per one outside line being used. This can save a significant amount of money in a small office, and more in the larger office environment. Table 10.1 shows a comparison of the use of outside lines on a three-to-one ratio. This does not include the costs for the key system, these will be shown in a later comparison. Suffice it to say, the comparison here is on the monthly recurring telephone company costs.

The obvious reason that an organization would choose a key system is the monthly recurring savings of $120.00 to $400.00, as shown in the table. However, the cost of the key system was not shown. The primary savings shown in this table are the cost of renting a telephone line for each user, as opposed to sharing the number of lines with the resulting savings. Now we can add the cost of the capital equipment, that being the key telephone system. A typical key system with installation, hardware, software, and warrantee will run an average of $500.00 per user. Therefore, a comparative analysis of the costs associated with the acquisition of the system is shown in Table 10.2.

TABLE 10.1   A Comparison of Line Rentals With and Without a
Key System

Number users	W/out key system	With key system	Difference
10 users @$20.00	$200.00	$ 80.00	$120.00 monthly
20 users @$20.00	$400.00	$140.00	$260.00 monthly
30 users @$20.00	$600.00	$200.00	$400.00 monthly

TABLE 10.2   Summary of Key
Telephone System Costs for Various
Number of Users

Number of users	Key equipment costs
10	$ 5,000.00
20	$10,000.00
30	$15,000.00

You can see from this table that the costs have been applied in a linear fashion. There might be better discounts involved as the system gets larger. Further, the system costs are associated with a basic telephone system, other features such as:

- Voice mail
- Station message detail recording
- Automated attendant
- Least cost routing
- Data and voice simultaneously
- Automatic call distribution

These features will all add on to the cost of purchasing a telephone system. The wiring is typically sold to a user based on the vendor's approach to installation. Many of these key systems for example, use a two-wire (one pair) connection to the desktop, others might use four-wires (two pair) and so on. In general, a good rule-of-thumb is to install a four-pair connection to every desk, whether you believe you need them or not. This will increase the cost of the wiring, but it is not that bad a deal. The cost of the wire is approximately $0.03–0.05 per foot for the added pairs. When all things are considered, the little add-on for the wires makes an insignificant increase to the overall system. However, over the life of the system, the extra pairs of wires will pay for themselves in no time. If a user needs a modem, or a fax machine for example, then the extra wires will be necessary. To have an installer come back and pull a whole new pair of wires after the fact, the cost will be from $250 to $350.00, mostly from the cost of the labor. This can also be disruptive in the workplace. Table 10.3 compares the equipment costs and the monthly rental costs for the telephone lines, to see just what the impact will be overall. Taking the comparative numbers from the original tables (10.1 and 10.2), you see that the cost structure is as follows:

Although this table might look like it is a break-even to use or not use a telephone system in lieu of the monthly rental of a line-per-user, there are some basic assumptions being made here. The equipment cost for the non-key system is for the purchase of basic analog single-line sets. No multi-line sets are involved in this scenario.

The monthly equipment costs are associated with a three-year depreciation cost on a straight line basis. In reality, the finance department would likely write the capital off over five years. Secondly, the finance department would also write the equipment off at an accelerated rate based on the IRS and financial policies accepted.

The one year differences will have to be spread over the number of years that the system is in place. After the payback period is achieved, the numbers change significantly.

Table 10.4 covers the same system costs over a five-year depreciation cycle, with the system life being a ten-year period. In this case, after the fifth year, the equipment costs go away, and only the monthly recurring costs remain. These are simple calculations—any increases in the line costs are not included and no factoring for maintenance is built in. The assumption here is to look at the overall differences of having a system or not. The same assumptions from Table 10.3 are used: the number of users, the cost of equipment at $60.00 per single line set (written off, in this case, for over five years), the line costs held constant over the five- and ten-year life cycle, etc.

Using this table results in a much different picture of the financial impact of using the key telephone system. Based on an example of 30 users,

TABLE 10.3 Summary of Costs for Key System vs. Non-Key System

Number users	Without key system		Total 1 year cost	With key system		Total 1 year cost
	Line cost	*Equip. cost*	*Deprec. @ 3 years*	*Line cost*	*Equip. cost*	*Deprec. @ 3 years*
10	$200	$ 600	$2,600	$ 80	$ 5,000	$2,660
20	$400	$1,200	$5,200	$140	$10,000	$4,980
30	$600	$1,800	$7,800	$200	$15,000	$7,400

TABLE 10.4 Five and Ten Year Comparative Costs

Users	Non-key costs	Line costs	Total		
		Monthly	5 years	10 years	
10	$ 600	$200	$12,600	$24,600	
20	$ 1,200	$400	$25,200	$49,200	
30	$ 1,800	$600	$37,800	$73,800	

Users	Key costs	Line costs	Total		Difference	
			5 years	10 years		
10	$ 5,000	$ 80	$ 9,800	$14,600	($2,800)	($10,000)
20	$10,000	$140	$18,400	$26,800	($6,800)	($22,400)
30	$15,000	$200	$27,000	$39,000	($10,800)	($34,800)

the difference in raw costs over the five-year period are $10,800; over ten years, the key equipment will save $34,800. These are significant savings over the life of the equipment when considering that the entire organization might only be 30 people. The savings will obviously pay for themselves over the five- and ten-year period. Remember also that the key system includes the features and functions of a telephone system, and will include multi-line sets. The non-key system comparison includes only single-line sets and no added features. Each of these will change the equation and show a better payout for the key telephone system. Having drawn this to a conclusion, you can now see why an organization might well consider the installation of a resource-sharing arrangement such as a key telephone system. However, there is more to it than just a financial calculation.

Just as covered in Chapter 9, there are other advantages to the use of the systems approach. One for example, is the use of the intercommunications channel, called the *intercom*. The intercom frees up the outside lines for incoming customer calls when two workers in the same office need to communicate. Using the intercom, party A calls party B on the internal communications paths in the system. Some systems have a fixed number of intercom paths, whereas others have an unlimited number. This is a function of the manufacturer equipment. When using the alternate system (the non-key telephones) the choices are far more restrictive. If user A wants to talk to user B, then the two outside telephone lines associated with these two individuals are busy. The customer calls coming into the building for either party will be blocked until such time as they terminate their conversation. The customer might get frustrated and take his or her business elsewhere if this is a constant occurrence.

Moreover, the ability to add features and services in the key telephone system are more readily accommodated than in the single-line telephone set environment. Users will ultimately like the features that can be achieved in the key system such as:

- Call transfer to another party

- Call hold by pressing a feature button

- Three-way calling, or conferencing calling

- Consultation hold, where a call can be placed on hold while the user confers with another user in the system

- Speed dialing on an individual or a system basis

- Speakerphones

- Hands-free dialing

Some of these features are displayed in Figure 10.2 where the buttons are allocated for the use of the specific features. No one telephone set or telephone system offers everything that the user wants; in the way they want it. However, strides have been taken over the past to accommodate what users have asked for.

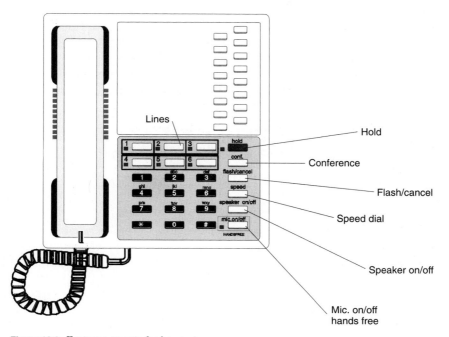

Figure 10.2 Features on sets for key system.

Vendor Interfaces

The telephone interfaces shown in Figure 10.3 are differing ways that a key system can be connected to the outside lines. In many cases, the telephone company will bring in individual two-wire analog circuits (lines) and terminate these in an RJ-11C jack. This will be the case for smaller systems of less than ten lines. If, however, the system is larger and more connections are required, the telephone company might well be asked to terminate the lines in an RJ-21X. This is a call that can be made by the vendor who will be requesting the service for their customers. Personally, there is no major difference between the connections other than some not-so-obvious benefits.

For example, using the RJ-21X consolidates the RJ-11s into a single block that will support 25 two-wire connections. This is a little cleaner than having 25 individual RJ-11s installed on a wall. The convenience for testing and troubleshooting at one spot on the wall has some benefit.

However, the individual RJ-11s can be used in lieu of the larger block. Although these might take up more wall space, if the telephone system ever fails, a user can go into the telephone equipment room and unplug one of the circuit connections to the key system, plug in a single-line telephone set

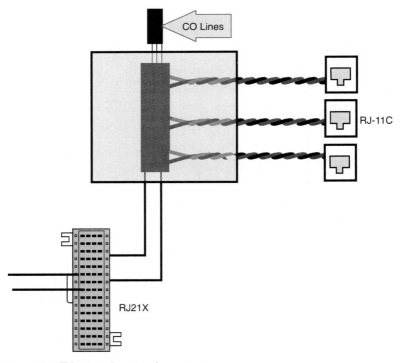

Figure 10.3 Telco interfaces to a key system vary.

(2500 set) and make calls to the repair functions for the telephone company or equipment vendors. Further, if critical calls must be made, this can be done for many users to allow them to enter a special access method into the lines.

These systems have been around for a long time, so there is no mystique in dealing with this connectivity. Vendors have pretty well gotten these systems down to a science. Users also have them under control. There are no major management issues in dealing with a key system, except when the system is failing. This is something that is usually rectified quickly by the vendor and telephone company when problems arise. The interface is a two-wire analog connection similar to the residential connection. The main problems a user will have might be power related. We always recommend that a back-up battery be provided with the installation so that a critical component of the system's performance can be preserved with the supplemental power. The back-up battery adds some minor costs, possibly a few hundred dollars at worst, but buys a lot of comfort and protection for the end user.

Key Players in the Key System Marketplace

The marketplace is quite a large one for the installation of these systems. In many cases, the manufacturers will sell these systems directly to the end user, or they might have a distribution agreement with several installation and maintenance companies in the area. The variableness in this market is extensive. We have had situations where we had the manufacturer and two or three distributors in an area competing for the business. The competition is fierce when this happens. In some cases, the distributors offered the system less expensively than the manufacturer's representative. This says a lot for the market, because it is almost as soft as the PBX market. In the list below, we look at some of the key players in the key telephone market. This ranking is on the basis of the number of systems sold, or the number of lines delivered. This in no way is designed to place a value on good, better or best. It is strictly a numerical value of total systems. However, there will be no surprises in this picture. The two top players are names that we are all familiar with and have the majority of the market.

- Northern Telecomm Inc.
- AT&T
- NEC
- Inter-Tel
- Telrad
- Siemens
- Tie Systems

- ComDial
- Hitachi
- Iwatsu

The top three players all combined control approximately 70% of the total U.S. marketplace. This again is not a reflection of the quality or ranking on better systems, merely the percentage points of systems sold. Many of the other systems are equally good, or fit a specific niche market. Their systems are equally feature-rich, price-competitive, and well-supported.

When dealing with the purchase of the key telephone system, you will be better equipped if all of the known benefits and features are discussed. The growth capacity of the system before a major upgrade is required is a beneficial piece of information. Never assume that the best deal is placed on the table on the first proposal. As a matter of fact, we prefer to do the following when considering a key system, a PBX or any other piece of equipment:

- Conduct a needs assessment: Determine what the needs of the organization are, to meet the core business functions of the office, plant or whatever.

- Determine what the buying motivations are: Is the system being considered to allow growth, replace obsolete or broken equipment, or just a feature need?

- Conduct user surveys to see if any hidden needs exist.

- Document the needs and the buying motivation.

- Send out a request for information (RFI) to gather information about the players in your area, such as the systems they offer, the average number of systems they have installed and the typical sizes.

- Pre-qualify the bidders that you will use based on the responses to the RFI.

- Determine the future needs as they pertain to the system growth.

- Develop a request for proposal based on the consolidation of all the information gathered.

- Obtain budgetary approval in advance for some value of the system intended for purchase. Allow some discretionary funding for added features or growth that you didn't anticipate. Never buy a system that only satisfies today's needs, make sure that plenty of room to grow is available.

- Release the RFP and place a target date on when the responses will be due.

- Analyze the information and begin to select a short list of possible suppliers that you will feel comfortable with. This includes checking references of other customers using a system similar to the one you are considering.

- Negotiate the contract, be firm. Make sure all costs are shown and documented.

- Allow plenty of time for the whole process.

These steps are generic enough that the process can be used straightforward for any system or service being considered. You can never do too much homework to evaluate a system or a supplier. However, you can get caught into a "paralysis by analysis" syndrome, where the process takes too long. This whole process for a key system can be done in a matter of weeks before getting to final negotiation. Nowhere is it written that such a study should take months to years. Technological advances and the product life cycles are such that you can only decide on the technology based on the information that is available today. Things will change at a very fast rate.

Chapter

11

Voice Processing

Introduction

Voice processing encompasses a variety of technologies that can be categorized into two major areas: directing calls, and message manipulation. As we will see, while each of the applications in this chapter focuses on one or the other of these two major areas, each application also includes elements of the other area. Moreover, as time goes by and the vendors of these products enhance their offerings, they become more and more alike. So bear in mind that these descriptions are of the essential functions present in each application; in fact, when you go out to buy, you might have a difficult time distinguishing one from another.

The technologies presented in this chapter include the following:

- Automated attendant
- Automated call distribution system
- Voice mail/messaging systems
- Interactive voice response

Other technologies that are still evolving fall under the voice processing category. These will include, for example, voice-to-text conversion and text-to-voice conversion.

Each of these technologies was designed to automate the processing of information, with one major limitation in mind: human beings. The major pitfall in the expansion of our telephony and telecommunications networks has been the inability of humans to keep pace with faster and more reliable communications. Organizations attempting to provide improved services with fewer people have implemented technologies to take up the slack. Many of these technologies have enhanced the price/performance ratios, but have subsequently caused user frustrations. No technology is good if the implementation is for the wrong reasons, or if the acceptance isn't there.

But before getting into the details, let us discuss two methods that allow human beings to interact with some of these technologies.

Control Alternatives: Touch Tone or Voice Recognition

Most voice processing technologies depend on Touch Tone (AT&T) input from callers, if any input is to be collected at all. As described in Chapter 4, dual-tone, multi-frequency (or DTMF, the technical term for Touch Tone signals) tones are specifically designed to be carried well by the analog telephone network. (Pulses, on the other hand, are generally not transmitted beyond a telephone's closest central office; it is impractical to use them for controlling voice processing.) DTMF telephone buttons can also be pressed by human beings far more quickly than can numbers be dialed on a rotary telephone.

An obvious alternative to any kind of fingered control is voice recognition. After all, people generally prefer speech rather than pressing a keyboard. And in fact, some systems now do implement limited voice recognition. But voice recognition technology is not quite ready for prime time in the area of general fielding of calls from the public. The problem is vocabulary.

Voice recognition systems generally can either recognize many words spoken by one (or a limited set of) user(s), or just a few words spoken by many users. Both approaches are being built into production voice processing systems. For example, voice transcription systems are now in use by some medical examiners, an ideal application: a very small user population, and a precisely definable vocabulary. Another example: some telephone companies are replacing some operator functions with voice recognition. Again, a relatively small vocabulary is required. Have you heard this on the telephone lately:

- "If you are making a long-distance call, please say 'operator' for operator assistance, 'credit card' for calling card service, or 'collect' for collect calls, now."
- "The number you called is not answering, please call back later."
- "The number you called is busy, to leave a message . . . "

These are forms of interactive limited vocabulary. But they work well. If you mumble and do not say the word correctly, you will hear the response that "We did not understand your request, please say it again clearly."

Increasingly sophisticated algorithms and more powerful computers are driving up the meaning of "many" and expanding the definition of "just a few" in the first sentence of the previous paragraph. As time goes by, voice recognition will be seen on more and more systems. But to minimize the vocabulary that must be recognized, the inputs will initially be limited to little more than what callers could have entered on their keypads; that is, numbers, letters, and a small number of special characters. But today, most systems accept control from Touch Tones only.

Automated Attendant

An attendant, in telephone lingo, is an operator for a private telephone system. At your organization, if you have Centrex or a PBX and dial "0," you will reach your attendant if you have one. But if you have an automated attendant, there might no longer be a human in that role. Or if there is an operator there, the operator might be the last resort.

The use of automated attendants to supplement normal operators was intended to assist customers in cutting right through a telephone system, if they knew who they wanted to speak to. Further, if customers didn't know who they wished to speak to, but knew the department, they could expedite their connections without human intervention. How often have you called an organization, to be greeted by a human with "Good afternoon, ABC Company. Please hold." The operator had more calls to handle than time to do so. Thus, you were put on interminable hold.

Automated attendants were designed to take the rote or redundant answering function away from the operator, so that quality time could be spent with customers who legitimately needed assistance. Some examples of this might include items contained in Table 11.1.

TABLE 11.1 Applications for Automated Attendant

Organization	Application
TV station	Use the Automated Attendant for newsworthy items such as weather storm conditions, parade information, repetitive information regarding TV news and community interest
Radio station	Contest information, record requests
Mortgage companies	Current rates on mortgages, information on applications
Colleges/universities	Student activities, registration, class schedules and fill rates
Movie theaters	Schedules, costs, specials

However, too many organizations implemented the system incorrectly. Instead of giving their customers a choice (if you want the automated attendant dial 555-1234, if you want to speak to an operator dial 555-1000), these users installed the system without warning. Customers who called on Friday afternoon and spoke to a human (as they were accustomed to doing over the years) called in on Monday morning to be greeted by a machine. This shock to the customers' egos and psyches created frustration and a reluctance to deal with this company. Thus, a system designed to improve upon customer service and relations was instead destroying both. Good installations, coupled with an understanding and concern for the caller would have prevented the problems.

Automated attendant systems fall into the "directing calls" category identified above. Automated attendants usually perform some or all of the following functions:

- Answer the telephone (usually only calls incoming from the outside).

- Announce options to the caller. These might include numbers of departments that can be selected, hours of business, and other information useful to callers.

- Accept Touch Tone input from the caller to direct the call, either to other information provided by the automated attendant, or to a human attendant.

- Forward the call to a human being if no Touch Tone input is received.

Organizations typically use automated attendants to reduce payroll costs for live attendants. Some have eliminated live attendants entirely; others reduce staffing for the function but allow callers to "escape" the Touch Tone system and reach a human attendant. A frequent application is as a "night answer" attendant. In this mode, live attendants answer the telephone during business hours; after hours, callers can select an internal extension if they know one.

Another possible use is to provide the daytime functional equivalent of direct inward dialing (DID) without implementing the latter with the telephone company. With this approach, callers must dial an additional four digits—but no additional trunks, DID or otherwise, need be acquired. The use of DID might not be practical if the key system or PBX is not equipped with these cards, or if an older system is not upgradeable. This is a supplement to using the DID trunks that are more expensive. Depending on the calling population, this can be a very effective supplement to a live operator. Callers who know the extension they want simply enter it; others either select the live attendant from the prompts, or time out and speak to the attendant.

Note that an automated attendant is not itself a PBX; it fits logically and architecturally into the same role as a live human attendant. Figure 11.1 illustrates an automated attendant configuration. Although from

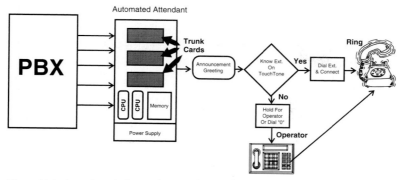

Figure 11.1 An automated attendant configuration.

callers' points of view the automated attendant is in front of the PBX, in practice it might be integrated into or reside behind it as shown in the figure. When the automated attendant is active, the PBX initially routes all calls that would normally go to the human attendant (usually all calls, unless DID is implemented or the automated attendant is present to handle overflow only) instead. The automated attendant attempts to determine the caller's real destination. It then uses the normal PBX call transfer capabilities to route the call to the correct extension, or the operator if appropriate. The conversation with the automated attendant might go something like this:

AA: *Thank you for calling XYZ Company! If you know the extension of your sales representative, please enter it on your Touch Tone telephone at any time. For a list of department extensions, please press the pound sign. If you would rather speak to an operator, please press zero, or just stay on the line and one will be with you momentarily.*

Caller: (Presses #)

AA: *For sales, press 31. For accounting, press 32. For installation scheduling, press 33. For customer support, press 34. To hear this list again, press the pound sign. To go back to the first greeting message, press 39. If you would rather speak to an operator, please press zero, or just stay on the line and one will be with you momentarily.*

Caller: (Presses 32)

AA: *Your call is being transferred.*

Note that in this example, no internal extensions would begin with the digit 3.

The sample dialog illustrates some important design considerations:

- Unless it is impossible or you are certain that it is not necessary, always give callers an out to a human operator. If you do not, you risk angering two groups of callers: those whose telephones do not generate Touch

Tones (and there are many of these still), and those who refuse to interact with any kind of telephone answering machines.

- Be friendly—it costs you nothing and will please some callers while offending few or none.

- Keep it short. The longer any one message (or the entire set through which callers must proceed), the higher the percentage of callers that will hang up (also known as balking or reneging). We have been through one system where the menuing function took us through twelve layers. (We only called that number once!)

- Keep it simple. Some calling populations are very technical (many of those calling technical support for a local area network vendor, for example). But more are not. A lowest common denominator approach, simple but not insulting, is usually justified.

- Give people information on how to repeat messages, and never leave them in a position where unless they remember the prompts, they are stuck.

- Do not use telecommunications jargon. You understand it; your callers do not. For example, do not refer to the attendant as such; he or she is "the operator."

- When possible, use obvious codes. For example, try to use 0 for the operator. Also, try only to assign a single use to a single code. In the example, the pound sign (#) always provides the department list, no matter where in the dialog it is pressed.

When providing a list of alternatives, provide the more specific choices first, then the more general ones. Not unreasonably, people will often select the first choice that sounds as though it might fit. "Customer support" can cover a lot of ground. If the department list had started with customer support, many callers who might better have been routed to one of the other departments would select it instead.

If your system allows it, allow callers to enter choices at any time; do not make them wait until a message finishes. People who listen to the end will not notice either way; but those who move more quickly and already know the choices from frequent use will appreciate the flexibility of the dialog.

A Few Important Points Not Directly Illustrated in the Dialog

When a machine answers a call, callers often immediately drop into a "find-fault" mode; many will be extremely easy to irritate. (Some will already be irritated, a price one pays to use this technology. As evidence that this technology can be a two-edged sword, there is at least one mail-order hardware vendor that includes in its advertising the fact that a human always answers their phone, "No annoying machines!") So consider designing your dialogs to be as inoffensive as possible, even to the point of blandness.

If your system is to be operational both during the day and after business hours, use different dialogs for different conditions. One of the more frustrating things you can do to callers is to inform them that they can escape to an operator, and then fail to do so because the operator(s) went home.

It is truly amazing how lousy some recorded greetings sound, some heard when calling billion-dollar companies, and what a bad impression such recordings make on callers. Use a professional studio containing a professional announcer and professional recording equipment to produce your messages, or your announcements also can fall into the lousy category. Remember that this is the doorway to your organization. First impressions are long lasting.

Dialogs must be designed and programmed (written out) much like computer programs. A poor flow, either of words or choices, will also reflect badly on the called organization.

Although it is important to list choices in an order that will cause the correct choice to be made, saving the most frequent choice to last is one of the surest ways to irritate people. If a large proportion of callers will correctly want a particular choice (e.g., customer service), consider providing a different telephone number for those people to call in the first place.

Try to incorporate a bypass through the prompts if the system allows it. As callers get used to the system, they will not want to listen to every offering. Rather, they'll try to dial the sequence immediately, expecting to get the party they always call.

If you make a material change in the choices, and your system has been in use for some time, announce early in the first menu that the choices have changed. Otherwise, your experienced users will blithely key their way down incorrect paths.

By itself, an automated attendant normally does not take messages or, without direction from callers, route calls other than to a default live answering position. But as mentioned above, the lines among voice processing systems are somewhat blurred. Do not be surprised if you find a voice mail system or ACD that encompasses all of the above functions.

Automatic Call Distributor

Automatic call distributors (ACDs) are the grandparents of voice processing systems. Although automated attendants are a fairly recent development (they have been around for a decade, more or less), ACDs have been around for several decades. When you call an airline, if you do not immediately reach a human being you will reach an ACD. If you respond to a television advertisement that states "operators are standing by," those operators will be reached via an ACD.

ACDs are used extensively for high volumes of incoming calls. This is a load balancing system to let the agents on the receiving end of the calls spread the workload over an even distribution of the agents. Other systems used in the early days only used a hunt system. The hunt would always start at the first agent and hunt down the line until it found an available agent. Unfortunately, if you were in the lead group of numbers, you would be swamped all day long, whereas an agent at the end of the group might only get an occasional call. This was an uneven distribution and caused a lot of stress and frustration in the workforce. The ACD rectifies this by leveling the load by number of calls, amount of time on the line, or other parameters set by the administrator. Typical users of ACDs include telemarketing organizations, the airline industry, travel-related industries (auto rental, travel agencies, hotels), etc. The purpose of the ACD is also to organize the smooth and orderly flow of incoming calls into the organization, for example: by customer, by region, or by product line. Primary responsibilities are:

- Prioritize the calls
- Route the call to the appropriate agent
- Deliver any announcements, queue
- Direct busy calls to announcements
- Handle messages for call backs

Because the machine is doing so much in terms of routing the calls to appropriate agents, or overflow to other groups of agents, companies can have far more incoming lines than they have people to answer them. This might or might not have some impact on the customer service function, because frustrations can arise. Think of how you feel when you get to an airline and are told that "All agents are busy, your call will be handled in the order of receipt, please hold on."

The ACD system can be configured in a stand-alone mode (see Figure 11.2), where all calls bypass the organization's PBX and go directly into the ACD for processing. This is a system similar to a PBX in that it has its own trunk cards, station (or agent) cards, CPU, printer, memory and power supply. The system can be simplex or redundant, depending on the finances of the organization. These systems can be as simple or as sophisticated as the company is willing to buy.

The ACD can also be a part of the company's PBX system, either integrated or as an adjunct to the PBX manufacturer's product line. The integrated system uses the same architecture as the PBX and the same CPU, memory, power supplies, etc. The agent's terminal is merely an extension off the PBX. There's some merit to having a single product for additional coverage, space considerations, etc. Again, this depends on the customer's willingness to integrate with a sole source. However, the PBX manufacturers are in the business of making PBXs and the ACD is a spin-off of this product line. Therefore, the system might not be as powerful or as fully featured as a stand-alone system designed strictly for the ACD function.

Another variation of the ACD function in PBXs is *uniform call distribution* (UCD), a basic system with less functionality than the normal ACD. Its function is to uniformly distribute calls to a group of agents, but without the high-end services of a full-fledged ACD.

ACDs can also be used with Centrex service (Figure 11.3). The LECs will offer a UCD function, as the ACD capability of a central office is somewhat limited, or as shown in the figure, the customer can run Centrex lines, 800

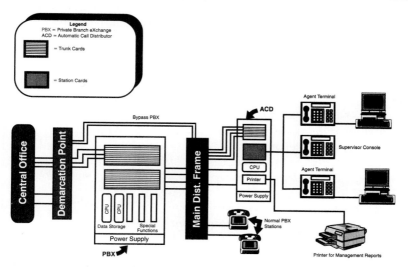

Figure 11.2   The ACD system configured in a stand-alone mode.

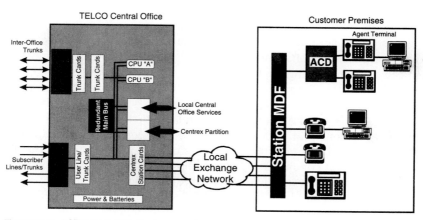

Figure 11.3   ACDs used with Centrex service.

lines, 900 lines, etc., directly through the Centrex into a customer premises system. The variations do not limit your ability to serve your customers.

Keep in mind the customer service side of the equation. Too few agents can create ill will toward your organization, too many will waste money. Thus, a traffic engineering calculation is required to come up with the right mix of agents and lines for the maximization of service (see Chapter 8).

Like automated attendants, ACDs fall into the "directing calls" category. But although automated attendants typically are used to replace one to six live attendants and interact with callers, ACDs typically provide access to dozens or hundreds of live "agents" and do not give callers any choices (although the ACD might make choices itself). An ACD routes calls based primarily on characteristics of the call other than input from the caller. These characteristics might include on what line the call arrives, the time of day at the answering location, what agents (not "attendants" in this case) are busy, and so on. A typical call to an ACD can go something like this:

ACD: *Thank you for calling SuperSoft customer support! All of our support people are serving other callers at the moment, but please don't hang up! Calls are answered in the order they arrive, and the average wait time at the moment is five and one half minutes.*

(music on hold for 30 seconds)

ACD: *SuperSoft also operates a support bulletin board at area code 212-555-7654, with modem speeds up to 14,400 bps at settings No parity, 8 data bits, 1 stop bit.*

(music on hold for 30 seconds)

ACD: *The probable wait time is now about four minutes.*

(more music on hold)

ACD: *Your call is being transferred.*

You get the idea. The above monologue illustrates some key characteristics and capabilities of some ACDs.

Unless integrated with other systems such as voice response units, ACDs are not prepared to accept input from callers.

Because the main purpose of an ACD is to smooth calling loads to make the most efficient use of agents' time, a key purpose of an ACD script is to keep callers from hanging up. One way to do this is to provide useful information to the callers; for example, information on how soon they might speak to an agent (a recently developed capability of some of the more advanced ACDs on the market), or other information related to the company's services.

The preparation of an ACD's response sequences is called *scripting.* Scripting is every bit as critical to a successful implementation as is the dialog in a television commercial. Usually, the process begins with the construction of a flow diagram (Figure 11.4). An automated attendant passes on a call and "forgets" about it. An ACD might also function as a PBX (or with one), and might track, retain, and report on numerous details about calls, including:

- The originating numbers
- How long each caller had to wait before an agent was assigned ("queue time")
- How many callers give up ("balk" or abandon) before being assigned to an agent
- The agent to whom each call was passed
- How long each agent took to answer each call
- How long the calls took to handle

As can be seen from the above list, an ACD is a management tool as well as a sophisticated call director.

ACDs are now being used with the automatic number identification (ANI) feature that allows the caller's number to be displayed to the agent when the call arrives (Figure 11.5). This service is typically offered with 800 Megacom (AT&T) services and other names for the bundled 800 services from Sprint, MCI, and others. The jury is still out on the caller identification issue around the country. Some states have allowed it, whereas others have ruled it as an invasion of privacy and therefore, disallowed it. On business

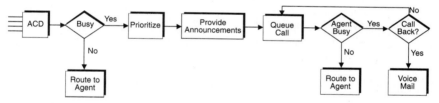

Figure 11.4 A scripted flow chart for an ACD.

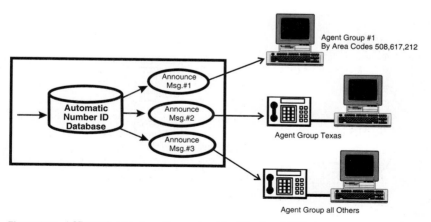

Figure 11.5 ACDs with an automatic number identification (ANI) database.

calls there doesn't seem to be as much dispute regarding this service. When future network technologies are fully deployed, this service might be available on every incoming call.

Using the ANI feature the caller's number can be routed to specific messages based on who is calling, from where they are calling, etc. Priority can be given to special clients using this service.

Another slant on the use of ANI with an ACD is to create a computer to PBX/ACD interface, an example of Computer Integrated Telephony (CIT). See Figure 11.6 for a graphic on this integration. With the customer's number delivered (inband) with the call, the system can route the caller id information to a computer via an X.25 packet switching link. The computer reads the incoming telephone number, looks up a database of numbers, searches and retrieves a customer database file, and delivers the call and a screen of customer information to a computer terminal simultaneously. This allows the agent to get right into the customer's inquiry, saving from 20–30 seconds per call by eliminating the exchange of who's calling, agent looking up customer info, etc. Many organizations have felt that the efficiencies gained by the use of the Computer to PBX Interface (CPI) have paid for themselves in customer service, improved agent performance, and better morale.

ACDs are not appropriate for all businesses. Typical requirements that can justify acquisition of an ACD include:

- Large volumes of incoming calls
- Callers will speak to anyone in a particular group (e.g., order takers, customer support personnel, travel agents)
- Need to fairly assign incoming calls among a group of agents

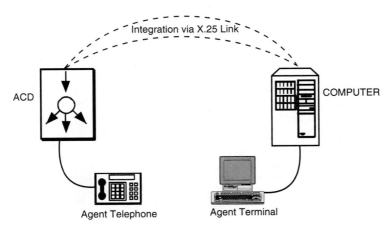

Figure 11.6 A computer to PBX/ACD interface.

Although large communications environments will often acquire a dedicated piece of hardware built and programmed for use as an ACD, several PBX manufacturers have come out with add-on software packages that allow their PBXs to function as low- or mid-range ACDs while performing their normal PBX functions.

Voice Mail

With fewer people in the office, organizations began to seek a new technology to enhance their operations. A company in Dallas, Texas, offered a new system called voice mail. The purpose of voice mail was originally designed to reduce the busy tones in the office. Users were absent from their offices more frequently. The old method of a secretary taking the name and number of the caller was not meeting the needs of the organization. Frequently, the called party diligently returned the call, only to find the calling party was away from their desk. This game is called telephone tag. However, some of these messages were information only; they did not require an interactive conversation. Thus, voice mail could provide a means of leaving one-way information, without the telephone tag. Further, the messages could be delivered at any time, at the convenience of the recipient. Figure 11.7 illustrates some of the ways voice mail can enhance productivity:

- It can be used as a remote dictation machine, although this is not its primary mission.

- It can be used to deliver a message to a recipient that is unavailable (or on the telephone) when the caller makes his or her initial call attempt.

- It can be used to remind one's self of important tasks; the reminders can be triggered at a specific time, as can other messages.

- It can be used to deliver identical messages to a target group with a single action.

- It can be used to dial out and deliver a message remotely to someone not even on the system!

Users have been somewhat reluctant to accept voice mail. From an internal perspective, the system can be effective. However, from an external standpoint, it can be a point of irritation. Users sometimes begin to hide behind the system, screening their calls before answering. Coupled with this problem is the response mode. If a caller has taken the time to leave a detailed message, and expects a return, the called party owes them a return call. Human nature states that the called party will not return the call if they are busy, or if the call is of an unpleasant nature. Thus the frustrations begin.

Voice mail does have benefits. It can gain additional hours of availability for your users, especially when dealing with callers from other time zones.

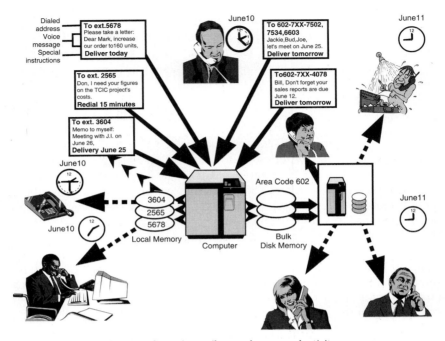

Figure 11.7 Some of the ways that voice mail can enhance productivity.

Users can be satisfied that, although they are not in the office, they can still receive information. Thus, the technology does have its positive side. As long as the users are properly trained on the use of the system, and the administration provides a follow-up on the appropriate use of the system, it can work and save your organization both time and money.

Voice mail, like automated attendants, is based on separate equipment. But voice mail falls into the message manipulation category mentioned above. Voice mail systems were originally designed to do for voice what electronic mail or messaging does for typed communications. That is, using Touch Tone pads users can:

- Dictate messages to the system
- Specify to whom those messages will be delivered
- Listen to, forward or delete messages directed to them
- Build and use distribution lists for messages
- Modify various characteristics of their "mailboxes"
- Provide a verbal buckslip (routing slip) attached to another voice mail, that is forwarded to another on the system for action and comment

To be most productive, voice mail systems require a fairly high degree of integration with their "host" PBXs or Centrex. Perhaps for this reason, most of the high-end voice mail systems on the market are sold by PBX manufacturers (e.g., Rolm's Phonemail, AT&T's Audix). In particular, it is important that the voice mail system be able to indicate to users when they have new messages. For users with indicator lights on their telephones, the voice mail system must be able to tell the PBX to "light the light." For other users, the PBX should provide a "stutter" (intermittent) dial tone to indicate new messages when the handset is raised.

Figure 11.8 illustrates a typical voice mail installation. Like OPX's, voice mail systems are normally connected on the line rather than the trunk side of a PBX (although they will also usually have some of their own external trunks to allow users direct access from off-site).

As with other telecommunications configurations, the number of trunks serving a voice mail system is a critical success factor. More than with most sytems, the usage pattern on voice mail systems is not a smooth one. Instead, users tend to check their voice "mailboxes" at similar times of the day: on arrival at work, when returning from lunch, and shortly before departure at the end of the day. The first of these, because it is often the same for the entire user population, often generates a peak demand for circuits far above the expectations of the system designer. Consider the reaction of your user population if 30% of them try to reach the system at the same time . . . and only 10 or 15% succeed.

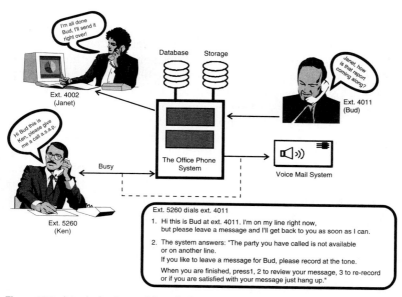

Figure 11.8 A typical voice mail installation.

Low end PC-based voice mail systems (typically limited to one or two physical telephone lines) are now available. An additional circuit card, along with appropriate software is required to do the voice processing; however, such systems can be implemented for little more than $100 (on top of the cost of the PC, of course).

Large systems are configured similarly to a PBX or key system. To avoid the access problem described above, a traffic study is conducted on the inbound and outbound traffic. Then, using the traffic engineering tables from the Poisson distribution, the number of lines and ports can be determined. Typical configurations are in four trunk/line configuration increments. Systems come in 4, 8, 16, 32, 64, etc., line configurations.

An additional key factor is the message retention capacity. High end systems are normally configured by the number of users they will support. This drives assumptions as to the number of hours of messages that must be stored. Because these systems digitize voice and compress it on hard disks, large capacities are used for the storage of voice messages. These drives are usually described in hours of storage. Examples would be 8, 16, 32, 64, etc., hours of storage, although these can also be in hundreds of hours.

Because all of the voice messages are digitized (although usually at less than the 64,000 bits per second of "toll quality" voice—the messages are compressed, sometimes down to as little as 9600 bps), a key consideration is the amount of disk space. So whoever is configuring the system must estimate an average message length, then decide how many messages the average user will retain (or will be allowed to retain) in his or her "mailbox." Knowing the bits per second at which messages are stored on a particular system will then allow a calculation of required disk resources. For example the figures in Table 11.2 represent a sample of the calculations used.

Doesn't sound like much, does it? But consider that at the standard 64,000 bps for "toll quality," the total would instead be 720 megabytes. Compression helps! This calculation was much more important a few years ago, before the cost of disk storage dropped to its current levels. (A one gigabyte disk can easily be acquired for less than $1000.) But understanding it can both assist you in selecting a configuration for a voice mail system, as well as illustrate the impact of some of the choices you can make when configuring the system. Note also that when a vendor touts the number of hours its system stores, and justifies the cost based on that storage, that if you can discover the digitization rate you can calculate how much disk storage you are actually getting for your money. Those could be very expensive disks.

TABLE 11.2 The Typical Calculation for the Hours of Storage Required

200 users retaining 5 messages each, averaging 90 seconds, digitized at 9600 bits per second requires $200 \times 5 \times 90 \times 9600 = 864,000,000$ bits, divided by eight, divided by one million, gives 108 megabytes.

Vendors will, of course, be happy to recommend a configuration to suit your requirements. But understanding that one of the key considerations is disk storage capacity should arm you in your discussions.

There are a few pitfalls to watch for in the implementation of voice mail systems. Some organizations have implemented voice mail systems as their primary answering positions. Like automated attendant systems, voice mail systems have the potential to irritate (or infuriate) callers who either do not ever want to talk with a machine, or who just take exception to the structure of menus on your system. Organizations have lost customers this way, so be careful! What is even worse than the structured messaging service is to wind up in the voice mail loop. You have probably already been exposed to this one:

Caller: Dials called party number and gets "Hi this is Janet, I'm not available to take your message right now, but if you leave a name and number I'll get back to you as soon as I can. If this is an emergency, please dial "0" and ask the operator to connect you to Bud who is covering for me."

Caller dials "0" and asks for Bud.

Caller gets "Hi this is Bud, I'm not available to take your call right now, but if you'll leave a name and number, I'll get back to you as soon as I can. If this is an emergency, please dial "0" and ask the operator to connect you with Don who is covering for me."

Caller dials "0" and asks for Don.

Caller gets "Hi this is Don. I'm not available to take your call. If this is an emergency dial "0" and ask the operator to connect you with Janet who is covering for me."

Caller hangs up!

Some organizations initially target these systems at selected individuals. Such implementations almost always are abysmal failures, because unless everyone has the facility, almost no one will use it. Few have the patience to ask themselves, "does this particular employee have voice mail?" before every message. The problem is similar to what might happen if the office mail room only delivered to selected individuals; everyone would deliver their own mail to ensure it got there. So if you want to put in voice mail, give it to everyone.

If you do give it to everyone, watch out! With a good kickoff, good training, and a good implementation, voice mail might take off to the point where your original traffic estimates err on the low side. So be sure to get a system that is not near its maximum configuration, as you will likely need to add disk capacity or trunks to keep users from receiving busy signals.

Interactive Voice Response

Interactive voice response (IVR) refers to technology supporting the interaction of users with the system (similar to voice mail or automated attendant), but also implies that information will be provided to the caller by the system itself.

The line between IVR and the previously described voice processing technologies is a fuzzy one; after all, an automated attendant often can provide telephone numbers to a caller. But typically, an IVR system is used to provide information to callers unrelated to the IVR system or technology itself. The data is normally either accessed directly in another system, or loaded into the IVR in advance from that other system.

For example, many mortgage companies have implemented IVR to allow clients access to their own mortgage records. A "typical" dialog follows:

IVR: *Thank you for calling Millie's Mortgage Machine! For current mortgage rates, press 1. To access information about your own account with us, press 2. To repeat these choices, press 9. To speak to an account representative, press 3 or just stay on the line.*

Caller: (Presses 2)

IVR: *Please enter your account number, followed by a pound sign.*

Caller: (Enters account number and a #)

IVR: *Thank you. Please enter your personal identification number, followed by a pound sign.*

Caller: (Enters PIN and #)

IVR: *Thank you. Your account is current through October 31st. As of November 1, your remaining balance is $83,432.21. Your monthly payment is $997.92; at 9%, you have 132 payments remaining. To have a balance payout statement mailed to you, press 1. To hear your account information again, press 2. To return to the previous menu, press pound sign. To hang up, press 8. To repeat these choices, press 9.*

Many of the points of voice processing systems also apply to IVR systems. The above monologue illustrates some additional key characteristics and capabilities of IVR systems.

Many systems require entry of a PIN. Any system capable of providing confidential information to callers should have at least this level of security. Normally, PINs are a minimum of four digits. Some service providers allow users to set their own, other organizations assign PINs to authorized users. Either way, the systems' organization must administer the PINs, an administrative requirement sometimes initially overlooked when planning a new IVR system.

Whereas an automated attendant or an ACD is primarily dedicated to providing communications between a caller and another human being, an IVR system's main reason for existence is often to directly provide the caller automated information. In fact, many organizations install IVR units to reduce the number of agents required "on the phones." Even if the initial setup cost of the system is tens of thousands of dollars, the savings in salaries can easily result in a payback within one year.

IVR systems always incorporate computers. Therefore, in addition to simple retrieval of data, they can also perform operations or calculations upon that data. In the above example, most of the numbers were probably calculated during the call rather than retrieved from storage. In another popular application, computers at banks can trigger electronic funds transfers (in effect, pay bills with paperless automated checks) under control of a caller's telephone call. The computer, probably interacting with another computer, adjusts the caller's balance as part of each such transaction. You can bet that both account numbers and PINs are required for such calls!

The integration of computers and telephones is sometimes referred to as Computer Integrated Telephony, or CIT. Figure 11.9 illustrates a possible architecture for such an application. In the figure, the IVR units access the computer database as needed to satisfy callers' inquiries. The IVR is separate from the computer, a configuration that might appear less and less frequently as computer manufacturers begin to offer voice interfaces and processing capabilities on their own machines.

IVRs can also be used in combination with ACDs. See Figure 11.10 for a representation of this arrangement. As an example, an IVR can be used in a customer support application to optionally provide callers access to a recorded database of common questions and answers. If the caller elects to inquire against the database (using his or her Touch Tone telephone to specify the questions to be played), an agent might not be required at all. If the caller selects "Speak to a customer support person" from the automated IVR menu, then the call can be directed to an ACD for queuing and ultimate connection to an agent.

Some ACDs now have as an option built-in IVR capabilities. We expect an expansion of such combinations as ACD providers seek out new ways to be competitive.

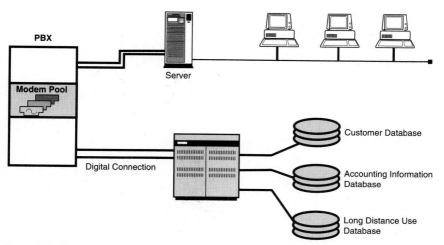

Figure 11.9 Computer integrated telephony is becoming more prevalent. Newer applications include LAN to telephone integration.

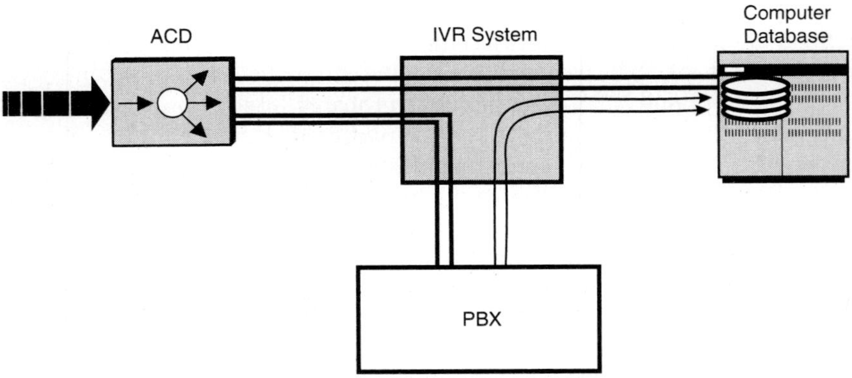

Figure 11.10 Access to computer systems through ACD and IVR integration is now a possibility.

Data Communications

Everyone who has ever had to deal with data communications has shuddered at least once. Telecommunications engineers, data processing personnel, and vendors alike all throw data communication terms around as though they were going out of style. The interesting point is that many of them really don't understand what they are talking about. Many think "If I learn the buzzwords, everyone will think I know what I'm talking about." Nothing could be further from the truth; these folks make complete fools of themselves in front of knowledgeable professionals. However, there isn't really a mystique associated with the use of data communications. Although some complexities do exist in this technology, the basics are fairly straightforward. If you can overcome the initial parameters of setting up a data transmission, the rest can be fairly well assimilated.

It is important to understand that the data world grew out of the voice world. The use of the analog dial-up network is where it all started. In order to communicate from a terminal or computer, or whatever, you merely have to put the pieces together in the proper order.

That order is:

- Select and deal with the transmission media
- Use communicating devices that will present the proper signal to the line
- Set up or abide by already accepted rules (protocols)
- Use a pre-established alphabet that the devices understand
- Ensure the integrity of information before, during, and after transmission
- Deliver the information to the receiving device

This chapter de-mystifies the pieces involved in all data communications processes. The later chapters focus on specific technologies, using the concepts and terminology introduced here.

Concepts

Like learning computer programming, learning data communications technology is a nonlinear process. That is, whatever starting point one chooses, one almost has to use terms that will be defined elsewhere. The usual solution to this problem is iterative teaching: teach a basic set, then go back and both use and expand upon those concepts, refining them as one proceeds. Our basic set begins with a discussion of some important concepts that permeate the world of data communications. Those concepts include:

- Standards
- Architectures
- Protocols
- Error detection
- "Plexes"
- Multiplexing
- Compression
- Standards

A *standard* is a definition or description of a technology. The purpose of developing standards is to help vendors build components that will function together or that will facilitate use by people by providing consistency with other products. This section discusses what standards are, why they exist, and some of the ways they could affect you. Specific standards will be mentioned in other sections where applicable.

There are two kinds of standards: de facto, and de jure.

De facto means *in fact*. If more than one vendor "builds to" or complies with a particular technology, one can reasonably refer to that technology as a standard. An excellent example of such a standard in data communications is IBM's Systems Networks Architecture (SNA). No independent standards organization has ever "blessed" SNA as an "official," or de jure, standard. But dozens, if not hundreds, of other vendors have built products that successfully interact with SNA devices and networks.

Note that de facto standards rarely become standards overnight. SNA was available for some time before vendors other than IBM could or would provide products that supported it.

Moreover, some technologies become standards because the creating vendors intend them to become standards (e.g., Ethernet), while others become standards in spite of the creating vendors (e.g., Lotus 123 menu structure).

De jure means *in law*, although standards do not generally have the force of law. In some parts of the world, when a standard is set, it in fact becomes law. If a user or vendor violates the rules, the penalties can be quite severe. A user who installs a nonstandard piece of equipment on the links could be subject to steep fines and up to one year in prison. These countries take their standards seriously. In the U.S., no such penalties exist, we are more relaxed in this area. But a standard is a de jure standard if an independent standards body (i.e., one not solely vendor-sponsored) successfully carries it through a more or less public standards-making procedure and announces that it is now a standard.

You might well ask, "What's the difference? And who cares, anyway?" But understanding which technologies fall into which of the categories, if either, can be a decision factor in what to buy. Generally, de facto standards are controlled by the vendors that introduced them. For example, Microsoft Windows is a de facto standard; many vendors provide programs that comply with and operate in this environment. But if Microsoft decides to change the way a new version of Windows works, Microsoft in theory can obsolete all of those programs. (In practice, Microsoft is most unlikely to do this, at least intentionally, because much of its market power stems from the fact that all those other products are built to its standard. Were Microsoft to make such a change, it is likely that those other software providers would look elsewhere for a target operating system. Microsoft's stock would drop precipitously, to say the least!)

One reason that vendors often build to de facto standards is that the standards-making process tends to be somewhat lengthy. A minimum of a four- to twelve-year period to come out with a new or revised standard is not at all uncommon in the industry. With product cycles under one year in some areas of the communications industry, waiting for finalization of a standard before introducing a product could result in corporate suicide. Ethernet is a good example of a standard that started as a de facto standard. Intel, Xerox, and Digital Equipment Corporation (DEC) introduced Ethernet with the intent of making it a de jure standard. But that process took years, and the final result was slightly different than the technology originally created by the three vendors. Nonetheless, many networks were created based on Ethernet before the 802.3 de jure standard was finalized, bringing profit to its creators and operating environments to their customers.

In the real world, the standards-making bodies rely in large part on vendors to develop the details of new and revised standards. In fact, most standards bodies have vendor representatives as full participants. It is a fascinating, political process with much pushing and pulling to gain advantage in the market. The vendors participate for several reasons, not the least of which is to get the jump on competitors that are not as close to the process. Other reasons include the ability to state in marketing materials that they contributed to or were involved in testing of a new standard, as well as the opportunity to influence the actual details of a standard to favor technology that they are most familiar with.

To be fair, it should be stated that the primary goal of most of those involved in the standards process is to define a good and useful standard. But when the process produces a "dual standard," as in the case of the Ethernet and Token Ring local area network standards, one can presume that the "best" was compromised somewhat in favor of what could be agreed upon.

Even de jure standards (usually identifiable by virtue of having unintelligible alphanumeric designations such as X.25, V.35, V.42 bis, etc.) change in ways that significantly affect the market—and you. One set of standards that affects thousands of users is the set of modem standards, discussed below. But beware a vendor that trumpets compliance with a "new standard!" The vendor's claim might be legitimate, but if no other vendors have product available in the same "space," the company might simply be hyping its own product in hopes that it eventually will become a standard. Or, there might be a standard under development but not yet approved. In the latter case, if that standard changes before final approval, the vendor's current products will instantly become "nonstandard" without changing in any way!

In the recent past, the standards committees were working on a new modulation technique to speed up data communications. The standards committees were locked in discussion with the rules to be applied. Yet, at the same time, every modem manufacturer began producing a new modem that was advertised and sold as compatible or compliant with the new V.Fast or V.34 standard. It was not yet a completed standard, but the manufacturers wanted to get their products on the shelves as quickly as possible and corral their piece of the market. So they produced a product with a disclaimer that offered a free or minimum cost upgrade to the V.34 standards if the standard changed. This is a classic example of the industry leaders setting the pace before the standards get completed.

Architectures

As with constructing a building, an overall design is needed when designing a communications environment. For a building, that design is described by its architectural drawings. A communications architecture is a coordinated set of design guidelines that together constitute a complete description of one approach to building a communications environment.

Several communications architectures have been developed. Some of the most well known include IBM's SNA and DEC's DNA. Architectures are covered in much greater detail in Chapter 14. But for those already familiar with the Open Systems Interconnect (OSI) model described in Chapter 14, most of this chapter (excluding codes, that reside at the presentation layer) addresses the physical and link layers.

The data communications architectures were modeled after the voice architectures. This is understandable, because data was merely a logical extension of the dial-up voice network. Devices are therefore constructed to fit into the overall voice network operation. Data equipment is designed and built to mimic the characteristics of a human speech pattern.

Protocols

Protocols are key components of communications architectures. Architectures are guidelines on how environments connecting two or more devices can be constructed, so most components of a given architecture in a network will be found on each communicating computer in that network. Protocols provide the rules for communications between counterpart components on different devices.

More detail, with examples, is provided in the discussion of the OSI model in Chapter 14. However, there is one aspect of protocols that also applies to hardware: whether they are synchronous or asynchronous. These key characteristics are covered in the following section.

Transmission Protocols (Synch vs. Asynch)

All lower-level data communications protocols fall into one of the two following categories: synchronous or asynchronous. The words themselves are based on Latin roots indicating that they either are "in" or "with" time (synchronous) or "out of" or "separated from" time (asynchronous). The underlying meanings are quite accurate, so long as one understands to what they must be applied.

All data communications depend on precise timing, or clocking. Chapter 5, analog vs. digital transmission, covered how voltage levels are sampled in the middle of a bit time in order to maximize the odds that the sample value will be clearly distinguishable as a one or a zero. But how does the equipment determine precisely when the middle of a bit time occurs? The answer is clocking; equipment at both ends of a circuit must be synchronized during transmission in order that the receiver and the sender "agree" regarding beginnings, middles, and ends of bits during transmissions. There are two fundamentally different ways to do this clocking: asynchronously and synchronously.

Simply put, asynchronous transmissions are clocked (or synchronized) one byte at a time. Synchronous transmissions are clocked in groups of bytes. But the differences in how these two approaches work go beyond the differences between individual bytes and groups of bytes.

Asynchronous communications is also called start/stop communications and has the following characteristics:

Every byte has added to it one bit signaling the beginning of the byte (the "start bit") and at least one bit added at the end of the byte (the "stop bits"). Bytes with seven data bits typically also include a parity bit, whereas eight-data-bit bytes usually do not. Thus, generally speaking, the total bits actually transmitted for every asynchronous byte equals 10 or 11. To get seven usable data bits, we must transmit approximately 10 to 11, or strictly speaking, we use a 30–35% overhead. See Figure 12.1 for the layout of a data byte in an asynchronous form. This was a special concern when data communications were initially used in the late 50s and early 60s. Then the cost per minute of a dial-up line was $0.60 to $0.65. Using that value, 30 cents of every dollar were spent just to provide the timing for the line. This amount of waste concerned everyone.

This also makes nominal speed calculations for such connections easy: dividing the rated speed of the circuit (e.g., 9600 bits/s) by 10 bits per byte gives a transmission speed in characters per second (e.g., 960 cps). As a rule, we divide the bits/s by 10 to get the nominal speed of an asynchronous circuit. ("Nominal" here means best case; in the real world, circuits rarely deliver 100% of their nominal capacity. But it's a starting point for capacity calculations.)

The bytes are sent out without regard to timing of previous and succeeding bytes. That means that none of the components in a circuit ever assume that just because one byte just "went by" another will follow in any particular period of time. Think of a person banging away on a keyboard. The speed and number of characters sent in a given period does not indicate in any way how many or how quickly characters can be sent in the succeeding similar period.

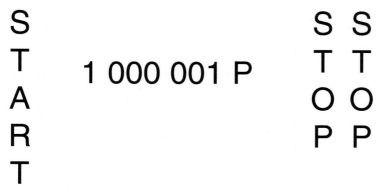

Figure 12.1 To send seven usable bits of data, we must use one start, one parity, and one or two stop bits.

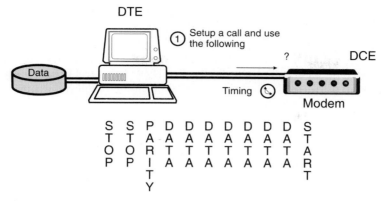

Figure 12.2 The data terminal equipment controls the timing to the data communications equipment (modem) as the bits are sent in an ASYNC protocol—a start bit and one or two stop bits help to set the timing.

Clocking is controlled by data terminal equipment (DTE). For example, when a personal computer is used to dial into CompuServe, clocking on bytes going toward the service is generated by the sending PC, see Figure 12.2. That first start bit reaching the modem begins the sequence, with all succeeding bits in the same byte arriving in lockstep at the agreed-upon rate until the stop bit is received. Then clocking stops until the beginning of the next byte arrives. Any intervening devices (especially modems) between the communicating DTEs take the clocking from the data sent by the originating DTE for any given byte.

Most PC and minicomputer terminal communications employ asynchronous techniques. The default communications ports on PCs (the "serial" or COM ports) only support asynchronous communications. To use synchronous communications on a PC, a special circuit board is required.

Synchronous communications have the following characteristics:

- Synchronous communications transmit blocks of data rather than individual bytes (characters).

- Individual bytes do not have any additional bits added to them on a byte-by-byte basis, except for parity.

However, bytes are sent and clocked in contiguous groups of one or more bytes. Each group is immediately (with no intervening time) preceded by a minimum of two consecutive synchronization bytes (a special character defined by the specific synchronous protocol, of which there are many) that begin the clocking. All succeeding bits in the group are sent in lockstep until the last bit of the last byte is sent, followed (still in lockstep) by an end of block byte. This layout of the synchronous characters (SYN) is shown in Figure 12.3.

Clocking is controlled by data communications equipment (DCE). Specifically, on any given circuit one specific DCE component is optioned (i.e., configured) at installation time as the *master device*. When the circuit is otherwise idle, the master generates the same synchronization character mentioned above on a periodic basis to all other DCE devices in order that all DCE clocks on the circuit are maintained in continuous synchronization.

Except in cases where smaller numbers of bytes (fewer than about 20) are sent at a time, synchronous communications make more efficient use of a circuit, as can be seen from Table 12.1.

Generally speaking, all circuits running at greater than 2400 bits/s actually operate in synchronous mode "over the wire." This is done simply because building modems to reliably operate asynchronously at higher speeds over analog circuits is much more difficult than taking this approach. Asynchronous modems that run faster than 2400 bits/s actually incorporate asynchronous to synchronous converters; they communicate asynchronously to their respective DTEs, but synchronously between the modems as shown in Figure 12.4. This doesn't normally impact performance: When smaller groups of characters are sent, there is time to include the additional overhead for synchronous transmission. When larger groups are sent, the reduced overhead of synchronous transmission comes into play. In practice,

S Y N	S Y N	S O H	S T X	Data up to 512 Bytes	E T X	E O T	S Y N
8	8	8	8	4096	8	8	8

Figure 12.3 The layout of data transmitted synchronously. In this case, the block size is 512 bytes.

TABLE 12.1 A Comparison of the Utilization of the Circuit

Data bytes	Asynch bits	Synch bits	Synch savings
1	10	32	–220%
5	50	64	–28%
10	100	104	–4%
20	200	184	8%
30	300	264	12%
100	1000	824	18%
1000	10000	8024	20%
10000	100000	80024	20%

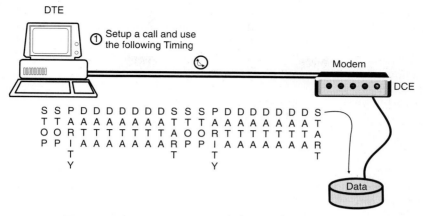

Figure 12.4 The data communications equipment (modem) will set up the communications synchronously. Blocks are used to create the synchronous transfer.

these higher-speed modems actually communicate between the modems at even higher than their rated speeds; the extra bandwidth is used for overhead functions between the modems.

Error Detection

We mentioned at the beginning of the chapter that ensuring the integrity of the information was one of the key responsibilities of a data communications environment. This does not mean that the data must be kept honest. Rather, it means that we must somehow guarantee with an extremely high degree of probability that the information is received in exactly the same form as it was sent.

More precisely, there are two tasks required: Detect when errors in transmission occur, and trigger retransmissions in the event that an error is detected. It is the responsibility of protocols (see Chapter 14) to trigger and manage retransmissions. Here we discuss some of the various approaches that have been developed to detect errors in the first place.

All of the code sets used in data communications (see the following) are designed to use all of their bits to represent characters (letters, numbers, other special characters). A not-obvious implication of this fact is that every byte received in such a code set is by definition a valid code. How can we detect whether the received code is the code that was sent?

The answer is to somehow send some additional information, some data about data ("data about data" is sometimes described as metadata) along with the primary data. All error-checking approaches depend on sending some additional data besides the original application-related data. The additional data is created during the communications process, used to check the underlying data when it is received, then discarded before the information is passed to its final destination.

In order of increasing reliability, the major methods used to detect data communications errors include:

- Parity bit, or vertical redundancy checking (VRC)
- Longitudinal redundancy checking (LRC)
- Cyclic redundancy checking (CRC)

Suppose you are my rich aunt, and that I am living in Paris (to further my cultural education, of course) and have run out of money. I've called you (collect, of course) to request that you electronically transfer some money into my account at the Banque de Paris. In a fit of generosity, you have decided to send me $1000.00. If the network used is not perfectly reliable and appropriate error detection methods are not applied by the transmitting financial service, a change in a single character—i.e., changing the period after the first three zeroes to another zero—could result in your sending me considerably more money than you intended: a total of $1,000,000.

What are the chances of my getting my inheritance early in this way? Not very high, given the odds that only the period would change, and only to a zero (out of either 126 or 254 other possibilities). But consider the probability of detecting the error, assuming that it has occurred. Using parity bits, the likelihood of detecting this kind of error (which requires several bits to be wrong at one time to change an entire character) is about 65%. A somewhat better method, longitudinal redundancy checking, would up the odds of detection to about 85%. But the cyclic redundancy checking method improves the odds of detecting and correcting such a multi-bit error to 99.99995%.

Because networks used to send monetary amounts generally use CRC techniques, it doesn't look as though I'm going to get rich because of their errors. But let's examine these methods in a bit more detail anyway.

Parity Bit / Vertical Redundancy Checking (VRC)

The parity bit approach to error detection simply adds a single "parity" bit to every character (or byte) sent. Whether the parity bit is set to zero or one (the only two possibilities, of course) is calculated by the sending digital device, and recalculated by the receiving device. If the calculations match, the associated character is considered to be good. Otherwise, an error is detected.

This is much simpler than it sounds. Two approaches are typically used: "even" and "odd" parity. Other forms of parity exist, such as mark or space parity. It makes no difference which is used; the only requirement is that the sending and receiving devices use the same approach. To illustrate using even parity, consider the ASCII bit sequence representing a lowercase letter "a": 1100001. Because we are using even parity, we require that the total number of "1" bits transmitted to the receiver to send this "a," including the eighth parity bit, be equal to an even number. Their position in the underlying byte is irrelevant. If we count the 1s in the seven-bit pattern, we get three, an odd number. Therefore, we set the parity bit to one, resulting in 11100001, an eight-bit pattern with an even number (i.e., four) of ones. (The parity bit is sent last. In our illustrations right-most bits are sent first, so we show the parity bit being added at the left.) See Figure 12.5 for an ASCII illustration of the word "Hello," complete with even parity bits. As you can see from the illustration, the bytes are represented as vertical sets of numbers; thus, "vertical redundancy checking": if we orient the digits vertically, we add a vertical bit that is redundant to help check the correctness of the underlying byte. The vertical orientation is arbitrary, of course. However, when we illustrate longitudinal redundancy checking, you will see that there is a method in this display approach.

Having gone to the trouble to describe parity bits, we must confess that they are of limited usefulness in data communications. Parity checking will catch 100% of errors where the number of bits in error is an odd number (1, 3, 5, etc.) . . . and none of the errors where the number of bits in error is

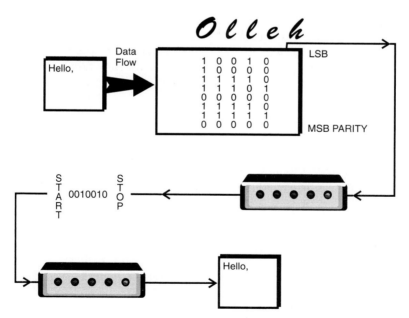

Figure 12.5 The vertical redundancy check shows the flow of data. The parity bit is added after each character is generated.

even. Put another way, if an error occurs (and communications errors rarely affect only a single bit), there is only about a 65% chance that parity checking will detect it. (The probability is better than 50% because there are somewhat more one-bit errors than any one type of multi-bit error, whether the numbers of the latter are odd or even.)

Parity checking is used extensively inside computers. There, it makes sense because it is entirely plausible that errors would occur one at a time (if they occur at all). Parity checking does well in this environment. Also, some networks (e.g., CompuServe) still have users set their communications software to use parity checking. But even CompuServe uses a more sophisticated protocol for file transfers. Some more sophisticated error detection protocols are described in the following paragraphs.

Longitudinal Redundancy Checking (LRC)

The concept of LRC follows directly from VRC, only more so. This example uses eight-bit bytes rather than seven-bit bytes. But LRC checking needs to operate on a group of bytes, rather than one at a time. For this example, it doesn't really matter what the bits represent, so let us create a set of eight eight-bit bytes. As you will see, although the bit patterns do not matter for the example, the number of bytes used does (See Table 12.2).

If we only use VRC (odd parity) as described above, we produce Table 12.3. But of course, we said that VRC only catches about 65% of errors, hardly acceptable. But what if we apply odd parity checking across the bytes in addition to vertically? In that case, the completely filled-in Table 12.4 would be generated.

TABLE 12.2 Setting Up for Vertical Parity Checking

1	0	1	0	1	0	1	0	
0	0	1	0	0	0	1	0	
1	1	1	0	1	1	1	0	
1	0	1	0	0	0	1	0	
0	1	1	0	0	1	1	0	
1	1	1	0	1	1	1	0	
0	1	1	0	0	1	1	0	
1	0	1	0	1	0	1	0	

TABLE 12.3 The Parity Bit is Inserted in VRC

1	0	1	0	1	0	1	0	
0	0	1	0	0	0	1	0	
1	1	1	0	1	1	1	0	
1	0	1	0	0	0	1	0	
0	1	1	0	0	1	1	0	
1	1	1	0	1	1	1	0	
0	1	1	0	0	1	1	0	
1	**1**	**0**	**1**	**0**	**1**	**0**	**1**	

TABLE 12.4 The Vertical and Longitudinal Redundancy Checks are Inserted Here

1	0	1	0	1	0	1	0	**1**
0	0	1	0	0	0	1	0	**1**
1	1	1	0	1	1	1	0	**1**
1	0	1	0	0	0	1	0	**0**
0	1	1	0	0	1	1	0	**1**
1	1	1	0	1	1	1	0	**1**
0	1	1	0	0	1	1	0	**1**
1	1	0	1	0	1	0	1	**0**

Adding both the horizontal and vertical checking, together referred to as *longitudinal redundancy checking*, improves the odds of detecting errors to about 85%. Not bad, although I wouldn't want to trust my money to such a transmission. But there is another disadvantage to LRC. Using eight-bit bytes for every eight data bytes, an additional two LRC bytes must be transmitted. That works out to 20% added overhead for error checking (two LRC bytes divided by the 10 total bytes transmitted in the set), not counting any degradation due to time required to compute the check bytes. This is not an efficient error-checking mechanism. In fact, considering that error-checking is only one of several sources of transmission overhead, it is abysmal.

LRC does have one advantage over CRC checking (the approach discussed next): the computational resources required to calculate the LRC bytes are far lighter than those required to calculate a CRC (unless the CRC checking is implemented with hardware). In fact, until recent generations of PCs became available with their vastly more powerful CPUs, CRC checking for asynchronous data communications in the PC environment was not practical because of its computationally intensive nature. Now, however, it is routinely used. Read on to see why.

Cyclic Redundancy Checking (CRC)

Although no practical error checking algorithm can guarantee detection of every possible error pattern, CRC comes close. A complete explanation with examples of how CRC works would (and does, in several data communications textbooks) require several pages of somewhat hairy binary algebra. Rather than put you through that, we'll describe here some of the method's key characteristics and indicate how this method is used.

Like the previously described approaches to error detection, CRC relies on on-the-fly calculation of an additional bit pattern (referred to as a *frame check sequence*, or FCS) that is sent immediately following the original block of data bits. The length of the FCS is chosen in advance by a software or hardware designer based on how high a confidence level is required in the error-detection capability of the given transmission. All "burst errors," or groups of bits randomized by transmission problems, with a length less than that of the FCS will be detected. Frequently used FCS lengths include 12, 16, and 32 bits. Obviously, the longer the FCS, the more errors will be detected.

The FCS is computed by first taking the original data block bit pattern (treated as a single huge binary number) and adding to its end (after the low-order bits) some additional binary zeros. The exact number of added zeros will be the same number as there are bits in the desired FCS. (The FCS, once calculated, will overlay those zeros.) Then, the resulting binary number, including the trailing zeros, is divided by a special previously selected divisor (often referred to in descriptions of the algorithm as P).

P has certain required characteristics:

- It is always one bit longer than the desired FCS
- Its first and last bits are always 1
- It is chosen to be "relatively prime" to the FCS; that is, P divided by the FCS would always give a non-zero remainder. In practice, that means P is normally a prime number.
- The division uses binary division, a much quicker and simpler process than decimal division. The remainder of the division becomes the FCS.

Specific implementations of CRC use specific divisors; thus, the CRC-32 error-checking protocol on one system should be able to cooperate with the CRC-32 protocol on another system; the CRC-CCITT protocol (which uses a 17-bit pattern, generating a 16-bit FCS) likewise should "talk" to other implementations of CRC-CCITT. Selection of a specific P can be tuned to the types of errors most likely to occur in a specific environment. But unless you are planning on engineering a new protocol, you needn't worry about the selection process; it already has been done for you by the designers of your hardware or your communications software. See Figure 12.6 for an illustration of the CRC creation process.

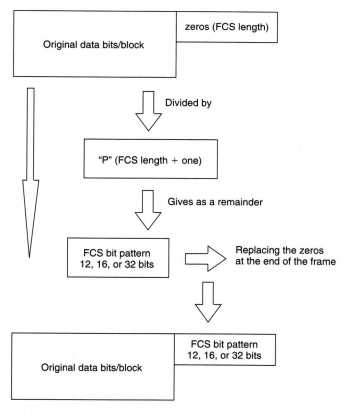

Figure 12.6 The creation of a CRC cycle shown.

CRC is typically used on blocks or frames of data rather than on individual bytes. Depending on the protocol being used, the size of the blocks can be as high as several thousand bytes. Thus, in terms of bits of error-checking information required for a given number of bytes of data, CRC requires far less transmission overhead (e.g., CRC-32 sends four eight-bit bytes' worth of error checking bits to check thousands of data bytes) than any of the parity-based approaches.

Although binary division is very efficient, having to perform such a calculation on every block transmitted does have the potential to add significantly to transmission times. Fortunately, CPUs developed in the last few years are up to the challenge. Also, unlike with most other "check-digit" types of error correction, the receiving device or software does not have to recalculate the FCS in order to check for an error. Instead, the original data plus the FCS are concatenated together to form a longer pattern, then divided by the same P used as a divisor during the FCS creation process. If there is no remainder from this last division, then the CRC algorithm assumes that there are no errors. And, 99.99995% of the time, there aren't.

Plexes—Communications Channel Directions

The next area is the directional nature of your communications channel. Three basic forms of communications channels exist.

One way (simplex)

This is a service that is one way and only one way. You can use it to either transmit or to receive. This is not a common channel for telephony (voice), because there are very few occasions where one person speaks and everyone else listens. Feedback is one of the capabilities that we prize in our communications, which would be eliminated in a one-way conversation. Broadcast television is an example of simplex communications.

To design an efficient data communications application using a simplex channel can require quite a bit of ingenuity. A good example is stock ticker tape radio signals. Bearing in mind that in a true simplex system (such as this one is), there is absolutely no feedback possible from the receiver to the transmitter. How then does someone using such a signal get useful, timely information? After all, unless one is simply gawking at the symbols as they go by, it is not practical to wait for on average half of the symbols to go by in order to find out that the particular stock in which you are interested just went up or down a bit.

The answer is a combination of communications and computer technology. The applications that implement this technique memorize locally (on a PC) the entire repeating communications stream once, then accept each new symbol/price combination received as an update to the local "database." The user inquires against that local database, getting what appears to be instant information, even though it might have been received several minutes ago. Naturally, the computer must be set to continually receive; otherwise, the user will have no assurance that the data is even remotely current.

Another approach requires the user to specify in advance to the software a set of symbols to collect. As that stock information goes by, the program snags only their information for retention and local query. This is not any faster than the previous approach, but it does require less local storage capability.

Two-way alternating (half-duplex)

This is the normal channel that is used in conversations. We speak to a listener, then we listen while someone else speaks. The telephone conversations we engage in are normally half-duplex. Although the line or medium (air, in this case) is capable of handling a transmission in each direction, most human brains can't deal well with simultaneous transmit and receive.

Many computer and communications configurations use half-duplex technology. Until a few years ago, most leased-line multidrop modems were half-duplex. One of the key differentiators among such modems was their *turnaround time*; that is, how quickly a pair of such modems could reverse the channel direction. This was measured in milliseconds—the fewer, the better. Entire communications protocols were built around this technology (e.g., IBM's Bisynchronous Communications—BSC). All-block mode (e.g., IBM 3270s) terminals still operate in half-duplex mode only, even if full-duplex facilities are available. This simply means that at any given time, the terminal is either sending to its associated computer or receiving from it, not both at once. This does not cause a problem because the entire system is designed around this behavior, and it works quite well. Of course, almost half of any given circuit's raw capacity (if one adds the capacity of the two directions together) is wasted. (IBM's more recent SDLC protocol-based front end processors can talk to one terminal and receive from another at the same time, minimizing this waste, but the individual terminals are still functioning as half-duplex devices.)

With some technologies, half-duplex can be used so effectively that the one-way-at-a-time characteristic of the circuit is invisible; it appears to be full-duplex (see below). An excellent example of such an approach is local area network communications. Most actual LAN technologies, including both Ethernet and Token Ring, are actually half-duplex "on the wire." But the information moves so quickly, and the responses are so fast, that the path appears to an observer to be full-duplex.

Two-way simultaneous (duplex) or full-duplex

True full-duplex communications make maximum use of a circuit's capacity—if the nature of the communications on that circuit takes advantage of it. In data communications, a circuit is implemented and used in full-duplex mode if a device can send to a computer and receive from the computer at the same time. Although we mentioned that human conversation is typically half-duplex, there is an exception to this: conversation among teenagers, who seem to be able to speak and listen (somewhat) at exactly the same time.

A common example of full-duplex communications is that seen when one dials into an on-line information system such as CompuServe or Prodigy. Most such systems allow users to continue typing at a keyboard even while the service is sending information to the user for display on the screen. Were the connection not a full-duplex one, this simultaneous bi-directional communications would not be possible.

One of the points that confuses some people is that the terms *simplex*, *half-duplex*, and *full-duplex* can refer to varying levels of a communications architecture. If the three levels are considered to be three points on an increasing scale of capability, one can say that a given level of a communications architecture must rely on lower levels with at least the capability of that given level as presented in Table 12.5.

This is a pretty table, but what does it mean? It means that a wide-area analog data circuit built to handle full-duplex communications (requiring the telephone company to support simultaneous communications in both directions, and use of full-duplex modems) can fully support half-duplex or simplex communications. But if a similar circuit is implemented with half-duplex modems, then full-duplex communications on that circuit will not work, although simplex will. A citizen's band radio provides a half-duplex channel—two directions, but only one way at a time. Simplex would work— one sender could lock down a key and just keep sending—but full-duplex would be impossible.

Air is a full-duplex channel. But a simplex signal such as the output from a stereo speaker system has no trouble traveling this full-duplex channel.

Compression

Although compression is not exactly a modulation technique, it does (usually) produce faster transmissions. To understand how compression works, con-

TABLE 12.5 Comparing the Capability and Directionality of the Circuit

This capability level	Requires all lower levels to have at least the following capability	But can also function without impairment on top of levels with the following capabilities
Simplex	Simplex	Half or full-duplex
Half-duplex	Half-duplex	Full-duplex
Full-duplex	Full-duplex	(no additional levels)

sider first how human beings communicate. Most human communication is inherently redundant. That does not imply waste; rather, human beings use that redundancy as a continual cross-check on what information is really being sent and meant. For example, in face-to-face conversation much more information is being sent than just the words. Facial expressions, tones of voice, limb positions and movement, overall carriage of the body, and other less obvious cues all contribute to the information stream going between two people having a conversation. But much of the information is duplicated. For example, anger can be communicated by the words themselves; but it can also be conveyed by tone of voice, facial expression, involuntary changes in the color of one's complexion, the stress in the voice, arm movement, and other cues. If some of these items were removed, the message received might be just as clear but the total amount of raw information might be reduced.

In data communications, *compression* is a technique applied either in advance to information to be transmitted, or dynamically to an information stream being transmitted. The underlying technology is essentially the same in both cases: removal of redundant information, or expression of the information in a more compact form, is used to reduce the total number of bytes that must pass over the communications medium in order to reduce the time the medium is occupied by a given transmission to a minimum.

A detailed discussion of compression techniques is beyond the scope of this book. But in this section we will describe two very basic approaches in order to help understand the fundamentals of the technology. Later, during the discussion of modems, we will briefly identify and describe the power of some compression techniques that are often built into such devices.

The simplest form of compression is the identification and encoding of repeating characters into fewer characters. For example, consider the transmission of printed output across a network to a printer. A typical report contains a very high number of blank characters, often occurring consecutively. Suppose every such string of four or more consecutive blanks (which are of course themselves ASCII characters) is detected and replaced with a three-byte special character sequence encoded as follows:

- The first character is a special character (one of the nonprint characters in the ASCII code set, for example) indicating that this is a special sequence.

- The second character is one occurrence of the character that is to be repeated, in this case a blank.

- The third character is a one-byte binary number indicating how many times the character is to be repeated. With one byte (using binary format), we can count up to 255, high enough to get some real savings!

How much can we save? Look at Table 12.6. We'll assume that on average, blanks occur in 10-byte consecutive streams, a very pessimistic assumption.

TABLE 12.6 A Quick Analysis of Compression Benefits

Total characters in print stream before compression	Blanks in print stream before compression	Total characters in print stream after compression	Percentage savings
1000	10	993	1
1000	100	930	7
1000	500	650	35
50000	2000	48600	3
50000	5000	46500	7
50000	20000	36000	28

As can be seen from the table, the savings depends on the number of occurrences of the character to be repeated. In practice, this is a reasonably powerful technique. In the example, we addressed only blanks. In practice, any character except the special character would be fair game.

But what about if we actually want to send that special character? After all, unlike print jobs, many transmissions must be able to handle every possible code; there are none left over that are "special." No problem—we add the following rules:

- We'll never try to compress multiple occurrences of the special character.

- Every time we encounter the special character as input during the encoding process, we'll simple send it twice. If the receiving hardware sees it twice, it knows to drop one of the occurrences.

With this approach, we only have to select a special character that is unlikely to occur frequently. If it then does occur frequently in a particular transmission, our compression algorithm doesn't break; it just will be very inefficient for that one transmission.

Note how the overall redundancy is squeezed out of a transmission using this approach. But just as in human communications, eliminating redundancy increases the risk that some information will be misinterpreted. In human communication, if the reddening of an angry person's face and other visual cues were not visible (e.g., if the conversation were on the telephone) and the speaker was otherwise very self-controlled, the listener might misinterpret angry words as being a joke; after all, many American sub-cultures routinely use "affectionate insults" without anger. Unrecognized anger is a very serious loss of information. In data communications without compression, omission of a single space in a series of spaces might cause a slight misalignment on a report, but will most likely not seriously distort its meaning. If compression is used and the binary count field is damaged—i.e., changed to another binary digit—dozens or even hundreds of spaces or other repetitive characters might be either deleted or added to the report, seriously compromising its appearance and perhaps distorting its meaning. Consider the havoc to be wrought in a horizontal bar chart! The error-checking techniques discussed earlier become much more important in a system using compression!

Another more sophisticated method of compression requires pattern recognition analysis of the raw data rather than just detection of repeating characters. Again, some special character must be designated, but now it presages a special short code that represents some repetitive pattern detected during the analysis. For example, in graphics displays capable of showing 64K (65,536) colors, every screen pixel has associated with it two eight-bit bytes (which together can represent 65K different values) indicating the color assigned to that pixel. If someone sets her screen to white on a blue background, the two-byte code for blue is going to appear thousands of times in the data stream associated with that display. The repeating one-byte compression algorithm described above will not detect anything to compress. But if analysis shows that a two-byte pattern occurs many times in succession, a more sophisticated approach might assign a specific character (preceded by the special character) to represent precisely twenty (or some other specific number of) consecutive occurrences of that two-byte sequence. The savings can be considerable, but they again depend upon the characteristics of the data being transmitted.

A third, very computationally intensive approach to compression has been designed especially for live transmission of digitized video signals. Unlike most other compression methods, this approach does not involve movement of representations of the entire digitized data stream from one point to another. Video signals consist of a number of still picture frames composed each second (visualize 30 photographs per second, in the highest-quality case). Although the first picture must, of course, be sent in its entirety, special equipment and algorithms must then continuously examine succeeding video frames to be transmitted, identifying which pixels have changed since the last "picture" was taken. Then, information addressing just the changed pixels is sent to the receiver, rather than the entire new frame. The receiving equipment uses this change information combined with its "memory" of the previous frame to continuously, locally build new versions of the picture for display. This approach is particularly fruitful for pictures that in large part remain static; for example, video conferencing. In video conferencing, usually the only moving features of the picture are the human beings. The table(s), walls, and other room fixtures stay still, therefore requiring transmission only once. Frequently, only the lips move for long periods of time.

One other compression-related concept is worth mentioning: lossy vs. lossless compression. What? I hear you ask. Does lossy mean what it sounds like? Would we ever tolerate transmission that loses information? The answer, for some applications, is yes. Moreover, you have probably settled for information loss when working daily with computers. And it caused you no hardship at all. If you use a personal computer with a VGA screen, but display any type of graphic that inherently has Super VGA (SVGA) level resolution, your VGA screen loses the additional definition in the image that is visible only when an SVGA controller card and monitor are used. And in fact, this example, while not involving compression as such, demonstrates precisely the type of situation where lossy compression would be tolerated: transmission of video images. Some compression algorithms used for transmission of video images lose some of the resolution of those images. However, if the received image is acceptably precise, the maintenance of the speed of the moving image might be more important to keep up.

Multiplexing

The paths available for moving electronic information vary considerably in their respective capacities, or bandwidth. If a company requires many paths over the same route (for example, many terminals each requiring a connection to one distant computer), it often makes sense to configure one large-capacity circuit and bundle all the smaller requirements into that one big path. The process of combining two or more communications paths into one path is referred to as multiplexing. There are three fundamental types of multiplexing, all of which have significant variations. These main types are:

- Space division multiplexing (SDM)
- Frequency division multiplexing (FDM)
- Time-division multiplexing (TDM)

SDM

SDM is the easiest multiplexing technology to understand. In fact, it is so simple that it would hardly rate its own special term, except that it is the primary method by which literally millions of telephone signals reach private homes. With SDM, signals are placed on physically different media. Then those media are combined into larger groups and connected to the desired end points.

For example, telephone wire pairs (which of course are also used for data communications), each of which can carry a voice conversation, are aggregated into "cables" with hundreds or even thousands of pairs (see Figure

12.7). The latter are run as units from telco COs out to wiring center locations, from there splitting out to individual customer buildings.

The biggest advantage of SDM is also its biggest disadvantage: the physical separation of the media carrying each signal. It is an advantage because of the simplicity of managing the bandwidth; one only must label appropriately each end point of the medium. No failures (other than a break in the medium) can affect the bandwidth allocation scheme. But because there is a direct physical correlation between the physical link and the individual communications channel, a provider has some difficulty in electronically manipulating the path to achieve efficiencies of technology or scale. For example, probably the largest single factor blocking the migration of the overall telephone network to all-digital technology is the embedded base of copper wire (and analog amplifiers) supplying telephone service to millions of homes and businesses.

FDM

FDM is inherently an analog technology. It achieves the combining of several digital signals onto (or into) one medium by sending signals in several distinct frequency ranges over that medium.

One of its most common applications is cable television. Only one cable reaches a customer's home, but the service provider can nevertheless send multiple television channels or signals simultaneously over that cable to all subscribers. Receivers must tune to the appropriate frequency in order to access the desired signal. Figure 12.8 demonstrates the combining of signals on a cable television coaxial cable.

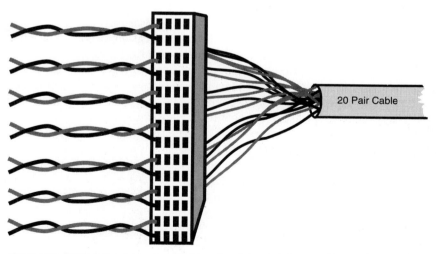

Figure 12.7 Individual pairs are bundled together into much larger cables on a one-to-one ratio.

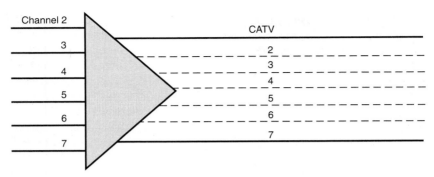

Figure 12.8 Frequency division multiplexing breaks the whole bandwidth into separate full-time channels.

Certain modems have built-in FDM capabilities. One can modify, using controls on the modem, how much bandwidth each connected digital device will be provided.

TDM

Time-division multiplexing is a digital technology. It involves sequencing groups of a few bits or bytes from each individual input stream, one after the other, and in such a way that they can be associated with the appropriate receiver. If done sufficiently quickly, the receiving devices will not detect or care that some of the circuit time was used to serve another logical communication path. The really high-speed communications technologies covered in later chapters all without exception use some form of TDM, so it is worthwhile to try to understand it.

Consider an application requiring four terminals at an airport to reach a central computer. Each terminal communicates at 2400 bits/s, so rather than acquire four individual circuits to carry such a low speed transmission the airline has installed a pair of multiplexers (see below), a pair of 9600-bits/s modems, and one dedicated analog communications circuit from the airport ticket desk back to the airline data center, as illustrated in Figure 12.9.

The time-division multiplexers work together to merge the data streams onto the 9600-bits/s circuit in such a way that each terminal appears to have a dedicated 2400-bits/s circuit. The multiplexer has enough buffer (or storage) space that as any of the four clerks presses a key, that keystroke is stored locally (in the multiplexer or "mux") until the time slot assigned to that clerk's terminal comes along. If the clerks are named Ellen, Joe, Susan, and Allan, their information flows as shown in Table 12.7.

As an example, we'll assume that Joe is inquiring as to the available of space on XXXXX Airlines flight 243. As part of this inquiry, he types the letters XA243. At the same time, Susan is seeing if there is space at the

Compris hotel in Chicago, so she types `Compris`. The other two clerks are out to lunch; their terminals are inactive. Those letters will be interspersed among letters typed by his colleagues. TDM assigns time slots to each configured device, whether or not the slots are used. If the data stream is to be interleaved one byte at a time, the data stream toward the computer resulting from this typing might look as shown in Figure 12.10 (the information on your right goes toward the computer first).

This is not really a form of compression because the same number of bytes is sent as is typed. In fact, as can been seen from the figure there is a certain inefficiency here. Those four-byte blocks are being pumped out by the multiplexer no matter whether anyone types a character or not. The first position in each four-byte block is reserved for Ellen's characters, while the fourth position is reserved for Allan. Because they are both at lunch, their blocks are empty—and wasted. Joe is a slower typist than is Susan, so even though he types fewer characters, his last one goes at the same time as Susan's last; some of his slots are empty. He "missed the train."

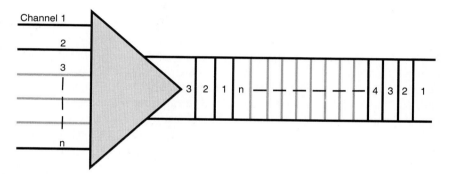

Figure 12.9 In time-division multiplexing, the entire channel is allocated for short periods of time.

Byte	From
1	Ellen
2	Joe
3	Susan
4	Allan
5	Ellen
6	Joe
7	Susan
8	Allan

TABLE 12.7 The Flow of Information for the Various Time Slots Used

In real life, many more slots would be empty because the capacity of the line is 240 characters per second—per clerk! Few can type that quickly. Also in real life, multiplexing is more commonly done at the bit level rather than at the byte level. This would be harder to illustrate understandably, so we used bytes in the example. But the principle is identical.

Next, we modify the example to assume that the characters typed (XA243 and Compris) are coming back from the computer instead of being typed. In such a case, they will be coming at "wire speed"; the computer can pump them out far faster than the line can absorb them, so we are now talking about wasting some time (Figure 12.11).

Were the multiplexer configured for only two clerks' terminals, the information from the computer could appear much more quickly (Figure 12.12).

This in effect is what a variation on a TDM, called a *statistical time-division multiplexer (STDM)*, does. The sending STDM analyzes the data stream on the fly to determine which ports or "tail circuits" are active—that is, how much service they require. If they are inactive, or less active (slow typists?), they are provided less of the time slots (or bandwidth). This allocation changes dynamically depending on the traffic pattern. Again, this is not compression because all the information is sent. Nonetheless, it seems like compression because extending this technique allow over-committing the line. If 10 terminals are connected to a 9600 circuit via an STDM, each terminal can be set to 2400 bits/s. So long as not all terminals are busy full speed at every moment, the STDM can make it look as though each terminal has its own 2400-bits/s circuit, even though the aggregate bandwidth required to support these settings without using an STDM would be 24,000 bits/s!

Paradoxically, STDM techniques are used more at lower speeds than on the really fast multi-megabit circuits. At very high speeds, the equipment is so busy just doing TDM functions that the extra computer power required to do the on-the-fly analysis for STDM would be just too much.

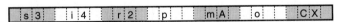

Figure 12.10   The use of TDM and the data time slots being used for two users.

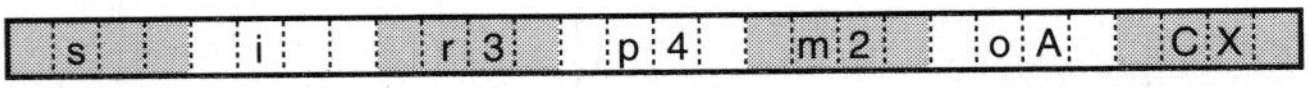

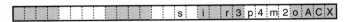

Figure 12.11   The time wasted on the circuit can be compensated for using TDM techniques.

Figure 12.12   The TDM is now configured for only two users so the data will be less wasted.

Codes

Chapter 5 covered digital transmission. The concept of a "bit"—an electronic expression of a one or zero—should now be clear. However, how do you get from ones and zeros to transmitting your resume over wires?

The alphabet must be built up from sets of ones and zeros. Specifically, we use one or more sets of codes or alphabets, standard definitions of patterns of ones and zeros that we will agree to mean letters, numbers, and other symbols that we wish to transmit and receive.

The alphabets most frequently used are either American Standard Code for Information Interchange (ASCII) which is fairly universal, or Extended Binary Coded Decimal Interchange Code (EBCDIC), which is an IBM alphabet. These two code sets, or alphabets, are used to convert a series of ones and zeros into an alpha or numeric character in our alphabet.

ASCII

One character (also known as a byte or octet) must be represented as a consistent bit pattern by both sender and receiver. As we use a keyboard (standard typewriter keyboards are known as *QWERTY keyboards* because of the sequence of the first row of alphabetic keys), we create a stream of combination of letters, numbers and symbols.

When ASCII (usually pronounced "ass-key" with a mild accent on the first syllable) was originally defined for use by the United States government, the bit pattern was defined to be seven data bits long. With seven bits, it is possible to differentiate 128 different patterns. So, to recreate these typed characters with ones and zeros, we can use a combination of up to 128 possible ASCII characters (Figure 12.13).

Using the combinations in this table, we should be able to transmit just about everything we presently understand in our vocabulary. And, in fact, for many years, the seven-bit ASCII was used for most non-IBM mainframe communications. But there were two factors that caused this form of ASCII to become less popular.

First, most computers handle data in eight-bit chunks (rather than seven) to represent characters. The terms *byte* and *octet* almost always refer to eight-bit patterns, not seven. Second, while seven-bit ASCII can indeed represent all the English letters and numbers, with some symbols left over for special characters and control information, there are many other characters used in written communication that cannot easily be expressed in a 128-character code set. Accented characters in the Romance languages, character graphic drawing symbols, typographic indications in word processors (bolding, underlining, etc.) are just a few examples of symbols difficult to handle with standard ASCII.

Bit 4	Bit 3	Bit 2	Bit 1		Bit 7 → 0	0	0	0	1	1	1	1
					Bit 6 → 0	0	1	1	0	0	1	1
					Bit 5 → 0	1	0	1	0	1	0	1
				Col Row	0	1	2	3	4	5	6	7
0	0	0	0	0	NUL	DLE	SP	0	@	P	\	p
0	0	0	1	1	SOH	DC1	!	1	A	Q	a	q
0	0	1	0	2	STX	DC2	"	2	B	R	b	r
0	0	1	1	3	ETX	DC3	#	3	C	S	c	s
0	1	0	0	4	EOT	DC4	$	4	D	T	d	t
0	1	0	1	5	ENQ	NAK	%	5	E	U	e	u
0	1	1	0	6	ACK	SYN	&	6	F	V	f	v
0	1	1	1	7	BEL	ETB	'	7	G	W	g	w
1	0	0	0	8	BS	CAN	(	8	H	X	h	x
1	0	0	1	9	HT	EM	)	9	I	Y	i	y
1	0	1	0	A	LF	SUB	*	:	J	Z	j	z
1	0	1	1	B	VT	ESC	+	;	K	[	k	{
1	1	0	0	C	FF	FS	,	<	L	\	l	¦
1	1	0	1	D	CR	GS	-	=	M	]	m	}
1	1	1	0	E	SO	RS	.	>	N	^	n	~
1	1	1	1	F	SI	US	/	?	O	_	o	DEL

Figure 12.13 The ASCII code set.

Extended ASCII

Extended ASCII is a superset of ASCII. Extended ASCII is the code used inside virtually all non-IBM computers, including personal computers. It is an eight-bit code set, doubling the possible distinguishable characters to 256. The seven-bit ASCII codes are present in extended ASCII in their original form with a zero prepended to the base seven bits. But another 128 characters are available with the same base seven bits as the original ASCII, but with a one prepended instead.

But whereas ASCII is a standard, extended ASCII is, well, not quite standard. While the original 128 characters communicate well from vendor to vendor, even in extended ASCII, every application defines its own use of the additional 128 characters. For example, you can easily write out ASCII text from most word processing programs, with the result being readable by most other word processors. But if you attempt to read a document created by a word processor in its native form with another, different word processor, you will only be successful if the latter specifically contains a translation module for material created by the first.

Nonetheless, the lion's share of non-IBM communications is now conducted using extended ASCII. Virtually all personal computers use it, including IBM's. If you set your communications protocol to "N,8,1" (no parity, eight data bits, one parity bit), those eight data bits are encoded in using extended ASCII.

EBCDIC

IBM, not a company to follow the herd, realized early on that 128-code ASCII did not contain enough patterns for its requirements. So, IBM created an entirely different code set, one twice as large as the original ASCII code set. IBM's eight-bit 256-character code set, used on all of its computers except personal computers, is called "extended binary coded decimal interchange code." (Most people pronounce it "eb-suh-dick" with a mild accent on the first syllable.)

Like ASCII, certain of the characters are consistent wherever EBCDIC is used. But other characters vary depending on the specific communicating devices. In Figure 12.14, the white space can be used differently depending on the EBCDIC dialect in use.

Unicode

256 codes might seem to be all anyone would need. But consider the requirements of Chinese, which has thousands of characters. Or Cyrillic, which, although not having a terribly large number of characters, does not overlap any of those defined in ASCII or EBCDIC. Another code set, called Unicode, is now being implemented in some products. Unlike the eight-bit extended ASCII and EBCDIC code sets, Unicode uses 16 bits, or two bytes, per character. While still just ones and zeros, this allows up to 65,536 (2 to the 16th power) separate character definitions. Of course, each character takes up as much storage and transmission time as two eight-bit characters. But Unicode is a truly international code set, allowing all peoples to use their own alphabet if they wish.

Modulation

How does the transmission process work? How does the data get onto the voice dial-up telephone line? We use a device to change the data. This device, known as a *modem*, changes the data from something a computer understands (numeric bits of information) into something the telephone network understands (analog sine waves, or sound). A modem generates a continuous tone, or carrier, then modifies or modulates it in ways that will be recognized by its partner modem at the other end of the telephone circuit. The modems available to do this come in variations, each one creating a change in a different way. Remember that the modem is a contraction for *modulation/dem*odulation. In order for the communications process to work, we need the same types of modems at each end of the line, operating at the same speeds. These modems can use the following types of modulation scheme (or change method):

Figure 12.14 shows the EBCDIC code set. Columns are defined by bits 8 7 6 5 (high-order four bits); rows are defined by bits 4 3 2 1 (low-order four bits).

Bits 4321 \ 8765	0000	0001	0010	0011	0100	0101	0110	0111	1000	1001	1010	1011	1100	1101	1110	1111
0000	NUL	DLE	DS		SP	&	-									0
0001	SOH	DC1	SOS				/		a	j			A	J		1
0010	STX	DC2	FS	SYN					b	k	s		B	K	S	2
0011	ETX	DC3							c	l	t		C	L	T	3
0100	PF	RES	BYP	PN					d	m	u		D	M	U	4
0101	HT	NL	LF	RS					e	n	v		E	N	V	5
0110	LC	BS	EOB	UC					f	o	w		F	O	W	6
0111	DEL	IL	PRE	EOT					g	p	x		G	P	X	7
1000		CAN							h	q	y		H	Q	Y	8
1001		EM							i	r	z		I	R	Z	9
1010	SMM	CC	SM		¢	!		:								
1011	VT				.	$	,	#								
1100	FF	IFS		DC4	<	*	%	@								
1101	CR	IGS	ENQ	NAK	(	)	_	'								
1110	SO	IRS	ACK		+	;	>	=								
1111	SI	IUS	BEL	SUB	|	¬	?	"								

Figure 12.14 The EBCDIC code set.

Amplitude modulation (AM)

Amplitude modulation represents the bits of information (the ones and zeros) by changing a continuous carrier tone. Figure 12.15 illustrates amplitude modulation. Because there are only two stages of the data, one or zero, you can let the continuous carrier tone represent the zero and the modulated tone represent the one. This type changes the amplitude (think of amplitude as the height or loudness of the signal). Each change represents a one or a zero. Because this is a 3-kHz, analog dial-up telephone line, the maximum amount of changes that can be represented and still be discrete enough to be recognizable to the line and the equipment is about 2400 times per second. This 2400-cycle changes per second is called the *baud rate*. Therefore, the maximum amount of data bits that can be transmitted across the telephone line with AM modulation is 2400 bits/s. Most amplitude modulation modems were designed to transmit 300–1200 bits/s, although others have been made to go faster.

As we chose to look at the different ways to change the ones and zeros, generated by the computer, into its analog equivalent, we had another choice in the process. Voice communications (or human speech itself) is the continuous variation of amplitude and frequencies, so we could choose to use a modem that modulates the frequency instead of the amplitude. This uses a modem as follows:

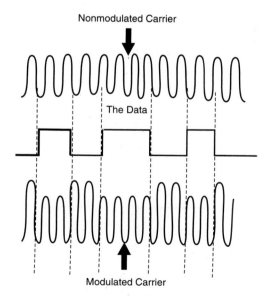

Nonmodulated Carrier

The Data

Modulated Carrier

Figure 12.15 The concept of amplitude modulation.

Frequency modulation (FM)

Frequency modulation is provided by an FM modem. This modem repre-sents the ones and zeros as changes in the frequency of a continuous car-rier tone. Because there are only two states to deal with, we can represent the normal frequency as being a zero, and slow down the continuous carrier frequency when we want to represent a one to the telephone line (Figure 12.16). The modem uses the same baud rate on the telephone line as the amplitude modulation technique, that being 2400 baud, or discrete changes per second. These modems modulate one bit of information per cycle change per-second, or a maximum of 2400 bits/s. You will note that the baud rate and the bits/s rate are somewhat symmetrical. Although both AM and FM modems are designed around what was once considered pretty fast transmission rates, we have continually been unsatisfied with any rate of speed developed. We want more and more, faster and faster.

Because we (as humans) and our offspring (the computers) are never satisfied with the speed of transmission over the telephone line, we de-manded faster. Throughput was expensive under the old dial-up telephone network. Therefore, we asked for additional speed to get more throughput and less cost. The engineers came up with a new process which modulates on the basis of phases.

Phase modulation

If we can change the phase of the sine wave as it is introduced to the line, at positions of 0, 90, 180, and 270 degrees, we can encode the data with more than one bit of information at a time. A phase modulation technique allows us to transmit a di-bit of information per signaling-state change. This gives us 4800-bits/s throughput.

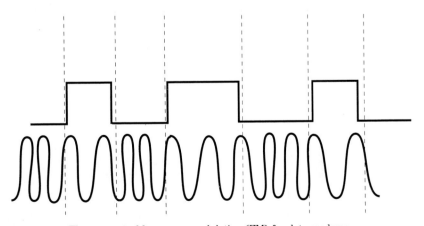

Figure 12.16 The concept of frequency modulation (FM) for data modems.

Bits	Phase
00	0
01	90
10	180
11	270

TABLE 12.8 The Four Phases in Phase Modulation Cause a 90° Shift When a Di-Bit Is Received

The di-bit represents the information as shown in Table 12.8.

As you can see, two bits of information at 2400 baud gives us the 4800 bits/s. This was a step in the right direction. But, we wanted more; so a combination of phase modulation and amplitude modulation was developed.

QAM

Called *quadrature with amplitude modulation (QAM)*, this combination allows us to use up to 16 possible steps of phase and amplitude modulation. QAM is mostly used with 4 bits of information per baud rate thereby producing a 9600-bits/s throughput across an analog telephone line. Theoretically, the system should be able to produce 38,400 bits/s of information (4 phases at 4 bits per phase at 2400 baud = 38,400 bits/s).

However, in practice, line rates can rarely support sustained throughput across the telephone network at this speed. Also, the telco limits the bandwidth to 3 kHz, and we use 2400 baud. If we try to send more data at a higher baud rate, the band limitation (bandpass filters) will strip the frequencies that go beyond the filters. This is a function of the telco equipment on the line at the CO. Thus, we usually have to settle for analog data transmission at slower speeds.

The driving force behind improving modulation techniques has always been to increase the speed possible over an analog circuit. Why? In addition to the obvious reason (i.e., accomplishing the task more quickly improves productivity), the cost of communications over dial circuits is directly proportional to the amount of time those circuits are in use; going faster saves money. But, when the absolute best available modulation technology is in use, one has not yet necessarily squeezed the absolute best transmission volumes out of a circuit. One can push it even further by using compression.

Devices

DTE vs. DCE

A basic of data communications is that every communicating device is either a terminal-type device (*data terminal equipment*, or *DTE*; also sometimes referred to as *data circuit terminating equipment*, or *DCTE*) or a communications-type device (*data communications equipment*, or *DCE*). DTEs use communications facilities; their primary functions lie elsewhere. DCEs provide access or even implement communications facilities. Their only role is moving information.

You might presume that some cosmic requirement is satisfied by this overall categorization. But in fact, the nitty-gritty reality of cabled communications is the primary cause of this division. Although we will get into cabling standards later, consider the following situation:

Two devices, A and B, must be connected. The connection will be via two wires, numbered one and two. Let us assign wire number one to transmit the data, and number two to receive. But wait! If A transmits on number one, and B also transmits on number one, and if they both receive on number two, neither will listen to what the other is sending. It would be as though two people each spoke into opposite ends of the same tube at the same time, with neither putting their end to his or her ear.

The solution seems simple: have A transmit on wire one, B listen on wire one, A listen on wire two, B transmit on wire two. But now let us introduce device C. How should C be built? If it is configured as is A, it cannot communicate with A; if it is configured as is B, it cannot communicate with B.

But devices are not configured at random. For example, terminals do not usually connect directly to terminals; printers are never connected directly to printers; and so on. Perhaps if one broad category of devices usually connects not to another in that category but rather to a device in another category, we could standardize on only two default configurations. This is what was done with DTEs and DCEs.

Categories and examples of DTEs

The grouping is not perfect. Some devices routinely connect to both DTEs and DCEs (e.g., multiplexers). Such devices do not clearly fall into either category, and must be configured depending on the specific installation requirements. But for the most part, any data-processing device which can communicate falls cleanly into one of these two categories. Most of the devices covered in this book are DCE devices. Here are some examples of the DTE devices to which we provide data communications services:

- Computers (mainframe, mini, midi)
- Terminals (CRTs, VDT, teletype)
- Printers (lasers, line, dot matrix)
- Specialized (bar code readers, optical character recognition)
- Transactional (point of sale equipment, automated tellers)
- Intelligent (personal computers)

The remainder of this section will focus on DCEs and ways of connecting components.

Modems

As mentioned earlier, *Modem* is a contraction of two words (modulator and demodulator). The role of the modem is to change information arriving in digital form into an analog format suitable for transmission over the normal telephone network. Naturally, modems work in pairs (or sets of at least two), and a second modem at the other end of the communications path must return the analog signal to a digital format useful to terminals and computers.

Dozens of manufacturers make modems, and hundreds of modem models exist. The market can be divided in many ways:

- Speed
- Supported standards
- Leased line vs. dial-up
- Two-wire vs. four-wire
- Point-to-point vs. multipoint (multidrop)
- With or without compression capability
- With or without error correction capability
- Manageable or nonmanageable

Entire texts could be (and no doubt have been) written just on this topic.

Chapter

13

T1 and the T-Carrier System

Evolution of the T-Carrier System

In the early 1960s, the Bell System began to introduce and use a new digital technology in the network. This was necessary because the older carrier and cabling systems were rapidly becoming strained for capacity. The demand for newer and higher-speed communications facilities was building among their customers, as well as within their own systems.

When this digital technology was being introduced, it was deployed in the public network as a means of increasing the traffic capacity within the telephone company on the existing wire pair cable facilities as interoffice trunks. The older systems, including the N-carrier system, used a two- or four-wire connection through an analog multiplexing device to deliver either 12 or 24 analog channels. This was still an inefficient use of the line capacity, and the analog service was both noisy and required expensive line treatment equipment. Therefore, the telephone company introduced its newer technology to overcome the limitations of the existing plant and transmission services.

Some of the problems with the analog systems were also related to the circuit quality. Anyone who can remember the older analog network as it used to work knows that a call placed across the country from East to West Coast was significantly different. The static and noise on the line made the call sound more like it went to the moon and back. Because that's all that was available, that's what the user became accustomed to.

The use of older analog systems was coming to an end within the telephone company networks. The telcos had to find a way of improving the utilization of the cable plant on an interoffice basis to overcome problems with under-utilized pairs of wires while the continued installation of inefficient systems was expensive and bulky. The average length of these wires between their offices was approximately 6.5 miles. As calling requirements continued to grow, the telcos needed to increase the traffic handling capacity on these interoffice routes. Yet, they were in a quandary. First, they didn't want to continue running bulky cables between offices because there simply wasn't enough space. Second, costs of maintaining the cable plant were escalating. Something had to be done quickly. Figure 13.1 shows the use of wires between the telco offices. These wires provide the inter-office communications from telco to telco. The end user or customer was kept on the old analog twisted pairs of wire. From a user's perspective, the changes were invisible, except that some call improvement was evident.

Analog Transmission Basics

Before going much further, a brief description of the dial-up telephone network would be prudent to understand the need for the digital architecture that was introduced through the use of the T-carrier system. The telephone system was designed around providing analog dial-up voice telephony. Everything was based on voice communications services on a switched (nondedicated) basis. A user could use the telephone to connect from their

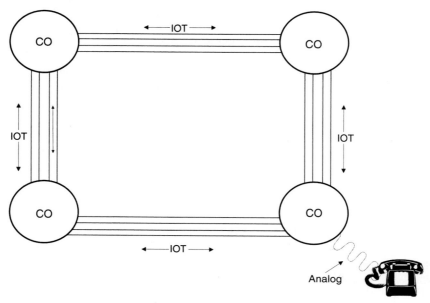

Figure 13.1 Bell used high-capacity services between their offices. Users were still kept on twisted pair analog.

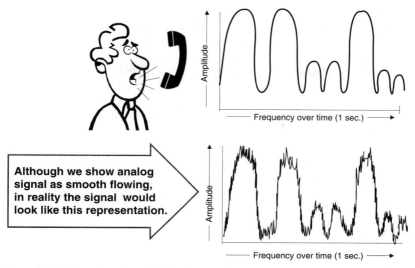

Figure 13.2 A representation of the analog wave being created.

set to another user on the network either through an operator connected or dial-up (a later evolution) addressing scheme.

Because voice was the primary service provided, the telephone set (Figure 13.2) has evolved into a device that took the sound wave from the human vocal cords and converted that sound into an electrical current, which was represented by its analog equivalent. The human voice produces constantly changing variables of both amplitude (the height of the wave) and frequency (the number of cycle changes per second). As these constantly changing variables of amplitude and frequency are produced, the telephone set converts the sound pressure into an equivalent electrical wave. This electrical energy will be carried down a pair of wires. As the electrical energy is introduced to the wires (using twisted pairs of copper for this reference) certain characteristics begin to work on it. First, resistance occurs on the wire impeding the flow of electricity, reducing the strength of the created signal. Second, the wires act as an antenna, drawing in noise (such as static, cross talk, or electricity from other conversations going on other pairs adjacent to yours, etc.). This loss of signal strength coupled with the introduced noise continues to distort the signal.

To overcome these problems, analog amplifiers are used to boost the signal strength back up to its original value. The amplifiers are normally placed on circuits greater than 18–20,000 feet in length. Figure 13.3 represents how the electrical signal on the wires is attenuated or weakened. As the signal is boosted back to its original strength, the noise begins to accumulate causing further distortion of the call. Figure 13.4 represents the analog am-

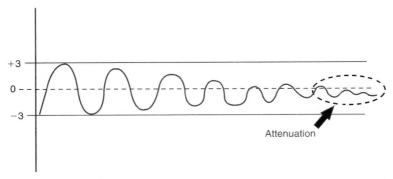

Figure 13.3 As the electrical signal is introduced, attenuation begins immediately. If the signal must travel great distances, it could run out of strength and disappear.

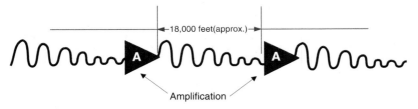

Figure 13.4 As the signal gets weak, the telco uses analog amplifiers to boost the signal and keep it going.

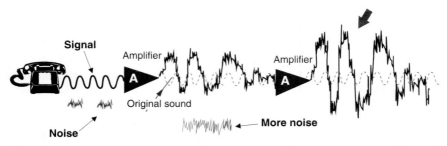

Figure 13.5 As the signal and noise are amplified, the noise takes on a cumulative effect until the transmission becomes virtually useless.

plifier on the line to boost up the signal. To overcome the loss on the line and to ensure the signal gets to the other end, the telco used analog amplifiers. These amplifiers were used on circuits to increase the strength of the analog transmission. However, they were prone to failures, and had other qualities in amplifying the signal which were undesirable. Further, Figure 13.5 is a representation of the problem encountered with the amplification process. As the signal is amplified, the signal, along with all the noise on the wires, are amplified together. This allows the signal to travel further, but after each amplification process, accumulation of the noise also occurs. A good deal of the problem stems from the fact that the amplifier can't distinguish the analog signal from the noise. Therefore, it amplifies everything.

The Evolution to Digital

Thus, the T-carrier system was born. The telcos were looking to enhance the quality of calls and better utilize the cable facilities. The T-carrier system allowed them to increase the call carrying capacity while taking advantage of the unused transmission capacity of their existing wire pair facilities and improve the quality of transmission by migrating away from the older analog techniques. The evolution to T-carrier was important for a number of reasons:

- It was the first successful system designed to utilize digitized voice transmission.
- It identified many of the standards that are employed today for digital switching and digital transmission, including a modulation technique.
- The transmission rate was established at 1,544,000 bits per second.
- The T-carrier technology defined many of the rules, or protocols, and constraints in use today for other types of communications.

As mentioned earlier, the analog transmission capabilities were inefficient. In many cases, the telephone companies used a single pair of twisted wires to carry a phone call. This meant that as uses of the inter-office facilities grew, the cable plant grew exponentially. As more central offices (or end offices) were added, the need for meshing these together grew. Interconnecting each office with enough pairs of wire to service user demands was becoming a nightmare. Figure 13.6 depicts the need for cabling or inter-office trunking needs to support the growing demand for services. As each new end office was added, the wiring systems had to grow to serve customer demands.

Remember that the telephone companies saw this carrier system as a telco service only. As the end-user population grew, the telcos still held back on deploying this digital capability to the user. The telcos used the higher efficiencies to support the end user digitally, from end-office to end-office or through the network. The user was still relegated to a single conversation on a twisted pair using analog transmission. Thus, this system operated analog on the local loop, but digitally on the inter-office trunks (IOTs). In metropolitan areas, the telcos were reaping the benefits of the T-carrier system. Figure 13.7 is a representation of the telephone company deployment of analog/digital transmission capacities.

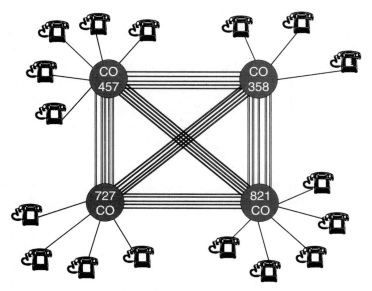

Figure 13.6 As more central offices were added to the network, the need for wires grew exponentially.

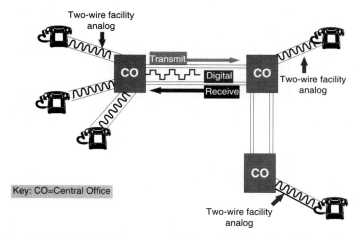

Figure 13.7 The telephone companies reaped the benefits of interoffice trunking, using digital transmission.

Some immediate benefits were achieved: the quality of transmission improved dramatically; and the utilization of their existing wire facilities increased. A single four-wire facility on twisted wires could now carry 24 simultaneous conversations digitally at an aggregated rate of 1.544 Mbps.

As digital switching systems were introduced into the dial network, they were designed around the same techniques employed in the T-carrier system. The digital switching matrices of the #4 and #5 electronic switching systems (ESS) developed by Western Electric and the DMS systems (10, 100, 250) as developed by Northern Telecom Inc. utilize pulse code modulation (PCM) internally, so that digital carrier systems and channels could be interfaced directly into the digital switching systems. By adhering to the standards set forth, the operating telephone companies and the long haul carriers could build integrated digital switching and digital transmission systems that:

- Eliminated the need to terminate the digital channels or equipment to provide analog interfaces to digital switching architectures.

- Avoided the addition of quantizing noise that would be introduced by another digital-to-analog conversion process. Whenever a conversion of the signal from analog to digital, or digital to analog is necessary, the risk of errors and noise increases.

These techniques were used to provide lower cost, better quality dial-up telephone services. However, this same technology underlies the idea of full end-to-end digital networking services that is the basis of integrated services digital networks (ISDN), the future and the higher-end broadband services that are emerging.

Analog-to-Digital Conversion

Prior to covering the T-carrier fundamentals, it would be appropriate to discuss the analog-to-digital conversion process. Digital architecture dictates the use of a digital bit stream; the analog wave must be converted into a useable format; and digital pulses are represented by a "0" or a "1." As the analog signal is being transmitted to the network, it must be converted from a wave of varying amplitudes and frequencies to a digital format, represented in the form of 1s and 0s (or presence and absence of a voltage).

The analog wave must be sampled often enough, and converted to create a stream of 1s and 0s that is precise enough to be recreated at the distant end. That result will produce what sounds like the original conversation. According to the Nyquist rule, a digital signal should be created by sampling the analog wave at twice the highest frequency on the line. For an analog circuit, delivered by the local telephone company, the frequency range on a twisted pair of wires is represented in hertz (Hz). To maximize the utilization of the older carrier systems, the telco delivered a usable bandwidth of 3000 Hz (3 kHz), as compared to 4000 Hz (4 kHz) capable to be carried on the line. The difference between the 3-kHz and 4-kHz services is in how the line is filtered. The human voice produces understandable information in the range of 300 to 3300 frequency changes per second. Therefore, the telco knew that all they had to provide to the user was 3 kHz of usable bandwidth. But, for separation between conversations and to minimize crosstalk, etc., they allocated 4 kHz per analog line. Using NYQUIST's rule, the sampling of an analog wave at twice the maximum frequency on the line meant that a minimum of 6600 samples per second should be sufficient to recreate the wave for the digital-to-analog and analog-to-digital conversion. However, to produce the quality of the higher range of frequencies of the human voice the number of samples is rounded up to 8000 per second.

The sample now had to be created into a bit stream of 1s and 0s, as already stated. To represent the true quality of tone, inflection, etc., of the human voice, enough bits need to be used to create a digital word. Using an 8-bit word creates enough different points on a wave to do just that. The wave is divided into 256 possible combinations of the amplitude at the moment of the sample itself. Figure 13.8 shows how the sampling would be accomplished, whereby using 2 states and 8 bits creates 28 or 256 points on the analog wave. As each sample is taken, the amplitude of the wave is what is being sampled.

Once the sample has been taken and the digital equivalent is created into an 8-bit word, the digital (or square) wave can be transmitted. The 1 represents the presence of a voltage and the 0 represents the absence of a voltage (Figure 13.9).

This conversion uses *pulse code modulation* (PCM), a technique to create the sample. Using PCM and the rules established, the transmission of

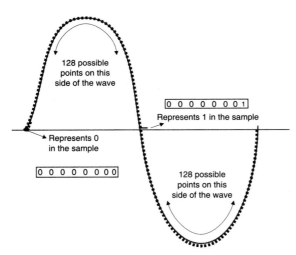

128 possible points on this side of the wave

0 0 0 0 0 0 0 1
Represents 1 in the sample

Represents 0 in the sample

0 0 0 0 0 0 0 0

128 possible points on this side of the wave

Figure 13.8 With 256 possible points being sampled, the quality of the digital transmission when converted back to analog should be exact.

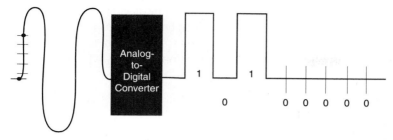

Figure 13.9 A sampling of the analog wave yields a transmission of 8 bits in a stream. Two pulses (voltages) are required in this sample in the 6th and 8th positions.

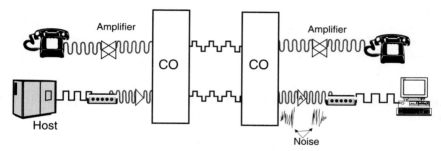

Figure 13.10 The customer still had to contend with analog transmission characteristics on the last mile.

the digital equivalent of the analog wave results in 8000 samples/SEC × 8 bits per sample = 64,000 bits per second for the basis for a voice transmission in PCM mode. Whether the end user transmitted voice, data, or facsimile, the copper wires were used to carry the analog wave to the central office, where the wave was converted to digital for transport across the telco network. At the far end, the receiving telco converted the digital signal back to its original analog equivalent for delivery to the customer. All this served to gain better transmission quality for the telcos, but left the customer with the last mile to contend with. The customer still received an analog transmission to their premises (Figure 13.10).

As the telcos were reaping the benefits of this transmission capability, the users began to request this service. More user demands created some upheaval within the Bell system. This technology was perceived to be a telco-only service. Yet, users wanted (demanded) end-to-end digital transmission to improve their throughput and quality. Because analog transmission was noisy, users were limited in what they could move across the network. Specifically, the constant growth in computer technology and the deployment of terminals and printers around the country, left the end user with limitations and restrictions that were becoming intolerable. Analog leased line or dial-up services left the user with data communications capabilities that were limited to 9.6 to 19.2 Kbps (leased) transmission, as described in Chapter 12. Not only were these services slower than what the user required, they were more expensive.

The Movement to End Users

Ultimately, the telcos had to install digital transmission to the customer premises. Originally these were in the form of lower-speed digital data services (DDS) at speeds of 2.4, 4.8, 9.6, and 56 Kbps digital. Newer introductions added 19.2–64 Kbps and high-capacity (HICAP) T1 services at 1.544 Mbps.

In order to deliver this digital capability to the customer, a special assembly, or individual case basis (ICB) tariff arrangement was initially used. This emergence didn't occur until the mid-1970s. At first, this was done with some reluctance, but as the movement caught on, the deployment was both quick and dynamic. The telcos had to modify the outside plant to accommodate the end user needs. First, to do this they had to engineer the circuit. Next, they had to remove all of the analog transmission equipment from the line. Next, they had to check the circuit (now a four-wire circuit) for splices, taps, bridges etc., which would all introduce loss and noise thereby impairing the transmission. These had to be removed or cleaned up. The next step was to provide digital equipment on the line called *regenerators* or *repeaters*.

With analog transmission, amplifiers are used on circuits over 18,000 feet. However, with digital transmission, repeaters are used at whatever intervals necessary to ensure quality (typically 2000 to 6000 feet). In digital transmission, the presence or absence of a voltage, the signal can deteriorate very quickly. Therefore, the telco needed four times the amount of repeaters compared to the old analog amplifiers. These repeaters, or regenerators ensure the proper signal is moved across the line. The signal is repeated to prevent it from falling below a threshold, where it can't be distinguished between signal and noise. Figure 13.11 is a representation of how the signal actually gets recreated rather than amplified and produces the quality we have come to expect. As the signal moves across the wires, the same problems are inherent; resistance on the line depletes the signal strength, and noise is prevalent. The digital regenerator is placed close enough to get around this problem. Therefore, the end user could move to an effective use of a four-wire circuit, at 64 Kbps. The first deployment of this service was for voice, but data quickly surpassed the demand for new digital transmission. It should also be noted that as T1 became more readily available, the primary use (75%) was to consolidate WATS lines onto a single digital trunk. This digital transmission capability was designed around the voice dial-up network, therefore, the service was a natural fit.

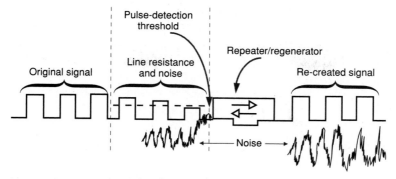

Figure 13.11 A representation of how the signal is recreated.

T1 Basics

To characterize a T1 by its operating characteristics, the T1 is:

A four-wire circuit

Because this technology evolved from the old twisted-pair environment, four wires were used. Two wires are used for transmit and two are used for receive. Other facilities can be used, but for now think of it this way.

Full duplex

Transmission and reception can take place simultaneously. Many customers derive other uses, such as one way only for remote printing, file transfer etc.; or as a two-way for alternate service, such as voice communications.

Digital

This is an all-digital service. Data, analog facsimile, analog voice, etc., are all converted to digital pulses (1s and 0s) for transmission on the line.

Time-division multiplexing

The digital stream is capable of carrying a standard 64-Kbps channel, which 24 channels are multiplexed to create and aggregate of 1.536 Mbps. Time division allows a channel to use a slot ½₄ of the time. These can be fixed time slots made available to the channel.

Pulse code modulation

Using the example, the analog voice, or other service is sampled 8000 times per second; an 8-bit word is used to represent each sample, thus yielding the 64-Kbps channel capacity.

Framed format

As the pulse code modulation scheme is used, the 24 channels are time-division multiplexed into a frame to be carried along the line. Each frame represents one sample of 8 bits from each of the 24 channels. Added to this is a framing bit. The net result is a frame of 193 bits. There are 8000 frames per second, therefore a frame is 125 microseconds long. Framing accounts for 8 Kbps overhead (1 bit × 8000 frames). Adding this 8 Kbps to the 1.536 Mbps covered in 4, yields an aggregate of 1.544 Mbps.

Bi-polar format

T1 uses electrical voltage across the line to represent the pulses (1s). The bi-polar format serves two purposes. It reduces the required bandwidth from 1.5 MHz to 770 kHz, which increases repeater spacing. And the signal voltage averages out to zero, allowing dc power to be simplexed on the line to power intermediate regenerators. Think of it as an alternating current (ac) version of a direct current (dc) line. Every other pulse will be represented by the negative equivalent of the pulse. For example: The first pulse will be represented by a positive 3 volts (+3 V), the next pulse will be represented by a minus 3 volts (–3 V) and so on. This effectively yields a 0 voltage on the line, because the +– +– equalizes the current. This bi-polar format is also called alternate mark inversion (AMI). The "mark" is a digital 1. Alternate ones are inverted in polarity (+,–).

Byte-synchronous transmission

Each sample is made up of 8 bits from each channel. Timing for the channels is derived from the pulses that appear within the samples. This timing keeps everything in sequence. If the devices on both ends of the line do not see any pulses, they lose track of where they were (temporary amnesia).

Channelized or nonchannelized

Generically, T1 is 24 channels of 64 Kbps each, plus 8 Kbps of overhead. This is considered *channelized service*. However, the multiplexing equipment can be configured in a number of ways. For example, the T1 can be used as a single channel of 1.536 Mbps; two high-speed data channels at 384 Kbps each, and a video channel at 768 Kbps. These examples can be mixed into a variety of offerings. The point is that the service does not have to be channelized into 24, but can be nonchannelized into any usable data stream needed (equipment allowing of course).

Any other suitable medium (fiber, digital microwave, coax, etc.) can also be used. The T1 is still treated as a four-wire circuit. When the other media forms are used, the T1 will be suitably taken from the transmission mode and converted back to the appropriate interface. For a four-wire circuit, the carrier will normally terminate the four wires into a demarcation point (Demarc) or network interface unit (NIU). Individual circuits can be terminated into a recommended jack—RJ48 or RJ68, now called *smart jacks*. These jacks serve as the interface to the four wires. In a larger environment or where multiple circuits are involved, other methods of termination can be used.

For example, an RJ2IX or a BIX block can be used for terminating a T1 on a main distribution frame in a customer location. From the RJ48, a 15-cable (using a sub-miniature 15-pin DIN connector) will be extended to the customer premises equipment (CPE). Usually, the CPE is a channel service unit (CSU). Figure 13.12 is a representation of the connection from the central office to CPE.

The customer premises equipment uses time-division multiplexing to carries the multiple voice and data conversation across the line. Time-division multiplexing is somewhat efficient, in that it allows time slots to be dedicated to each of the conversations to be carried on the line. Using this fixed time slot, a conversation will be in the same bucket (or slot) at each sample time.

Remember that the equipment using the NYQUIST rule operates at twice the highest frequency of the line. Therefore, each individual conversation is sampled at 8000 times per second. There are 24 (typical) paths being simultaneously multiplexed on to the T1. Each of these paths, therefore, carries one sample ($\frac{1}{8000}$ of a sec.) of voice, data, fax, video, etc.

The time slot is always present, and if no traffic is being generated between points across the line, the time slot goes unused. The slot is empty. For this reason, time-division multiplexing is inefficient. Attempts to get

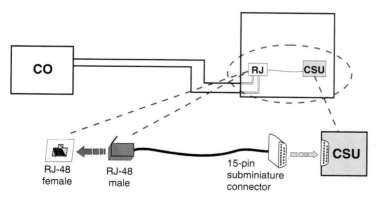

Figure 13.12 A representation of a 4-wire circuit terminating into an RJ48. This is probably the most common method of delivery. The RJ48 is an 8-conductor modular jack.

the most amount of service from a T1 would be thwarted because of the fixed time slot problem. Figure 13.13 is a representation of this time-division multiplexing scheme (TDM or Mux).

Framed Format

T1 uses some very specific conventions to transmit information between both ends. One of these is a framing sequence that formats the samples of voice or data transmission. It's easy to think of a frame in terms of the 8-bit samples of the 24 channels being strung together in a logical sequence. After the 192 bits of information are compiled, a framing bit is added, creating a frame of 193 bits of information.

The framing bit can be equated to a pointer or address. Because the line is moving bits of information at 1,544,000 per second, it would be very easy to skew left or right and get information out of sequence. Therefore the extra 8000 bits of information (1 bit per frame, 8000 frames per second) create a locator for the equipment to lock in on. This pointer allows the devices to read a pattern of bits to know which frame is being received/transmitted and the location of each channel thereafter.

Framing has undergone several evolutions over the years. The first use of T1 service was for voice so the information was easier to use. The evolution of framing followed a sequence as outlined on Table 13.1.

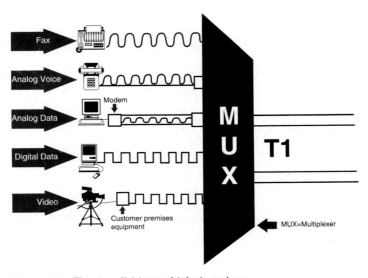

Figure 13.13 The time-division multiplexing scheme.

TABLE 13.1 Framing Evolution

Framing	Use	Pattern/use
D1	Voice or analog data	Alternating 1s & 0
D2	Voice or analog data	12 frame sequence Superframe
D1D	Voice or analog data	Upgrade compatibility for D1
D3	Voice or analog data	Superframe format and sequence bits
D4	Voice & data (digital)	Superframe
D5[1]	Voice & data (digital)	Superframe
ESF	Voice & data plus maintenance	Extended Superframe (ESF)

[1] D5 really doesn't exist as a standard. However, in the electronic switching systems this framing was termed as a software version of the D4 format.

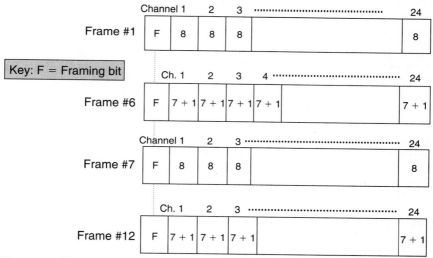

Figure 13.14 The robbed bit signaling framed format.

The D4 superframe uses a framing pattern for voice and data. The framing pattern in the D4 superframe is a repeating 12-bit sequence (1000 1101 1100) that allows signaling bits to be "robbed" in the 6th and 12th frames of the superframe. The 8th bit from each sample in frames 6 and 12 is used to provide signaling. This pattern of the framing format is shown in Figure 13.14.

Table 13.2 shows the actual framing of a data stream in the D4 framing format. The framing bit pattern repeats every 12 frames. 12 frames make up a super frame.

Once again, this requires that each of the 24 channels is robbed of the least significant bit (LCB) in the 6th and 12th frames. This has an impact on the data capacities that always limits the user to 56 Kbps of effective throughput (resulting in a net overhead of 192 Kbps given up to provide signaling).

Bi-Polar

When encoding the voice and data streams onto a digital circuit (the T1), as mentioned earlier, up to 1,544,000 bits of information can be transmitted. The 1s are the presence of a voltage, and are represented by 3 volts per pulse.

As a result, the bi-polar concept was introduced. The first pulse (a "1") is represented by 3 volts (positive), followed by the next pulse being represented by a negative 3 volts. This would serve to bring the overall line volt-

TABLE 13.2 Superframe Format. The 6th and 12th Frames Are Used to Denote the Presence of Signaling Bits

Frame #	Bit #	Framing bits Term frame F_1	Framing bits Sig frame F_1	Bit use in each time slot Traffic (all channels)	Bit use in each time slot Sig	Signaling bit use options T	Signaling bit use options Signaling channel
1	0	1	-	1–8	-		
2	193	-	0	1–8	-		
3	386	0	-	1–8	-		
4	579	-	0	1–8	-		
5	772	1	-	1–8	-		
6	965	-	1	1–7	8	-	A
7	1158	0	-	1–8	-		
8	1351	-	1	1–8	-		
9	1544	1	-	1–8	-		
10	1737	-	1	1–8	-		
11	1930	0	-	1–8	-		
12	2123	-	0	1–7	8	-	B

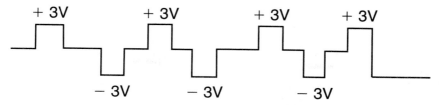

+ 3V   + 3V   + 3V   + 3V

        − 3V   − 3V   − 3V

Figure 13.15 The alternating voltages shift from positive to negative, etc., yielding an average of "0" volts on the line.

age down to 0. Figure 13.15 is a representation of the voltage/pulses on the line. Bi-polar does the following things:

- Reduces the bandwidth necessary for transmission from 1.5 to 0.772 MHz.

- Allows the use of isolation transformers for inexpensive connections.

- Allows simplexed dc voltage to power intermediate repeaters. This can be as much as + and −130 volts.

Byte Synchronous

When transmitting on the T1, timing and synchronization come from the bits transmitted. Because each sample is made up of 8 bits, a byte is formed. The transmitters and receivers synchronize based on the pulses in the byte format. If a string of zeros (0) is transmitted in several frames, then synchronization will be lost. The devices get amnesia as to where they were. Consequently, conventions on the transmission of the byte were established. These conventions are called the "1s density rule." What this means is that in order to maintain synchronization, a certain amount of 1s must be present.

For voice communication, this doesn't pose a problem because voice generates a continuous change in amplitude voltages which when encoded into an 8-bit word (byte), the presence of voltage (1s) is covered. However, in data transmission strings of 0s are possible (even probable) when refreshing or painting screens of information. Hence the *1s density rule* comes into play.

Simply stated the "1s density rule" states that: in every 24 bits of information to be transmitted, there must be at least three pulses (1s); and no more than 15 zeros (0s) can be transmitted consecutively. A more stringent requirement set by AT&T in the implementation of their digital transmission was that in every 8 bits of information, at least one pulse (1) must be present.

Think of how this affects data transmission. If your data must have at least one pulse per byte, you have to change the way you deliver data to the line. A technique known as *pulse stuffing* is used to meet these conventions. The 8th bit in every byte was stuffed with a 1. This limited the data transmission rate to 56 Kbps, because 7 usable bits plus a stuff bit had to be transmitted. Figure 13.16 is a representation of the ones density rule and pulse stuffing.

To overcome this bit stuffing, yet meet the one's density requirement, a technique was developed. This is known as *bi-polar 8 zero substitution* (B8ZS—pronounced as BATES). At the customer location B8ZS is implemented in the channel service unit (CSU). As data bits are delivered to the CSU for transmission across the line, the CSU (a microprocessor controlled device) reads the 8-bit format. Immediately recognizing that a string of 8/16/24 zeros will cause problems, it strips off the 8-bit byte, and substitutes a "fictitious byte."

- 8 zeros will be stripped off the data stream and discarded.
- The CSU will then insert a substitute word (byte) of 00011011.

However, to let the receiving end know this is a substitute word, and not real data, 2 violations to the bi-polar or AMI convention are created. Remember that the bi-polar convention states alternating voltages will be used for the pulses.

In the 4th and 7th positions, violations will be created that act as flags to the receiver that something is wrong. See Figure 13.17 for a graphic representation to this substitution.

Channelized vs. Nonchannelized

Up to now, discussions have been on using the T1 for 24-channel capacity at 64 Kbps per channel. For the average user, this was the norm. However,

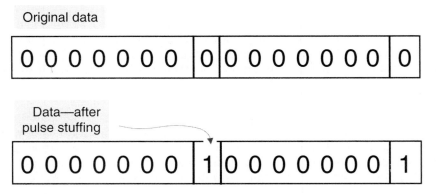

Figure 13.16 When strings of zeros must be transmitted, a 1 is stuffed into the 8th position (LSB). This yields a 56 Kbps throughput on the line.

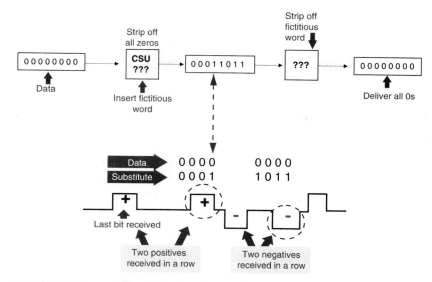

Figure 13.17 The receiving CSU noting that two bi-polar violations occurred in the 4th and 7th positions recognizes this as a "fictitious word." It strips off the substitute word and replaces it with all zeros for delivery to the computer, terminal, printer, whatever. Thus, the ability to transmit all usable data at 64 Kbps clear channel becomes a reality.

as newer uses for the T1 service became evident, it is possible to configure the capacity to meet the need:

- A single channel of 1.536 Mbps is possible for point to point video conferencing.
- High-speed data on a channel of 512 Kbps, plus 16 channels of lower-speed data and voice at 64 Kbps, are possible needs.
- Any other mix of services might be required.

To accommodate the mix of needs, the end-user (CPE) equipment can be super-rate or sub-rate multiplexed. This is fine, except that you must ensure that the provider (LEC, AAC, IEC) knows that you are using the service for something other than the conventional 24 channels at 64 Kbps.

If confusion exists, the carrier or provider could do some reconfiguring in the network (at their end) that could literally disrupt service for the throughput being used. Consider if a 512-Kbps data channel was somehow rerouted through a telco office, and was then brought back and forced into a single 64-Kbps channel. This would obviously not work. Hence, all parties should know what uses are being made of the transmission capacity.

Digital Capacities

One of the by-products of the T-carrier system was that the transmission rate employed became the standard building block for a multiplexing hierarchy. In high-speed digital transmission systems, a carrier combines many lower speed signals into an aggregate signal for transport. To simplify this process and to hold the line on the costs of equipment, standard rates were defined.

Each of the transmission rates were assigned a number that identifies the rate and configuration for the signal. Each time a higher speed is created, the lower speeds are combined and extra bits (bit stuffing) are added to come up to the signaling rate. These extra bits allow the equipment (multiplexers) to compensate for distances, clocking, etc., which cause transmission delay cycles.

The designations for the digital transmission are called *digital signal X*. Many times people will transpose or intersperse these designators. For example, referring to a T1 as a DS1 is often heard in the industry. To clarify this point, T1 is the first level of the T-carrier system. DS1 is the multiplexed digital signal, first level, carried inside the T-carrier.

The digital hierarchy for North America includes the following:

DS0

Although originally not a formally defined rate, the DS0 is a 64-Kbps signal that makes up the basis for the DS1. 24 DS0s combined produce the DS1. The standard 64-Kbps pulse-coded modulation signal (PCM), is the basis for all of the future networks employing digital signaling. Included in this is the integrated service digital network (ISDN).

DS1

The digital signal level-1 is a time-division multiplexed (TDM), pulse-code modulation aggregate of 1.544 Mbps, regardless of the media used to carry the signal.

DS1C

This is a digital signal equivalent to 2 DS1s, with extra (stuff) bits to conform to a signaling standard of 3.152 Mbps.

DS2

The composite of 4 DS1s multiplexed together yielding an aggregate rate of 6.312 Mbps. The Bell system used a DS2 capability to deliver subscriber loop carrier (SLC-96) to customers. A total of 96 DS0s could be carried across the DS2.

DS3

The 44.736 Mbps aggregate-multiplexed signal is equivalent to 28 DS1s or 672 DS0s.

DS4 / NA

The 139.264 Mbps aggregate-multiplexed signal equivalent to 3DS3s or 2176 DS0s.

DS4

The 274.176 Mbps aggregate multiplexed signal equivalent to 6 DS3s or 4096 DS0s.

See Table 13.3 for a summary of these bandwidths. These rates are based on the ANSI T1.107 guidelines. From an international perspective, the CCITT has set a hierarchy of rates which differs from the North American standard. A comparison of these rates is shown in Table 13.4. As can be seen, differences between the North American standards and the international standards exist.

However, other rates are evolving for transmission capacities. The standard 64 Kbps is derived from using pulse-code modulation. In several vendor products, 64 Kbps for the transmission of voice and/or analog data is considered too much bandwidth to carry traditional voice or analog data. Therefore, lower speed capacities are derived at 32 Kbps.

Signaling

Signaling comes into play when dealing with voice and dial-up data services. In the traditional arena, signaling is provided on a dial-up telephone line,

TABLE 13.3 ANSI T1.107 Rates

Designator	Capacity (Kbps)	Equiv. DS1	Equiv. DS0	Stuff bits
DS0	64	—	1	
DS1	1,544	1	24	8,000
DS1C	3,152	2	48	64,000
DS2	6,312	4	96	136,000
DS3	44,736	28	672	1,504,000
DS34/NA	139,264	84	2096	5,056,000
DS4	274,176	168	4032	14,784,000

TABLE 13.4 International CCITT G.702

Designator	Capacity (Bps)	Equiv. DS1	Equiv. DS0
E1	2,048,000	1	32
E2	8,448,000	4	128
E3	34,368,000	16	512
E4	139,264,000	64	2048
DS0	64,000	-	1
DS1	1,544,000	1	24
DS2	6,312,000	4	96
J1*	32,064,000		
DS3	44,736,000	28	672
J1*	97,728,000		
DS4	139,264,000	84	2016

* J designators are for Japan's network/also these capacities will be used for the BISDN rate H21.

across the talk path. This is referred to as *in-band signaling*. You might recall that the need to find bits to send between transmitter and receiver was accomplished in the 6th and 12th frames. Bit robbing, or stealing the 8th bit in each of the channels (1–24) in these two frames allows enough bits to signal between the transmit and receive ends. These ends can be customer premises equipment (CPE) to central office for switched services, or CPE to CPE for PBX-PBX connections, etc.

The most common form of signaling on a T1 line is 4-wire E&M signaling types I, II, or III. It would be safe to say that the easiest implementation, acceptable to all vendors and carriers alike, is 4-wire E&M Type I.

Signaling is used to tell the receiver where the call or route is destined. The signal is sent through switches along the route to a distant end.

The common types of signals are:

- On hook
- Off hook
- Dial tone
- Dialed digits
- Ringing cycle
- Busy tone

4-wire E&M is used for tie lines between switches. Occasionally, other services are bundled onto a T1 circuit. These could include:

- Direct inward-dialing (DID)
- Direct outward-dialing (DOD)
- 2-way circuit
- Off premises extension (OPX)
- Foreign exchange service (FX)

With these other services, the type of signaling might differ. DID/DOD on a PBX might use ground-start trunks, which requires a ground to be placed on the individual circuit to alert the central office that service is requested, etc. Regardless of the type signaling used or the services used, signaling requires bits of information. To overcome the robbed bit signaling that limits data to 56 Kbps, another opportunity exists. *Common-channel signaling,* or CCS is a method to get the clear channel capability back.

Remember that there are constant demands on the use of T1 for clear channel capacity. The use of bit stuffing to conform to the "1s density rule" for timing was overcome by using B8ZS. Now, to overcome bit robbing for signaling, the use of CCS can apply.

By dedicating the 24th channel in the digital signal, 23 clear channels can pass 64,000 bits of information. The choice is how much to give up. Table 13.5 is a decision table to help decide this process.

This common-channel signaling technique is also called transparent mode for signaling. A single 64-Kbps channel (#24) is given up rather than 8 Kbps per channel [[×]] 24 (192 Kbps) for robbed bit. At some later date, as the carriers deploy newer digital dial-up services under the auspices of ISDN they will be using CCS7. In the T1 world, this means the primary rate interface (PRI) at 23B+D. Simply stated, 23 bearer (B) channels at 64 Kbps

TABLE 13.5 Decision Matrix for Deciding on Robbed Bit or Common Channel Signaling

If (switched)	Then	O/H given up (Kbps)
Voice only	Robbed bit	192
Voice and analog data	Robbed bit	192
Voice, analog data & 56 Kbps digital data	Robbed bit	192
Digital voice or digital data @64 Kbps	CCS	64
Digital data @64 Kbps or greater	CCS	64

clear channel, plus a data channel for signaling at 64 Kbps (D). The choices are not always obvious, but understanding the requirements helps to steer the decision better.

Clocking (Network Synchronization)

Any digital network synchronization between sender and receiver must be maintained. As the DS1, the digital signal level 1 of 1.544 Mbps, is delivered to the network, it is likely that it will be multiplexed with other digital streams from many other users. All of these signals will then be transported as a single signal over high-capacity digital links (DS3 and above). Further, if the line is directly interfaced to a digital switching device, then synchronization must be maintained between the customer's transmitter and the switch.

Synchronization is imperative in a digital transmission system. If the timing of arrival or transmission is off, then the information will be distorted. Regardless of whether voice, data, video, or image traffic is present, the presentation of a digital stream of 1s and 0s is contingent upon a timed arrival between the two ends.

There are a number of ways to synchronize a digital network, but the issue must definitely be addressed. Some of the ways to address the synchronization deal with levels of synchronization.

The levels are:

- Bit
- Time slots for time-division multiplexing
- Frame

Bit synchronization

As a digital stream of 1s and 0s is delivered to the line, the timing (or clocking) of the bit is important. The transmitter should be sending bits at the same rate the receiver can take them in. Any difference, faster or slower, could result in lost bits. Therefore, the bits must occur at a fixed time interval.

Time slot

Whenever several links are connected or routed, through a network processor, switching system, or end mode, the potential for lost bits or degradation of the link increases exponentially. Using the pulse-coded modulation technique, eight bits are encoded from each sample of information. These eight bits are then assembled and placed into a time slot. As signals and links are processed through a network, it is the eight-bit pattern that is routed from time slot to time slot. Should a slippage or mismatch occur, multiple streams of information will be lost.

Frame synchronization

After the data stream of 192 bits of information is assembled (8 bits × 24 channels) an extra overhead bit is added to let both transmitter and receiver know the boundaries of the frame (think of it as a start/stop bit sequence). When dealing with a T1, the bit sequencing is easier to derive. However, at multiplexing schemes above the T1 rate, additional bits are inserted in the data stream to maintain a constant clocking reference. The use of these overhead (or stuff bits) brings both transmitter and receiver up to a common signaling speed to maintain frame synchronization.

Potential synchronization problems

When a digital system is scheduled to receive a bit, it expects to do just that. However, clocking or timing differences between the transmitter and receiver can exist. Therefore, while the receiver is expecting a bit that the transmitter hasn't sent, a *slip* occurs. There will most likely be slips present because of multiple factors in any network. These can come from the two clocks at the ends being off or from problems that can occur along the link.

Along the link, problems can be accommodated. The use of pulse stuffing helps, but other methods also can help. Each device along the link has a buffer capability. This buffer creates a simple means of maintaining synchronization. Pulse stuffing is done independently for each multiplexer along the way, which enhances overall reliability of the network. However, pulse stuffing has its negatives too. The overhead at each multiplexer is basically a penalty. Further, at both ends the location and timing of each stuff bit must be determined, then signaled to the receiver to locate and remove the stuff bits. When de-stuffing occurs, a timing problem known as *jitter* can cause degradation of the signal. When passing through multiple switching sites, the signal must be de-stuffed from a received signal, then re-stuffed to a newly transmitted signal. This is expensive to do in both equipment and in overhead.

Obviously when slippage occurs or if a problem exists in the network buffers, the re-transmission of a frame or frames of information will be required. For voice, this isn't too bad, but for data transmission, this can result in errors that can render the data unusable. The result will be re-transmission requests (NAKS) that will affect the throughput of the link and potentially increase the burden on other systems to detect and correct the errors.

Performance Issues

Once a decision is made to use digital transmission capabilities, and to consolidate 24 or more voice, data, video, etc. circuits on a single T1, the performance of the T1 becomes an issue. The placing of "all your eggs in a single basket" is an issue to be dealt with.

The first users of this technology were risk takers. If 24 analog voice and data lines are being used, and one fails, 23 are still working. However, when 24 lines are digitally encoded on a single circuit, and one fails, they all fail. This is a concern to many users. Further, it should be noted that failures will occur. A T1 is based on the four-wire circuit in an error-prone network. Equipment failures, cable cuts, flooded cable vaults, etc., all contribute to circuit failures. If total circuit failures were the only problem, the risks would be minimized. But, cases can occur that cause distortion, high bit-error rates, timing slips, and loss of framing sequence (to name a few), which can impair or disrupt transmission. Combining these situations can leave the potential user's knees weak.

A further statistic is also one for concern, that being 90% of all errors/failures, will likely occur at the local loop (or last mile from central office to customer premises). This is not to imply that the local exchange carriers (LECs) are doing a poor job of provisioning these circuits. It is merely a statement of fact because the local loop is exposed to more of the external problems mentioned above.

Considering average performances, the use of T1 technology is based on 7 days per week by 24 hours per day. Table 13.6 is a summary of total hours and percentage comparisons at availability times. You can see the impact of a few percentage points, and your exposures with these statistics.

Although these numbers are dramatic, the actual performance is completely different based on the problem, customer, location, etc. Situations have been recorded where services have been out for 6 to 7 consecutive days because of variable circumstances. However, many of these outages are extreme, yet rare.

To ensure that availability of the circuit, and the call-carrying capacity of the channels, the carriers, equipment vendors, etc., have all adopted to a new set of standards. This new standard is known as *extended superframe format* or ESF. Before delving into ESF, however, an understanding of the older format is appropriate.

TABLE 13.6 Total Hours Available @24 × 30=720

Performance %	Hours down/month	Hours of yearly downtime
99.5	4	48
99.0	8	96
98.5	11	132
98.0	15	180
97.0	18	216
95.0	36	432
90.0	72	864

D3/D4

First the D3 and D4 framed formats were designed to format the channelized information of a T1. Because D3 was a voice only (analog data input) and fixed format for input, the use of T1 was designed around tie lines, WATS lines, and dial-up analog data. The evolution to the D4 frame was to accommodate voice, data, and image. Both of these formats had their limitations. The superframe format (SF) used provisions to allow for signaling (robbed bit) and ones density for timing (stuff bits). These two capabilities, combined with the framing bits, were constantly taking channel capacity from the T1. All of the user information was designed around a 56-Kbps data stream. The robbed or stuffed bits were designed around control. Yet, when looking at the potential for maintenance and diagnostic capabilities, no spare capacity exists for the data channels to do this, short of giving up additional overhead. Given the pricing structure and the other forms of overhead, this option was not acceptable.

Maintenance Issues

Whenever a user of T1 with D3/D4 framing experienced problems on the line, the sequence of events became elongated and disruptive. The sequence would occur in the following order:

- The user would experience some hits on the line (or downtime) that affected data throughput. These hits would be in the form of loss of framing, loss of synchronization, bi-polar violations, etc. Regardless of the problem, disruptions on the circuit for seconds or longer occurred. The user could see lights flashing on the channel service unit (CSU).

- The user would notify the appropriate carrier (either local-exchange or inter-exchange carrier) of the problem. Further discussions would evolve, regarding the errors, etc.

- To test the line, the carrier would ask the customer to get all users off the circuit and give the line to the carrier for testing. In many cases, this might be an immediate situation; in other cases, this was a scheduled event. Whether or not this was scheduled for a later time, the issue at hand was the need to give up 24 channels of voice/data/image, etc., for a couple of hours for testing.

- The carrier would then test the line. A loopback arrangement would be used in some cases. The carrier would loop the circuit from the exchange office to the CSU or demarcation point. A pattern of 1s and 0s were transmitted to the customer premises, and immediately shipped back around to the originating office. As the data was shipped out, a comparison of what was received back was made. If any errors were evident, the necessary fixes could be either done on the spot or scheduled.

- However, oftentimes the carrier conducted the tests and detected no problems. A technician might then be dispatched to the customer's premises to provide further testing. Equipment using test patterns [quasi-random signal source (QRSS)] was used. Again, the possibility of no errors being detected existed.

- The carrier would then give the circuit back to the customer. When challenged as to the nature of the problem, the carrier's response was no trouble found (NTF). Obviously, the customer had problems accepting this answer. The feeling that the carrier was either hiding something or just didn't know what was going on prevailed.

- Once the circuit was placed back in service (2 hours later), the customer might see errors again. At this point, the cycle would start all over again. Usually with the same end result—NTF. Consequently, the working relationships and confidence levels between the customer and provider become very strained.

Error Detection

The easiest way to describe errors on a T1 is the process where any logical one is changed to a zero (0); or a logical zero is changed to a one (1). These logic errors are the source of the corrupted data on the line. The errors can be caused by noise from a spike of electrical energy on the line (i.e., lightning) or induced by electromechanical interference (i.e., a large motor in a building). Further types of errors can be caused by cable losses, flooded cables, rodent damage, or crosstalk on the line, to name a few. Equipment can introduce errors from the digital repeaters, multiplexers, or DACS on the line. Unfortunately, the process of detecting these errors can be tenuous. Any pulse (a 1) or absence of a pulse (a 0) transmitted along the line is subject to noise that can change the actual data. When traveling through equipment such as multiplexers, repeaters, DACS, CSU, DSU, etc., the pulses are accepted as valid. The equipment, therefore, cleans up the valid pulses, but because the electronic systems can't differentiate between valid and erroneous pulses, it cleans up the entire data stream. Consequently, the receiver at the distant end accepts the clean, well-timed input as correct information. Hence, errors are introduced and undetected.

You might think that an error on the line could be easily detected by the equipment because certain format problems might also be evident. For example, an error (an extra pulse on the line) might violate the bi-polar format convention, or a dropped pulse might violate the 1s density rule. Clearly, these would be easy to detect. However, if the signal must pass through equipment along the way, that equipment might clean up the errors. They do have the ability to log an error on the inbound side, then reformat the signal on the outbound side and make it appear to be valid.

To further understand this process, look at both errors where a pulse is inserted and where a pulse is dropped.

Errors of Omission/Commission

The errors are either dropped pulses (omitted) or inserted pulses (committed). An error where a pulse is omitted should in turn create a bi-polar violation. If the number of pulses omitted is even, the bi-polar violation won't be created. Only an odd number of errors causes this violation. Figure 13.18 is an example of how this will work with an even number of removed pulses.

A error occurring with an odd number of pulses being dropped would be detected. Figure 13.19 is a representation where three pulses are dropped, therefore, causing a bi-polar error.

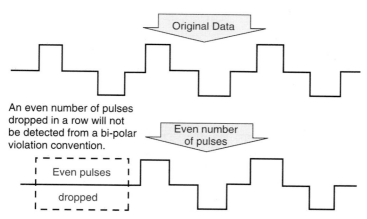

Figure 13.18 An even number of pulses being dropped will not be detected from a bi-polar violation convention.

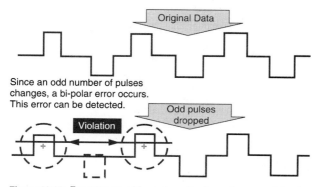

Figure 13.19 Because an odd number of pulses changes, a bi-polar error occurs. This error can be detected.

When pulses are introduced (committed) a similar problem occurs in detecting the errors. For example, when an odd number of pulses is introduced, a bi-polar error will occur, but when an even number of errors occur, the creation of a bi-polar violation might occur, but is not guaranteed. Figure 13.20 is the commission of an odd number of pulses which will create a bi-polar violation.

However as mentioned earlier, when an erroneous signal passes through an equipment along the link, the signal is cleared up and put into a format correct state. Figure 13.21 is a sampling of how this would work.

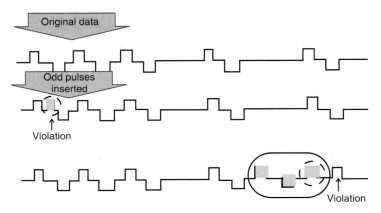

Figure 13.20 When an odd number of pulses is inserted, a bi-polar violation will occur either with the first pulse inserted, or with the first valid pulse after the commission of the pulse.

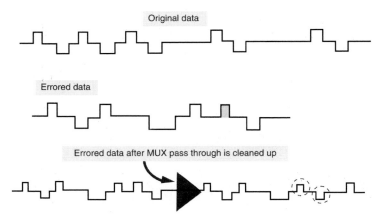

Figure 13.21 As the data with errors and bi-polar violations passed through MUX, CSU, DACS, DSU, etc., the electronics cleans up the signal so that it comes out properly formatted.

The detection of errors, therefore, becomes less obvious in the transmitted data stream due to these problems. A format error (bi-polar violation) occurring implies that a logical error has occurred, thus the data is erroneous.

ESF—A Step to Correct the Problem

In the early 1980s, AT&T suggested the implementation of extended superframe format as a means to provide nondisruptive error detection and perform nonservice affecting diagnostics on T1 circuits. This was a step to correct the problem of lost customer confidence and the disruptive performance testing. As a service provider, the interexchange carriers write tariffs to provide service within certain guidelines of availability. ESF helps to address this problem.

Essentially, because overhead on the circuit is at issue, customers are unwilling to give up more of the channel capacity for maintenance and diagnostics. ESF uses the 8 Kbps of overhead normally assigned for the framing format and thereby does not affect the data bandwidth.

What does ESF do?

Extended superframe format as implied by the name, extends the superframe from 12 consecutive frames of information to a repetitive 24 frames of information. The 8 Kbps overhead originally used strictly for framing is now sub-divided into three functions. Table 13.7 is a summary of the overhead being shared to support ESF.

ESF takes advantage of newer technologies to use the 8 Kbps framing overhead and uses only 6 bits (2 Kbps) for framing synchronization. Six additional bits are used for error detection by employing a cyclic redundancy check 6 (CRC-6). This leaves 12 of the framing bits for a facility data link communications channel (4 Kbps). Table 13.8 is a comparison of the framing bits used for SF and the newer use for ESF.

Using this shared facility, the carrier has the option to perform the three functions necessary to support the customer needs. These being:

- Framing synchronization
- Error detection
- In-service monitoring and diagnostics

TABLE 13.7   The original 8 Kbps of Overhead in SF Format Is Now Sub-Divided to Provide Three Separate Functions

Original SF format	Revised ESF format
8 Kbps for framing *only*	2 Kbps for framing
	2 Kbps for error detection (CRC)
	4 Kbps for facility data link

TABLE 13.8 Frame Bit Definitions

Frame #	Bit #	FPS	FDL	CRC
1	0	-	m	-
2	193	-	-	C1
3	386	-	m	-
4	579	0	-	-
5	772	-	m	-
6	963	-	-	C2
7	1158	-	m	-
8	1351	0	-	-
9	1544	-	m	-
10	1737	-	-	C3
11	1930	-	m	-
12	2123	1	-	-
13	2316	-	m	-
14	2509	-	-	C4
15	2702	-	m	-
16	2895	0	-	-
17	3088	-	m	-
18	3281	-	-	C5
19	3474	-	m	-
20	3667	1	-	-
21	3860	-	m	-
22	4053	-	-	C6
23	4246	-	m	-
24	4439	1	-	-

Framing

As already stated, the fundamental purpose of the bit pattern is for framing synchronization. Framing is important because its loss can impact performance by the incorrect synchronization. (Channel 1 could wind up connected to channel X [2-24] if the framing is incorrect.) Errors might cause a loss of frame and disrupt the circuit; and far end and/or intermediate equipment could be thrown into a loss of frame condition, thereby a loss of data. While the devices along the circuit are reframing data, throughput is disrupted for a period of time. Depending on the equipment in use, this loss could be quite substantial (approx. 200 ms). Newer equipment performs a reframing operation in less time (approx. 10–20 ms). This helps to improve performance. With ESF using only 1 of 4 bits, the delay could increase significantly unless the proper equipment is purchased.

CRC6

Cyclic redundancy checking is designed to detect errors that occur on the line. Because the network is error prone, a foregone conclusion is that errors will occur. Using CRC6 an entire ESF (24 frames of information or 4632 bits) is checked for accuracy. This check is a fictitious number created by the CSU performing a mathematical computation on the 4632 data bits. When the math is calculated using a prime number polynomial, a 6-bit pattern is the end result. In the next ESF (24 frames), the result of this mathematics is transmitted to the distant end.

At the distant end, the receiving CSU calculates the exact same mathematical operations on the ESF data, then computes an answer. When the next ESF is received, the result of the first calculation is contained in the CRC bits (6 bits) of the overhead. When the two results are compared, the answers should be the same. If, however they differ, an error has occurred. The use of CRC-6 allows for an error detection efficiency of 98.4% or simply stated, 63 of 64 errors will be detected.

The Facility Data Link

The 4-Kbps overhead set aside for facility data link control is a synchronous communication channel. This channel can serve multiple purposes, one of which is for the exchange of information between equipment devices along the circuit. Under the AT&T publication (54016), a set of standard message formats has been specified for communicating across the data link to the storage devices. The standard conforms to a basic X.25 (BX.25) link procedure and uses an AT&T standard known as *telemetry asynchronous block serial protocol (TABS)*. The carrier or end user can communicate with the remote equipment (CSU) using a maintenance message format. Some of the maintenance messages allowed under the AT&T technical reference (54016) include:

- Send 1-hour performance data reports—this will include the existing status of the link, how long has it been since the last check (up to 24 hours), errored seconds, and failed seconds in the present 15-minute cycle, overall 24-hour cycle and the past four 15-minute cycle performances.

- Send 24-hour errored seconds (ES) performance data—the CSU will send specific error events logged within the past 24 hours, in 15-minute increments.

- Send 24-hour failed seconds (FS) performance data—the CSU sends the 24-hour historical information of errored events logged in 15-minute increments.

- Reset registers—empty the buffers and start counting errors all over again.

- Send errored ESF—the CSU will send the current count of errors accumulated (up to a maximum of 65,535).

- Reset ESF register—empty the buffer and start at 0.

Benefits of ESF

The reasons for the use of ESF are to increase the circuit availability and improve performance. Once again, putting 24 channels (circuits) on one T1 puts the user at risk if performance is poor. Just about all of the major carriers (inter-exchange) have implemented this enhancement. Because all of the interexchange carriers internetwork; that is, they connect to each other, the use of the AT&T standard (54016) is the norm.

Through the use of ESF, vendors and carriers can see the improvements. When a burst of errors occurs, the carrier needs only access the CSU from the facility data link and query for status, etc. The problem of old, the "no trouble found" situation, tends to fade into the background because the CSU has buffers to collect the errors for later retrieval. Further, with the facility data link, the carrier can monitor real customer data traveling across the line nondisruptively. Errors occurring on the line can be seen and diagnosed real time. This all leads to a better track record. In fact, many of the carriers will contractually guarantee circuit availability between 98.4% and 99.5% because of the proactive capability gathered in ESF.

This seemed to solve the problem for users and carriers alike. Despite the passive collection capability and the on-line diagnostics, the parties all began to create a working rather than an adversarial role.

Problems with ESF

Unfortunately, one player not mentioned in this scenario and yet who is an essential ingredient in the total circuit is the local exchange carrier (LEC). An integral piece is the "last mile," or the circuit running from the LEC to the customer premises. ESF was designed to function as a part of the customer premises equipment (CPE). However, the LEC by mandate cannot poll the CSU that stores the error information. Under the guidelines of the modified final judgment (MFJ) or what we know as divestiture, the LEC cannot go beyond a clearly defined demarcation point. This demarcation is the line termination in the RJ48, RJ68, or smart jack—where the LEC stops and customer takes over.

This point, therefore, excludes the LEC from benefits derived from ESF as defined by the AT&T technical reference (54016). To overcome this problem, the American National Standards Institute (ANSI) came up with a proactive approach to ESF: ANSI's standard T1. 403 uses the CSU to monitor performance on the T1 line, then puts together an activity report every second and transmits this ESF report back out across the network.

The Open Systems Interconnect Model (OSI)

The three-letter acronym *OSI (open systems interconnect model)* appears in several places throughout this book. No one concept in this industry has been more misunderstood, which is what happens when several different operations are merged into one single industry.

We first had a telephone industry. Later we started to call it a *telecommunications network*—especially as newer services were being supported on this one infrastructure. The merger started when we introduced the data processing industry to the telecommunications industry. The data processing industry had already become entrenched in their own jargon and concepts. While the data processing function was localized to the mainframe and all transactions were performed in the computer room, no problem existed. However, as all things evolve, so did this industry. Soon, the movement was from the computer room to the desktop where users wanted connectivity to the mainframe. This was accomplished through the use of specialized wiring systems. Nothing new here, other than these special wires were very expensive. The wires had to carry the data from the terminal to the computer and also do the reverse. This was accomplished by using age-old communications techniques, by converting the data into electrical energy and carrying it down the wire; and the data was represented in its digital form without notice.

Then it happened: not only did we need connectivity from the desktop to the computer room, we also needed it from a desktop across town or across the country. Now the process was going to change the way we did business.

The data communications industry became the next battleground. The data processing folks all thought that data processing and data communications should be similar enough that they should control the deployment. The telephony people felt that this computer stuff was a pain anyway, so they were content to let it reside in the data processing arena.

But some problems existed with this whole situation. To send the data across town or across the entire country, it would be necessary to use a device called a modem. Recall that we talked about this in an earlier chapter, but the modem was used to make the data look like voice. As long as it looked like voice, the data folks wanted to send it back to the telephony department, especially when they didn't understand or care to understand how this analog network functioned.

So the marriage of the three techniques began back in the late 1950s and into the 1960s. The evolution was one of a slow migration because the telephone monopoly controlled the delivery of the transport system. The data processors were looking for a way to make the network digital to avoid the digital-to-analog and the analog-to-digital conversions. This seemed like too many steps, and as each step was taken the risk of errors was introduced into the data stream.

Things rolled along nicely, but the data processing evolution kept going from the mainframe environment to that of a distributed computing slant. Organizations were cropping up that would offer smaller scale computers (midi and mini) that could handle very specific functions and off-load this process from the mainframe. Once again, this started the ball rolling. Many of the computers in use at the time were not compatible with each other. This led to a new set of predicaments. As the smaller computer companies were competing with the mainframe world, two major players were doing battle—IBM and DEC. Each had their own hardware and software platforms. Now organizations began to move computers into the buildings from different manufacturers. No problem, they were very specific application processors, so there need be no conflict. That was true, but a user who needed access to the mainframe and the specialty machine now required two separate sets of wires run to the desk and two different terminals on top of the desk. This sounds familiar from the historical section of this book where the interconnection of various telephone companies would not work and the user needed two or three phones on their desks to inter-communicate. The battleground was one of compatibility rather than technology. The problems were only just starting.

First, let's take another look at this from what the managers of the computing systems were facing. In Figure 14.1, the organization had a mainframe that was based on an architecture created by the manufacturer—in this case, IBM. The hierarchy of this architecture allowed various connection arrangements, based upon the device's level on the planes of the archi-

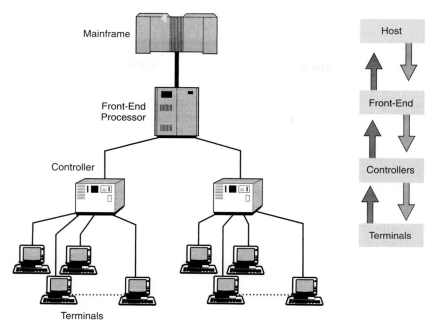

Figure 14.1 The hierarchy in a mainframe environment.

tecture. This figure shows how the transfer of information among and between users would take place. Things worked pretty well because IBM had everything in place: they delivered the hardware, the software, the connections and the talent to make everything work together. This was all based on an architecture called systems network architecture (SNA). See Figure 14.2 for the architecture in its stack. This is a seven-layered architecture that IBM created to make everything work in harmony. Using this structure, IBM would fix something that did not work, for a fee, of course.

Later, as the specialty machines appeared, a separate approach was used by the provider. This provider was created as a result of spin-offs from IBM. The founders of this particular company (DEC) were upset with the strategy and direction that IBM used. So, they created their own off-shoot company to offer lower-end machines to the customer, but at a more reasonable pricing and open environment. DEC of course, being an off-shoot of IBM, used a lot of similar approaches. Let's face it: When an engineer or manager leaves one place of employment, it's tough to leave behind all learned information, so the cross pollination of ideas and goals took root. In the design and roll out of their products, DEC introduced an architecture (shown in Figure 14.3) that just happened to be a seven-layered architecture. Called *digital network architecture (DNA)*, these seven layers were based on

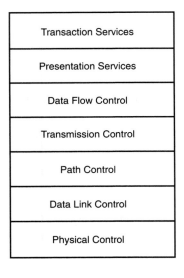

Figure 14.2 The SNA architecture.

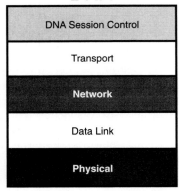

Figure 14.3 DEC's DNA architecture.

DEC hardware, software, connections, and talent to make everything work in harmony. Sounds familiar doesn't it?

Progressive end users who had embarked on a multi-vendor approach in meeting their computing needs would come up with two separate architectures to work with. The third approach to this is that many users wanted to be connected to both systems. Ultimately the end user wanted to be able to access these two platforms, hardware, software, applications shared data sets, whatever; but they wanted to do it from a single device. What's more, many of the specialty applications that the users wanted required the data that was stored in the mainframe. This meant that the users were re-keying the data from printed reports that came from the mainframe. This left a new set of problems on the table.

- The timeliness of the data, because it had to be re-keyed it could be obsolete before it ever got entered.

- The accuracy of the information was questionable because data entry errors are not uncommon.

- The cost of the information was now more expensive because the input was done at least twice.

- The frequency of the input was becoming exponential because as soon as the data is printed out from the mainframe, it could be obsolete as a result of changes made by other users. Unfortunately, this was changed in the mainframe, but the update in the specialty machine was not automatic.

Therefore, the scenario got more comical. Here's a scenario to make this concept easier to understand. This is a hypothetical situation—all names and companies used only for purposes of this example. Do not try this yourself, as this was tested by professionals who are skilled at ducking the issues.

Let's use an example of an organization who has two suppliers of their computing platforms. One is an ABC mainframe, the other is a DEF BAX. The company has users attached to both machines, and has received complaints of the connections. Users are getting confused when they are trying to log onto sessions on both machines. The MIS manager decides to go to the vendors and ask if there is a way to provide transparent connectivity. The dialog goes something like this:

IS Mgr.: Hello ABC, I've got this ABC mainframe where the users are attached using a 333X terminal. I also have a DEF BAX system and my users are using a VL2000/3000 type terminal to connect to the BAX.

ABC: Yes, what can we do for you?

IS Mgr.: Well what I'm trying to do is let my BAX users log onto the mainframe and view or manipulate the data on the ABC host. When the need dictates, I'd also like for my users to pull the data from the host and move it over to the BAX where the user can modify or delete the data. Then when they are done, they should be able to save it back to the original data set on the host.

ABC: So you want your users to access the data on the host, use it; modify it or whatever. Then you want that data saved on the host so that others can see it and use it. Is that correct?

IS Mgr.: That's it exactly. But I also want the user to be able to use the same keystrokes on their terminal that my 333X terminals use.

ABC: OK! There's no problem, we have the perfect solution for you.

IS Mgr.: You do! That's great, this is exactly what I was hoping to hear. What do I have to do?

ABC: It's simple. All you have to do is sell all those DEF terminals and the BAX, then buy all ABC terminals and hosts. You'll have full transparency and everyone will enter the exact same keystrokes.

That's not exactly what the IS manager was hoping to hear. So off to DEF. The dialog will be somewhat the same.

IS Mgr.: Hello DEF. I have these two systems, one a BAX using VL2000/3000 terminals serving a special application. I also have this ABC mainframe that is serving some of my back-office functions like accounting, billing, inventory control and the like. What I'd like to do is provide seamless and transparent connectivity between the two computers and let any user access any data from the terminals at their desk. This can be a VL2000/3000 or a 333X. Do you have a way of handling this for me?

DEF: You have a BAX and an AB who?

IS Mgr.: An ABC host.

DEF: Oh this is easy. We have these requests all the time and we help our customers through this arrangement very easily. As a matter of fact, we can do it all for you.

IS Mgr.: You can? This is great. I knew you folks would come through for me. What do I have to do?

DEF: It's really quite simple. Sell all those other devices, buy more BAXs and VL2000/3000s. It will work transparently and everyone will love it!

Once again the dilemma. What was the IS manager to do? Fortunately others had already gone through this problem and a new market had emerged. The third-party manufacturers of "black boxes."[1] The box was not black, it was the reference that no one really knew what it did, it was a protocol converter. All the user knew was that if a user on one system typed the normal way on a terminal, the commands and data that was typed went into this box as DEF and came out the other side as ABC, and vice versa. These protocol converters were not inexpensive, but they sure were a more attractive option than the others that the IS manager got. So, the IS manager bought a third-party converter and attached it to both machines. With a connection to each machine the converter sat in the middle. This is shown in Figure 14.4, where the two hosts are connected to the converter. All is solved. Or is it?

After getting the installation done, things appear to be working fairly well. Then one day things start to go awry. The IS manager calls the ABC folks and asks if they have done anything different with the system or code. The converter worked fine yesterday, but today nothing seems to be functioning. ABC, of course, would respond that they have done nothing. They

---

[1] The reference here is generic to a device that converts the information from one system and format to another. The use of the black box in no way refers to The Black Box Company, an organization in Pittsburgh, PA. These are highly skilled and qualified personnel that do an excellent job for their customers.

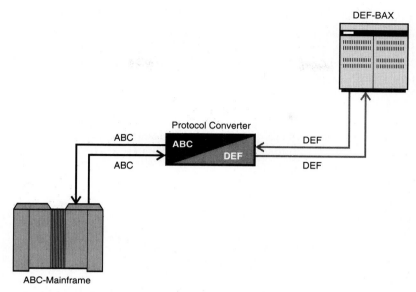

Figure 14.4 A "black box" was used to create transparency.

would, therefore, suggest it must be those other folks. Now, the IS manager goes to DEF and gets the exact same response. So, it must be those others. Lastly, off to the third-party supplier. They, of course, suggest that they have done nothing different so it must be those others. The result of all this is shown in Figure 14.5. This is called the finger-pointing routine. Point and hide is really what happened.

Even if nothing went wrong, things might not be exactly the same as what the end user accesses at the various machines through protocol converters. The follow-on to this is that the code sets and command lines might be different. Specifically in a 333X world, the end user might go through the following keystrokes to perform a function:

- To repaint a screen with a blank accounting form, press the CMD key, which is located at the upper left-hand corner of the keyboard using the left index finger.

- At the same time, press the F9 key using the right index finger.

- Voilà the new screen appears with the blank form ready to go.

To perform the same function from a VL2000/3000 device in the same host do the following:

- Press the ESC key with the left index finger. Because there is no CMD key, a remapping of the keyboard must take place. The ESC is located in the upper left-hand corner of the keyboard.

- At the same time, press the F9 key with the right index finger.
- At the same time, press the shift key with the left thumb. The shift key is on the lower left side of the keyboard. And at the same time press the ALT key with the right thumb. The ALT is located at the lower right hand side of the keyboard.
- And at the same time press the space bar with your right big toe. The space bar is located at the bottom center of the keyboard.
- Voilá you now have the screen repainted with the blank form ready to be filled in! See Figure 14.6 for what this might look like.

Figure 14.5 When a problem occurred, the finger pointing started, leaving the IS manager at a loss.

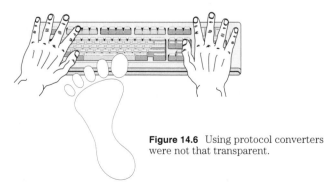

Figure 14.6 Using protocol converters were not that transparent.

Obviously, by now you have recognized that this is strictly in jest. There are past ills in the industry, but they have not been this bad. The issue was that transparency in the connectivity, communications, and use of the data was not to be had easily. The result was that the users put pressure on their coordinating committees in the industry to come up with a resolution. In 1978, the International Organization for Standards (or as we abbreviate it ISO) was asked to come up with a solution that would allow the transparent communications and data transfer between and among systems regardless of manufacturer. This was originally pointed toward the data communications devices. Thus, a new committee was formed to evaluate the way that this might happen. Clearly, everyone in the industry held their breath. What form of solution will they come up with? Because this is a committee made up of users, vendors, and manufacturers of the systems, and standards members, what could we expect? The result was to come several years later with what we now have come to know as the ISO's OSI reference model. No, they did not just reverse the letters of the organization, this stands for the open systems interconnect reference model. It is a seven-layer architecture. Just like the other two architectures that already existed (SNA and DNA). Two of the member companies of the committee were IBM and DEC. This seven-layered architecture is shown in Figure 14.7. The seven layers will be covered in more detail later, but are listed:

- The application layer sits at the very top of the model. This is the direct interface for the application to request a service. This layer provides communications services to the end user.

- The presentation layer sits below the application layer and provides a service to the application layer. This is where format, code conversions, data representation, compression, and encryption are handled for the application.

- The session layer, located under the presentation layer, is responsible for the establishment and maintenance of connections to a process between two different users or systems. The control of the direction of data transfer is handled here.

- The transport layer is right below the session layer. The transport layer controls the connection for error recovery and flow control. It is responsible to assure error-free data delivery end-to-end in a cost-efficient manner.

The next three layers deal with the network-specific issues of the communications process. Where the upper four layers deal with the end-to-end communications and data transfer, the bottom three layers are concerned with the node to node connection. This portion of the network may be provided by a single entity (such as a carrier, PTT, or third-party communications company).

- The network layer sits just under the transport layer and is responsible for the switching and routing of the connection. Also, this layer is responsible to take the data that is to be shipped from node to node, not end user to end user, and break it into smaller pieces to accommodate the transmission system. Congestion control, alternate routing services, and such connectivity are handled here. The network layer provides the establishment of the connection, transfer of the data, releasing the connection, and in some cases, the transfer of data in a connectionless service.

- The data link layer is responsible for the delivery of the information to the medium. It will create frames of information if so required and sends the frames along to the wires. Additionally, this layer is responsible for the reliability of the information and error checking at the node to node basis.

- The physical layer that sits below the data link layer is the actual electrical or mechanical interface to the physical medium. In this layer, we have the physical pieces and the necessary components based on the dependency of the medium. This layer is responsible for the transmitting of the bits onto the guided medium (wires), specifying various physical portions, such as the voltage levels, the pins to use in placing the voltage onto the link, whether the signal will be electrical or photonic (fiber) or modulated onto a nonguided (radio) based service. You will see how all of these pieces tie together in later text.

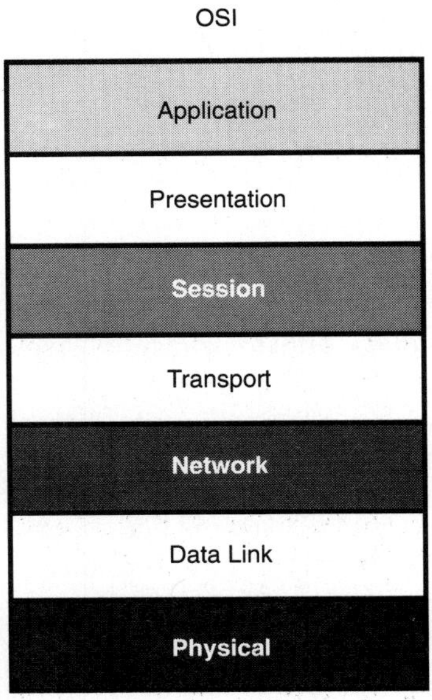

OSI

Application
Presentation
Session
Transport
Network
Data Link
Physical

Figure 14.7 The OSI reference model.

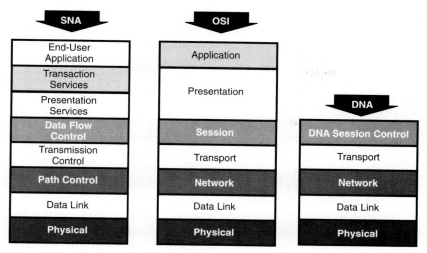

Figure 14.8 Comparing OSI, DECNET, and SNA.

Even with all of the effort that was placed on creating this set of protocols and the entire architecture, many industry and end-user groups were somewhat frustrated. They wanted a single solution to provide the transparent communications between systems of different manufacturers, and what they got was still another seven-layer architecture. They were amazed that the standards committees would do such a thing. In reality, this took a while to sink in, but it was the best way to provide the solutions. The seven layers are independent of each other, even though they provide services to the upper layer and they rely on their lower layer, there is still some independence. Why this is important is that as any protocol suite in the seven layers the one layer needs to be changed, but the rest of the architecture remains intact. A single solution would not have done that.

Figure 14.8 compares the three seven-layered architectures, the OSI as a base, the SNA, and the DNA to see how they line up against each other. In fairness to both DEC and IBM, their architectures were in place before the OSI model was created; therefore, they cannot be expected to be the same. DEC, over the years, has changed the architecture of their DecNet protocols to be more consistent with the OSI. However, IBM has basically done nothing over the past 18 to 20 years to change. IBM felt that they already had the single largest and most widely accepted architecture installed at all of the Fortune 1000 companies; therefore, there was no need to change what already worked for them. So, even though the architecture was developed in the 1978 time frame, it was not totally accepted until 1983, and there has been very few implementations of it.

Part of the reason was the cost of moving over to a new architecture. The U.S. Government wanted a standard applied to all of their purchases in the past. In an effort to get something to happen in the OSI acceptance and implementation, the GSA specified a modified version of the OSI reference model. If you wanted to sell something of a computing or data communications nature to the government, you had to comply with their modified version. It was called the *Government Open Systems Interconnect Profile (GOSIP)*.

Using this model as a base reference has been the foundation of many of the changes that have occurred over the past 20 years. Although very few manufacturers have gotten to full openness in their systems, they have attempted to use the OSI as a model for their future compliance. Again, this will become clear as you go further through this chapter. You must remember that there was little incentive for any manufacturer to comply fully with the OSI model because the openness could potentially jeopardize future revenue streams. Not to pick on any vendor, but when you bought a computer system using SNA from a large computer manufacturer, it was also an unwritten rule that most if not all future software and hardware purchases would be from that same vendor. If you bought from them, you were assured that it would work seamlessly into your existing hardware and software platforms. If you went the third-party route and something did not work, you could expect little support from IBM. Thus, their future revenue streams were aligned to the proprietary nature of their architecture. If a completely open attitude and platform was introduced, it would work with any hardware you acquired.

Now back to this complex model. The seven layers always confused the steel hearted who had a background in the data-processing or data-communications industry. Further, these folks took 5- and 10-day workshops, only to leave even more confused than when they went in. How can we show you in one chapter of text how this works and keep it simple? We assume that you are likely a novice or getting into some new portion of the industry, and that this is your reference manual. So, let's try it with our usual method: a story. We can hope that by the time this works through, you will have a basic understanding of the function of the model and each of the layers as they play into the industry. Here goes!

Figure 14.9 shows a situation where two people (Don and Bud) are thinking about moving to a new area in a suburb of a major city. There is no subdevelopment yet, but there is a lot of property that can easily be broken down into perfectly sized lots all with a great view.

These two players meet with a builder who owns the land and ask for their respective lots, which just happen to be right next to each other. The builder agrees to build both Bud and Don's houses at an agreed-to fee.

Now that the agreements have been reached, both parties decide that they'd like their houses rather quickly. So, they both go back to the builder and ask that the process be escalated. The builder, of course, will balk at the

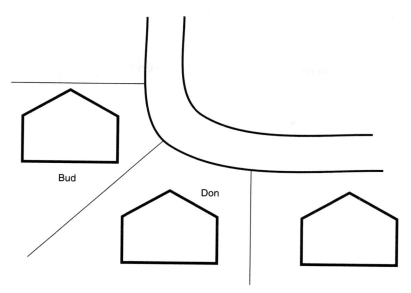

Figure 14.9 Building on new sub-development areas.

idea of doing this too quickly. But, perseverance wins out and the two potential homeowners get the builder to cave in and escalate the construction.

The result is that in two weeks, the houses are ready to go. Now Bud and
Don can move in. Immediately, a couple of things become apparent. Because
their houses were escalated, the builder did not have time to notify the local
telephone company of the impending move-in and escalation. So, the two
owners can move in but, they have no form of telephony service. Bud and
Don are neighbors and here they are living next door to each other without a
means of communicating with each other. What options do they have?

Immediately, they both call the local telephone company and are told that
the construction of facilities into their new area will take 6 to 12 months.
This is too long to be without a communications method between next-door
neighbors. So, they consider their options and come up with the following
choices (Table 14.1). The option is shown in the first column, the pros and
cons are shown in the other two columns. These are not all inclusive, but we
have limited this to a few options.

In the figures shown from the table results, there were benefits and
losses associated with the communications options. So, Don and Bud decide to model after the service that they really wanted; a telephone. They
aren't going to go out and build a central office. After all, they really just
want the convenience of speaking with each other. So, they decide to build
on a model. Let's use the OSI model and make sure that this all ties in while
we provide transparent communications.

TABLE 14.1 Summary of Communications Options

Option	Benefits	Deterrents
Use a megaphone and yell out the window from one house to the other (see Figure 14.10)	■ Easy to install ■ Relatively short distances ■ Can be implemented immediately	■ What if one isn't home? ■ What if Bud wants to talk to Don at 2:00 a.m.? ■ Loss of transmission due to distances ■ Noisy for other neighbors in the area ■ Total loss of privacy
Throw pebbles at the window, when the receiver hears the pebbles on the window, open the window and yell out (see Figure 14.11)	■ Pebbles are readily available ■ The method of signalling each other is straightforward	■ Given the distances, the throw will take some effort. This can cause damage ■ In inclement weather this can be inconvenient and awkward
Use a flashlight and send the signals in Morse Code	■ Easy to use ■ Quick communications	■ Don doesn't know Morse Code ■ During daylight hours the sun may diminish the light ■ Batteries can fail
Go next door and visit. Host our conversations during the visit.	■ Easy to accommodate ■ Doesn't cost anything	■ Weather may make this difficult ■ What about those 2:00 a.m. conversations? ■ The inconvenience will outweigh the benefits

Bud comes up with the idea that they need to send their information across a medium. The earlier options used the airwaves as the medium, but this was limited by weather conditions and so on. Therefore, both neighbors agree to use a physical medium. Using a medium, they decide to install a piece of everyday common household string between both of their houses. This is shown in Figure 14.12 where the string is run in a direct line of sight between both parties. The string is laid out and the two parties are connected with the medium.

Given the physical string is now attached, can the two parties communicate? The answer is somewhat obvious that there is still something missing. So they decide that they have to use the model to handle their next step. Don comes up with the ideas that they need to address the physical medium with some electrical or mechanical interface (ala layer 1 of the OSI). They decide that electrical is not applicable because they do not have an electrical conductor in the string. So, all they need is a mechanical interface. Don suggests that they can attach a can on both ends of the link to build this interface. Using everyday household tin cans on both ends of the string, they now have addressed the bottom layer of the OSI model (Figure 14.13).

Figure 14.10 A megaphone isn't a good choice, but it could work.

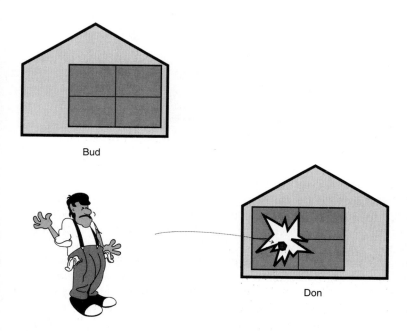

Figure 14.11 Throwing pebbles (or rocks) may not be a good alternative.

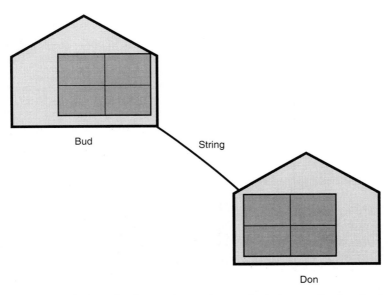

Figure 14.12 A piece of ordinary string produces a physical connection (medium).

Figure 14.13 Two tin cans are used to satisfy the mechanical interface (layer 1 OSI).

A problem has come up with this arrangement. To make this all work properly and transparently, the concept is that everyone build the connection in the same way. Bud has attached the string through the side of the can, whereas Don has attached the string to the bottom of the can. Although this might work, when you pull the string tight, the can at Bud's end is in an awkward position. This means that the two have to agree how they will connect the two cans to the string to make this work more efficiently. The agreement is that they will both take an empty can, punch a hole through the bottom and attach the string from the outside of the can, through the hole in the bottom, and knot it off inside the can. Then, they will pull the string tight to get a good connection. Now, the bottom layer of the OSI is handled.

The strings are now attached the same way, so can the parties now communicate? Well maybe! Bud decides to test it out so he picks up his tin can on his end and starts to talk to Don. Don is preoccupied with the television set and does not hear that the communications is being used. The whole conversation is wasted thus far. So, they need a means of using the link. They build a protocol that will solve this problem. First, the two parties agree that they should have an alerting mechanism to let the other party know that a conversation is requested. Let's use a technique that states when Bud wants to talk to Don, or Don wants to talk to Bud, the originating party needs to grab the can on his end, and yank on it three times. This will cause the can on the other end to rattle around. When the called party sees the can bouncing around then he will know that the other end wants to talk. To use the data link (layer 2), they establish a protocol called the *yank and rattle*. One end yanks the string causing the can on the other end to rattle. Now, they have overcome this obstacle.

We can skip by layer 3 of the model for now. Is it obvious why? Look at Figure 14.14, and it should become fairly clear. The strings are attached to bath parties directly. The connection is a point-to-point private line. There are no switching or routing decisions to be made because the connection always goes to the same location. Therefore, the network layer doesn't apply per se, so it will be transparent. But now they have the necessary connection and the alerting mechanism, so they should be able to communicate without much ado. Correct? Maybe not, but see what happens.

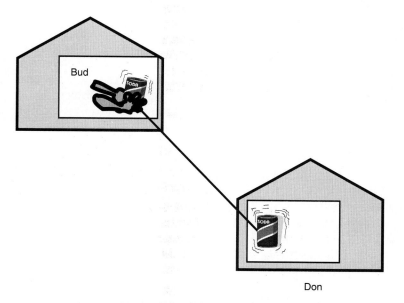

Figure 14.14 The point-to-point connection doesn't need a network layer. No switching decisions need to be made.

Bud decides now that he wants to talk to Don. So, Bud picks up the can and yanks three times. This causes Don's can to bounce around on the other end, which alerts Don that Bud wants to talk to him. Kind of like a ringing phone isn't it? So Don jumps up and grabs the can, picks it up, places it to his mouth and says "Hello." At the same time, Bud on the other end of this connection is yelling into the can for Don to pick up, "Don. Hello Don, are you there?"

Are these two parties communicating? No, they are not. Both ends have the can to their mouths, no one is listening. If both parties are talking and no one is listening, then the information transfer is null. So, a new set of protocols are required. Let's introduce the transport layer in the OSI model now. Because this whole idea is Bud's, they agree that whenever Bud wants to speak to Don, or whenever Don wants to speak to Bud, after the can has been rattled a couple of times, Don will always pick the can up and place it to his ear. Bud on the other hand will always start the conversation off by placing the can to his mouth. Now the problems already encountered are handled. So we should be able to communicate without any further complications. Right? Here is the transmission process as has been defined.

Bud: Yanks the string.

Don: Picks up the can on his end and places it to his ear because that's the rule.

Bud: "Hi Don! Can I borrow your lawn mower today?"

Don: Takes the can from his ear and places it to his mouth and begins responding. "No way, you told me last week that you were going to finally buy your own."

Bud: At the same time Don is responding. "I know I told you last week that I was going to buy my own but I haven't had a chance to get to the store yet."

Now both parties are talking at the same time. After a given amount of time, they will probably both be listening at the same time. The communications flow is not working properly. So, they have to add a new set of protocols to allow the orderly transfer of the information. The problem stems from this being a half-duplex link. There is only one transmitter/receiver on each end. There are two possible solutions. These solutions are shown as follows:

Run a new string between both parties. On this new string, they will attach two new cans, set up just like the first connection. Each can will be labeled T or R. On Don's end, the T can will be connected to the R can at Bud's end, and vice versa. Now when an alert comes in, each party will pick up the two cans. The T can will be placed in front of both Don's and Bud's mouth. The R can will be placed in front of each party's ear respectively (Figure 14.15). Now there are two separate transmission paths, one for transmit and one for receive. This is probably the best option, but Bud and Don are reluctant to do this because this adds to the complexity of maintaining two sets of string between their houses, and requires twice as many cans on each end. This will take up too much space in their homes.

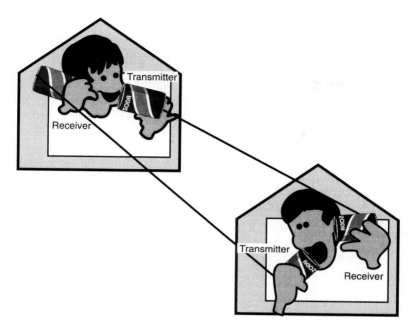

Figure 14.15 Two separate cans used allow simultaneous transmit/receive, not the flip.

The alternative to this communications dilemma is to use a set of rules (protocol) that will define the session layer. The control of the data transfer between the two ends. The session layer will work like this. Because this is Bud's idea, whenever Bud or Don yanks on the string, Don always listens first, and Bud always talks first. The problem occurs when the transmitting end is done sending and wants to await the reply, the two parties need to know how to time this out properly. So Bud will talk first (did you ever notice that Bud talks a lot?). When Bud finishes his statement he will use an "over" protocol. When Don hears the word "over" accentuated, he will take the can from his ear and place it to his mouth. Bud will at the same time take his can from the mouth and place it to his ear. Then Don finally gets a chance to talk. When Don finishes, he will say "over" and the process will reverse back to the beginning. And so on . . .

Now, the two parties have the ability to communicate with each other. Essentially, they can send their information back and forth as they had planned. Now that they have established this whole scenario, you have probably noticed that a couple of layers were missed in the OSI model. These would be handled transparently by nature of Don and Bud working with the same communications protocols. However, let's go back and pick those layers up to see how they will play out under our scenario. The three we have yet to address are the network, presentation, and application layers. So we shall complicate the issue by introducing the following add-on needs.

Just as Don and Bud got the communications working by the rules established, a new neighbor moves in next door to Don. This is going to happen all around the neighborhood, and we should be prepared. So Helen moves in next to Don. Helen is neighborly and decides that she might have a need to speak to Don and Bud from time to time. Therefore, the rules that we already had in place can be used to provide the connectivity between all three parties. See Figure 14.16 for what is about to be discussed. To set up communications between Helen and Don, a string can be run between the two houses and the cans attached to the string. The rule will be: because Don was here first, Don will always speak first, Helen will listen regardless of who rattles the can. The over protocol can also be used to provide the session layer and so on. Now Helen also needs to talk to Bud, but as the figure shows, in order for this to happen they need to run the string from Helen to Bud. This is not a problem for Helen or Bud. As a matter of fact, we can even color code the cans (red for Don, Blue for Bud).

But there is a problem with Don in this scenario. In order for Bud and Helen to connect the string to each location, the string must go in one side of Don's house, pass through and exit the other side to connect to Bud. Don will therefore have to leave his windows open on both sides of his house at all times for this to work. Don doesn't like this idea, so he offers to become a network layer relay point. The process, as shown in Figure 14.17 will work this way. When Bud wants to speak to Helen, he will pick up the one can that is associated with this communications system. He will then yank the can, causing Don to pick up and immediately listen. Bud will then say "Don, this is a call for Helen." Upon hearing this, Don who has two cans and strings, will grab the other can that is connected to the string to Helen's. Don will "yank" the can causing Helen to pick up the can and listen. Don speaks first on this connection so this is fine, he has a can to his ear from Bud and the other can to his mouth lead-

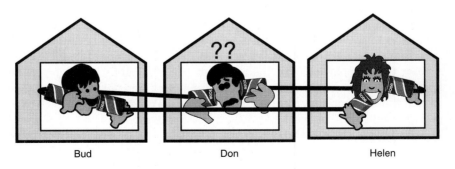

Bud   Don   Helen

Figure 14.16 Adding a third party may complicate the process (for Don).

Figure 14.17 By using the relay point, we satisfy the bottom three layers of the model.

ing out to Helen. When Don knows that Helen is on the link he will say "Stand by for a message from Bud." At this point, Bud will speak to Don, who in turn will automatically relay everything he hears to Helen. Now the routing of the information is accommodated. The network layer is met, of sorts.

A problem exists with this whole scenario that isn't as obvious. Helen only speaks German. Bud only speaks English. Now the problem: If Bud speaks English into the can connected to Don, how do we make the format or language conversion into something that is useable and understandable to Helen? Luckily for them, when Don was in school he took German 101. He didn't excel in this but he has a working knowledge of the language. Now when Don hears Bud say "Good Morning, Helen" in English, Don will relay the information out through the other can in German as: "Guten Tag!" And when Bud finishes and says "over," the process switches. Bud takes the can from his mouth and moves it to his ear. Don takes the one can connected to Bud and moves it from his ear to his mouth, at the same time he takes the can connected to Helen and moves that one from his mouth to his ear.

Helen in turn moves her can from her ear to her mouth and begins to speak or respond. When Helen responds with "Morgen, wie geht's?" Don hears it and converts it to Bud as "Morning. How are you?" And so it goes. The presentation layer has just been addressed to accommodate the format or protocol (language) conversion so that transparent communications can take place. The only layer in this case is the applications layer that has not been addressed. What is the application in this regard? Chit-chat, I just called to say hello!

By now, you should have an understanding from this simplistic comparison how the process works through the OSI model. Further, the importance of this all taking place transparently should be obvious. If we had to go through the trials and tribulations of information transfer every time we placed a call, this would be far too cumbersome. The next time a connection to another neighbor is needed, the same rules can be applied.

We already have a simplistic network here, so let's draw it to a conclusion by introducing a couple of other players. Mary moves in behind Bud. There will be a need for Mary to communicate with Bud, Don, and Helen. Rather than trying to run strings all around the neighborhood we'll just use our relays and rules already in place. So Mary connects a string to Bud. Communication to Bud is simple. To reach Don the call will go through Bud to Don; to reach Helen, the call will go though Bud and Don to ultimately get to Helen. One added complication is that Mary only speaks French. Therefore, when Mary says "Bonjour" to Bud, he repeats the information to Don as "Good Morning." Don then sends it to Helen as "Guten Morgen." Finally. Roy moves in behind Don. We now run a connection between Don and Roy no problem. Roy only speaks Pig Latin, but that's okay because Don can deal with it. Now when Mary wants to speak to Roy she calls through Bud in French, where Bud converts to English and sends it on to Don. Don receives the message in English and converts it to Pig Latin for Roy. The connection is sent through and routed/switched depending on the intended target location. Helen, as you recognize, is left out of this conversation, but she has no need to be connected. The result is Figure 14.18, where the connection is routed and switched transparently between and among the neighbors. What has also happened is that Don becomes very similar to a central switching system and service provider. Now for voice or data, the telecommunications network is built to handle any user to any user, regardless of the language (format and protocol) or origin of creation (manufacturer). This is what the OSI model was designed to do for us.

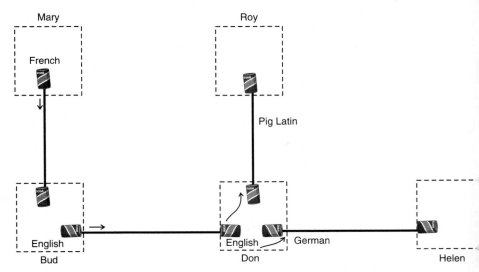

Figure 14.18 The layers can all be met through this network.

Is it here? Partially! Work continues on the standards, especially in how this can all be achieved. Secondly, many manufacturers are implementing OSI compatibility in different ways. That's okay, but it might lead to some kludges. The issue is not how the arrangement is created, but the opportunity to have a universal set of rules and protocols that everyone can build to. That is what the OSI reference provides.

That description is probably more than what most of us need to know about this. However, we should at least be aware of just what it was meant to do for the industry as a whole. If we substitute our players in the example above, then we can have an organization using a computer platform manufactured by Company Bud. This is connected to or through a computer manufactured by Company Don and ultimately sharing and swapping information with a computer manufactured by Company Helen , and so on . . . The communications network might well be used as a point-to-point, point-to-multipoint, or a switched dial-up service. Protocols and services have been applied, such as ISDN, X.25, switched 64 Kbits/s and leased-line T1, so that the connectivity arrangements adhere to the tin cans and string. Yes, some of the cans might be larger, or rounder than others. As long as the same rules for electrically and mechanically attaching to the string (wires, light beams, radio waves), then this should all work.

Other Network Architectures

Using the OSI as the reference is fine. It will be the basis of all future development. But there are still other architectures that were in existence or sprung up around the wait for the ubiquitous implementation of OSI. We might never see the final end, but we will see variations and improvements over the OSI as it applies to vendor specific architectures or proprietary applications. Here are some of the more common models:

- SNA, which is now over 20 years old and has had very few modifications from the original architecture.
- DEC Net. As covered, this is also twenty plus years old, but has gone through several iterations with an attempt to create openness.
- TCP/IP, a set of protocols that work on a different architecture developed back in the late 1960s and deployed more in the early 1970s for the government. Now it is the protocol for the internet and the primary set of protocols that LAN, MAN, and WAN users are implementing for openness and robustness in their networking needs.

SNA

We have already discussed the SNA world earlier in this chapter. Therefore, the continued discussion of this architecture will concentrate on the subtleties of the SNA to OSI comparison. In 1974, IBM introduced its architecture called *Systems Network Architecture (SNA)*. SNA defines a structure and all of the protocols required to implement a network in which a wide variety of computers, software, and terminal devices can interact and provide a very high degree of network efficiency. Some of the characteristics that were introduced with the SNA concept include:

- Application software uses a standard interface for communications, leaving the software network and device transparent and independent.
- All of the data links can use the same set of rules called *link control procedures*.
- Applications can interexchange data with each other.
- All applications can be intertwined so that they can allow multiprocessing capabilities.
- Regardless of the location, different devices can access any application on any host.

Various techniques exist for local area data transport through recommendations and implementations by the vendors. Remember that the installations were merely recommendations to move data across a cable system between any two stations on the localized data transmission facility. Networking, whether local area, metropolitan area, or wide area, requires higher level protocols to manage the transfer of information and make it useable and understandable. Without these higher level protocols, nothing is guaranteed.

The functionality of SNA is to augment and supplement the local area network. Imagine that the LAN actually existed under a different name prior to industry acceptance of the term *LAN*. In the earlier days, this was implemented in the hierarchical networking arena provided by SNA. This was introduced in 1974 and changed on a limited basis only. IBM felt that they had an infrastructure in the major organizations that did not require change. The goal of SNA was to introduce the structure and protocols required to implement a network in which a wide variety of computer hardware, software, and terminal devices could interact and provide a high degree of efficiency and utilization of the network. SNA represents several characteristics of the architecture such as:

- Application software uses a standard communications interface, and maintains network and device independence.
- All data links in SNA use the same link control procedures.
- Applications can exchange data between and among each other, allowing the use of multiprocessing services.

- Different types of devices can access applications in different hosts, regardless of the location.

These characteristics of the SNA function to work cohesively in any implementation from IBM. SNA, therefore, details the specifications for the devices and nodes on a network, and the logical path between these nodes. Both paths and nodes are arranged in the hierarchy using several layers of protocols. This is organized to facilitate the distribution of processing under a common control for the nodes. Further, it is structured to provide flexibility and robustness in routing on the paths.

SNA Components

Each device in an SNA network from the individual dumb terminal to the host computer functions with a level of control, depending on its position in the hierarchy. It will then operate under the control of the next higher level device. Terminals, for example, function under the control of the cluster controller. The controller functions under control of the front end processor and the front end functions under control of the host. This is the true hierarchical concept IBM introduced 20 years ago.

The network addressable unit (NAU): A network addressable unit is any segment, code, or etc. that emphasizes the device to a network, program, or device. There are basically three different types of NAUs as follows:

- The systems services control part (SSCP)
- The physical unit (PU)
- The logical unit (LU)

The SSCP

The SSCP functionally resides in the communications access method of a mainframe computer, or the control program in a midrange computer. See Figure 14.19 for a representation of this control part. This function controls the addressing and routing tables, name service to address translation services, routing tables for the network, and all instruction sets that deal with these entities. The SSCP establishes the communications connection between nodes in the network, provides the informational flow control and queuing services for network efficiency and selects the route for the nodes to communicate with each other.

The Physical Unit

IBM defines the physical unit as a single device on the network. See Figure 14.20 for a representation of the PU. In a host computer and a front-end

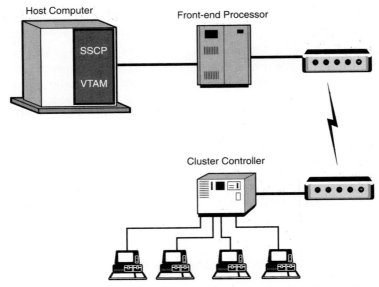

Figure 14.19 The SSCP works as the communications software and manages the network. It will control the setup and tear-down of connections between users.

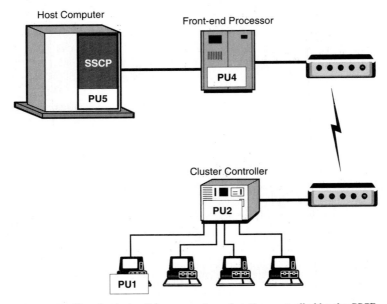

Figure 14.20 The physical unit is a supervisory function controlled by the SSCP.

communications processor, this is implemented in software. In a terminal device (less intelligence applied in these), the physical unit is implemented in firmware.

IBM defines the PUs as follows:

- Type 1 is a dumb terminal or a printer.
- Type 2 is a cluster controller (3274) function for the terminals or a batch terminal.
- Type 3 was under study.
- Type 4 is a communications controller such as the front end processor (37X5).
- Type 5 is a host computer with a system services control part (SSCP).

In SNA, the SSCP controls the physical unit. Each PU is treated as a physical entry point on the network between the network and one or more logical units.

The Logical Unit

The logical unit is the basic communications entity in SNA. The LU is a port where end users access the SNA network. See Figure 14.21 for this access method. Two LUs communicate through what is called an LU-LU session, a temporary connection where data can be exchanged based on some mutually agreed upon protocols. The layers of SNA shown a little differently in Figure 14.22 than in previous versions of the architecture (not to change, but to clarify the architecture) build upon each other from bottom up. Every SNA node contains the bottom two layers shown in this figure. These two layers—the data link control (DLC) and the path control (PC)—combine as what is called the *path control subnetwork*.

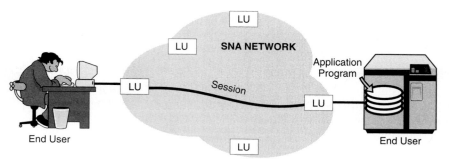

Figure 14.21 The logical unit is a port where end users access the SNA network.

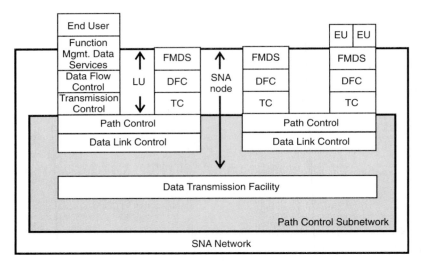

Figure 14.22 The SNA architecture shown differently.

This is the basic method for the movement of information from a source to a destination node. Sometimes this can involve passing through intermediate nodes. The LU, which resides in the SNA nodes, consists of the upper three layers in this figure and use the data transport capability of the path control subnetwork.

The different types of LU are:

- LU0 are sessions for special applications where IBM or the end user defines the parameters.

- LU1 is a session between a host application and a remote batch terminal.

- LU2 is a session between the host application and a 3270 display terminal.

- LU3 is a session between a host application and a printer in the 3270 display family.

- LU4 is a session between a host application and a word processor, or between two terminals.

- LU5 is under study.

- LU6 is an intersystem communication, a session between two or more different applications, usually in two or more different systems. The more common of this LU is an advanced program to program communications (APPC), a session between an application program in a host and an application program residing in an intelligent workstation or terminal. This is also called the *LU6.2*.

The first SNA networks were simple tree networks, using the hierarchy shown. All terminals and controllers were attached to a single host computer. LUs in the terminal and controller communicated only with LUs in the host, but never with each other. All of these arrangements were purely hierarchical. Later versions of SNA allowed multiple hosts to reside on an SNA network and communicate as peers on the network. However, LUs in the terminals and controllers still only communicated with host LUs, never with each other. The difference was that they could communicate with LUs in multiple hosts instead of a single host.

Later advances in VLSI integration changed the computer networking strategies. Dumb terminals became intelligent workstations. Personal computers and mid-range computers were introduced that changed the complexion of the network. Because these devices contained their own processing and storage capability, it was a natural evolution to allow for distributed processing. SNA evolved with the market and adapted more of a peer to peer processing capability and the Advanced Peer to Peer Networking strategy.

Digital Network Architecture (DNA)

Digital Equipment Corporation's DNA is the model and architecture for the functions of DecNet. DecNet is a group of communications software and hardware that enables DEC's operating systems and components to function in a network with other DEC systems and is open enough to communicate with systems manufactured by others. DecNet is a group of computers with equivalent DEC software and communication hardware that connect with physical channels or lines. Each computer implemented in a network with DecNet software is called a *system*.

DNA defines standard protocols, interfaces, and functions that allow DecNet systems to share data and access various resources, programs, and functions. The DNA currently has two separate stacks of protocols; one supporting proprietary protocols and interfaces on a DEC environment, and the other supporting the protocols defined in an open systems environment (OSI model). Using a layered approach the DNA is functionally grouped into services and functions based on the layers associated with DecNet. Functionally, DNA is broken down into the following:

- User functions
- Network functions
- Communications functions

Using a slightly different view of DecNet compared to the OSI model than what was shown earlier, the DNA is shown in Figure 14.23. Two separate

OSI	DNA	Function		
Application	User	■ File Transfer ■ Down-Line Loading ■ Virtual Terminal ■ Remote Resource Access ■ Remote Command File Submission		
	Network Mgmt			
Presentation	Network Application			
Session	Session Control	Task to Task		
Transport	End Communications			
Network	Routing	Adaptive Routing		
Data Link	Data Link	DDCMP Point-to-Point Multiport	X.25	Ethernet
Physical	Physical			

Figure 14.23 The DecNet comparison shown differently.

stacks are shown here, the first is the DEC layer, and next to that is the function served by this layer. The verbiage on the DNA is different than that of the OSI, indicating that DecNet is not 100% compatible with the OSI reference. This has been demonstrated when DEC introduced their DecNet Phase V. This met with only moderate interest in the industry and limited implementations.

DEC has introduced a family of products and services over the years to allow access to and from DEC computing platforms and other manufacturer's platforms.

Internet Protocols (TCP/ IP)

While the industry and standards bodies wrestled with openness in the communications and computing arenas, another evolving set of protocols emerged. This is also a layered architecture, and is comprised essentially of four layers. A result of a government contract to look at the internetworking of computers in the event of a national disaster, the protocols emerged from the original Advanced Research Projects Agency Network (ARPANET). Developed by Bolt, Brenack, and Newman of Cambridge, this simple but robust set of protocols were geared to link systems in an open architecture. The transmission control protocol with internet packeting protocol (TCP/IP) continued its evolution from the original ARPANET to what we know as the Internet. The TCP/IP architecture is shown in Figure 14.24. This is a four-layer stack that deals with the equivalent seven-layer architecture of OSI, SNA, and DNA.

Actually, the TCP/IP is fast becoming the widest accepted set of protocols in the industry. Included in all forms of the UNIX operating systems and now the protocol of choice in many LAN to WAN environments, it is the "middleware" for interconnectivity. The two most prevalent features of this protocol stack are its ability to work with just about any environment because of its ability to use application program interfaces to emulate most services on other architectures, and the ability to packetize the data and send it out.

IP

Internet packeting (or internet protocol) performs a packetization of the user data to be sent between and among the various systems on a network. When a large file is sent down the protocol stack, the IP function is responsible for the segmentation and packetization of this data. Then, a header is placed on the packet for delivery to the data link. The routing and switching of this data is handled at the IP (network) layer. To be somewhat simple in our analogy, IP is also a dumb protocol. When a packet is prepared for transmission across the medium, IP does not specifically route the call across a specific channel. Rather, it just puts the header on the packet and lets the network deal with it. Therefore, the packets going out the door can take various routes to get from point A to point B. This means that the packets are in a datagram form, and the packets are not sequentially numbered as they are in other protocols. IP makes its best attempt to deliver the packets to the destination network interface; however, it makes no assurances that:

- The data will arrive.
- That the data will be error free.
- That the nodes along the way will concern themselves with the accuracy of the data and sequencing, nor will they come back and alert the originator that something is wrong in the delivery mechanism.

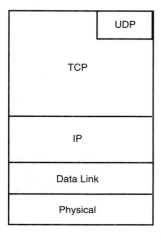

Figure 14.24 The TCP/IP protocol stack (architecture).

This might sound strange in a networking environment, but it allows the robustness to be achieved. The nodes along the network merely read the header information and route the packet to the next logical downstream neighbor. This means that if anything gets corrupted on the network, the node will not know where to send the packet, so it will throw it away. The network will not send a message back to the originator and let it know that the packet did not get delivered. Another possibility is that the nodes along the network can send the packet out on the basis of the best route (based on some costing algorithm, or least amount of points to pass through). Yet in the transfer of the packets, if one of these routes is busy or broken, the packets will be rerouted around the problem to another node. This has great possibilities, but it can lead to a problem.

It is possible that in the IP routing of a packet that the packet can be sent along the network in a loop (Figure 14.25). The routes might set up in a loop fashion and the packet will be out on the network like a spinning top. This doesn't cause concern if it is only one packet, but if this were a general occurrence, the network could get quite bogged down with packets spinning around. Therefore, IP has a mechanism in its header information that allows a certain amount of "hops" or what is called *a time to live on the network*. Rather than let an undeliverable packet spin around on the net-

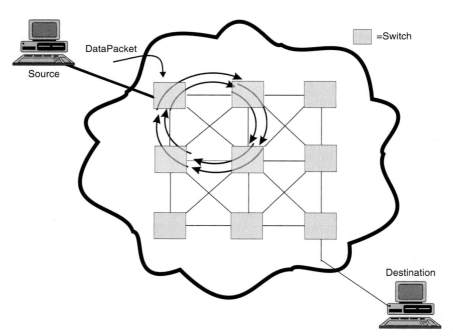

Figure 14.25 IP can send data into a loop, meaning it will not reach the destination computer. The switches could send the data around and around.

work, IP has a counter mechanism that gets decremented every time the packet passes through a network node. If the counter (usually 16 hops or less) expires or gets set to 0, the node will discard the packet. As you might imagine, if the network can throw away packets, we should be alerted of this when it happens. This is not part of the IP protocol, it does not come back and tell the originator that this has occurred. Additionally, packet number four (just an example) was spinning around the network for several hops and finally got discarded. Yet packets five, six, and seven go through the network without ado. Thus, the later packets arrive at the destination, but four is gone. IP does not have a mechanism to know that this has happened, and no numbering sequence to put everything back in order. Why would anyone favor such a dumb protocol? It is this robustness to deliver the packets across the network and throw away bad or undeliverable packets that helps to improve the efficiency of network utilization.

TCP

Nary a network manager or administrator would accept the unknowing status of whether data gets delivered across the network. Therefore, working with IP is the transmission control protocol (TCP). TCP provides the smarts to overcome the limitations of IP. Here, the controls will be put in place to ensure that the reliable datastream is sent and delivered. At the sending end, TCP puts a byte count header on the information that will be delivered to the IP protocol layer. This is encapsulated as part of the data in the IP packet. The receiving end, when it gets the packets, is responsible to put the data back into its proper sequence and ensure its accuracy. If things are not correct, the byte count acknowledgment (ACK) or nonacknowledged (NAK) is sent back to the sending end. The sending end receiving a NAK will resend the bytes necessary to fill in the blanks. Further, TCP holds all of the later received bytes (packets five, six, and seven from the previous example) until it gets four re-sent. Thus, it will buffer the data at the receiving end. This makes the responsibility of the data reception and accuracy the end user (node) responsibility, not the network's responsibility. The reason for this is to prevent the network from getting bogged down if errors or loops are occurring. The network is after all a transport system, not a computer processing function.

Using TCP/IP, the network operates efficiently and improvements are being seen every day as the quality of the circuits are improved with all of the fiberoptic backbones being installed. Therefore, the movement in the past few years has been heavily toward TCP/IP.

15

Packet Switching Technologies (X.25)

We have spent a considerable amount of time throughout the past chapters covering the pros and cons of the dial-up telephone network. Keeping in mind that when the network was originally fashioned, it was designed as an analog voice telephone network. Restating this position, it was a network that was designed to carry voice communications only.

In the late 1950s, the thought of carrying data communication began to blossom. Newer needs to move information from a user to a mainframe computer system rapidly emerged. Therefore, the industry was challenged with a means of creating a device that would allow a terminal device in a remote office (across town, or the country) to communicate with a central computer system, a.k.a., the mainframe. This was not a major problem because many other systems were already in place to carry low-speed data-type communications (Telex, TWX, etc.). Dial-up modem communications on the telephone network were already moving about rapidly. Most users were in the mode of transmitting their data across the dial-up telephone network at relatively low speeds, such as 2400 or 4800 bits/s. Only in rare cases did 9600 bits/s come into play. These were rudimentary data transfers and host access application being used on the network.

A problem existed, however, with the data transfer. In our data communications chapter, we discussed the problems that could be experienced on the dial-up telephone network. Because the use of a modem implies that an analog data transmission is used, all of the problems inherent with analog communications were present. For voice the analog systems posed no major problem. If a human does not understand part of a conversation, the problem is easily rectified. We merely shout "What?" into the phone, and the conversation is immediately retransmitted. That is, the person on the other end repeats what was just said. If the line is so bad (e.g., static, snap, crackle, and other noises) then the two humans agree to hang up and replace the call, hoping for a better connection.

Data Communications Problems

In the data communications world things were different. Back in the early days of data transmission, transmission protocols were not as sophisticated as they are today. Consequently, if a data transfer was attempted, the overhead was significant. Refer back to the overhead in the data communications chapter of this book, and you see that with an asynchronous transmission, using an ASCII code sent over the dial-up telephone network used approximately 40%. This was a tremendous burden in the data transfer. When you apply a 2400 bits/s transmit mode with 40% overhead, the net data rate is only around 1440 bits/s.

Again, if you think back to the earlier chapters discussing voice communications, the cost per minute for a voice (or data) call was $0.50 to $0.60. This overhead on the call meant that approximately $0.20 to $0.24 per minute was automatically lost to transmitting nothing usable. Clearly the problem became more pronounced as the data needs grew. Adding to the problem, the network was highly unreliable for data transmission. You can only imagine how frustrating it was to set up a file transfer of 2 megabytes of information.

Transmission protocols that were rudimentary in their approach to handling data transfers were still limited. During the data file transfer, if the end user got 1.5 megabytes transmitted over the network, timing was critical. A 1.5-megabyte file transfer at 2400 bits/s would take a significant amount of time. As a matter of averaging this out, the connection would take approximately 83 minutes (without the associated overhead). In reality, the time could be 116 minutes (close to two hours). Now imagine that this 2-megabyte file got through the first 1.5 megabytes in 116 minutes, then the inevitable happens. The line fails for a number of reasons. Whether a poor connection existed or the network degraded over time, the result was the same. The call disconnected three quarters of the way through the transmission. Clearly today, this would pose no problem. By redialing the far end and establishing the connection, one could pick up where the transmission left off. Not so back then. If the call disconnected, the protocols could not pick up in the middle of a data set and begin where it left off. As a matter of fact, everything had to be scrapped and the whole file would have to be retransmitted. Not only was this frustrating, but expensive too! After 116 minutes of transmission at $0.50 per minute, you just scrapped $58.00.

The Data Communications Review

This scenario was more the norm, rather than the exception. Users were easily frustrated with the communications needs and dreaded using a modem. Clearly, the industry had to do something about the problem. In the late 1960s, the CCITT commissioned a study group to look at a resolution to this and many other problems. The study group arrived at several initial findings:

- The network is unreliable. Further, the use of data transmission across the analog network is at risk.

- The roll-out of digital services is still very limited, and the service is primarily a carrier service.

- Other systems, such as message store and forward, although more reliable than straight dial-up connections, still get congested and deliver messages at the customer demand. Therefore, the network can delay delivery for hours if not days based on the congestion factor.

- Other services do not offer any reliability of their services, regardless of the delay.

The resulting response to this whole study was the recommendation to use a packet switching transport system, that would guarantee the reliable delivery of data, break the data down into more manageable pieces, deliver the data at the convenience of the network, sequentially deliver the information, and recover from any failures that occur on the link without having to retransmit the entire file.

Packet Switching Defined

Thus, packet switching was born to accomplish these goals. To define packet switching then we will use the following as the guidelines:

Packet switching is a means of taking a very large file of information (data) and delivering it to a piece of hardware or software. From this interface, the hardware or software will break the information down into smaller, more manageable pieces. As these pieces are broken down, additional overhead will be applied to the original segment of data. This overhead is used for control of the information. Because the information is segmented, the packet service will insert the telephone number of the addressee, along with the segmentation number (aka, packet #1, packet #2, packet #3 . . . etc.) so that the data can be reassembled at the receiving end. Once the overhead is attached to the segmented data (now called a packet), the packet will be transmitted across a physical link to a switching system that reads the address information (telephone number) and routes the packet accordingly. This will establish a virtual connection to the distant end and each packet will be sent along the same route as the first packet. The system will use a connection-oriented transport, based on a virtual circuit.

That is quite a definition to work from. From this as a starting point, the rest should be fairly easy to break down and discuss. As a matter of fact we can packetize the definition and work at defining it piece by piece.

What Is Packet Switching?

As already mentioned, the use of packet switching is a means of breaking down the larger data files into more manageable pieces. Here's an example that might make the whole thing a little easier to comprehend.

The Packet Switching Analogy

For the sake of describing this concept, visualize a meeting with all of your peers in the industry. You have all been called into a local meeting to discuss a matter of extreme importance. The U.S. Postmaster General asked that you meet regarding a serious problem. The problem is that more and more people (like you) are using other types of delivery services for various articles (i.e., mail, packages, etc.). This, of course, detracts from the U.S. Postal Services (USPS) revenues. Therefore, the Postmaster General (PG) has an offer to get you back as a customer for your routine deliveries. The scenario plays like this: AU are the authors' comments.

PG: We know that you have some concerns about using the U.S. Postal Service. Therefore, in order to get you back as a customer, we are willing to hold the price of a postage stamp at $0.32! Yes the rumors are true that by 1997 the cost of a stamp will rise again, but if you'll sign an agreement here today, we will hold the rate at the $0.32.

AU: How many of you would sign an agreement to stay with the Postal Service for all of your delivery needs? Just as we thought, this was not enough to move you.

PG: OK, let's sweeten the pot. Not only will we hold the price of the stamp at $0.32; if you agree to give the USPS 100% of your mail, we'll lower the cost of a stamp to $0.25 and hold that rate for five years. This I guarantee.

AU: Some people will be moved to sign this agreement now! Many of us think in terms of business needs and money is always a decision factor. But, only a few of the audience will bite for this deal. This leaves a bunch more that have not committed. Now for the PG's next move . . .

PG: Okay folks, here's my last offer. I will drop the price of a stamp to $0.25 and hold it there for five years. Also, I guarantee that any letter you drop in a USPS mailbox will be delivered anywhere in the country overnight! Now how many of you will sign the agreement.

AU: By now you are worn down and will sign on the bottom line. This is an attractive offer. For five years, you get cheaper postage and overnight delivery guaranteed. How can you lose? So you go ahead and sign.

PG: *Now that you've all signed on the bottom line, we have a contract that you will use the USPS for all of your mailing and delivery needs. However . . .*

AU: Watch out! Here it comes.

PG: *In order to do all these things we contracted to do, I have to give you the conditions of this agreement. You never asked, so I can assume that overnight delivery at $0.25 per letter is what concerns you the most. To meet the terms of our agreement, the conditions are:*

All mail that you wish to send through the USPS must be placed in a standard #10 business envelope. This is shown in Figure 15.1. The envelope will be a white 9½" × 4" solid (no windows) business envelope. There shall be no other envelopes used: We don't want to see the holiday red or green; no Mother's Day orchid colored; no big seasonal envelopes for birthdays and Valentine's Day, etc.

In each of these business envelopes, you can only put one sheet of paper (standard U.S. business page of 8½" × 11"). You can print on it front and back, but still limit the contents to only one page per envelope. See Figure 15.2 for the standard page letter.

For those of you who have a major pamphlet, book or other document to mail, the solution is easy. If the pamphlet has two hundred pages, you merely have to send two hundred envelopes. This means that at each end you will hire a person to handle the process. On the sending

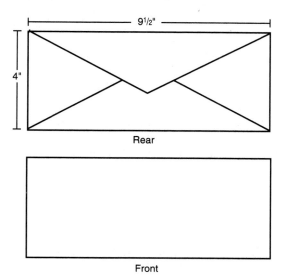

Rear

Front

Figure 15.1 The standard business envelope will be used.

Figure 15.2 The standard page for letters shall be 8 × 11 inches. You can print on both sides of the page.

end, the postal clerk in your organization will take the pamphlet apart and stuff one envelope at a time with one of the pages. This will require a total of 200 envelopes and stamps ($50.00). The mail clerk will then add some overhead to the envelope such as:

- *A "from" address will be placed in the upper left-hand corner of the envelope.*
- *A "to" address will be placed in the center of the envelope.*
- *A $0.25 postage stamp or meter imprint will be placed in the upper right-hand corner of the envelope.*
- *To be assured that everything works okay, in the bottom right-hand corner of the envelope, have the mail clerk label this as 1 of 200, 2 of 200, 3 of 200 . . .*

AU: How are you doing? Are you still with us so far? Does this sound like a workable solution just to save a few dollars? Will you really save that money, probably not! But let's continue, there's more.

PG: Once the letter is ready for delivery to the USPS, (see Figure 15.3 for the final envelope), your mail clerk has to drop these into the mail box or at the post office. Your job is partially done. At the mail box, a new set of sequences will begin:

- *When the postal employee comes along, the envelopes will be stacked up and checked (counted) to make sure that all of the envelopes are there. If the postal clerks find that one is missing, they will search around in the mail box for the missing envelope. If this proves futile, they will contact the mail clerk in your organization and ask that a new copy of the missing envelopes be generated. This of course means that the mail clerk must keep a copy of every page until we accept the envelopes and determine that none are missing.*

- *If none are missing or if this did exist and they were replaced, the mail clerk will receive an acknowledgment from the postal employee. The mail clerk can now discard all of the spares that were held (called buffering).*

- *Before the acknowledgment is given to the mail clerk, the postal employee will do a couple of things. The envelope count confirmed, the postal employee will open each and every envelope and read all of the information to ensure that the information contained inside is intact. If something happened to the page, as shown in Figure 15.4, we merely alert the mail clerk to give us a new copy. Secondly, as the information is read and validated, the Postal employee will make a duplicate of each page of information and store it in a holding bin.*

- *After these have been completed, the acknowledgment gets sent to the company mail clerk who empties the holding bin at the originating location (flushes the buffer). What also happens here, the USPS is now responsible for the letter and the integrity of the information. The letter will be passed on to the next location in the chain.*

- *At each step along the delivery trail (e.g., the mail box to the post office, to the plane, to the receiving post office, etc.) the information will be counted, opened and read, copied and accepted. The previous holder of the information will be given an acceptance, then destroy the copies. If anywhere along the line, the system breaks down, the holder of the stack needs only to go to the previous location and request a new copy.*

- *At the end point (your location), the process will end. A mail clerk at the receiving end will count all the envelopes, open and read them and accept them one at a time. But in this case, the mail clerk will finalize the process by putting the page in each envelope back in the original order until the pamphlet has been reassembled.*

- *Your organization has just received overnight delivery of the letters, with guaranteed integrity of the information and sequentially delivered in order.*

AU: By now you are saying to yourself "These authors have been sniffing too much laser printer toner while preparing this book." This whole scenario sounds too complex to be real, doesn't it? But, this exactly is how X.25 packet switching works. The process is a little simpler, but the concept is the same. How does this sound now?

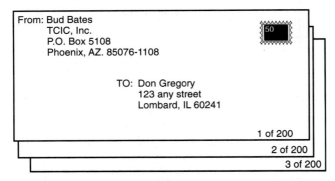

Figure 15.3 The completed business envelope has all the necessary information. Note the numbering system on the bottom right corner.

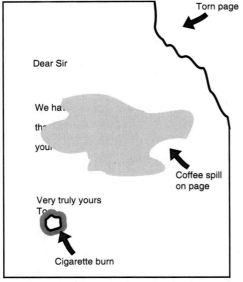

Figure 15.4 Many things can happen destroying the integrity of the printed page.

The Packet Concept

Packet switching works like the scenario described above. If an organization has large amounts of data to send, then the data can be delivered to a packet assembler/disassembler (PAD). The PAD can be a software package or a piece of hardware outboard of the computer system. The PAD acts as the originating mail clerk in that the originating PAD receives the data and breaks it down into manageable pieces or packets. In the data communications arena, a packet can be a variable length size of information, usually up to 128 bytes of data (one page in the example). Other implementors of X.25 services have created packets up to 512 bytes, but the average is 128. The 128-byte capability is also referred to as a *fast select*. The packet switching system can immediately route the packet to a distant end and pass data of up to 128 bytes (1024 bits).

Overhead

To this packet, the PAD then applies some overhead as follows:

- An opening flag that is made up of eight bits of information. Using a standard high-level data link control (HDLC) framing format covered in the data and the LAN chapters of this book, the opening flag is a sequence of eight bits that should not be construed as real data. (This is an envelope.)
- A 16-bit address sequence that is a binary description of the end points (the from address).

Control information consists of eight bits of information to describe the type of HDLC frame that is traversing the network (this is a notation on the envelope that describes the information inside). These can be supervisory, unnumbered or information fields. To be more specific these break down as follows:

- Information (I) is used to transfer data across the link at a rate determined by the receiver and with error detection and correction.
- Supervisory (S) is used to determine the ready state of the devices, receiver is ready (RR), receiver is not ready (RNR), or reject (REJ).
- Unnumbered is used to set parameters, such as set modes, disconnect, and so on.

Packet-specific information follows the HDLC information. The packet information will consist of the following information (this information is similar to the "To" address and the designation of the routing that will be used such as first class, book rate, etc.):

- General format identifier (GFI), four bits of information that describes how the data in the packet is being used; from/to an end user, from/to a device controlling the end user device, etc.

- Logical channel group number (LGN), four bits that describe the grouping of channels. Recognizing that only four bits are available, only eight combinations are used.

- Logical channel number (LCN) is an 8-bit description of the actual channel being used. The theoretical number of channels (ports) available is 2048. The logical channels are broken down by channel groups so that the numbers play out (Table 15.1). Although the number of logical channels can be 2048, most organizations implement significantly less ports or channels.

The next overhead elements consist of the packet type identifier (PTI), an 8-bit sequence that describes the type of packet being sent across the network. Six different packet types are used in an X.25 switching network. These packet types define what is expected of the devices across the network. The packet types are shown in Table 15.2. Note that the packet types have different uses, so their binary equivalent is shown for the benefit of separation.

The variable data field is now inserted. This is where the 128 bytes of information are contained in the packet. The 128-byte field is the standard implementation, but as mentioned it can be larger (as much as 512 bytes).

Following the information field is the CRC, a 16-bit sequence that will be used for error detection and/or correction. Using a CRC-16, the error detection capability will be approximately 99.999995%. This is to ensure the integrity of the data, rather than having to deliver information over and over again, the concept is to deliver reliable data to the far end.

The closure of the packet is the end of frame flag. In the HDLC frame format this denotes the end of the frame, so the switches and related equipment know that nothing follows. The switches then calculate all of the error detection and accept or reject the packet based on the data integrity.

Summary of Packet Format

Figure 15.5 is the actual HDLC frame format used for the packet of information. In this case, one can look at the fields and recheck through the sequencing information of what each field represents. Users who have become intimidated through this process can relax. This is done by the equipment that will be generating the packets across the network.

The Packet Network

Figure 15.6 shows typical network layout. Here, the cloud in the center of the drawing represents the network provided by one of the carriers. From our discussions in earlier chapters, the obvious network configuration is

TABLE 15.1 Summary of Logical Groups and Channel Numbers. Although the Number of Logical Channels Can Be 2048, Most Organizations Implement Significantly Less Ports or Channels.

Type	Logical group	Logical channel	Number of channels
PVC	0	1–255	0–255
PVC	1	256–511	0–255
Incoming only	2	512–767	0–255
SVC	3	768–1023	0–255
Two way	4	1024–1279	0–255
SVC	5	1280–1535	0–255
Outgoing only	6	1536–1791	0–255
SVC	7	1792–2047	0–255

TABLE 15.2 A Summary of the Packet Types and Identifiers

Packet type	Description	Packet type identifier
Incoming call	Call request	00001011
Call connected	Call accepted	00001111
Clear indication	Clear request	00010011
Clear confirmed		00010111
Data		xxxxxxx0
Receiver ready		xxx00001
Receiver not ready		xxx00101
Reject		xxx01001
Interrupt indication	Interrupt request	00100011
Interrupt confirmation		00100111
Reset indication	Reset request	00011011
Reset confirmation		00011111
Restart indication	Restart request	11111011
Restart confirmation		11111111

that each of the designated carriers has their own cloud. Actually, they all interconnect to the clouds provided by each other so that transparent communications can take place. Once the cloud is established, the next step is to provide the packet switching systems (called packet switching exchanges [PSE]). These are nothing more than computers that are capable of reading the address and framing information. They then route the packet to

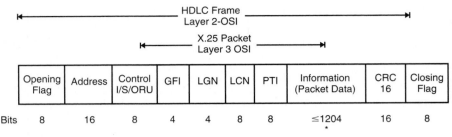

Opening Flag	Address	Control I/S/ORU	GFI	LGN	LCN	PTI	Information (Packet Data)	CRC 16	Closing Flag

Bits   8   16   8   4   4   8   8   ≤1204   16   8
                                                                                      *

I = Information
S = Supervisory
U = Unnumbered   * Variations can exist using the 4096 bits of data

Figure 15.5 An HDLC frame is used to carry the X.25 data to the network. Note that the packet is layer 3 and encapsulated into the data link frame.

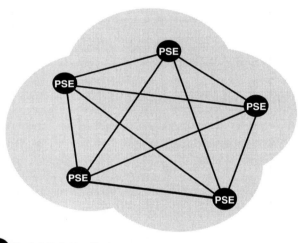

PSE Packet Switching Exchange

Figure 15.6 The typical X.25 network is a meshed network. The cloud represents the network.

an appropriate outgoing port to the next downstream neighbor. In many cases, these PSEs are connected to several other PSEs.

Therefore, a meshed network is provided by the carrier, as shown in the figure. The packet switches can select the outbound route to the next downstream neighbor based on several variables. The selection process can be the least used circuit, the most direct, the most reliable or some other predefined variable. This again is the magic that takes place inside the cloud. Now that the network cloud and the packet exchanges are in place, the next step is to connect a user.

The User Connection

Users sign up with the carrier of choice and let the carrier worry about the physical connection. As shown in Figure 15.7, the user connection into the cloud is through a dedicated or leased line. The carrier notifies the LEC and orders a leased line at the appropriate speed (in the leased line, this can be up to 64 Kbits/s on digital circuits). The original network connections back in the initial rollout of X.25 services were on analog lines at up to 9.6 Kbits/s. A modem was either provided by the carrier at the customer end, or the customer purchased and provided a modem. Now that the modem is attached at the customer end the circuit is terminated in a port on the computer (PSE). This is the incoming port that can be used as a permanent virtual connection or as an incoming only channel. This is part of the addressing mechanism inside the packet where the PSE reads the address from where the packets are coming. The incoming only channels are the channel numbers assigned to each customer.

The next step follows the connection where the customer must initiate the PAD function as shown in Figure 15.8. Remember that the PAD is the hardware or software installed to break the data into smaller pieces, attach the overhead, and forward the packetized data to the network. In the reverse order, the PAD is responsible to receive the packets, peel off the overhead and reassemble the data into a serial data stream to the data terminal equipment, whether it is a terminal or host computer. So the PAD function is crucial. If this is a software package that is performing the PAD function-

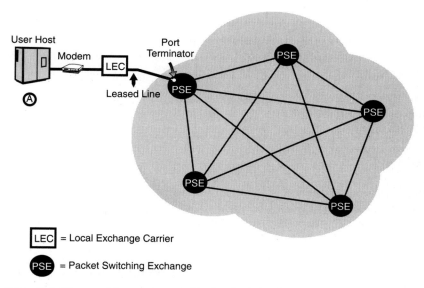

LEC = Local Exchange Carrier

PSE = Packet Switching Exchange

Figure 15.7 The user A is now connected to the cloud via a lease line.

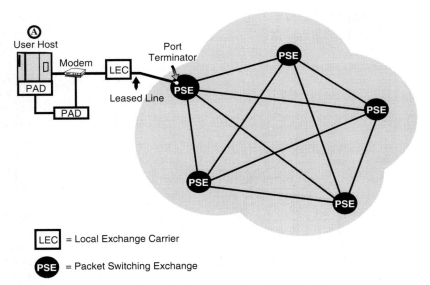

LEC = Local Exchange Carrier

PSE = Packet Switching Exchange

Figure 15.8 The user A can have a hardware or software PAD function.

ality, then it is the customer who must purchase (or license) the software and install it. If the solution is a piece of hardware, the options are different. The customer might buy the PAD and install it; or the carrier might provide and install it. This can be rented, leased, or sold by the carrier. Now the connection exists on one end of the cloud.

In Figure 15.9, another device is attached on the other end of the cloud. What happens now is the magic of the packet switching world. As packets are generated through the network from user A to B across the network several things happen, as shown in Figure 15.10.

- The data is sent serially from the DTE to the PAD (that acts as the DCE for the computer terminal).

- The PAD will break the data down into smaller pieces and add the necessary overhead for delivery.

- The PAD then routes the packets to the network PSE.

- The receiving PSE (PSE 1) sees the packet coming in on a logical channel so it remembers where the packets are coming from. After analyzing the data (performing a CRC on the packet) and verifying that it is OK, the PSE will send back an acknowledgment to the originating device (which is the DCE).

- The PSE then sends the packet out across an outgoing channel to the next downstream neighbor (PSE 2). This establishes a logical connection between the two devices (PSEs) from the out channel to an in channel at the other end. The logical channel is already there, it is used for the transfer of these specific packets. A virtual connection is also created, allocating the time slots for the packets from A to B to run on the virtual circuit.

- Back at PSE 1, the next packet is sent down from the originating PAD. This again is analyzed and acknowledged, then passed along. At the same time this is happening, PSE 2 is sending the first packet to PSE 3. Once again at each step of the way the packets are opened and a CRC is performed before the packet is actually accepted.

- At each PSE along the way, the packet is buffered (at PSE 1) until the next receiving PSE accepts the packet and acknowledges it (PSE 2). Only after the packet is acknowledged, does the first one (PSE 1) flush it away. Prior to that the network node (PSE 1) stored it, just in case something went wrong.

- Don't forget what happens at the receiving end (device B), where the packets arrive in sequential order, are checked and acknowledged. Then, the overhead is peeled away so that a serial data stream is delivered to the receiving DTE.

- This process continues from device A through the network to device B until all data packets are received. Every packet along every step of the link is sent, accepted, acknowledged and forwarded until all are through. Here is the guaranteed delivery, of reliable data properly sequenced. The logical link that is established between devices A and B are full duplex. The two devices can be sending and receiving simultaneously.

You can see from the packet delivery process that there is a benefit to the packet switching process. However, nothing is perfect. The overhead on the packet, the buffering of multiple packets along the route, the CRC performs at each node along the network and the final sequenced data delivery, all combine to present the risk of some serious delays. What you receive in the integrity and reliability can be offset in delays across the network. You must always weigh the possibilities and choose the best service.

Benefits of Packets

The real benefit to this method of data delivery ties back to the scenario painted in the beginning of this chapter. Remember the problems with the dial-up telephone network and the risk of sending ¾ of a file transfer, only to have a glitch in the transfer? Using the packetized effort, if a glitch occurs the network might have from 7 to 128 outstanding packets traversing the links. Therefore, instead of scrapping the entire file, the network automatically recovers and resends the packets that were lost or corrupted. This

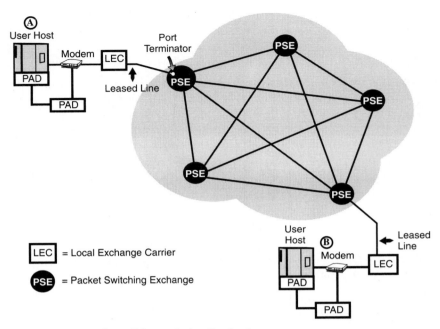

Figure 15.9 A second user B is attached to the cloud.

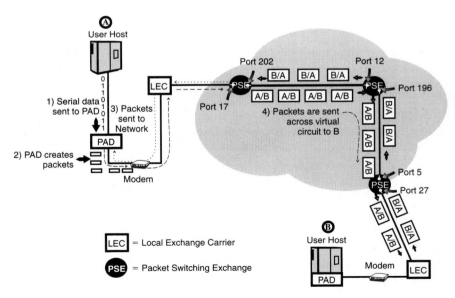

Figure 15.10 Serial data sent to the PAD get packetized (A/B) then sent across a virtual circuit to location B. B can be returning responses B/A on a duplex circuit.

means that the users would save time and money on an error-prone network. But, again the risk of congestion and delay on the network might cause others to look for alternative solutions.

Other Benefits

Above and beyond these benefits, other benefits can be achieved from the use of the dedicated link into the network. As Figure 15.11 shows, the link is not solely for one user at a time, nor is it for two specific locations. When the organization uses a packet switching network, the users might have the need to have multiple simultaneous connections up and running. Therefore, the PAD will act similar to a statistical time division multiplexer (statmux). The statmux capability allows multiple connections into the single device, and will sample each of the ports in a sequential mode to determine if the port has anything to send. If a packet has been prepared, the statmux will generate the call request (initiate a call) to the network. The connection will be created and the data will flow. This assumes that no problems are being experienced on the network.

As a new user logs on and generates a request to send data, the PAD will then set up this connection. Packets will then be interleaved across the physical link between or among the various users. One can imagine that not all devices are going to transmit at the same rate of speed, or will not have as much two-way interactive traffic. Therefore, the statmux function of the PAD interleaves the packets based on an algorithm that allows each device to appear as though a dedicated link is available to it. The use of a statmux

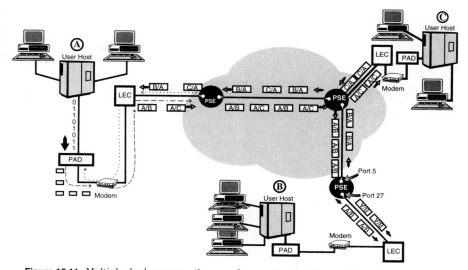

Figure 15.11 Multiple duplex connections can be running simultaneously on the same circuit.

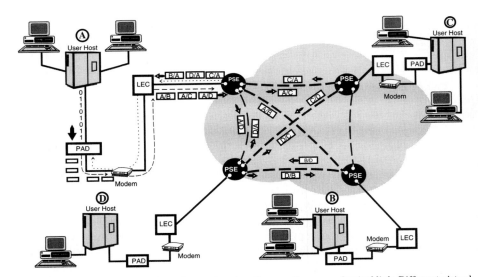

Figure 15.12 Packets are interleaved to various locations on the same physical link. Different virtual circuits are used to get the packet to its destination.

capability will be a benefit by using the expensive leased link to the maximum benefit of the organization. This requires less physical links and takes advantage of the dead time between transmissions, etc.

Figure 15.12 shows a series of sessions running on a single link, all interleaved. This also shows that packets are not specifically interleaved by order of A,B,C . . . etc. Instead, they can be interleaved based on the flow or delivery method used, such as A,B,A,B,C,B,A . . . Herein lies the added benefits of the packet switching world, so that the user achieves the data throughput necessary without having a dedicated resource that is only periodically used.

Advantages of Packet Switching

Packet switching is considered by many to be the most efficient means of sharing both public and private network facilities among multiple users. Each packet contains all of the necessary control and routing information to deliver the packet across the network. Packets can be routed independently, or as a series of packets that must be maintained and preserved as an entity. The major advantages of using this type of transport system are:

- Access shared among multiple users, either in a single company or in multiple organizations.
- Full error control and flow control in place to allow for the smooth and efficient transfer of data.

- Transparency of the user data.

- Speed and code conversion capabilities.

- Protection from an intermediate node or link failure.

- Pricing advantages, because the network is used by many the prices are based on per packet, rather than cost per minute.

Other Components to Packet Switching

Although the primary components of the packet switching world have already been discussed, there are other pieces. As the evolution of the X.25 standard continued from the early 1970s, several different implementations were enhanced. This could include devices or access methods that you should understand. These are summarized so that you can gain an appreciation of how these pieces all work together. Some of these added pieces include:

1. The ability to dial into the packet switching network from an asynchronous modem communication. Although packet switching (ala X.25) is a synchronous transfer system, the need for remote dial-up communications exist. Therefore, the standards bodies included this capability in one of the enhanced versions of the network. As shown in Figure 15.13, a dial-up connection can be made from a user. In this case, the asynchronous communications will be a serial data transfer to a network-based PAD function. The PAD will accept the serial async data, collect it, segment it into packets, and establish the connection to the remote host desired. In this case, the connection is now across the X.25 world synchronously moving packets across the network. This uses an X.28 protocol to establish the connection between the async terminal and the PAD.

2. The PAD-to-host arrangement is controlled by the X.29 protocol. Control information is exchanged between a PAD (X.3) and a packet mode DTE or another PAD (X.3). This is shown in Figure 15.14. In this case, as the communications is established between these devices, the X.3 PAD will set the parameters of the remote device. This could be the speed, format, control parameters, or anything that will be appropriate in the file transfer. The X.29 parameters can also provide keyboard conversion into network usable information.

3. The internetworking capability of an X.25 network uses a protocol called *X.75*. Although this should be user transparent, the network needs the X.75 parameters to provide a gateway function between two different packet networks or between networks in different countries. In each case, the gateway function is something that should only concern the network carrier. However, as more organizations install their own private network switching system, the need to interconnect to the public data

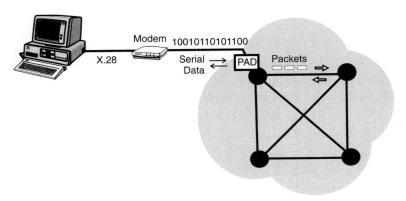

Figure 15.13 Asynchronous dial-up terminals can dial into the X.25 network through a PAD on the network.

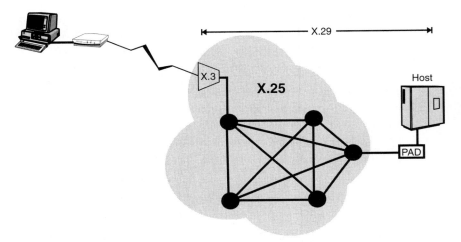

Figure 15.14 The X.3 PAD can set the remote host parameters to handle the ASYNC user.

networks rises. Therefore, the internetworking capability is moving its way closer to the end-user door. A representation of an X.75 interconnection is shown in Figure 15.15.

The X.25 Numbering Plan

The X.25 numbering plan takes advantage of the worldwide numbering plan designed by the CCITT (ITU-TSS). It is known as the *X.121* worldwide network numbering plan for the public data networks. This was a design to allow for public networks only. However, in most of the world this is acceptable because only one public network exists, controlled and operated by the Post Telephone and Telegraph (PTT) organizations.

In the U.S., however, the proliferation of both public and private networks is completely different. Looking at all of the players involved today in the carrier community, you can see that easily 300 plus networks could exist publicly. Couple that with the private organizations who can add packet switching to their networks and want to access public networks, the number can expand exponentially.

Applications for X.25 services

There are no single solutions that fit any transport system. Therefore, the variations of the X.25 capabilities are as numerous as the organizations using them. However, if you think in terms of the overall use, the following can be applied:

If a user organization has data to transfer between or among a limited number of locations and the data transfer will be continuous (such as 8-12-24 hours per day) or the files are constantly very large, the use of X.25 might or might not apply. In this case, the connection will likely be a leased line between two hosts (for example) and the data transfer will be constant.

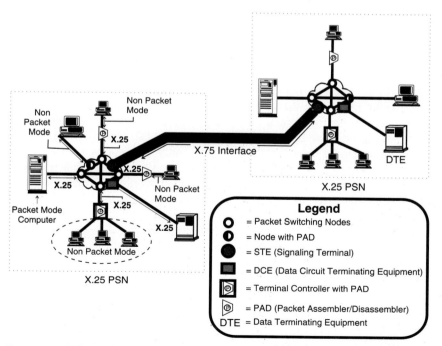

Figure 15.15 X.75 Interconnectivity protocols allow different networks to communicate.

The leased line is an expensive proposition unless the volume of data justifies the connection full time. A data transfer will use very little time to initiate a session between hosts or end-to-end devices because they are physically connected together.

If on the other hand, a user organization has multiple locations (10, 20, 30, etc.), and they only need to establish a connection occasionally during the day (twice or thrice as examples), and the data is batched in a large transfer of one hour of information transfer, then a dial-up circuit might be appropriate. In this case, the cost associated with a leased line would likely be unjustified. Therefore, a dial-up modem communications will be a better solution. The time it takes to generate a connection and handshake between the modems can be approximately 1 to 2 minutes. As a portion of the data session, the call setup is insignificant. Again the dial-up communications will be a possible solution.

If, however, an organization has many (50, 100, 200, 1000) locations that must pass information to a central computing platform, and the amount of data is relatively trivial, then packet switching might be the most appropriate solution. In this case, let's use a department store that has 100 locations spread around the country. The stores use a point of sale system (POS) to run the registers, inventory system, and other associated data processing functions. Each night, the store must communicate back to the home office with three pieces of information: how much the daily sales were, the amount of money deposited in the local bank for transfer back to corporate, and any items that are totally out of stock and need immediate replenishment. The entire data transfer here will be approximately 10 seconds.

It would not be prudent to spend the money to lease data lines to each of the 100 locations. The expense could never be justified. Even in a multipoint circuit environment this will be too expensive.

A dial-up connection would be inappropriate, because at 1–2 minutes of set-up and handshake it will take longer to establish the connection than the actual transmission of information when only 10 seconds of data will be transferred. Secondly, the amount of time lost to sequentially dial 100 locations with the associated timing and handshakes, would take far too long to get all the calls through in a reasonable amount of time. Therefore, the packet switching arrangement would be more appropriate.

Using the packetized data, several simultaneous devices can be generating packets of information to the host and transmitting the nightly information across the network. Different logical connections will be established and the packet reassembly process will take its due course to deliver the required information in the appropriate form.

An organization that has nightly or periodical information to transfer is a prime candidate for packet switching. Other forms of information transfer that could be served by a packet switching service are:

- Transactional processing, such as automated teller machines (ATMs).
- Process control information such as electrical distribution, oil drilling, and chemical production facilities.
- Database inquiries.
- Credit card approvals.
- Motor vehicle registration information.
- Signaling system seven (SS7) messages for call set-up and tear-down.
- File transfers where reliable and guaranteed data delivery is required.
- Money transfers, such as on the SWIFTE network.
- Others.

One can see that many of these activities can take full advantage of the packetization of the data streams and still conduct the normal day-to-day business activities.

Other Forms of Packets

As an industry accepted standard, X.25 is the widest installed and accepted form of data transmission. This is a standard that is accepted worldwide, and the interconnectivity solutions defined have created a useable form and format throughout the world. In later chapters of this book, we discuss other forms of data transmission. However, these are emerging forms of packet switching (fast packets) that are not ubiquitous. Other forms of transmission might or might not be in use in third world countries. Speeds are limited and facilities to transmit at various other means might not yet exist. Therefore the X.25 standard has become the industry norm.

Still though, other forms or variations of packet switching exist today. One is called *fast select*, which is a single packet transfer (rather than a sequentially numbered series of packets) where only one packet needs to be sent. This can be applicable in a Unix world where a single character is sent across the network to a Unix host. It would be inappropriate to have a rash of packets sent across the network when all that is needed is a sole packet. Another use of such a system is in the credit card approval. The single packet that gets generated is limited in the data needs, such as:

- The credit card number
- The expiration date of the card (optional)
- The vendor ID number requesting the approval
- The dollar amount of the transaction

In this case, because only a very limited amount of information is required, it can all be contained in a single packet. The network does not use the sequential transfer as it does in X.25, nor does the reliability check take place on the network. Instead, the packet is acknowledged by the POS equipment when the main credit clearinghouse sends back an acknowledgment. If something is corrupted in the transfer of the packet, the vendor must retransmit the entire transaction. This is a sporadic type of data transfer and works well for the most part. Delays do occur, but mainly as a function of the clearinghouse computers being busy.

Still another form of packet switching is used in the industry, called a *Datagram*. A *datagram transfer* is a form of packet switching. Typically, the datagram does not follow the same virtual circuit concept. Every packet is sent across the network and can take a different route to get from point A to B. Also, a datagram is not checked at every node along the route, only the header (or address) information is checked for the location of the destination. As each packet is received, it is scanned for the destination address then immediately routed along to the next node. No CRCs are done on the data. It is not the network's responsibility to guarantee the integrity of the data, that will be left to the receiving device using a higher level protocol. With this concept of a different path for each packet, the idea of sequentially numbering the packets is lost. The network treats every packet as its own entity (or as packet number one). If a packet gets lost or whatever, the network does not care. The sequencing and reassembly of the data back into its original form is handled by the receiving end's higher layer protocols. An example of this is transmission control protocol with internet packeting (TCP/IP). IP packets are datagrams that are segments of the original data stream. It works at the layer 3 of the OSI model, as covered in Chapter 14. As a network layer protocol, the only concern of IP is to break the data down into packets and make a best effort to deliver the packets.

TCP at the receiving end is responsible to reassemble the data back into the serial data stream, and ensure the integrity and sequencing of the information. TCP is not a network layer protocol, it works at the higher layers of the OSI model (transport and above). TCP/IP is the mainstream of the internet and has been widely adopted as the protocol stack of choice in many LAN and WAN arenas. Because of its robustness and the ability to deal with the packetization of the data, more organizations are using this as their WAN protocol. TCP/IP is not without its problems, but the advantages far outweigh the disadvantages. The industry as a whole and in particular the vendor community has recognized the importance of TCP/IP in their mainstream of products. Just about every LAN network operating systems (NOS) vendor now supports the use of TCP/IP. From a physical network and data link layer, TCP/IP will run on most any topology. It is supported by the hardware suppliers of network interface cards (NICs).

Local Area Networks (LANs)

What Are LANs?

The digital data communications technologies covered up to this point can function over essentially arbitrary distances. Whether one mile or 10,000, they will work about the same. But certain data communications functions would be much more useful (e.g., file sharing and transmission) if higher speeds could be used than are generally available on a wide area basis. What if some slick optimization and engineering techniques were used to radically improve communications performance, at least for short distances? The productized answers to these questions are local area networks (LANs).

In the early 1980s, a new device made its way to the desktop. IBM introduced the PC, a single diskette system that was functionally a terminal device that just happened to have its own intelligence. The single floppy diskette was sufficient to run the entire system and still allow the user plenty of storage space. The second version of IBM's desktop computing system (PC) was a dual floppy disk system, where one diskette was used to run the operating system for the machine, and the other floppy was used for the data storage. Not much exciting happened during this era, but there were rumblings of the single and dual diskette systems being too limited and awkward to use. The industry had no problems. Then, it happened!

The third evolution of the PC introduced by IBM had a hard disk. The disk operating system was stored on the hard disk so that the machine could operate without playing around with those floppies. This in itself was a major improvement, but this version of the PC was a bit pricey ($5000 to $6000). Further, management and information systems personnel had to reckon with a different situation. The hard disk had a total capacity of 10 megabytes. These professionals knew back then, just as you and I would know, that no one would ever need 10 megabytes of hard disk space all to themselves. Therefore, a new sharing of resources was needed. The first LANs were resource sharing capabilities so that five users could share that 10-megabyte disk. Hence, the LAN was born. Because it was hard disk sharing, this was typically performed in one department and was very localized. The actual LANs started out as serial port connections where users were daisy chained together to access the disk.

A little later on, the introduction of the laser printer (HP's wizardry) led to a new problem. One specific user in the department might have a laser while others had a dot matrix printer. In the early 80s, this dot matrix output was nothing to write (or print) home about. So, everyone wanted their own laser printer on their desk, rather than have the dot matrix. Clearly, management did not want to spend $5000 for every employee to have their own laser printer. The LAN was the resource sharing service, so the printer moved onto the network.

What is a LAN? Originally, the industry defined local area networks as data communications facilities with the following key elements:

- High communications speed.
- Very low error rate.
- Geographically bounded.
- A single cable system or medium for multiple attached devices.
- A sharing of resources, such as printers, modems, files, disks, and applications.

Let's take these one at a time.

High communications speed

At the time (the 1970s), use of communications speeds above one megabit per second on either a wide or a local area basis was exceedingly rare, other than locally connected terminals to a mainframe. So, "high-speed communications" meant in the millions of bits (or "megabits") per second. Even then, it was obvious that sub-megabit speeds would become a significant speed limit unless they could be vastly increased.

But why not simply use the existing WAN protocols for LANs? They certainly can be so used. The reason, in a word, is "overhead." One major source of overhead in WAN protocols is their routing capabilities. These range from relatively simple (e.g., IBM's SNA, that traditionally requires much human intervention to establish or change routes for network traffic) to extremely complex (e.g., Digital's DECnet, that can generally figure out where to send something without any assistance at all). But regardless of their sophistication, they add overhead not required in a local area network. So, LAN protocols eliminate routing overhead by functioning as though all nodes are physically adjacent. Another source of overhead is error checking, covered in the following paragraphs.

Most local area network technologies currently in use provide maximum communications speeds well above one megabit per second (Table 16.1). This is a summary of the typical speeds available from the various approaches. We look at the more common of these transport systems individually in later chapters. For the moment, the products that produce these speeds are what we are trying to get at.

Appletalk is the only true LAN in the table that falls below one megabit per second. It is nonetheless very popular due to its excellent ease of configuration (not to mention that it comes free with every Macintosh!).

Very low error rate

An "error rate" is the typical percentage of bits that can be expected to be corrupted on a regular basis during routine transmissions. As mentioned in the discussion of protocols, error checking ensures that such damaged bits are retransmitted so that applications never see the errors. But the data link protocols in use for wide area networks must contend with much higher error rates than would normally be experienced using LAN technologies.

Wide area analog communications typically experience somewhere between one error in 10^5 bits and one error in 10^6 bits transmitted. A "very low error rate" is at least two orders of magnitude better than this or one error in 10^8 transmitted bits. Several technologies meet this challenge.

In a typical WAN protocol "stack," error checking takes place at multiple levels of the stack. At each level, that error checking requires both additional bits in the transmitted frames and additional processing time at both ends (at the sending end, to create the error checking bits; at the receiving end, to determine if any errors occurred). If some of that error checking

TABLE 16.1 A Summary of Major Architecture Speeds

LAN name	Typical speeds
Appletalk	256 Kbps
Arcnet	2.5 Mbps
Token Rng (original)	4 Mbps
Ethernet	10 Mbps
Token Ring (updated)	16 Mbps

overhead can be eliminated because of fewer errors to catch, higher speeds can more easily be achieved—which, not just incidentally, is another of the prime characteristics of a LAN.

The physical elements of a LAN do not depend on the vagaries of the telephone companies' facilities. Those elements are all-digital, and are designed specifically to deliver high-speed communications with very low error rates. Thus, the higher levels of the protocols can be relied on to handle the few errors that do creep in. Most LAN protocols do little or no error checking, and this is one of the design elements that permits their higher speeds.

Geographically bounded

A local area network is bounded by some geographical limitation. (Otherwise, by definition it would not be local.) Engineering considerations and physics impose this limitation; vendors did not impose it by fiat. The technologies employed to meet the other requirements (especially the requirement for high speed at a low error rate) tend to preclude more wide-area transmission capabilities.

The geographic limitations of the various LAN technologies are expressed in different ways. There is usually a maximum distance from one node (or connected, network-addressable computer) to the next, expressed in tens or perhaps hundreds of meters. There is usually a maximum distance from one node to any other network node, typically hundreds or a few thousands of meters.

Network technologies that reach only up to the bounds of a single campus, or group of co-located buildings, satisfy these specific limitations. However, as you shall see, a number of new technologies and connectivity options have worked together to significantly mitigate the distance limitations, which applied to LANs as they were originally defined and designed.

Single cable system or medium for multiple attached devices

One of the less-obvious, but nonetheless key characteristics of a local area network is that all devices on it are viewed, and to all intents and purposes function, as being adjacent. This implies that no special routing capabilities are required for communications traffic from one device to another; just put the information on the LAN, and it will reach its destination. To put this into the ISO context, no routing functions should be required for a LAN to operate.

To accomplish this feat, all of the devices must reside, or appear to reside (from the point of view of an attached device) on a single cable system. Thus, each end node-to-end node link is also functionally a point-to-point link. No routing choices need be made if every machine you can talk to is right "next door." All signals from all devices propagate throughout that cable system.

Different types of LANs accomplish this function in very different ways. Some of those ways are covered in this chapter; later chapters focus specifically on Ethernet and Token Ring signaling.

A LAN by Another Name

The previous definition of a LAN provided focuses on the components used to provide connectivity among communicating devices. In fact, initially LANs tended to be configured to provide computer-to-computer communications on a more or less egalitarian, or many-to-many, basis. But the rapid development of network operating systems (NOSs) gave rise to a different, although related, definition of local area network.

What do users of LANs see when they use the network?

Configurations vary widely. However, when a user turns on his or her workstation, it normally automatically runs a number of programs up until a point when the user is prompted to enter an identification (i.e., "user id") and a password. This is called *logging in*. If you ask the user (with, of course, precise grammar), "Into what are you logging?" they will answer, "I'm logging in to the network, of course!"

Logging in to a bunch of wires? Well, not exactly. Because security and access to services is typically (although certainly not always) provided for a given user from a single file server (see the following for an explanation of "file server"), users often perceive and refer to that server as "the network" or "the LAN." Managers of those servers often go along with this usage in order to avoid confusion—or because they believe it to be the only definition anyway. For single-server shops, this causes no problems. In fact, not surprisingly, many of the NOS vendors also use the term LAN in this way. It is what they sell, so one cannot fault them for focusing attention on what to them is the key component of the environment.

But referring to a server as a LAN can cause quite a bit of confusion in certain cases. Consider the case of having multiple servers on a single-wiring facility. This is certainly a reasonable thing to do, particularly if they provide different types of services. Is each a LAN? Are they all, collectively, "the LAN?" If someone logs in to one of these servers from a different cable plant (connected using methods described later in this chapter), which LAN are they on?

Or, consider the individual responsible for design and implementation of the wiring plant, together with the connectivity devices described in the following paragraphs. That person will most likely refer to the wiring plant as the "LAN." In this usage, LANs provide connectivity among devices, including servers. This book generally refers to the wiring plant together with the active devices providing connectivity as the LAN. File servers are just one category of servers, albeit a very important one.

Why They Are Used

For many, the answer to "Why use LANs?" is "How could we work without them?" But business operated for centuries without LANs, and automated data processing functioned successfully for more than a decade without them. What changed?

Performance

One major change was the increasing popularity of minicomputers (not personal computers; the development of LANs predates that of personal computers). Mainframe communications architectures had been optimized for use of large numbers of terminals. In fact, a special category of communications hardware ("front-end processors") was developed in order to off-load the additional work necessary to manage such terminals.

In the minicomputer environment, however, front-end processors could not be used; the minicomputers were not built to handle the high input/output speeds necessary to service such processors. Instead, terminals were directly connected to *serial communications boards*, printed circuit boards with multiple RS-232 connections that plugged directly into the backplanes or buses of those minicomputers.

Minicomputers are designed to respond immediately to certain requests for services called *interrupts*. Every keystroke on terminals connected in this way generates an interrupt on the associated minicomputer. And "interrupt" is exactly what happens; all processing ceases for an instant while the single keystroke is processed. You can easily see that as additional terminals are added to a minicomputer, the amount of substantive processing accomplished (vs. that required merely to receive characters from attached terminals) decreases rapidly!

If the communicated information from the terminals could be presented to the minicomputers more efficiently (i.e., in groups without generating one interrupt per keystroke), the minicomputers could serve far more terminal users. A new input/output method was needed.

Wiring

Another issue, again in the minicomputer environment, was the snarl of wiring that communications was generating both in computer rooms throughout the world, as well as in buildings containing those same terminal users. Before local area networks, almost every logical (nonwide area) connection from a terminal to a computer required a separate set of physical wires from the terminal all the way back to the computer. Even if not every terminal needed access to every computer, wiring such as this was common. Figure 16.1 shows the spider web of wires that would be required

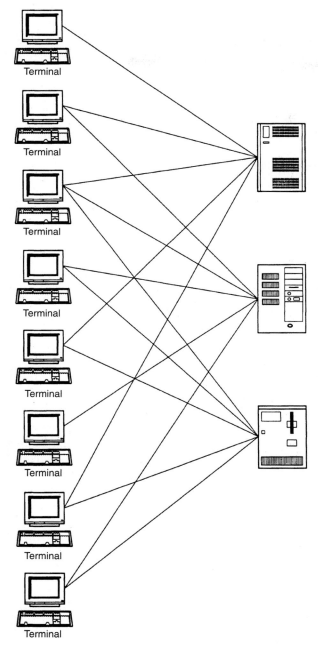

Figure 16.1 The typical wiring scenario in a multicomputing environment.

to link the various devices together. This is expensive to install, difficult to manage, and inefficient in the use of ports and facilities.

Data switches, or data PBXs, were alternatives, but they could only support lower speeds—and to configure them economically, they still had to be large and centrally located, much like the computers! Such devices helped, but were not an optimal solution. But what if a network device could be used to consolidate the terminal wiring? Terminal servers are LAN devices that can perform such a function (Figure 16.2).

This might not look all that much better than the previous drawing. But terminal servers can be located near the terminals, with one long cable back to the main computer system, instead of one per terminal. In a real business setting, they greatly simplify terminal wiring. A terminal server can allow any given terminal to have more than one session active simultaneously, either with one or several destination computers. If a new computer is added to the environment, only the authorizations change; no wiring other than the new computer's connection to the LAN is required. Best of all, the way LANs are implemented on minicomputers eliminates the interrupt-per-keystroke problem. Naturally, the sessions still need to be responded to; however, a much smaller percentage of the minicomputer's resources are devoted strictly to communications functions.

Not only do these terminals generate wiring snarls, but sites with multiple minicomputers were not uncommon. Suppose you had six computers, each with a need to reach all of the others on occasion. You could easily end up with a network wiring diagram like the one shown in Figure 16.3. The wiring nightmares were so bad that many organizations have joked that if the wiring was pulled out, the building would probably fall to the ground.

Imagine what a mess this would be if there were 8, 10, or 20 computers to connect! But if you used a common cable system instead, you might get something like the drawing in Figure 16.4. Clearly, this is providing some improvement in the process but the real question is: Is it easier to manage? You bet! Also, the number of components required, and therefore cost, decreases geometrically.

Initially, local area networks were developed to address these problems. Other issues, such as sharing of printers, were initially not of concern—that issue came later; connection of terminals to the minicomputers provided access to the printers on those computers as well as WAN communications resources on those computers.

Share Resources

Much of this progress occurred before personal computers were generally available. With the advent of personal computers, additional requirements surfaced. A general need appeared for resource sharing where the resources were not connected to a central computer. One of the two primary

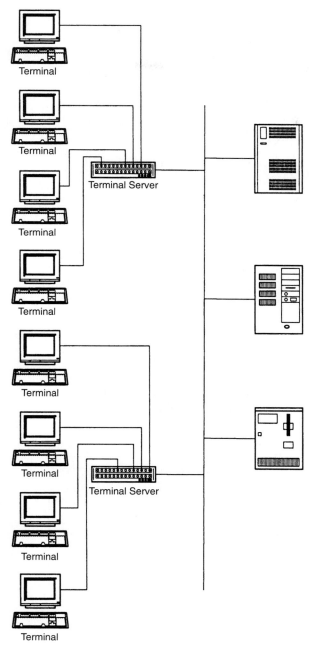

Figure 16.2 The use of a terminal server eliminates the wiring mesh.

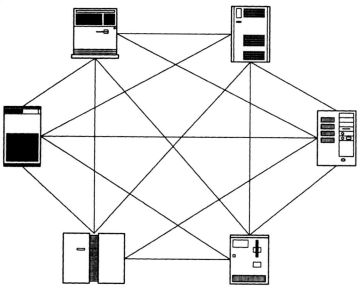

Figure 16.3 The minicomputer connectivity required the same form of wiring mess.

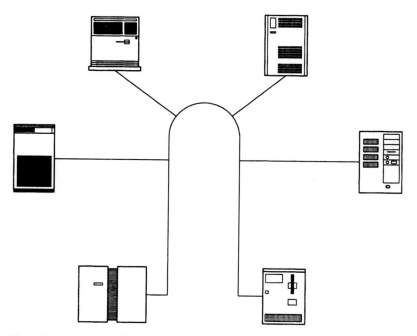

Figure 16.4 A single cabling structure might be a better approach.

resource categories first shared was printers. High-quality printers were initially quite expensive; management was understandably reluctant to provide one laser printer for each user of a personal computer. Printers could be connected in three general ways:

- To a central server, perhaps providing file services as well as print services.
- To a user's computer, allowing others use of that computer at the same time the user was using it.
- Directly to the LAN by themselves, if the specific printer had LAN capability.

The other major resource first shared was file services. File services provide a PC user the ability to reference files on another machine (the *server*) as though those files were actually located on the user's PC. Why not have them on the user's PC? There are several reasons.

- Initially, large disks were also very expensive; the cost per megabyte was lower for larger disk drivers, making sharing of such drives economically attractive.
- The larger drives were also much faster; network managers in some cases could actually provide better disk input/output performance across the network than the user would experience were the information on the user's local drive.
- Users do not back up data. Yes, there are exceptions, but a manager would be foolish to assume that his or her users fall into that category. If information is stored centrally, then it can be backed up under management control rather than relying on users to protect their data.

The observant reader might ask at this point, "why not just use a mainframe or minicomputer as before?" A very good question. Perhaps the answer is that the applications becoming available on PCs were so much more functional and attractive (remember "user friendly?") than those applications users had access to on larger computers. Also, PCs allowed many functions that hitherto could not be automated because of central IS departments' backlogs to easily be implemented using spreadsheets on PCs. Users were going to use their PCs anyway. So, ways had to be found to properly and economically support both the users and their PCs.

As network offerings matured, additional shared resources became common:

- Database servers
- Application servers
- Communications servers
- Backup servers

Database servers provide centralized processing capability optimized for efficient handling of databases. In practice, this typically means that they have large, very fast disk drives. Some mainframe environments also dedicate machines to handling databases. The difference here is that such a mainframe would typically cost well over $1 million; a minicomputer or PC in the same role will typically cost less than $0.5 million, sometimes less than $100,000.

One of the particularly useful capabilities of a LAN is to allow companies to provide individual processor configurations optimized for specific functions. The ability to provide such servers goes a long way to justify the extra management effort involved in managing LANs.

Application servers are machines dedicated to handling specific applications. Such a server might be marketed together with a specific application by a value added reseller, or VAR. Such a configuration can be used in either of two basic ways: via terminal access (as with mainframes, but for one application suite only), or via a client/server approach.

Communications servers allow centralization of external communications facilities and hardware. Modern communications server capabilities include the ability to provide both in- and out-bound modem pools (either separated or shared) along with routers and gateways (see below) that serve an entire user community. Without such servers, it is not uncommon to see a configuration involving one modem per PC user. Centralizing a modem pool allows acquisition of faster, better modems and more efficient use of them (assuming that only a percentage of the user population will be using a modem at any one time).

Backup servers allow off-loading of backup functions, which would most likely otherwise reside on file servers. Because the failure of a file server is typically the single-most disastrous possible occurrence in a LAN environment, it does not make sense to make the file server one's primary backup device (i.e., the unit with an installed tape drive). If the server fails (e.g., because of a failed power supply), it becomes somewhat difficult to restore the data to another machine.

LANs are not limited to these categories of servers; these are just the most frequently implemented. Other possibilities include scanner servers and facsimile servers. Additional functions will most likely evolve over time.

Distributed Systems

The term *distributed processing* predates LANs. However, LANs make it much easier to architect. What is distributed processing? It is not simply spreading function out among multiple computers, although many people believe that is all there is to it. Distributed processing has a few basic premises:

- Do the work close to where the results are required (minimizing communications costs, a key objective of distributed processing).

- Dedicate processor(s) configured appropriately to their specific functions.

- Use cheaper MIPS (a raw unit of processing power that is generally less expensive when purchased in smaller units).

(For robustness—i.e., the ability to continue key functions in the presence of failures—provide a certain degree of redundancy; i.e., do not have just one processor in any given function category.)

Of course, distributed processing is also architected in wide area networks (WANs), but allowing computers to communicate at high speeds among themselves is much easier to accomplish in a LAN environment.

Client /Server Architecture

Client/server design is, depending on one's viewpoint, either an extension or a subset of distributed processing design. Whereas distributed processing typically begins with the premise that "all computers are created equal"—at least insofar as their communications patterns—client/server design uses a more asymmetric model. But there are more subtle differences.

The basic premise of client/server design is similar to that of distributed processing: dedicate processor(s) configured appropriately to their specific functions. But in this case, a certain amount of predefinition is applied. Specifically, it is usually assumed that for a given application, precisely two processors will be involved: a desktop PC used by a human being (the client) and one other machine (the server). There are many ways the division of labor between these two can be assigned. Many books have been written on the subject, so it is not pursued here. However, we mention the general approach because in most cases, a critical characteristic of client/server design is high-speed communications between clients and servers, speed that can only be easily accomplished via LAN technologies.

Scaleability

A key benefit often provided by both distributed systems and client/server architectures is scaleability. Roughly speaking, scaleability is the ability smoothly to increase the power and/or number of users in an environment without major redesigns or swapping-out of equipment and software. In a mainframe environment, upgrade paths often looked like the chart shown in Figure 16.5.

With smaller individual machines linked together with LANs, capacity upgrades tend to look more like the next chart (Figure 16.6). In this chart, the capacity is added in smaller increments as user needs dictate.

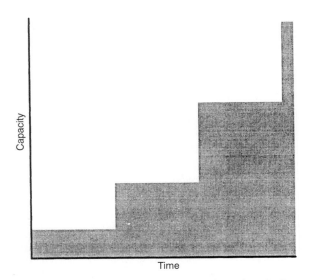

Figure 16.5 The function of capacity over time when dealing with the mainframe.

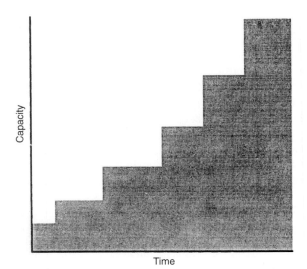

Figure 16.6 Capacity for smaller based machines over time.

The key difference is that with smaller machines, each upgrade is less expensive and one is able to use the capacity in the near future. Each upgrade is less financially traumatic, so when the need arises, it is usually easier to make the appropriate acquisitions. With mainframes, each upgrade is a big bite and it takes a while to fill them up once you have them. Because each upgrade costs a lot, justification of it usually takes a while, often resulting in demand exceeding capacity until the approval comes through for an upgrade.

Scaleability is more an argument in favor of many small processors, rather than specifically for LANs. However, the easiest way to connect those processors is with a LAN.

How They Work

A LAN environment includes, at a minimum, nodes and wiring. We first cover what makes up a node.

Node configuration elements

A LAN node is a computer that has a unique network address. That address consists of a sequence of letters and numbers (or possibly just numbers) that, just like your postal address, identifies the node so that messages sent to it can be delivered to the correct device. LAN nodes comprise certain basic elements (Figure 16.7). The key components of the LAN include:

- Processor (computer, possibly but not necessarily including keyboard and monitor, etc.)
- Software
- LAN network interface card (NIC)

The processor can be just about any computer currently available. The software and NIC, however, deserve more discussion.

LAN software for LAN nodes

Every LAN node requires a complete, functional implementation of the International Standards Organization's (ISO) Open Systems Interconnect (OSI) model. This does not mean that the specific software is the ISO standard software; rather, it simply means that all the functions described in the OSI model must be addressed, one way or another. With the exception of part of level 1 (the physical layer), all of the OSI functions are accomplished in software, so it should be no surprise that most LAN software involves not one but rather several software modules that collectively allow a node to communicate.

At the bottom layer of the protocol stack (of these software modules) is certain software that is built into the NIC (in read only memory, or ROM), part of the physical layer. This is generally invisible to the user—or even to the installer— but it is critical. When installed, the on-board NIC software

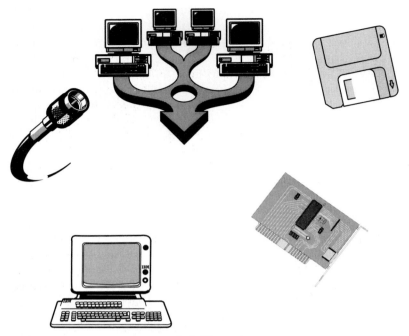

Figure 16.7 The basic components of a LAN.

executes in the same "address space" used by other software on the node; thus, certain address space might have to be set aside by the node in order that the software does not get stepped on by other programs. This might be handled directly by the node's operating system, or it might have to be addressed directly by the installer.

The next layer of software is the network driver. This software is the go-between or interface between the rest of the software on the node and the NIC. In the world of PCs, there are two prevalent "standards" for this software: Network Distributed Interface Standard (NDIS, from Microsoft) and Open Data Interface (ODI, from Novell). These standards define not how the software drivers operate, but rather the standard approach for the other software on the computer to interact with the NDIS or ODI drivers.

It is the responsibility of the NIC manufacturers to provide such drivers with their interface cards. Providing such standard software is much more difficult than you might assume. As with all of the other software on a node, the NIC interface software runs under the operating system supporting that node. Computers, especially personal computers, can run many different operating systems. A different driver implementing a given interface standard is required for each operating system. If a vendor supports both ODI and NDIS (and most do), that doubles the number of modules to be provided.

Above the level of the driver software is the province of Network Operating Systems (NOS), such as Novell's Netware, IBM's LAN Server, and Banyan's Vines. Because the driver standards insulate these environments from the hardware, all of these NOSs can support several different types of LANs. The NOS software on a node provides two major functions:

- It provides all of the communications functions specified in the OSI model above level 2.

- It intercepts and satisfies resource requests on the node for resources that would normally appear on the node, but instead are to be provided via the LAN (e.g., file access, communications access, etc.).

These resource requests can originate directly from the keyboard (e.g., a user types a file from a server disk to the screen), or from a program (e.g., a word processor program retrieves a document from a remote server). Such requests are normally satisfied so transparently that the user or requesting program is completely oblivious to the fact that a LAN function occurred.

LAN NOS software is typically priced on a per-active-user basis. At this writing, such prices vary widely depending on the NOS as well as the volume of such licenses. They can range from less than $100 per license to as high as close to $300 each. Fortunately, when this pricing structure is used, there is normally not any additional charge for the server software.

LAN network interface cards

Network interface cards (NICs) typically fit into an expansion slot on a computer. In some cases, however, the network interface is built into the computer in the factory. The latter approach usually results in a lower overall cost, but can reduce or eliminate the opportunity to later change or upgrade the network interface.

Functions performed by a NIC include the following:

- Provides a unique LAN address for the node that is built in by the manufacturer.

- Performs the link level functions with other nodes appropriate to the physical connection (e.g., fiberoptics) and protocol (e.g., Ethernet) used on the particular LAN.

- Accepts protocol frames of information (see below) at full LAN speed, buffering (storing) them on the card until the computer is ready to process them.

- Recognizes, examines the address information of, and ignores received protocol frames that are not addressed to the node in which it is installed.

- Responds to certain management signals on the network. These might be inquiries as to the status of or recent communications load imposed upon the card, or commands to stop communicating on the network. (The latter might arrive if the card fails in such a way as to begin placing spurious traffic on the network.)

Like NOS software, NICs vary widely in price. Factors affecting price include:

- How many NICs are purchased (volume discounts)
- Type of LAN to be supported (Ethernet, Token Ring, etc.)
- Management capabilities, if any
- Buffer sizes
- Speed
- Bus type

These last three, buffer size, speed, and bus type, bear more discussion.

All NOS node software functions in parallel with other software on a node. For example, the computer does not stop doing word processing when LAN communications occur. Thus, the node can on occasion get quite busy, perhaps so busy that frames arriving across the network might not always be dealt with promptly. But those frames continue to arrive at the NIC at the rated speed of the LAN (e.g., if Ethernet that speed is always 10 Mbps). What happens if the bits pile up in the NIC? If all of the storage area on the NIC devoted to this purpose (i.e., the buffers) fills up and more bits continue to arrive, some bits will be lost ("dropped on the floor" in the vernacular). Obviously, the more buffer area on the NIC, the less likely this is to happen.

There is an interaction between needed buffer space and the speed of the node: If the node's speed is well-matched to the LAN, then minimal buffers are needed. If the node is much slower than the LAN's rated speed, then larger buffers might be desirable. Buffers that are too small will result in retransmissions of the data from the sender, reducing the overall throughput of the LAN.

The speed of the NIC is also a factor. But, you ask, does not the NIC communicate at the defined speed of the LAN? It does, on the LAN side. However, the speed with which it communicates to its node (the *local speed*) is independent of the LAN speed. The part of the NIC doing local communications might or might not be as fast as the node receiving the information. If it is not that fast, then the NIC is a communications bottleneck. If overall performance is not satisfactory, this is one possible cause.

The type of bus in the node can also have a major effect on the local NIC speed. We are not speaking here of the LAN "bus," described below, but rather the internal architecture of the node. In the case of IBM-compatible personal computers, there are currently at least four different types of buses:

- Industry standard architecture (ISA)
- Microchannel (MC)
- Extended industry standard architecture (EISA)
- Peripheral component interconnect (PCI)

ISA is generally slower than all but the slowest LANs, no matter how fast the node processor or the NIC is. All of the other types can easily keep up with most LAN speeds, provided that the NIC internal speed is also fast enough.

Most other (non-IBM compatible) nodes can accept data more quickly than most LANs can deliver it. Therefore, if there is a bottleneck at the node, it usually will be either the LAN itself or the NIC.

Topologies

The word *topology* is used in at least two different ways when discussing local area networks. Unfortunately, speakers rarely identify which of the two meanings they are using. Even worse, they often muddle the two meanings together. We try to distinguish the two meanings by qualifying the term: *physical topology* and *logical topology*.

Physical topology constitutes the way one lays out the wires in a building. The major physical topologies include:

- Bus
- Ring
- Star
- Combinations of the above (e.g., tree, double ring)

Logical topology describes the way signals travel on the wires. Unfortunately, logical topologies share much the same terminology as physical topologies. Signals can travel in the following fashions:

- Bus
- Ring
- Star (or switched)

So, is there really any difference between physical and logical topologies? Emphatically yes! As you shall see, most of the logical topologies can be

used on most of the physical topologies. Thus, you can use bus communications techniques on a bus or star; ring communications techniques on a
bus, ring, or tree; and star communications techniques on a bus or star.
Confused? Read on—we sort the main variations out one at a time.

Physical topologies

The bus physical topology has the following key characteristics:

- All locations on the communications medium are directly, electronically
  accessible to all other points.
- Any signal placed on a bus becomes immediately available to all other
  nodes on the bus, without requiring any form of retransmission.

Physical bus topologies are usually drawn as a line with nodes attached
shown in Figure 16.8. Note that the bus is a single high-speed cable with all
devices connected to this cable.

But as you will see when we cover media, some buses do not require any
wires at all.

The ring physical topology has the following key characteristics:

- Each node is connected to only two other nodes.
- Transmissions from one node to another pass via all intervening nodes in
  the ring.

Complete failure of a node, together with its NIC, can (in theory) cause
the entire network to fail. We say "in theory" because in practice, ring NICs
are designed to fail "open"; traffic continues to pass.

Rings are typically drawn as shown in Figure 16.9. There are variations to
this ring but these are covered in greater detail in Chapter 18.

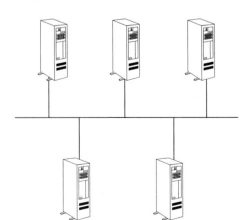

Figure 16.8 The layout of a bus
topology.

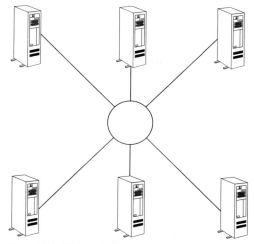

Figure 16.9 The typical physical ring as it evolved.

Some types of rings (e.g., FDDI) have additional active devices to construct the ring itself (Figure 16.10). This architecture is not a perfect circle as we think of it, but is still a ring.

The star physical topology has the following key characteristics:

- Each node connects to a central device with its own wire or set of wires.

- The central device is responsible for ensuring that traffic for a given node reaches it.

- Data might or might not pass through multiple nodes on the way to its destination, depending upon the specific capabilities of the central device.

Star topologies often are drawn something like the one shown in Figure 16.11.

Note that even though there is always some type of device at the center, star diagrams often do not show it. It depends on whether the focus of the diagram is the nodes or the way the network is built. In the latter case, the diagram would look like this variation shown in Figure 16.12.

Logical topologies

As mentioned earlier, logical topologies include bus, ring, and star (or switched). A logical bus operates with each transmission visible to every other node on the LAN. The easiest way to do this, of course, is with a physical bus; all devices connect to the same wire.

But consider the user of a star physical topology. If the central device instantly retransmits every frame it receives from one wire out to every other

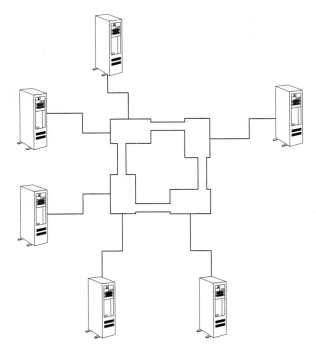

Figure 16.10 The FDDI uses a different type of ring. Not all rings are laid out in perfect circles.

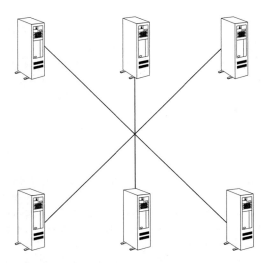

Figure 16.11 The typical star network layout.

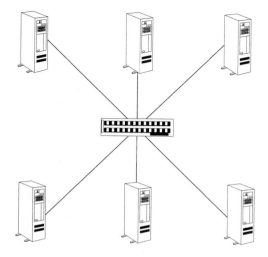

Figure 16.12 The star has some form of electronics that all devices connect to.

wire, how could the individual devices detect whether they are on a physical bus or physical star? Would they care? The answer to both questions is a qualified no. A certain amount of engineering goes into building such devices. However, Ethernet, for example, was originally designed to function on a physical bus. Nowadays, almost all Ethernet installations consist of physical stars.

Bus topologies require a method of handling collisions, times when two or more nodes attempt to transmit at the same or overlapping times. Different methods are used with different protocols.

A logical ring consists of a network where any given transmission passes from one device to another in a fixed sequence until it reaches its destination. It is naturally most obvious on a physical ring. However, it also can be implemented on a physical bus or a physical star. In the case of a physical bus, although all devices see all transmissions, each transmits on the common medium in a fixed sequence, sending packets only "downstream." An advantage of this tightly disciplined line protocol is that there can never be a "collision"; all devices know their turn and follow it. On the other hand, there can be quite a wait until one's turn arrives.

A logical star cannot be built on a physical ring or bus; it requires a central device to make intelligent choices as to the disposition of each frame, sending the frame only to the correct destination. Ethernet switches, a relatively recent development, operate in this fashion. Collisions cannot happen in this environment, although buffer overruns in the central device can occur if it is not quick enough to keep ahead of the frame arrival rate.

Mixed topologies—the real world

As we will see in later chapters, ideal designs as described above have never stopped engineers from building what will work. Token Rings are really stars, Ethernet is also usually a star, and the fiber distributed data interface (FDDI) is the first real ring that was widely implemented—and it uses the variation shown with devices building the ring. Here is a brief list of some of the more popular LAN protocols, with notes as to their topologies:

Ethernet and Token Ring. A later chapter discusses these protocols.

LocalTalk. This Apple protocol uses a bus physical and logical architecture. Its contention management approach is called *CSMA/CA (carrier sense with multiple access and collision avoidance)*. All devices listen continuously to media (carrier sense). Multiple devices connect to the same media (multiple access). When a device wishes to send a packet, it sends out a very small precursor packet requesting permission. Because the packet is very small (and is only sent out in the first place if the medium is idle), it is unlikely that it will collide with another packet. Presuming that it doesn't collide, all other devices are now on notice that the real packet is coming, so they do not step on it with their own transmissions. This is LocalTalk's approach to collision avoidance.

Manufacturing automation protocol (MAP). MAP uses a token-passing bus. The token-passing bus is designed to have the benefit of the bus (high bandwidth) and the predictability of the ring (token passing). Each device's transmissions are heard by all devices. However, every device transmits to the next in a fixed sequence, thus completely avoiding collisions—but lengthening the average time until any given device receives its turn to transmit.

Asynchronous transfer mode (ATM). ATM is both a late-model LAN and a WAN technology. It is designed to complement the LAN by performing non-contention LAN emulation. This means that the ATM will look like a star and provide dedicated bandwidth to each device on the network. In reality, the older LANs (called *legacy LANs* now) will not be sharing a single cable system and competing for the bandwidth as they do today.

Peer-to-peer vs. server-based. This is an area of network design that sounds as though it is a topology-related subject, but really is not. Peer-to-peer and server-based are two different approaches to designing network operating systems and applications. Whole volumes have been written on both approaches; we briefly summarize and distinguish between the two approaches here.

Peer-to-peer networks afford users great flexibility because every node, potentially, can function as a server. In practice, this means that relatively small shops can set up many or all of their computers to provide file and other services to all other computers. In principle, this is an egalitarian and flexible approach. In reality, it can quickly turn into a nightmare if not managed carefully.

The problem is that managing any server properly requires that a certain amount of time and expertise be focused on that task. Users, not unreasonably, tend to focus their efforts on their own jobs to the exclusion of network management-type functions. If their machines are configured as servers, and they do not manage them, who does?

Another problem with this approach is that server functions consume computer and input/output resources (i.e., time). If you as a user are also providing services to other users, your local performance will be directly affected by the load others put on your machine. Because a major benefit of personal computers is that the power on your desktop is yours alone, peer-to-peer configurations can quickly cause frustration because of degraded performance—unless each machine has more power than is needed solely to satisfy its assigned user.

Having given some of the more serious drawbacks of peer-to-peer environments, it should be stated that literally thousands of small businesses function very well using this technology. The key is that they are small enough that the disadvantages described above never become serious enough to outweigh the advantages. Some of the advantages of peer-to-peer NOSs include:

- Any system can be configured as a server, allowing very effective group access to any files desired no matter where those files reside.

- Typically lower cost per node, partly because of lower software costs and partly because of the elimination of the need to purchase high-powered dedicated servers.

- Reduced management costs, partly because some management functions typically do not get performed, but also because these systems are designed to be more easily managed than the dedicated server systems.

Server-based networks predominate. Their advantages and disadvantages are partly the flip side of those listed in the peer-to-peer discussion. But two should be emphasized:

Server-based networks can be and are typically tuned so as to support larger numbers of users (anywhere from 20 to 300) on a single server. The management effort involved to support such numbers is much less on a per-user basis than would be the case were a peer-to-peer approach to be used for such a large user base.

Although it seems a self-fulfilling prophecy, the very large numbers of installations of server-based systems (especially Netware) have resulted in an appropriately sized base of people expert in the configuration and management of such systems. Also, that same installed base has attracted many creators of after-market software products that complement and extend such environments, a far larger selection than is available for the peer-to-peer products.

Internetworking

Internetworking refers to the connecting together of two or more networks, either LANs, WANs, or a mixture of the two.

The creators of LANs designed them to be truly "local." This was not an arbitrarily imposed design restriction; by not requiring a LAN to be capable of wide area communications, the engineers could take advantage of a much more controlled environment. That control allowed use of techniques that, while not practical over a wide area, could deliver far higher communications performance than WAN communications of the day as well as reliability orders of magnitude faster than those same WAN communications.

But, people being what they are, demand soon materialized for the ability to provide LAN-type communications (that is, fast, lower error rate, and using typical LAN protocols) over areas wider than those originally envisioned as being supportable by LANs. And, engineers being what they are, they came up with not one, but several ways of accomplishing this. All of them involve the use of additional devices that, in varying ways, take a LAN signal and send it further than the original LAN specification allows.

Terminology is important here. We cover four types of devices that provide such extended LAN connectivity. The first type (repeaters) is considered to extend the reach of a given LAN. As such, using repeaters is not technically "internetworking" although we describe it in this section because it nonetheless involves extending the reach of LAN technology in a given environment. All three of the other technologies (bridges, routers, and gateways) are considered to provide connections among or between different LANs, thus providing true internetworking capabilities. The reason for this division between repeaters and the rest should become apparent as we cover the devices in more detail. We begin with the simplest of the devices, and go on from there to more sophisticated products.

Repeaters

Repeaters are relatively "unintelligent" connections between two LAN segments of the same type (where "type" is Ethernet, Token Ring, etc.). A repeater is sensitive to the traffic's content only at the ISO level 1 (physical) layer. An Ethernet repeater will transmit any traffic that can be sent on Ethernet, a Token Ring repeater will send on all Token Ring traffic, etc. Repeaters satisfy only three key functions:

- Distance
- Electrical isolation
- Media conversion

Bridges

Bridges are intelligent connections between two LANs of the same type. They operate at layer two of the OSI reference model (data link). The bridge is responsible for linking the two segments or LANs together and transparently passing frames across the link to each other. No form of choices are used, there are no format or protocol conversions or any other services other than providing the physical and logical paths between the two networks or segments.

The bridge can be intelligent, and providing filtering and forwarding services across the link. A filter is used if an addressable device is on the network that is sending the frame. This is so a frame of information does not get sent across the wires to another segment or network that has no need to see the frame. Forwarding takes place when the bridge sees a frame that is addressed to a node on the other side of the link.

Routers

Routers are switching devices that connect two LANs where multiple paths exist, or a single path exists on a dial/leased line basis. Working at the layer 3 of the OSI model (routing, or network), the router will be responsible to ensure that the data gets to its destination. There are several considerations about this decision process, being the router is more sophisticated; therefore, it is more complex to manage. However, this added complexity brings significant gains to the users who are looking to internetwork their LANs. Many large organizations now use routers in their backbone network to connect segments (departments, floors on a building, etc.) together with alternative paths to get from one LAN segment to another. The feeling is that the resiliency is worth the added complexity.

Gateways

Gateways connect two networks of different types (i.e., LAN to X.25 network). These operate at higher levels of the OSI model. ISO levels 4–7 (transport, session, presentation, application). In effect, the gateway handles everything above the network layer in the OSI model. The controlling portion of a session, the format and protocol conversion to make all things common, and the actual application interface reside in the gateway's domain for interconnectivity.

Benefits of using gateways

Cost saving:

- Initially—yes
- Including management—no
- Performance—definitely
- Flexibility/scaleability—definitely

Chapter

17

Ethernet

Ethernet was the first commercial approach to using a LAN on a bus topology. Ethernet is probably the most commonly used LAN technology today. Although certain recently-developed technologies are faster (including ATM, two varieties of "fast Ethernet," FDDI, 20-Mbits/s ARCnet, and 16-Mbits/s Token Ring, and the new 25-Mbit/s Token Ring/ATM, coming from IBM), Ethernet still enjoys continued popularity and growth for a number of reasons:

- For its speed, it is by far the least expensive networking approach.
- Its various standards supporting a wide variety of media are sufficiently well-defined that products from multiple vendors can be mixed and matched with a reasonable expectation that they will work well together.
- Ethernet equipment exists to allow virtually any intelligent device to connect to an Ethernet.
- A variety of interfaces still exist for dumb terminal interconnection through terminal servers.
- It is fast enough for the vast majority of applications in use today despite what we hear in the press or from the vendor community.

The Ethernet standard(s) define an approach to building a contention-based LAN on a bus topology. As originally defined and implemented by Xerox, Intel, and Digital Equipment Corporation (called the *DIX design*), to connected devices the bus appears to be a single straight cable system (although more recent implementations, in fact, can be built on different types of cable systems). The cable is available to multiple users at the same time, even though only one user can be sending data at a time. We present these concepts in more detail in the rest of this chapter.

A note on standards. Technically, the LAN protocol implemented most widely and still referred to as *Ethernet* is actually defined in the 802.3 standard, and is not precisely the same as either of the two "Ethernet" standards (I or II) defined by the Xerox-Intel-DEC triumvirate. However, most equipment available today can support both standards; also, in most cases, the devices can inter-operate on the same media. In any case, most people, when not attempting to be particularly precise, refer to 802.3 as *Ethernet*. We do also.

Concepts

Bus—CSMA/CD

The human ear is much better at picking out a single voice from among multiple simultaneous speakers than is any electronic receiver capable of selecting one signal from among many. Thus, every local area network using a bus (i.e., a shared medium) design must incorporate a contention management or arbitration mechanism that results in only one signal being present at a time so that on the shared medium, no matter how many devices, signals have to wait their turn. Otherwise, signals on the cable will conflict like voices on the floor of a commodity exchange—but with less profit. Ethernet uses a contention management approach called *carrier sense*, multiple access with collision detection (CSMA/CD) to ensure that devices "speak" politely, one at a time. CSMA/CD is designed to provide equal use of the network for all attached devices. CSMA/CD actually breaks down into three meaningful two-word terms:

Carrier sense

All stations listen to the cable continuously. Many non-Ethernet signaling approaches send a single unvarying signal (a carrier) and impose variations on it based on the desired information to be transmitted, whether data, voice, or music. Examples include AM and FM radio, as well as television. Ethernet is different in that Ethernet nodes do not send out any signal at all unless they have something to say—kind of like Mr. Ed, the talking horse. So, when stations monitor the cable, they are listening for the presence of a carrier signal, whether modulated or not. If a device that wants to transmit detects a carrier on the cable, then it knows that another device is either preparing to send, is sending, or has just finished sending information onto the cable. In any case, the waiting device will hold back until the carrier signal vanishes.

Multiple access

If no one is using the cable, any of the network devices attached to the Ethernet can transmit data onto the network at will; there is no central control, nor a need for any. This is a more powerful design element than might be apparent at first blush. Networks "degrade" (or experience reduced performance) in different ways, depending on their design. One factor that does not in and of itself degrade an Ethernet is the attachment of literally hundreds of devices (the theoretical limit of the number of addressable stations on the network is 1024, although most networks do not come close to this number). Because the wire is truly idle unless a device is transmitting, only active devices impose any load upon the network. Thus, a myriad of devices can be connected, so long as most of them are not transmitting at any one time.

Of course, there is a potential fly in this ointment: If all those attached devices need to transmit a lot of data, great congestion will ensue. But that might be acceptable for some networks—not the congestion, but the threat of it. Some networks are built to support primarily word processing, an occasional file transfer, perhaps some electronic mail. With such networks, the light load per machine allows many devices to be present without degrading the network. Networks that will experience much traffic at all per device need to be engineered more carefully, perhaps by being divided up into multiple bridged segments (described in Chapter 16).

Collision detection

In the event that two devices attempt to transmit at the same time, or even during overlapping time intervals, a collision will occur.

Think of two people simultaneously beginning to speak at a cocktail party. If they both continue to speak, neither will be understood. This might be fine for a New Yorker, who can talk and listen at the same time, but for this type of network, it can be devastating. On an Ethernet, if a collision occurs, the data is lost and each system must retransmit.

At the cocktail party (if the attendees are reasonably well-behaved), both speakers will stop as soon as they become aware of the problem. On an Ethernet, the first device that detects the problem will stop the regular transmission and send out a special jamming signal, which tells all attached devices that a collision has occurred. The "jam" signal will be heard by every device because all devices continue to listen to the network even if they are transmitting.

At the cocktail party, the speakers will each hesitate a moment, then try again. The one that hesitates the shortest time gets the word in edgewise, while the other will wait for the next opportunity. On an Ethernet, any device that was attempting to transmit at the time the jam signal arrives will assume that a collision occurred. It will therefore calculate and wait a semi-random amount of time. What is semi-random? The calculation is an attempt to generate a random number, but the (guaranteed unique) address of each device is a part of the calculation. Thus, the amount of time waited in such circumstances will always be different for each involved device, ensuring that the next attempt by the device that waits the shortest time will succeed. Well, almost.

The problem is that the only devices that apply the waiting algorithm are those that experienced the collision. Any other device that has occasion to transmit will only follow the standard Ethernet approach (listen, then transmit). If a new participant "decides" that it is time to transmit, it could do so just as the "fastest waiter" in the previous group finished its wait and went ahead with its second try. If this happens, another collision will occur.

You might well ask, "how does anything ever make it through an Ethernet?" We must emphasize that all of the shilly-shallying about described in the previous paragraphs happens in a small number of microseconds. And remember, unless devices attempt to transmit at almost exactly the same time, collisions will not happen; the second device waits its turn, then transmits.

Nonetheless, if you understand how the collision detection and avoidance mechanism works, it should be no surprise that as the number of communicating devices attached to an Ethernet increases, the number of collisions also increases. In practice, close management is indicated when steady traffic on an Ethernet exceeds about 30 to 40% of the 10-Mbits/s capacity of the network. To put this in perspective, however, the vast majority of Ethernets installed today run at less than 5% average utilization. This really means that an Ethernet with an average number of users on it will yield an effective throughput of approximately 3.3 to 4 Mbits/s. This assumes a mix of 30 to 1000 terminals, or 30 to 35 high-end terminals, 6 CAD workstations, or 2 minicomputers. In all of these examples, we are using averages; however, the authors know of organizations that have supported 1000+ terminals on large Ethernets. But, the difference was the type of work being performed. If all the work is word processing, as you might find in a legal business environment, then the demands on the network will be minimal. Yet, put 5 to 6 CAD terminals on the network and things will definitely be different. These 5 to 6 terminals will saturate the network and be collision prone.

Half-Duplex

If you have ever seen a fast typist using a terminal connected to a mini-computer via Ethernet, you might have concluded that Ethernet is a full-duplex protocol. However, Ethernet is actually a half-duplex protocol. It is so fast, though, that it is functionally full-duplex. Although no more than one frame can occupy the wire at a time—and in the case of a typist, it is possible that each 64-byte frame might literally include only one byte of data in each direction—the frames are moved back and forth so quickly that it looks to a human being as if the typing and the receiving of characters are simultaneous.

If two files were copied, one from node A to B, the other from node B to A, at exactly the same time, the flow of frames in both directions would so intermingle that again it would appear that the transmission was going in both ways simultaneously. But because at any one instant only one frame can occupy the wire at a time, the medium is actually only half-duplex.

Bandwidth

Ethernet has always been a 10-Mbits/s facility (except one no-longer-marketed variant—AT&T's Starlan, that ran slower, at 1 to 2 Mbits/s). There is an advantage to all implementations being the same speed: No significant speed matching is required when connecting multiple Ethernet segments, even those built using different media.

However, even though all Ethernets run at 10 Mbits/s, not all devices on an Ethernet actually communicate at 10 Mbits/s. This concept is difficult to grasp, but it is important. Every node connected to an Ethernet incorporates an interface capable of receiving from and transmitting onto the Ethernet at 10 Mbits/s—exactly. Otherwise, it does not comply with the Ethernet standard. But that interface card might not communicate to its own node at 10 Mbits/s. After all, a large percentage of existing desktop PCs cannot even accept data at 10 Mbits/s through their internal buses! Sometimes the interface card is not built to interface with the node at a high enough speed; in other cases, the card could go faster but the node cannot. This inability to keep up with the nominal speed of the Ethernet itself is one major source of performance problems, especially if the slow interface is found on a server.

As with all protocols, not all of the bandwidth on Ethernet is actually available for carrying application-related data. Some bits are required for overhead functions. Ethernet supports a variable length "frame" or packet size; the minimum length frame is 64 bytes, the maximum 1518 bytes. The variation results from changing the amount of data transmitted. In the Ethernet (actually, 802.3) protocol, 18 bytes in each frame is dedicated to overhead, as shown in Table 17.1.

TABLE 17.1 The Composition of an 802.3 Frame

Function	Bytes	Description
Destination address	6	Ethernet address of the node to which the frame is addressed
Source address	6	Ethernet address of the sending node
Length	2	Indicates the number of bytes of data in the next (data) field
Data	46 to 1500	The "payload"—the actual data being transmitted
CRC	4	Cyclic redundancy error detection bytes

Let's take these one at a time.

Destination Address

Both the destination and source addresses are 6-byte fields, usually represented in hexadecimal format; for example, BB-BB-BB-BB-BB-BB.[1] Every Ethernet interface on a LAN receives and processes all frames transmitted on the LAN. The main job of the receiving interface is to compare the destination address in each received packet with its own address. If the addresses match, the frame is delivered to higher protocol layers at the node

---

[1] In our normal numbering system, decimal or *base ten*, the digits zero through nine (0–9) are used to count. Note that even though we call it base ten, and there are ten digits, there is no single digit that represents the quantity ten. Each position (ones, tens, hundreds, etc.) in a number contains a decimal digit (that is, 0 through 9) indicating how many of a power of ten (100 = 1, 101 = 10, 102 = 100 and so on) goes into that number. Thus the number 264 is four 1s, plus six 10s, plus two 100s.

In the hexadecimal numbering system, or *base sixteen*, we also use the digits zero through nine (0–9) to count. But we need six additional symbols to represent the values 10 through 15 (there is no need for a single symbol for sixteen). Hexadecimal uses the capital letters A through F to represent the values ten through fifteen, respectively. Each position in a hexadecimal number contains a hexadecimal digit (0 through F) indicating how many of a power of sixteen (160 = 1, 161 = 16, 162 = 256 and so on) goes into that number. Thus the hexadecimal number 264 is four 1s, plus six 16s, plus two 256s or, in total as represented in the decimal numbering system, 612. The hexadecimal number "BB" is eleven 1s plus eleven 16s, or a total of 187 decimal.

The "natural" internal numbering system used in most data processing and communications equipment is binary, or base 2, using only ones and zeros. Network addresses (and many other items in data communications and data processing) are usually represented in hexadecimal notation because of the ease with which it can represent or be converted into 4-bit binary (base 2) numbers, as follows:

Hex digit	0	1	2	3	4	5	6	7	8	9	A	B	C	D	E	F
Binary value	0000	0001	0010	0011	0100	0101	0110	0111	1000	1001	1010	1011	1100	1101	1110	1111
Decimal value	0	1	2	3	4	5	6	7	8	9	10	11	12	13	14	15

Because it only takes 1 hexadecimal digit to represent any possible single 4-bit binary value (as illustrated in this table), and 1 byte equals 8 bits, it takes exactly 2 hexadecimal digits to represent the contents of 1 byte.

for further processing and eventual hand-off to an application. If the addresses do not match, the frame is "dropped on the floor"; that is, it is discarded.

Source Address

As mentioned elsewhere, every Ethernet node contains an interface device, sometimes built-in, other times provided as part of an add-on printed circuit board that implements the interface. Every single one of those devices, worldwide, contains (or should contain) a unique 48-bit Ethernet address, usually stored in programmable read-only memory (PROM). Every company that wishes to manufacture such Ethernet devices applies to the IEEE (in the past, Xerox performed this function) for a unique 3-byte code. As the company manufactures the interfaces, it "burns in" its 3-byte code into the first three bytes of the 6-byte address of that device. The other three bytes of the address are normally assigned sequentially, one per device, but can in fact be assigned; however the manufacturer wishes. The only requirement is that every device have a different 6-byte address. If properly followed, this system guarantees that every manufactured Ethernet device will have its own unique 6-byte physical address.

Apparently, at least one Far East manufacturer has "broken" this approach to guaranteeing address uniqueness. It acquired chip designs for Ethernet interfaces—and then proceeded to manufacture them identically, one after another, with identical addresses. If more than one of those interfaces were to be installed on the same local area network, they would be unable to communicate, either with each other or with other devices.

Other variants to this rule are that users were allowed to assign their own 3-byte extension on the addressing mechanism. This allowed a manager of a network to assign specific addresses to servers, sometimes in a sequence that made sense to the manager for the appropriate LAN. However, this led to some complications where the manager would assign two identical addresses to the devices on the network which led to conflicts. Most installations today are with the standard addressing that comes on the card because of the past ills in this process.

Length

Valid lengths for Ethernet frames are from 64 to 1518 bytes, inclusive. The length field in the frame indicates only the number of payload bytes, and that value can range from 46 to 1500, inclusive. Notice that a frame cannot contain fewer than 46 bytes of data. What happens if the sending device provides fewer bytes to be transmitted? The controller will add "pad" bytes to bring the overall frame size up to the minimum length of 64 bytes.

Why is there a minimum length? There is an interaction between minimum frame length and the ability to detect collisions. Each medium supported by the Ethernet standard has a defined maximum length for any one segment. One factor that goes into determining that length is the need to ensure that a device transmitting a frame onto the segment will still be listening if a collision occurs—and recognize that the collision happened to its own transmitted frame. Devices stop listening for their own collisions a very short time after they finish transmitting. Therefore, once a device begins transmitting, it must continue to do so for long enough for the most distant device to receive at least the beginning of the frame, plus the time required to receive a jam signal from that distant device if such occurs.

If the minimum frame size were smaller, then the defined maximum segment sizes would also have to be shorter. To allow longer than specified segment lengths, the minimum frame size would have to be increased, requiring additional pad characters and reducing the efficiency of the Ethernet protocol. The 64-byte minimum cannot be claimed to be "ideal"; rather, it is an engineering judgment call by the original designers of the Ethernet protocol. If installers adhere to the rules regarding segment lengths, no undetected collisions can occur.

Data

This is the frame's payload, the reason it is being transmitted in the first place. The data can be up to 1500 bytes, although it does not have to be arranged in 8-bit octets (bytes); actually, it can be any pattern of up to 12,000 bits (that is, 8×1500). In most cases, the actual number of application-related bytes carried will be smaller than this, because some of the 1500 bytes will be used by higher level protocols for their overhead bytes, sometimes many more per frame than are required by Ethernet itself.

CRC

The cyclic redundancy check digits allow the receiving node's Ethernet interface to determine whether a frame was received intact or not. If not, the frame is discarded. If the interface is one of the more expensive available, it might also keep track of the number of frames discarded in order to facilitate management and detection of network faults. Error checking, including CRCs, is covered earlier in this book.

Components

Speaking very generally, the components used to implement a LAN tend to be similar across implementations; there must be an interface built into (or added to) each node, each node must have driver software installed that communicates between the nodes (network) operating system software and the interface, and the interface must be connected to the network medium. However, when one delves a little deeper, there are significant differences, both among LAN architectures and even among variations in the same LAN architecture, such as Ethernet.

For example, Ethernet is probably supported over more different types of media (with truly standard implementations) than any other LAN protocol. And, the detailed piece parts, as well as design considerations vary greatly depending on the specific medium.

The rest of this chapter covers the various components that can be used to build an Ethernet. We also mix in some design information and rules. Before getting down into the details, though, we would like to indicate what happens to people who violate the rules. Actually, at most organizations there aren't network police. But there are many network installers and users who wish there were. Generally speaking, there are two kinds of consequences that can result when LAN design rules are violated. The network doesn't work. Obvious, right? What could be worse? What could be worse is the alternative, the network that does work . . . sort of.

The designers of the various LAN standards (not just Ethernet, of course) attempted to specify standards that would result in robust networks—networks that could continue to operate in the presence of certain kinds of problems. But TANSTAAFL ("There ain't no such thing as a free lunch"). If a network does have some kind of problem and continues to operate anyway, that operation is usually somewhat degraded. In Ethernet's case, for example, intermittent short circuits appear to users as collisions. No data is lost, but overall performance can slow. Out-and-out failures are often much easier to diagnose and repair.

10BASE5

This would be a good time to explain the naming "standard" used for the various Ethernet "substandards." Bearing in mind that these were defined by committees, here goes. Each term is made up of three parts. The first part, "10," indicates that the network runs at 10 megabits per second. The second part, "base," indicates that the network is a baseband network (see below for 10Broad36). The last part indicates some physical component of the cabling system. In the case of 10BASE5, it indicates that one segment can reach up to 500 meters (1640 feet) in length. We explain the variations in these terms, as we cover each of Ethernet's supported media.

Ethernet was originally defined to operate over a particularly heavy type of coaxial cable, one thicker than most individuals' thumbs. People in the industry often referred to it as the "orange hose"—a particularly apt name because of its usual color. A more conventional term for the 10BASE5 medium is thick wire.

10BASE5 Piece Parts

As with any LAN, every attached device must have a network interface card (NIC). In the case of thick wire, that NIC has a port called an *AUI interface*, a 9-pin interface designed to connect to a transceiver cable (sometimes also referred to as a *drop cable*).

The transceiver cable can extend up to 50 meters (164 feet), but usually is much shorter, somewhere between two and ten meters. This cable is not Ethernet cable; it is not coaxial, but rather multi-pin, similar to RS232, but with fewer wires. The transceiver cable plugs into, naturally, a transceiver.

Transceivers include the componentry for a combined transmitter and receiver. Minimum spacing between transceivers on the cable is 2.5 meters (8.2 feet). When using the orange Ethernet cable, there are black marks spaced on the cable to indicate the appropriate spacing between the taps. If you do not adhere to the distance limits, other problems can occur. Ethernet transceivers are also sometimes referred to as "taps" or sometimes "vampire taps."

Vampire tap is most descriptive. Thick-wire Ethernet taps can be attached to the orange hose without interrupting the network. They do this by inserting two sharp points into the cable, one deep enough to reach the core, the other stopping at the braid. (A special tool is usually used to pre-drill the holes.) The tap clamps onto the cable with the two points sunk into the cable. The design might seem somewhat baroque, but the ability to add or remove devices from an active network was a significant advance in network technology at the time Ethernet was first designed.

Not all networks—not even most other Ethernets—require separate transceivers, so it would be worthwhile to indicate their function here. Ethernet transceivers perform several important functions:

- Physical access to the main LAN cable, as described above
- Media conversion from coaxial to multistrand cable
- Electrical protection and isolation in both directions (LAN to NIC, and vice versa)

But they are essentially passive devices; the protocol "smarts" reside in the NICs.

Thick wire is the only Ethernet medium that must be configured as a physical bus. As mentioned above, thick wire can be configured in segments up to 500 meters long. Up to three segments can be connected in series using two Ethernet repeaters. If more segments are so connected, the (very small) delays unavoidably introduced by the repeaters might foul up the collision detection part of the protocol.

Up to 100 direct physical attachments can be made to one segment, although there are ways to attach far more network devices than that. Up to 1024 network addresses can reside on a segment, or set of segments connected by repeaters. As with the other rules, violators of this configuration rule might be punished by undetected collisions.

Barrel connectors are small passive devices used to connect two pieces of coaxial cable to make a longer one. Barrel connectors are not defined parts of the Ethernet standard. They can be used, but the rules for segment lengths simply apply to the aggregate length built up of smaller wire pieces; that is, a 500-meter length can include one or more barrel connectors, but it still is limited to 500 meters.

Terminators are also small passive devices. But unlike barrel connectors, terminators are defined Ethernet components, and are essential to proper operation of a 10BASE5 Ethernet. Every 10BASE5 segment must have precisely two terminators installed, one on either end. One and only one of those terminators must be properly grounded. We used to tell students that this was required to ensure that the bits do not "fall out of the wire." But actually, in a matter of speaking, terminators are intended to ensure that the bits do "fall out of the wire"; improperly terminated coaxial cable ends can reflect transmissions back onto the wire, interfering with themselves and/or later transmissions and generating unnecessary collisions.

Configuration of the Parts

Setup

Designing 10BASE5 networks is relatively simple. But installing them is less so. If only one segment is needed, typical configurations include those shown in Figure 17.1.

Note a few characteristics of the 10BASE5 cable layouts:

- No loop is created; the cable forms a line, however bendy.

- The cable is strung in such a way that all parts of the building are within reasonable transceiver cable reach.

- The cable is continuous and is most likely not configured to take advantage of any existing cable troughs, etc. In fact, it is probably placed in the ceiling.

- The 500-meter segment length limitation applies to the length of the cable on the reel; how it is laid out does not affect the length limitation at all.

- Everything that attaches to a 10BASE5 cable attaches via a transceiver that is clamped to the cable, or to something else that is attached via a transceiver. The cable ends never plug into anything but the appropriate terminator.

- If more than one segment is required, the segments connect as shown in Figure 17.2.

Because of the complexity of the layout, the differences in building designs, and the difficulty in using this thick wire coax, this has become a

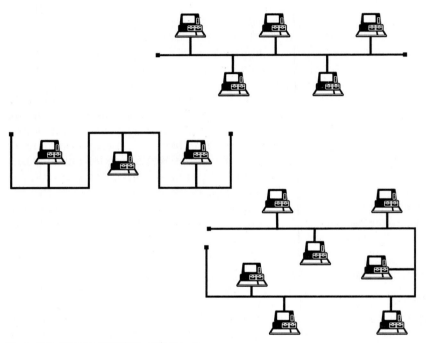

Figure 17.1 Typical 10BASE5 cable layouts.

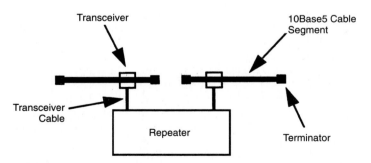

Figure 17.2 Connecting 10BASE5 segments.

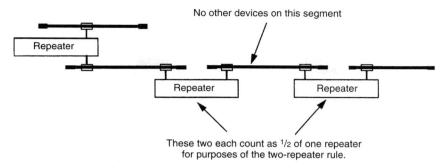

Figure 17.3 Three repeaters complying with the two-repeater rule.

legacy in the installation business. Very few new 10BASE5 networks are being installed these days, so we will not go into great detail on their possible layouts. But it is a good idea to expand upon the two-repeater rule here because it applies (sometimes in modified form) to other Ethernet topologies as well. The rule says, in effect, that "Thou shall not configure an Ethernet such that traffic must flow via more than two repeaters to get from any node to any other node." Seems straightforward, but there is at least one nonobvious subtlety. If two repeaters are configured so that the cable between them has no other drops (not even one), then for purposes of the two-repeater rule those two repeaters each count as one half of one repeater. Thus, the two of them add up to one. An Ethernet can thus have up to four repeaters in series without violating the two-repeater rule. Figure 17.3 illustrates a configuration with three repeaters.

The two-repeater rule is not as onerous as might be the case. It is possible to configure a 10BASE5 network with an almost arbitrarily large number of segments. For example, consider the "comb" configuration shown in Figure 17.4.

In Figure 17.4, no single packet must traverse more than two repeaters to reach anywhere on the network, even though seven repeaters are in the configuration. Each segment can still be 500 meters; there is opportunity with such layout to connect a great many nodes, if needed. Nodes can be placed on any segment although, in practice, the top segment, that is connected to all seven repeaters, is often designated as a "backbone" and reserved for repeaters and perhaps network monitoring equipment.

Changes

Changing connections to a 10BASE5 network is simple; simply unplug or plug in components as needed, adhering to the configuration rules. No special software configuration is required for new nodes; removed nodes require no close-out procedures removing them from any tables or anything

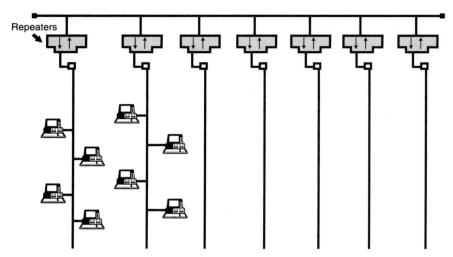

Figure 17.4 Comb 10BASE5 layout.

else. (This simplicity might or might not extend to the upper layers of the protocols used on the Ethernet.)

However, physical access to the main cable itself is required. In an office environment, the main cable is typically strung in the ceiling, with transceiver cables leading to individual offices. So, moves and changes can be quite an effort. For this reason and others, 10BASE5 is rarely installed anymore. See Figure 17.5 for how this can be a problem to access the cable and the needs to add a drop cable from the ceiling to the device.

10BASE2

10BASE5 wiring is extremely difficult to work with, and costly on a per-foot basis. With improving technology, companies determined that by applying some engineering smarts, they could design a new Ethernet standard (still 802.3) for use on a much lighter, flexible, and less expensive type of coaxial cable. 10BASE2, also often referred to as *thin wire*, or *cheapernet*, is specified to run on RG-58A coaxial cable. This cable is almost identical to cable that you probably have running into your home for cable television. But there are quite a number of differences from 10BASE5 in the way 10BASE2 is configured.

10BASE2 piece parts

As with the other 802.3 media, 10BASE2 requires a NIC in each node. However, the NIC is much smarter, and the "tap" much simpler (and less expensive) than in a 10BASE5 configuration. Also, there is no transceiver cable required or even allowed; instead, the "tap"—called a *T connector*—plugs right into the back of the NIC. The T connector splits the signal on the cable, allowing it to both continue on and also sending a copy directly into the connector on the NIC—a coax connector of the same type as is on the other two sides of the T. A T connector looks something like the one shown in Figure 17.6.

   T connectors have advantages and disadvantages, compared to transceivers. They are far less expensive, about $\frac{1}{10}$ the cost. They can be added or removed from a 10BASE2 daisy chain in seconds, much more quickly than adding or removing a 10BASE5 transceiver. However, unlike with 10BASE5, T connectors are inserted in line in the coaxial cable; if a T connector is to be added or removed, the entire cable will be down for those few seconds during which the change is being made. This is true even if the T connector is being added to the end of the cable. Once a T connector is added to the cable, unplugging it from or plugging it into its attached node does not disrupt the cable—so long as the T connector itself remains in the cable. For this reason, many installations preconfigure many T connectors in each daisy chain to avoid later disruption.

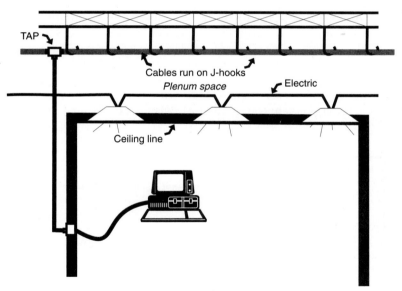

Figure 17.5 To move a station is very complex. It requires going above the ceiling. The drop behind the wall may not lend itself to removal.

Figure 17.6 The 10BASE2 T connector.

The minimum distance between T connectors is 0.5 meter (1.6 feet). One 10BASE2 segment can be up to 185 meters (607 feet) long, much shorter than a 10BASE5 segment. As with 10BASE5, every segment must end with a terminator, in this case one designed for 10BASE2. As you can see in the next section, there are differences in how the coax is terminated and in how it is laid out, when compared with 10BASE5.

Configuration of the Parts

Setup

10BASE2 is supported in two different kinds of configurations, and is also often used in a third not originally envisioned in the specification. The three configurations can briefly be summarized as follows:

- Star with daisy chains
- Pure star configuration
- Daisy chain bus

The original specification for 10BASE2 was for a star configuration, with each radiating spoke a daisy chain bus. Such a configuration might look like Figure 17.7.

A *multi-port repeater* (MPR) takes every signal it receives on one daisy chain, and sends it out over all of the other daisy chains to which it is connected. Up to 27 T connectors (some vendors support up to 30) can be installed on each daisy chain; the number of daisy chains is limited by the

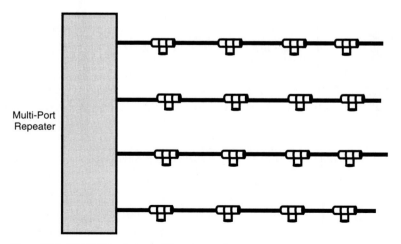

Multi-Port Repeater

Figure 17.7   10BASE2 using a multi-port repeater.

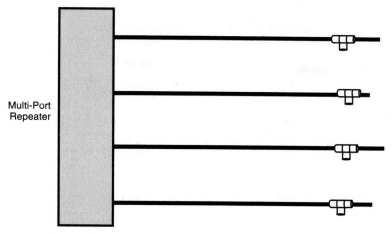

Figure 17.8 10BASE2 pure star.

number of ports on the MPR. The last T connector on each daisy chain must have a terminator on one side. The MPR is an active device; it plugs into ac power. As such, it is grounded through its power cord, and therefore provides an electrical ground for each attached segment. Thus, no separate ground is required (or indeed allowed) for the individual segments. Note that each daisy chain is plugged directly into a port on the MPR; the cable end at the MPR does not require a T connector.

This MPR configuration of daisy chains is very cost-effective, compared to 10BASE5. However, wiring and maintaining it is kind of a pain. Until the last few years, using this approach for wall-plate installations (vs. just running the cable along the floor) was very difficult and often subject to frequent outages as individuals interrupted connections, disrupting their entire daisy chain. More recently, special wall plates have come onto the market that simplify such wiring; however, they are quite expensive. 10BASE2 installations can also use a pure star configuration, as shown in Figure 17.8.

With only one T connector (and therefore only one Ethernet node) per cable, the per-port cost of a 10BASE2 pure star is much higher than the daisy chain approach; the cost of the MPR itself is spread across far fewer devices than with the previous configuration. Nonetheless, the star has two major advantages over both the daisy chain and 10BASE5:

- Because it is no longer a physical bus (although it is still a logical bus), it can use the physical communications paths installed in buildings for the purposes of telephone wiring because telephone wiring too is star-configured.

- Because each node is on its own cable, cable faults generally do not affect other nodes through the MPR.

Both of the two previous 10BASE2 configurations are officially supported by vendors. The following one is not. The pure daisy chain bus (Figure 17.9), without use of any repeater, was not contemplated by the specifiers of the 10BASE2 standard (perhaps because the opportunity to sell equipment to implement it is minimal). The biggest problem with it is that, although every Ethernet segment is supposed to be grounded, daisy chains without repeaters usually are not in practice grounded.

Because such networks are by definition very small (not to exceed 30 T connectors), the problems experienced by not grounding them properly usually do not cause major interruptions. Certainly, this is probably the least expensive type of Ethernet configuration on a per-port basis that supports more than two nodes. All of the cabling components together (except the NICs and software) can in most cases be assembled for about $10–$25 per Ethernet node, total.

Changes

Removing a node from any 10BASE2 network is simple; just unplug the T connector from the back of the NIC, leaving the T connector in the cable. Adding a node to either of the daisy chain configurations is more difficult. If dealing with pre-terminated coaxial cable lengths, adding a node requires at least one additional cable length and a T connector.

The problem is not so much in acquiring the materials as in getting at the portion of cable to which the new cable length is to be added. In addition, unless the new connectors are used (and these must be pre-installed), the entire daisy chain is down until the new T connector is inserted and connected.

The pure star approach is typically pre-cabled to every work location, just as is normal telephone wiring. With appropriate use of patch panels at a central location, moves and changes in such an environment are easier than moving telephones. (Normally, telephone moves involve telephone number changes; Ethernet moves require only connecting the new location up to any active MPR port.) Locations not requiring Ethernet connectivity simply do not have their coaxial cable runs connected to an active MPR port. If 10BASE-T wiring had not been invented, the 10BASE2 approach might have become much more popular.

10BASE-T

10BASE-T defines a standard for running Ethernet over unshielded twisted pair (UTP) wiring. The prospect of using existing, in-place (but unused)

Figure 17.9 10BASE2 daisy chain bus without repeater.

telephone wiring provided much of the impetus for the development of this standard. Many organizations had already been using some of the existing wiring in their buildings after the break-up of the Bell System in 1984. What actually happened was that Bell came along and told users that the in-house (station) wiring was theirs. Bell would no longer be in the in-house wiring business and would not maintain the wires as they had in the past. Realizing that the number of wires in the building just for the telephone systems were extraordinary, and the cost to install new wiring is expensive, many organizations saw the opportunity to use the existing spare telephone wires to run their higher-speed LAN connections. However, in practice, little of the wiring that was in place (when the standard came out) could meet Ethernet performance requirements. Even if these wires could meet the standards, the vendors required that they be certified. This meant that the vendor would send in a team of wiring experts who would test and certify each cable run. Reality revealed that it was more expensive to certify the existing wiring than to install all new wires. This proved to be a catch 22 for the end user.

Nonetheless, the much lower wiring cost of 10BASE-T (primarily because of the lower cost of telephone wiring as opposed to coaxial cable), as well as the availability of people who know how to work with UTP wiring, has made this cabling approach extremely popular. The vast majority of new installations today use 10BASE-T for horizontal (that is, to the desktop) wiring.

The possibility of running Ethernet over UTP excited many companies before the actual standard was developed. Several vendors, including Synoptics, DEC, and others came out with Ethernet-over-UTP variants before the 10BASE-T standard was finalized. They worked, but naturally it was impossible to intermix components from these vendors to build an Ethernet-over-UTP configuration. Most of those vendors did, however, have the integrity to guarantee free conversion, or at least plausible migration paths, from their proprietary hardware to hardware compliant with the 10BASE-T standard when the latter became available.

That promise does not mean, of course, that all customers bothered to take advantage of such upgrades when they became available. You might find yourself dealing with an existing UTP-based Ethernet installation with which you had no previous involvement. If so, we suggest you verify that the hardware is indeed 10BASE-T compliant rather than an older proprietary variant before blithely assuming that added 10BASE-T hardware will integrate properly.

10BASE-T piece parts

10BASE-T NICs are functionally similar to 10BASE2 NICs. That is, there is no separate transceiver as required with a 10BASE5 architecture. The connections are quite different, however. Ethernet UTP wire is terminated at each end with an 8-pin modular RJ45 jack (only 4 pins are used). The modular jack plugs directly into the NIC; no T connector is required. Each cable supports only one Ethernet node, as with the pure star thin net approach. See Figure 17.10 for a representation of the 10BASE-T environment. At least category 3 wire must be used (see the chapter on media for more on categories of UTP wire). Its length must not exceed 100 meters (328 feet).

A 10BASE-T hub (sometimes also called a *concentrator*) is almost always required. This is another version of the LAN in a box. The box or hub is the backbone network, whereas the horizontal run to the desktop is the drop. The only exception is if only two devices are to be connected; two 10BASE-T devices can be linked via a crossover 10BASE-T cable. 10BASE-T hubs are functionally identical to multiport repeaters, although some vendors provide sufficiently fast central hubs that the two-repeater rule can be slightly relaxed.

10BASE-T hubs (and, to a large extent, other hubs as well) come in two broad categories:

- Work group hubs, sometimes referred to as stackable hubs.
- Backplane chassis, sometimes referred to as concentrators.

Work group hubs are relatively inexpensive (as low as $200 to $300) devices that typically support 12 or 16 connections. If they incorporate SNMP management agents, the cost increases by several hundred dollars. Work group hubs, by definition, have very few options in their configurations. But their low cost and simple environmental requirements (they can be mounted in a rack or simply placed on a table or on the floor) make them extremely popular.

Chassis-based concentrators offer great modular flexibility. Such devices consist of a power supply, some control logic, and several slots into which interface boards can be inserted. Each interface board can support several 10BASE-T cables, often 12. Depending on the manufacturer, other boards might be available to support different types of networks out of the same chassis. Those other network types (for example, Token Ring) might be interconnected to the Ethernet board either via an additional interface board just for that purpose, or via an external device not inserted into the chassis.

Figure 17.10 The 10BASE-T connection.

10BASE-F (Fiber)

Newer uses of various cabling systems have always been explored. The fiberoptic medium has been catching everyone's attention because of the declining costs of the glass, and the electrical characteristics (that is: the freedom from electrical and radio frequency interference and the isolation from electrical components being non-issues) of the fiber. When users discovered that the fiber could be used inexpensively in the backbone network, the thrill set in. If they could use a high-speed medium that was impervious to the electrical, mechanical, and radio-frequency interference, then the network could be stretched to areas that were previously unavailable. Further, all of the grounding and bonding issues of copper cable were not requirements of the fiber network. When and if the Ethernet standard ramped up to higher speeds (100 Mbits/s Ethernet, for example) the fiber would support these higher speeds with ease. Thus, the backbone saw more implementations with the 10BASE-F installed.

The 10 Mbits/s standard was still used in the backbone network but to the desktop, UTP wiring was still the least expensive proposition. Therefore, a medium changer is required. In the closets where the LAN will run from floor to floor, or from closet to closet, fiber is used. At the hub ports exist for the connection from hub to hub via the fiber backbone. Inside the hub, the electronics are present to convert the fiber backbone to a copper station drop, or in better terms from the light to the electrical pulses needed for the copper. A typical configuration of the 10BASE-F environment is shown in Figure 17.11. This could be extended to the desktop, but as already stated the cost of putting fiber cards acting as transceivers in a PC or workstation is too steep for the average user. Therefore, this is done in a mixed environment.

10Broad36

Some networks have been built on the basis of a broadband coaxial cable in the backbone. This might be more prevalent in a campus area (such as a college, corporate park, or hospital) where multiple buildings are interlinked. The use of the broadband cable was an expedient to support the needs of the organization's voice, data, and video needs because all of these components were analog transmissions that were connected between buildings. Therefore, the broadband backbone cable (such as a CATV cable) was used when the Ethernet was introduced. Customers who needed to link multiple buildings together were constrained by the 500-meter distance limitations of the Ethernet baseband coax. Even with the two-repeater rule, the most that the cable could be extended was 1500 meters, just under a mile. Although this might sound like a lot of cable, which should reach just about anywhere, that one mile of cable gets used very quickly when it needs to be snaked through ceilings, up and down from the ceiling heights, and so on.

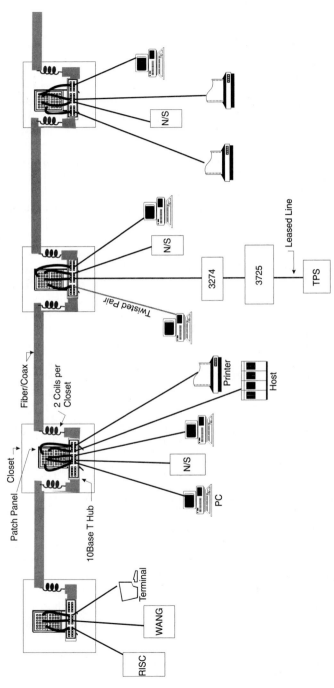

Figure 17.11 The 10BASEF uses fiber in the backbone, but converts to copper to the desk.

Therefore, the existing broadband cable offered a significant distance increase to the Ethernet users that had to exceed the distance allowances. Using a 10-Mbits/s transmission speed on a broadband cable, the distance limitations were increased to 3600 meters. This is a seven-fold increase in the distances that the cable could be run. So, the vendors came up with a solution to the distance and the medium needs for the end user. Broadband cable inherently is an analog transmission system. Special devices were attached to the cable called *frequency agile modems* that modulated the electrical signal onto the carrier. The cable is normally broken down into various channels of 6 MHz each. Using a bridging arrangement, two channels could be connected together as a 12-MHz channel and used to transmit 10 Mbits/s Ethernet on this coax. This allowed more flexibility in the overall networking ability because the Ethernet as stated in the beginning of this chapter has been implemented on a variety of media.

Digital Equipment Corporation (DEC) was a supporter of this form of connectivity in their DecNet on Ethernet from the onset. Others also came up with the necessary piece parts to configure and use the broadband cable systems for the 10 Mbits/s Ethernet standards. This has also been accepted as a connection under the IEEE standards committee. See Figure 17.12 for a cable using the broadband Ethernet connection.

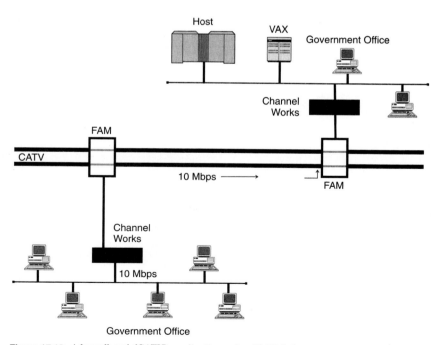

Figure 17.12 A broadband (CATV) application using 10 Mbits/s on coax.

A recent innovation from a spin-off of DEC called LANCity Inc., has produced a device or a concentrator called *channel works* that allows direct connection to a CATV operator's cable system to deliver 10 Mbits/s Ethernet connectivity across the cable. In this regard, the channel works product is supposed to be able to garner up to 83 separate Ethernets on a single cable. The primary application here has been to link multiple buildings in a metropolitan area network (MAN) for government, schools, universities, and hospitals where they have multiple buildings all across town. This has been a pretty well accepted innovation in many communities.

18

Token Ring LANs

Token Rings

Another access method that can be used in a local area network (LAN) is the token-passing concept. The *token-passing concept* is an access method that was initially developed to work on a physical and a logical ring. Remember, LANs are used in both physical and logical topologies. A *topology* is the layout of the physical wiring plan to provide a single shared cabling system. If you would think about some of the constraints and limitations of the BUS topology covered in the previous chapter, the issue is the collisions that can occur on a network. Two areas that seemed to rise in the discussion of the topology of the physical Ethernet were that the signal (information) has to propagate (travel) down to the ends of the cabling system. At the ends of the system are terminators (No! not Arnold Swartzenegger) that absorb, or remove, the electrical pulses from the cable. This meant that in order to send information and be assured that the information got where it had to go, the transmitting device could only use 50% of the time to generate the information onto the wire. The other 50% of the time allotted to this device was devoted to listening to the cable to make sure the signal made it all the way to the ends without colliding with some other device's data. This would appear to be somewhat wasteful, because only a 50% duty cycle is allowed. To overcome this situation, the use of a ring topology was introduced. Along with the topology, a collision avoidance scheme was also introduced. If a device wants to transmit, it must have a permission slip to do so. The permission slip is in the form of an electrical signal, called a *token*, which travels around the network constantly. Only one user on the network will be at a time. Only one token is needed.

The IBM Token Concept

In 1984, IBM announced plans to produce a set of network interfaces that adhered to the IEEE 802.5 token-passing concept. This access method was IBM's defense against the use of Ethernets. As you will recall, Ethernet was produced by Digital Equipment Corporation, Intel, and Xerox as the network topology and access method of choice. With the proliferation of LANs in the industry, IBM had not endorsed the Ethernet, yet had no counter offer—other than Systems Network Architecture (SNA) for a networking environment. To arrive at an IBM stamped LAN then, the Token-Passing Ring was announced. In 1985, the Token Ring LANs were beginning to appear. End users who had not migrated to the bus topology were waiting for the IBM LAN. They immediately began to jump on the use of rings. After all, "no one ever got fired by buying IBM." So, a new set of boundaries was established. Users were in one of two camps, either supporting or criticizing peer implementations of the LAN topologies. IBM created a new approach in defense against the Ethernet supporters. Many of the IBM systems would not work on an Ethernet, so immediately a new connectivity solution was introduced to support the big iron and the in-between hardware. The range of devices from the desktop (in either an ISA or an MCA architecture) to the mainframe computing platforms could all be introduced to the LAN.

IBM's version and the standardized IEEE version of tokens included the use of a deterministic-natured system. By using a single token on the network, only one device at a time will be allowed to transmit. Therefore, the risks and problems associated with the CSMA/CD LANs could be eliminated. Each device would get its fair share of use on the network. In order to do this, the devices are allowed a time to use the token as it arrives.

Initial Layout

In the very beginning days of the Token Ring, the cables were installed as a physical ring. In Figure 18.1, the actual layout of the cabling is shown. In this initial layout, each device installed on the ring is connected to the physical and closed ring of wires. Wires run to an inbound slot on a device [actually to the network interface card (NIC), which is now called a *Token Interface Card (TIC)*], then out to an outbound side of the device. What this actually means is that every device has a wire in/wire out connection. This is done on a 4-wire cable. As the devices are attached to the network, they are wired from their upstream neighbor and to the downstream neighbor. This continues around the LAN until the circle is closed.

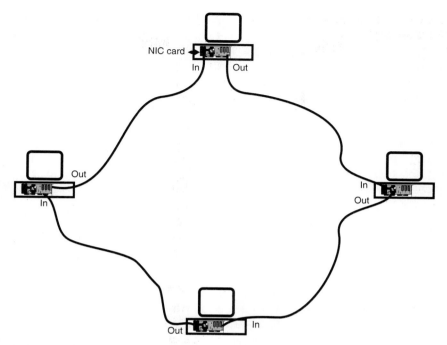

Figure 18.1 The physical ring connects each device with a wire-in/wire-out arrangement—a closed circle is created.

Problems Encountered

Nothing ever works the way we want it to all the time. Unfortunately, several problems arose with the use of this structure of wiring systems together physically. These include the following issues, but are not limited to these:

Lack of power

When the devices are wired together, the electrical signal (or data) runs down the wires, through one device to the next. This means that the electricity comes into the TIC card, gets read, and then regenerated to go out to the cable. However, this implies that the TIC card must be active at all times. When users on the ring finished their daily activities, or if they did not come into work, a problem was discovered. In both of these cases, the user's device (PC) was not powered on. The TIC card that is installed inside the PC in an expansion slot draws its power from the PC. However, if the PC isn't on, the TIC card is dead, which means that the LAN is down (Figure 18.2). This left many a LAN administrator in a quandary because this was

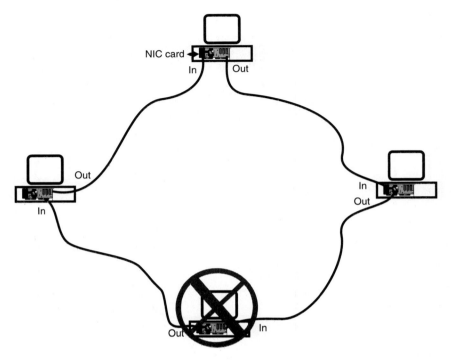

Figure 18.2 If a PC is powered off, the network doesn't work.

not initially obvious. Many hours were spent as administrators ran around the building looking for the cause of the problem on the LAN.

Lack of connection

After an end user has waited for the connection to the LAN for what they perceive as forever, the LAN administrator finally gets the wires installed, the PC configured, and the software loaded. This usually happens off-hours or weekends. At any rate, on the following business day, the user comes into the office. Initially, the end user is ecstatic about finally being connected to the LAN. However, this wears off quickly and the user then decides that the device is not located in a convenient location. Therefore, the user picks up the PC and moves it across the room to another location in the office or work area. Unfortunately, as shown in Figure 18.3, this also pulls the wires out of the wall and the jack. The broken wires render the LAN unusable because electrical continuity is lost. The LAN must be wired as a closed circle so that the electrical current will flow around the wires consistently.

Constant changes

The last scenario involves the dynamics of the networking environment. After a LAN administrator painstakingly sets up the network, the inevitable happens. A new user must be added to the network. Sometimes this can be quite simple; yet other cases require a whole new layout of the wiring system to reach the new user. In any case, however, the result is the same. The LAN administrator has to cut the connection and run the wires from the existing termination on a PC to the new one added to the network. This means that the LAN is out of service or "down" for the duration it takes to get the new wires run and connected. Only when everything is put back into the physical ring will the network be back "up" again. This is shown in Figure 18.4 where a new user must be added.

Obviously, the results of these scenarios left one conclusion: the LAN is down a lot! Something had to be done to preclude this from continually happening. IBM decided to create the "LAN in a box" concept.

The Solution to Physical Problems

To overcome the downtime and the lost credibility, the evolution to a centralized network began. IBM developed a modification to the physical structure of

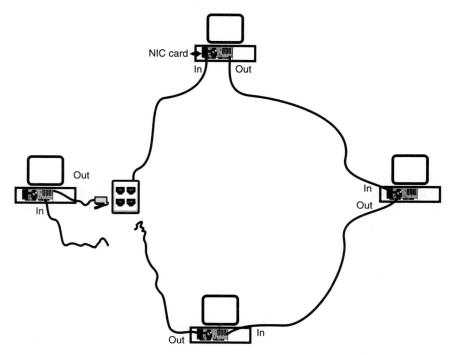

Figure 18.3 If a user moves the PC, the wires can be broken. The LAN is down.

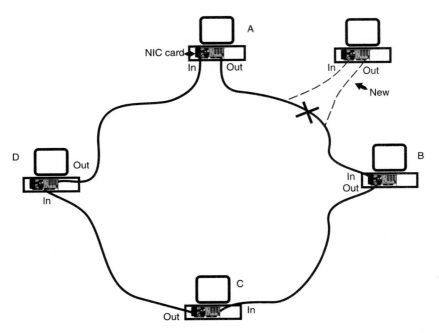

Figure 18.4 If constant changes are required, the wires between A and B must be cut and run to E. The LAN is down.

the ring. Using a version of a star network, the multi-station access unit (MAU) was introduced. In Figure 18.5, a multi-station access unit was installed in a star configuration in a telephone closet. The ring now was a physical star and a logical ring. Physical defines the structure of the wiring, whereas logical defines the signal (energy) flow. Using a basic telephony concept, wires were no longer pulled from device A to B to C to D to A as they had been. Instead, they were all pulled from location X. A single cable run from X to A, X to B and so on was used. This allows for some other unique characteristics.

Additional capabilities of the newer Token Ring networks allow for the self-healing of a cable or device problem. Using the physical ring from MAU to MAU, if the cable gets broken (Figure 18.6), the electronics running the ring are intelligent enough to close off the ring and reverse itself. This creates a far better approach to keeping the LAN up and running than what was used in the past. If the cable gets cut or disconnected, the LAN senses the open, drops a shorting bar and loops the transmit and receive wires together. The resultant network is shown in Figure 18.7. Although this is not a perfect solution, it does help to keep the uptime needs of the organization more to the design and intent. Because the ring closes and reverses itself, designing the cable lengths is very important. What used to be a short connection between MAUs has now just doubled in length because the signal

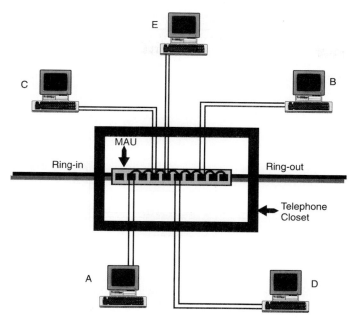

Figure 18.5 The MAU uses a physical star, but a logical ring. This "LAN in a box" helps to overcome the downtime problem.

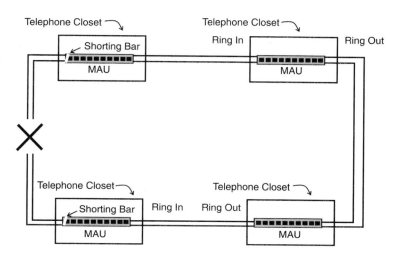

Figure 18.6 If the physical ring (MAU to MAU) is cut, the electronics in the MAU can close the ring by dropping a shorting bar.

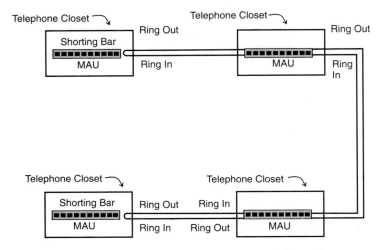

Figure 18.7 The resultant network is a closed loop around the problem.

must run on a longer cable to get from device to device. Distance limitations are still a factor, as far as the signal must run on the wires. By reversing the wires, the cable distances are doubled. This merely means that the LAN administrator must be cognizant of this limitation.

The original Token Ring topology was delivered at 4 million bits of information per second (4 Mbits/s). Distances of wires run from device to device could not exceed a total length of 1600 feet. Newer rings operate at 16 Mbits/s, but the distance was cut in half, or a total cable run of 800 feet. Too many organizations realized that this distance limitation was in the way. To solve this problem, IBM included a repeater function in the MAU so that each port on the MAU (Figure 18.8) would regenerate the signal and overcome the risk of signal loss. The MAU typically had only eight ports, as shown in the figure, plus a ring-in and a ring-out port. This allows multiple MAUs to be hooked together in a ring fashion and allows for more than eight devices to be clustered into a single logical ring.

Further, each of the ports in the MAU also includes electronics in the componentry that have shorting bars. A shorting bar allows for a single (or multiple) device to be unplugged from the network without causing the network to go down. The original, as you recall, required that all devices be physically operating and plugged together in order to work. Now users can unplug a device, or power it off without risk to the network.

Cable Types Used

When IBM introduced the Token Ring, a second part of the network design was the introduction of the structured wiring plan. To ensure that the networks would operate in an office or factory environment, different types of cables were introduced. The cable types allowed for various environmental and operational conditions to exist. This was not an inexpensive cabling solution, but if one wanted the IBM seal of approval, the IBM cabling system had to be used. In Table 18.1, the variations of the cabling are shown. This wiring included the typical twisted pairs similar (but different) to telephone wiring. Telephone wiring existed in buildings for years and many LAN administrators thought that a simple installation would be to use the existing telephone wires. This led to devastating results. Cables reflect energy. They also act like antennae and draw in electrical energy. Unless the wires are protected from each other (bleed off of energy) and from other sources of disturbances in a building (such as electric motors, electrical welders, fluorescent light ballast, etc.) the results will be degradation of the network and corruption of the data. Shielded wires were introduced by IBM to preserve the integrity of the data and prevent the influx of noise onto the cable. Thicker wires were also used. Instead of the traditional telephone wire, which is very thin (26 AWG), IBM used a thicker wire (22 AWG). In some cases, they also recommended a fiber solution to carry the higher speed data. The cabling design included both stranded and solid conductor copper cables, shielded and unshielded, and varying gauges. What IBM was attempting to accomplish was a one-stop shopping arrangement for all of an organization's cabling needs. Typical of IBM, many felt that this was overkill.

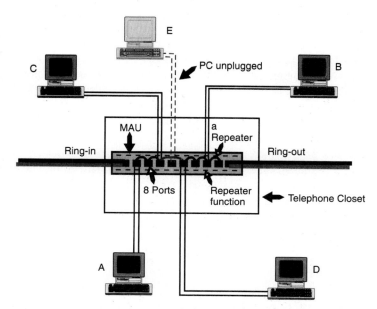

Figure 18.8 A repeater function is included in the MAU to overcome the distance limitations. Unplugged PCs are bypassed with shorting bars in the MAU.

TABLE 18.1   A Summary of the IBM Cable Types

IBM type	Number of wires and type
1	Two pairs of shielded solid core 22 AWG
2	Two pairs of shielded solid core 22 AWG and four pairs of unshielded solid 22 AWG
3	Four pairs of unshielded solid core 22 AWG
5	Fiberoptic cable multiple strand
6	Two pairs of shielded, stranded 26 AWG
8	Flat cable to go under carpets
9	Two pairs of shielded solid core 26 AWG

Further, the pricing for such a cabling system left many organizations with sticker shock. Many organizations were quoted installation prices in excess of $1 million. This cost was greater than anyone ever could have imagined—especially when everyone was accustomed to using the less-expensive telephone wiring.

Although this was a very expensive and a very bulky wiring scheme, there are organizations that installed this system nearly 10 years ago, and still supporting the original and the upgraded capacities and speeds on the same wires. To prevent problems with radiation and cross talk, IBM also used a device called a *media filter*. The purpose of the media filter was to reduce or eliminate the interfering emanations from the wires that would cause corruption of the data. These, again, have worked for the duration.

Shielding of the cables when higher data rates are being used is important. However, the shielding is dependent on being properly grounded. The ground wire should be connected to a proper building ground. The building ground should have a good solid earth ground. This is a possible problem for all forms of buildings, and subsequently can cause problems on the LAN. If the LAN wiring is not properly grounded, disruptions are likely to occur. If the building ground is suspect, it should be checked by a certified engineer and problems should be corrected. Old ground rods corrode or deteriorate, losing the proper depth to maintain a good ground. Installers like to use a cold water pipe to tap the ground, but fail to recognize that many buildings now use plastic pipes not copper, hence no ground. One must be very specific in requesting a proper ground.

Speeds

As already mentioned, IBM's first version of the Token Ring operated at 4 Mbits/s. A lot of these rings are still in use today, nearly 10 years after the announced product. The industry and the Ethernet proponents scoffed at the slower speed network when they heard of the limited speed. They perceived the situation as IBM was so desperate to get a network out into the marketplace that they short-changed the user. What most people did not recognize was the significance of the deterministic, rather than the CSMA/CD network access control. A well-installed Ethernet operating at 10 Mbits/s will support up to 1024 nodes and yield approximately a 40% net throughput. This equates to about 4 Mbits/s of effective utilization. In actuality, the Ethernet yields about 3.3 Mbits/s. The main reason for this difference is the propagation time, the potential for collisions and other timing delays that are prevalent on the bus. If collisions begin to occur, the back-off algorithm used by each of the NIC cards prevents either of the two conflicting devices from transmitting until a period of time passes. If continued collisions occur, the back-off algorithm used by the NIC cards begins to exponentially get longer and longer. This means that more devices are waiting to use the cable than those that can use the cable at a given time.

Conversely, the deterministic nature of the ring using a token-passing access control does not have the same problems. There shall be no collisions, because only a single token exists and only one device can transmit when it controls the token. Consequently, the ring can be more effectively used. The effective throughput of a 4 Mbits/s token-passing ring is approximately 3.3 Mbits/s. This parallels the performance of the Ethernet even though the raw speeds are different. Further, IBM's second version of the Token Ring Card introduced a higher rate of speed. Operating at 16 Mbits/s, the faster speed using a deterministic access control method yields approximately 12 Mbits/s. These speed comparisons are shown in Table 18.2 as a quick reference in the event this issue ever has to be reviewed. Because two camps always exist, pro or con on Ethernet or Token Ring, this one point is usually a neutralizing factor. Therefore, we wanted it to be clearly understood and visible for future reference.

Media Access Control (MAC) Layer

As with any LAN, the Token-Passing Ring requires certain access control methods. To summarize from earlier discussions, the LAN really operates at

TABLE 18.2 Comparing Effective Throughput on LANs

Access method	Raw speed	Effective throughput
Ethernet CSMA/CD	10 Mbits/s	3.3 Mbits/s
Token-passing ring	4 Mbits/s	3.3 Mbits/s
Token-passing ring	16 Mbits/s	12.0 Mbits/s

the bottom two layers of the OSI model, that being the physical layer and the data link layer. The data link layer is subdivided into two parts: the media access control and the logical link control functions. A Token Ring therefore can be shown (Figure 18.9) as it stacks onto the model. This comparison shows that the following apply:

- The physical link deals with the physical structure of the cabling system and the access method used at the physical layer. This is layer 1, as shown in the figure.
- The data link sub-layer, called the *media access control layer*, shows the access control in support of the IEEE and the ITU model for access control.
- The logical link control brings the various topologies together in a common format. The LLC represents the upper portion of layer 2 of the OSI model.

In each of the cases, the other forms of LAN access methods are shown in the same layers of the model. Layer one deals with the forms shown in Table 18.3.

The Frame

The Token Ring uses three different types of frames in its ongoing operation. These three frames are determined by the service that they provide on the network. They are:

- The token
- The frame
- The abort message

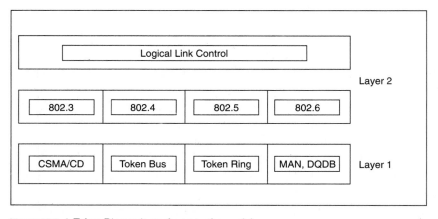

Figure 18.9 A Token Ring as it stacks on to the model.

**TABLE 18.3 Comparing the
Standards and the Topologies**

Topology/class	IEEE standard
Ethernet	802.3
Token-passing bus	802.4
Token-passing ring	802.5
DQDB, CATV services	802.6

We will look at each of these and their purpose so that the differences can be clearer. In many ways, the Token Ring is considered the most popular network concept (true or not) by users who support this infrastructure. The ills of the past Ethernet concept were addressed on the ring so that less confusion will preside on the ring, and troubleshooting, maintenance, and diagnostics can all be simpler. The word "simpler" might frustrate many a network administrator, but the intent is to simplify the detection of errors and the resolution of problems. We (Don and Bud) both differ in our opinions about which of the two network strategies is better, and why. Obviously, the intent of this book is not to steer any user into a specific product or service, but toward a better understanding of what each does. From this perspective, the control placed on the Token Ring yields a better management scheme. The network facilitates the use of the cabling system in the way it was designed. Now back to the frame formats. Our intent is not to bog the discussion down with a lot of bits and bytes. But if you don't understand what the ring is doing and what the reasoning is behind it, then the whole situation becomes a moot point. Again the differences are significant in the way they operate; the functionality that they serve is similar.

The Token

The token is made up of 24 bits of information (3 octets). The 3 octets define the start and stop sequence within this frame. A token will traverse the network constantly, even if no traffic exists. Refer to Figure 18.10 for the frame format of the token as we continue through this discussion. The first octet in the token is the *start delimiter* (SD). It is made up of a combination of good bits and violation bits (in the way we produce digital pulses, in this case, called *Manchester Coding*). The reason for the good and bad bits is to differentiate a token from real information. Therefore, the start delimiter is a very unique set of ones and zeros that can be differentiated by all devices on the network as being the beginning of a token. These eight bits are always formed in the same way.

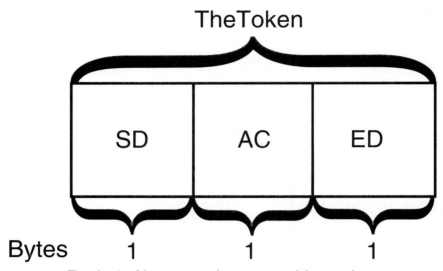

Figure 18.10 The token is a 3-byte sequence that moves around the network.

Skipping to the third octet in the token is the *end delimiter (ED)*, which contains a combination of good bits and bad bits. Much like the start delimiter, the end delimiter is used as a specific sequence to signal the end of a token. Again, this is a unique combination, so there is no confusion about what is being presented to the network.

Now back to the middle octet in the token. This octet is called the *access control (AC)* byte, which really equates to the working portion of the token. The eight bits are made up of a series of working bits that define what is happening inside the token (Figure 18.11). Bits one through three in the access control byte are marked as priority bits (P). Each device (or node as they are called) on the network has an assigned priority by the system administrator. Because there are three bits, a total of eight priorities can be used on the network, working from 000 through 111. In order of succession, the 000 will be the lowest priority device and the 111 will be the highest priority device. In order for a node to transmit information on the network, it must have the token (of course, the token must be free). The device must have a priority of equal or greater priority than the token in order to use it. This will allow for certain devices to have a higher priority to send information than others. An example of this might well be that the server on the network will have a higher priority than a regular low-end user who only performs word processing functions. This is not a cast system, but a means of providing service and controls on the network so that one device cannot take over the network.

The fourth bit is called the *token bit (T bit)*, which signals what is present, a token (T = 0) or a frame of data (T = 1). Two possibilities exist as the token comes around, it is either a free token or a busy token containing user information. This token bit can establish the ability to let the server use the token more often than an end user, thereby allowing all others on the network to gain access to the server and receive their information. You can only imagine how the performance would be viewed if the server is waiting to send a file, save a file and so on. Everyone on the network would think that the network is performing slowly, and in reality it would be. Regardless of the reason, if the user does not get instantaneous response, then the network is slow. There are no other options.

Bit number 5 in this AC byte is used as a *monitor bit (M)* for network token control. Think of this bit as a traffic cop on the network. A device on the network is assigned the responsibility to monitor what happens on the network. When a token is sent around the network, it would not be wise to let it circle around and around continuously. Therefore, the monitor station is responsible to watch that this doesn't happen. The transmitting station (sender) sets the M bit in the token to a 0. As the token goes by the active monitor, it sets the M bit to a 1. If the monitor station receives an incoming priority token or frame with the M = 1 set, it knows that the transmitting station did not take the information off the network after a round trip. The monitor station would then take the token off and clean it up. Then, it would issue a new token to be used on the network. There is still only one token on the network at this point. To prevent any major problems on such a network, the monitor station is set up to handle the problems with tokens on the network. Thus, if something happened to the monitor, one or two neighbors will be designated as the standby monitors to take over the responsibility (Figure 18.12).

Bits 6, 7, and 8 of the AC byte are allocated as reservation bits (R). Remember the priority levels? Well, if a server or any other device has information to send across the network and it has a high priority, it can re-

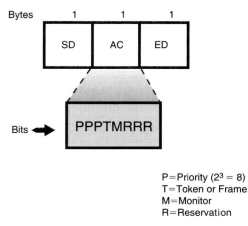

Bytes

SD	AC	ED

Bits ➡ PPPTMRRR

P=Priority ($2^3 = 8$)
T=Token or Frame
M=Monitor
R=Reservation

Figure 18.11 The access control byte format.

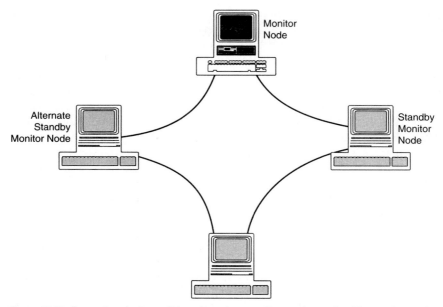

Figure 18.12 One or two devices will be selected as standby monitor nodes. The monitor node is usually a high-priority node.

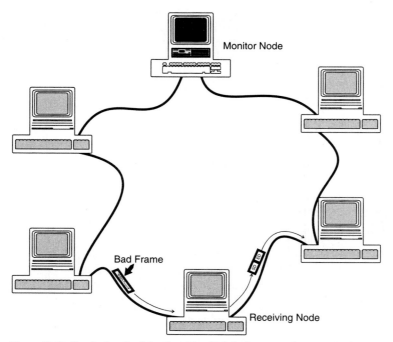

Figure 18.13 By placing the "start" and "end" delimiters together, a node detecting a bad frame will alert all other nodes that a problem exists.

serve the token for its use on the next pass around the network. These three bits are assigned for the use so that a single device cannot control the network and send frame after frame of information, denying access to all other stations on the network. Therefore, as the token goes by, the NIC inside the node will set the reservation bits and request the next available token. Certain constraints can be allowed on the network, where a single node might only be allowed one use of the token, then it must release it. Other higher priority devices might be allowed to use the token two or three times before relinquishing the control. It is through these bits in the access control field that this is implemented so that other devices have a fair share of network usage. This was a well-thought-out plan for the network.

The Abort Sequence

Figure 18.13 shows the abort sequence that is used in the Token Ring network. In the event that problems exist on the network, the nodes are designed to recognize the problem and discard the token. The abort sequence includes the ability to read the start delimiter and the end delimiter, place them together and issue it to the network by the detecting node. This alerts all nodes on the network that the problem exists. Things that can go wrong, will go wrong. This can include such problems as data corruption, lost tokens, time outs, address problems, and so on.

The Frame

Above and beyond the token format, the ring will carry a frame of information, known as the *data frame* (Figure 18.14). This frame is a variable length based on the data inside. This is comprised of the following pieces:

- The start delimiter (SD) in the token frame.
- Access control (AC). These constitute the first two bytes of the data frame.

Following these two bytes comes:

- The frame control (FC) byte indicates the type of data in the frame.
- Next comes the destination address (DA) of the token, a definition of where the frame is being sent. The destination address is 6 bytes long (48 bits).
- The source address (SA) is the indicator of where the frame came from, or who sent it. This address is also 6 bytes long.
- The data follows. In this field, the actual information being sent is provided. It can be just information, or it can also contain such information as

MAC or LLC information or routing information. This allows all physical and data link overhead to be contained inside the data packet transparently. The data field is a variable amount of information that can be up to 4048 bytes on a 4-Mbits/s network and up to 16,192 bytes on a 16-Mbits/s network.

- The frame check sequence (FCS) is a CRC-32 error detection pattern that checks the validity of the information in the FC, DA, SA, data, and FCS bytes. Using a 32-bit CRC, the error-detection capability is better than 99.99995%. If an error occurs in the transmission of the frame, if a single bit error occurs, the frame will be tossed out. This is fairly stringent, but must be used to ensure the integrity of the data on the network.

- The end delimiter (ED) is from the original token. This is a single byte of information.

- Last is the frame status (FS) byte, which indicates the status of the actual frame. In this byte control, mechanisms are used to determine what has happened since the data was initially sent by the transmitting station.

In Figure 18.15, the various bits used in the FS byte are shown. There are reasons for this technique. First the figure shows that the byte is comprised of a sequence that mimics the following:

ACxxACxx where:

- The A bits are used by the receiving device to indicate that the destination address was recognized.

- The C bits are used by the receiving device to indicate that the information (data) was copied.

- The x bits are not used and therefore ignored.

- The A and C bits inside the FS byte are important for the overall control of the network.

Because the transmitter is using a one-way cable transmission system, the information is sent out onto the wire. The sender must know that the data arrived at the desired location. However, because this is a one-way system, the only indication to the sender is if the receiver sends back a "return receipt requested" notification. That is where these bits come in to play. If the sender does not receive the frame back (in the full circle) with the A and C bits set properly, then the message must be retransmitted. If, however, the A and C bits are properly set, the transmitter knows that everything worked according to plan. Therefore, the transmitter will remove the

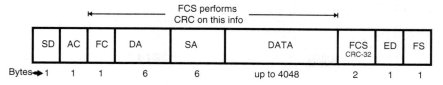

Figure 18.14 The data frame layout in a Token Ring (802.5) network.

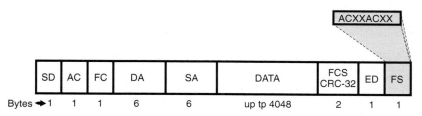

Figure 18.15 The frame status byte includes the A and C bits. A is used to designate a recognized address (to a receiver) and C is used to designate the data has been copied.

information from the network, free up the token and send it back onto the network for another device's use. Note that there are two A and C bits in this byte. The frame status byte is not checked by a CRC, therefore as a check and balance on this arrangement, the two A and two C bits must be identical. In Table 18.4, the combinations of the A and C bits inside the frame status is shown. If an A or C bit is set to 0 it is not recognized or used, whereas if set to a 1, it means that everything is okay.

To summarize the use of a Token Ring network, the following scenario will hopefully get the point across simply. We recognize that many readers are seeing this information for the first time and that many of the framing formats and control sequences are complicated and confusing. Sometimes the information needs to be read and re-read. Therefore, we attempt to use the network as best we can through an analogy. The process is fairly straightforward, but the concept can get quite complicated if you do not have a basic understanding of just what is taking place. So, fasten your seat belts and enjoy the ride around the ring.

Case Example

We wish to set up a Token Ring, so we establish the number of users who need to be connected. As we begin the process, we use an older format of physical and logical ring. Therefore, we find that we have five users that need to share the resources on the network. In Figure 8.16, we run the ca-

TABLE 18.4 A Summary of the Frame Status Bits Being Used

Pattern of frame status byte	Explanation
00xx00xx	Nothing was recognized or copied. The station may be out of order, turned off or the frame is discarded.
10xx10xx	The receiver recognized its address and responded, indicating that it is on the network. However, the receiver did not copy the information. The data CRC may have been corrupted, etc.
01xx01xx	If this happens, you have a major problem. The addressee did not respond to the frame of data but somehow it copied the frame. This is theoretically not possible, or there could be a promiscuous device that set the C bits but received someone else's data.
11xx11xx	The addressee recognized that the information was for it, and the data was in fact copied. Everything worked the way it was intended to.

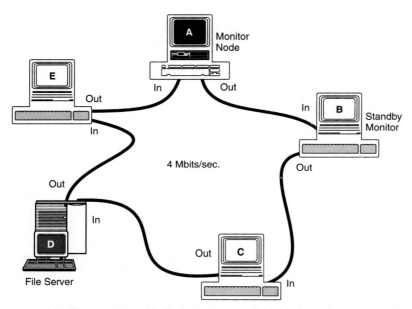

Figure 18.16 The physical and logical ring in place. A 4-Mbits/s ring is created and devices/nodes are tagged as functions require.

bles from device to device. Here, we have a wire connected to the in port of the NIC card at station A. The second wire is connected to the out port of the NIC in device A, but the out wire runs to the in port of device B. At device B, we run the out wire from B to the in port at C. From C, we run the out port wire to the in port on device D. At D, we run the out port wire to the in port wire at device E, and at E, we connect the out port to device A.

Now we have a physical ring. The circle is closed and continuity exists between and among all devices.

Next, we install the LAN drivers in each of the PCs to basically let each PC know how to use the NIC card that has been installed in the expansion slot. The card has several components to include: the electrical connection to the backplane on the PC, a chipset that operates at 4 Mbits/s, buffers to hold the data, a framer that will take the necessary data to be transmitted around the network in chunks. (Remember that in a 4-Mbits/s Token Ring, the frame can be as large as 4048 bytes.) Next, we select device A as the monitor station and B as the backup (or standby monitor). We begin by assigning the connected device D as a file server. Now we are just about ready to begin the process.

As each device is powered up, the 6-byte address is recognized as being the card's location. So the card's CPU recognizes its own address. At the physical layer we now have full connectivity.

When we load up the network operating system (NOS), we also build a database of the network devices attached to this ring. This will also be the time that we grant certain rights and privileges to the individual PC, including the services that can be accessed, the priority of the individual PC, its address and the upstream and downstream neighboring information.

Next we begin to operate the network. User A wants to send information to the file server. Therefore, user A begins by using a save command on a file in word processing. The PC and its operating components (whether in DOS, Windows, or other) send the information to the NIC card for delivery to the file server. The NIC card stores the data in a buffer. Next the NIC CPU formats the data into a frame of information. In Figure 18.17, the data is being prepared into a frame (envelope). In this envelope the NIC stuffs 4048 bytes of information inside. For all intents and purposes, this is two pages of information. So the two pages of the information have been inserted into the empty envelope.

In Figure 18.18, the addressing information is placed on the frame. We now have a frame just about ready to go. There is more information that must be attached to this packet though, but that is the permission slip referenced earlier in this chapter. The permission slip comes from the network.

The NIC monitors the incoming port to see if anything is coming down the wires. It waits for the token to come down the path to collect the data. Now the NIC sees the token coming in on the in port and grabs the information. In a very quick analysis, the NIC determines that the token has other information attached to it. Unfortunately the token comes by, but the token is busy. This is a service where only one token at a time will be available to use. Therefore, the NIC just passes the busy token along through the out port back onto the network so that the information can get to its designated address. The NIC now waits for the next pass of the token.

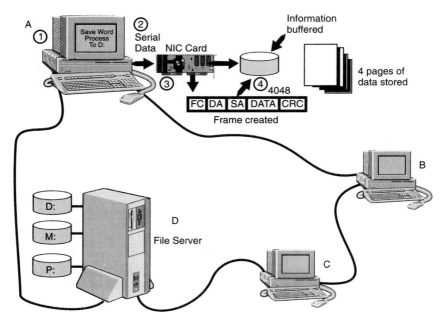

Figure 18.17 The process of preparing a frame for transmission. The frame is formatted by the NIC and stored in a buffer until the node can send to the server.

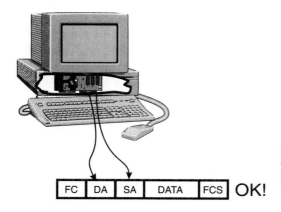

Figure 18.18 The NIC inserts the addressing information and performs a CRC on the information in the frame.

Here it comes again. This time the NIC grabs the information and determines that the token is free (Figure 18.19). The NIC does several things:

- The NIC grabs the free token, and makes it busy by changing a bit field in the AC byte.

- The NIC then appends its data to the token starting with the frame control, the destination address, the source address, the information field (4048 bytes), and the frame check sequence.

  Behind this information are the end delimiter and the frame status bytes.

- Now the NIC sends the frame out to the network. As the frame is moving down the wire, the next NIC located at device B sees the token coming and grabs it.

- NIC B then reads the token and sees that this is a frame so it checks the destination address. Realizing that the frame is not for B, it then immediately sends the frame directly back on the network to its downstream neighbor. The process continues through C.

- As C sends the frame out to the network to D, things change. The addressee is the file server at D. So when D gets the frame, it reads the information and sees that the destination address is D.

- D now makes a copy of the frame and stores it in memory on the card. Now D sets the A (address recognized) and C (frame copied) bits in the Frame Status byte to 1.

- Leaving the entire frame intact, and having modified the A and C bits, D sends the frame back out to the network where it will go through device E then on to device A (the originator).

- When A receives the frame back, it analyzes the frame and sees that the A and C bits have been set to 1 (return receipt requested). Therefore, A strips off the frame field sets the token as idle and sends the free token out to the network for the next station waiting to transmit to use. A node typically only gets one use of the token, then must wait for the next pass before using the token again.

- The process restarts all over again depending on the source and destination addresses used.

As you see, this can be an efficient use of a network service. The larger the network gets, however, the more the devices will be waiting to transmit rather than transmitting. Growth and latency on these networks will slow the network down by one bit time per node attached. Therefore, when IBM introduced its souped-up Token Ring network at 16 Mbits/s, the frame size increased by four-fold and two tokens were allowed to exist on the network. This should be helpful in keeping things moving quickly.

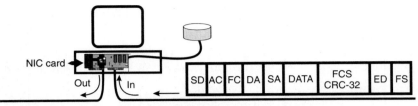

① The NIC sees the inbound frame, examines it and sees it's busy.
Sends the entire frame back out to the network.

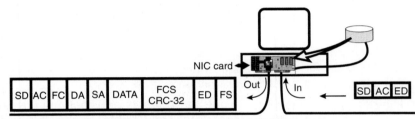

② The NIC sees the free token on the next pass so it inserts the buffered
frame into the token and sends it out to the network destined for the file server.

Figure 18.19 The NIC grabs the free token and appends the data frame. Note the sequence.

19

Baseband versus Broadband

When LANs were first emerging in the late 1970s, they were called *local nets*. These local networking services and techniques would be the foundation of the LANs of the future. As with any other emergence, two lines were formed as to how these services might be provided. From our earlier discussions in this book to date, we said that there were really two factions in the industry: the voice and the data communications factions.

Whenever a voice person got involved in the picture, all data needs were satisfied by the traditional data communications techniques. This meant that the voice analog dial-up or leased line services were installed. Following that, a modem was used to convert the digital signals of a computer into the analog signals of the voice world for transmission. This was the same for local as well as long-haul communications needs. Most voice people felt comfortable with the use of modulation techniques on the link.

But whenever a data communications person got involved, the picture changed. Most of what was installed in the data world evolved around the single largest player in the industry, IBM. IBM's strategy was to use a coaxial cable to allow the high-speed communications needs for computing platforms. This included the hierarchical structure of their SNA world. We've already covered most of these concepts in past chapters, so by now it should be getting comfortable to mention the acronyms. At any rate, the coaxial world was introduced to overcome the limitations of the noisy twisted pair wiring that the voice folks always used. This coax was traditionally what was used in a baseband arena. A single cable was extended to the end-user device (3270 terminal, 3274 controller or other). Nothing else needed to be done, because IBM would directly apply the electrical voltages on the coax for propagation down the wire to the next hierarchical device. This was a baseband application of the cable. So a nonmodulated approach was used. Electricity was directly applied to the cable and it propagated (ran) down to the ends unchanged.

Another faction that cropped up in the industry came from the radio frequency side of the house. The use of radio-based systems were also prevalent in the industry prior to this era. So the radio folks looked at the need to send simultaneous signals down a medium to an end device. They began to use radio frequency techniques (RF) to carry multiple data signals. Where localized communications was the only requirement, the RF folks decided to apply some of the characteristics of the CATV business in the local data communications business. Local nets were set up with the use of a broadband communications capacity, the equivalent of CATV, to support the need for multiple data streams on a single medium; all at the same time. The voice half of the business took sides with this approach because a modulation technique was used, ala modems.

Baseband and Broadband

Baseband and broadband are two different types of signaling that can be used on a wire. We are speaking here of the basic electrical wave forms that are injected onto the medium. But do not tune out at this point just because you are not an electrical or radio engineer (or even if you are)! We cover these methods in the context of benefits and choices, rather than instructing the reader on the design and construction of the associated components. The emergence from the local nets into a formalized LAN arena introduced a new battleground between these factions in the industry: those that felt comfortable with the use of a baseband service (the data communications groups) and those who felt more comfortable with the use of a higher-speed communications channel capacity (the voice communications, broadband RF, and electrical engineers). The stage was set for the next set of arguments. In the early 1980s, this was a hot battle, with both sides claiming to have the best solution, and that the other side did not know what they were talking about.

Many an organization had to take this decision to a very high level of management for approval. When you think about that concept, why was management bothered with these complex decisions? All management wanted was a communications solution that met the organizational needs without spending money frivolously. Here, management was asked to decide on a technical decision that they were not qualified to make. Hence, they brought in outside consultants to assist with the decision process. Unfortunately, the external consultants were equally split in their recommendations. We have seen where management brought in a consultant at the recommendation of the baseband side of the business. Then, a management consultant was brought in from the broadband side of the business. Management now had two recommendations each different, so a third (independent) consultant was brought in to be the tie breaker. What a waste of time and money!

Let's look at some of the differences between these two concepts. But first review the fundamental motivations that were used in determining whether to use a LAN or other connectivity arrangement.

Motivations Driving the LAN Decision

If you go back to when these decisions were being wrestled with, there is some historical perspective that will help you to understand the argument in its base form. When local nets and the ensuing LANs were being researched, the fundamental needs were clear:

The number of terminal users in the late 70s and early 80s were limited. Many very large organizations had a total population of terminals that were based on a 50:1 or 100:1 ratio. For every 50 or 100 total employees, there would only be one terminal device. This might be a shared device among users, or a centralized departmental function (data entry) where all others brought their data to this terminal user for input/output.

The industry was touting that by the mid-1980s, the terminal population would be closer to 10:1 and closing in on a 1:1 basis. The office automation and future office would mandate that the deployment of such terminal devices would be required for every "knowledge worker."

The alternative to massive wiring jobs to support this increasing population of users was to use a form of backbone system with easier connectivity to a central computer, rather than using "home runs" to the host from the user terminal. However, the use of a LAN to replace or simplify the cabling nightmare and the expanding population goes far beyond this. Here is a group of other reasons for installing a LAN on a single-shared medium:

Data rates

The original data rates for terminals were normally handled by local cable runs or dial-up communications. The newer terminal devices that will use bitmap overlays, graphical user interfaces, and the density of the data will require much higher data speeds than what was available. The LAN is designed to support much higher throughput speeds than traditional dial-up communications and local cable attachments.

Interconnection

Most terminals at the time were hard wired back to the host, or controller, or were connected by a dial-up line. The use of a LAN moves away from the concept of a central computer control into a distributed control architecture. This will allow a server to support users locally, and only connect to the host when necessary. Interconnectivity to do this is a function of the LAN. Figure 19.1 shows the other problem with interconnectivity. A user might have a need to access three different computers, therefore, a terminal attached to each was required. The lack of desk space and the cost of such a solution is exorbitant.

Integrated resources

Voice, data, text, and video were all on separate networks and wiring schemes (Figure 19.2). The LAN was touted as the first solution to integrate these all onto a single infrastructure. Although this is still not a reality, except in limited tests, the concept of a single access to any service was a motivation to consider LANs (Figure 19.3).

Compatibility issues

Various access and protocol stacks were used. The LAN was to provide a common interface for access to any connected service. The operative word is common communications among various vendors. LAN interfaces were to be highly intelligent so that it did not matter what the need, so long as the common path could be established.

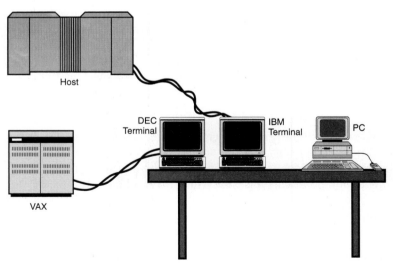

Figure 19.1 The need for a separate terminal connection to each computer was expensive and robbed all desk space from the user.

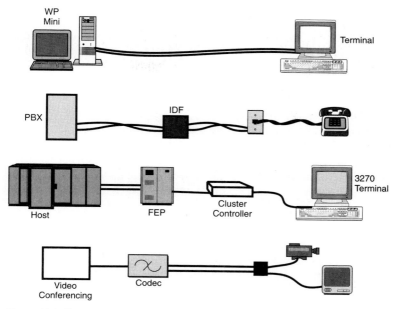

Figure 19.2 Four separate wiring systems were used to provide appropriate services. This is an expensive nonintegrated solution.

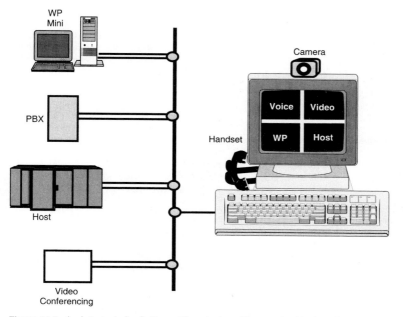

Figure 19.3 An integrated solution with a single cable was the ideal goal of a LAN.

Cost implications

As already mentioned, the cost of separate systems and wiring solutions were exorbitant. The LAN single connection was to provide this at a much more reasonable rate to the organization (Table 19.1).

Table 19.2 represents the cost comparisons of the alternatives. In this table, we are using example prices only because the actual connections and services go beyond what we can address for every organization. The numbers are generic, but roughly approximate to that of what was found in an organization back then. We looked at the cost of wiring a voice connection to the PBX, a data terminal via coax to the host, a text-based terminal to a word processing system (specialized or mini-computer) and a video connection. Then, we compared this to the LAN connection costs that were offered back then.

TABLE 19.1 The Primary Motivations for Considering the LAN

Reason	Discussion
Data rates	The original data rates for terminals were normally handled by local cable runs or dial-up connections. The newer terminal devices that will use bitmap overlays, graphical user interfaces, and the density of the data will require much higher data speeds than what was available. The LAN is designed to support much higher throughput speeds than traditional dial-up communications and local cable attachments.
Interconnection	Most terminals at the time were hard wired back to the host, or controller, or were connected by a dial-up line. The use of a LAN moves away from the concept of a central computer control into a distributed control architecture. This will allow a server to support users locally and only connect to the host when necessary. Interconnectivity to do this is a function of the LAN. In Figure 19.1 we see the other problem with interconnectivity. A user may have a need to access three different computers, therefore a terminal attached to each was required. The lack of desk space and the cost of such a solution is exorbitant.
Integrated resources	Voice, data, text and video were all on separate networks and wiring schemes as shown in Figure 19.2. The LAN was touted as the first solution to integrate these all onto a single infrastructure. Although this is not still a reality, except in limited tests, the concept of a single access to any service was a motivation to consider LANs as shown in Figure 19.3.
Compatibility issues	Various access and protocol stacks were used. The LAN was to provide a common interface for access to any connected service. The operative word is common communications among various vendors. LAN interfaces were to be highly intelligent so that it did not matter what the need, so long as the common path could be established.
Cost implications	As already mentioned above, the cost of separate systems and wiring solutions were exorbitant. The LAN single connection was to provide this at a much more reasonable rate to the organization.

TABLE 19.2   A Comparison of Costs for a Single versus Multiple Cabling System Connection

Needed connection	Separate connection	LAN connection
Voice, twisted pair to the PBX	$ 500.00	
Data terminal, coax cable 327X device	$1000.00	
Word processing connection to a specialized system or mini-computer on a coax or twisted pair	$ 450.00	
Video connection on a coax CATV cable	$ 500.00	
Integrated on a LAN, voice, data, video and word processing		$ 800.00
Total	$2450.00	$ 800.00
Differences		($1650.00)

This was an appropriate comparison, but a problem emerged with this philosophy. The various systems being offered in the LAN arena would not support all of these services on a single cable system. The reasons for this argument are:

Voice, although a baseband system, is a two-way simultaneous conversation, full duplex. Even though we saw that voice operates primarily as a half-duplex service, the need to be fully duplexed is always there. Conversations are built around the need for two people (or more) to interact as they would in a face to face conversation. The LANs would not specifically allow this in the Ethernet and Token-Passing Ring installations. They are baseband.

Both voice and video require a constant bit rate of transmission. They are not a data communications transport. The voice and video were not easily packetized or placed in frames like data can. Data and LANs deal with more variableness in their need to transmit, but can suffice with the packetization or framed format of transport. If an integrated cable is used, various streams will be required on the cable simultaneously, in a full and half-duplex method. This was not an available service at the time because baseband LANs are primarily one-way transmission systems.

As a shared cable system, the use of the cable by a single user at a time on the entire bandwidth was not appropriate. Voice does not need 4, 10, or 16 Mbits/s. Video can use this much, but we have already compressed this into much lesser bandwidth requirements. Hence, the baseband LAN was only perceived as a single transport for data, not voice and video. This underpinned the benefits of a fully integrated solution.

Three ways were available to provide the "integrated solution"; these included the following:

The PBX

Integrated PBX solutions were being touted for their robustness in the business environment. Because most organizations already had a telephone system, the PBX manufacturers were pushing their wares as the solution to the voice, data, and other needs. However, the PBX was limited in its ability to handle the higher speeds that are integral to the LAN desires and data transport. Digital PBXs were still relegated to the 64-Kbits/s digital or a 9.6-Kbits/s analog dial-up data communications connection. The PBX players were quick to state that the average LAN transmits at 4 or 10 Mbits/s, one way. However, the digital PBX can handle much higher aggregated speeds than a LAN. Voice is two way, so the PBX has a two-way simultaneous connection. Table 19.3 shows how the PBX suppliers were touting their bandwidth capacities.

Although you might argue the numbers, the PBX makers also conducted studies and found that most computer systems were limited to 64-Kbits/s data transfer rates, and that the average user needed no more than that. We have to give them credit for trying! PBXs introduced one added risk that everyone jumped on right away. The system is a centralized architecture, which meant that all of the associated risks of a single box on a network are associated here. Further, this was a digression from the distributed architecture that the LAN players were trying to accomplish.

Baseband cable systems

As mentioned in earlier chapters, an Ethernet and Token Ring normally use baseband signaling. Baseband signaling is actually rather simple in concept.

TABLE 19.3 The PBX Manufacturer's Comeback to Their LAN Service

Item	PBX	LAN bus	LAN ring
Number of users that can be simultaneously wired	10,000	1000 in Ethernet	260 in Token Ring
Maximum speed per connection	64 Kbit/s	10 Mbit/s	16 Mbit/s
Two way benefits	64 Kbit/s	not avail	not avail
Total number of users communicating at once	5000[1]	1	1
Number of users times aggregated bandwidth	640 Mbit/s[2]	10 Mbit/s	16 Mbit/s
Ability to handle video	Yes	No	No

[1] This assumes that 5000 users can simultaneously call the other 5000 users in the system meaning that 50% of the people can call the other 50%.

[2] This number is derived from the following: 5000 users transmitting and receiving at 64 Kbit/s is 128 Kbit/s times 5000 connections or 640 Mbit/s. This is arguable but they were there in the running and hyped what they had to offer.

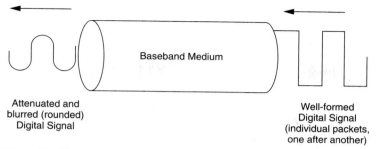

Attenuated and
blurred (rounded)
Digital Signal

Well-formed
Digital Signal
(individual packets,
one after another)

Figure 19.4 The square wave degrades over distances limiting the overall LAN length.

Devices using this technique place a digital (square waveforms—that is, ones and zeros) signal directly on the attached medium (the cable). Only one device can communicate at a time on a baseband system. There is no means for allowing two devices to originate a signal on the medium during overlapping time intervals. If two devices did try to transmit simultaneously, both digital signals would instantly be garbled beyond recognition. In the case of a bus network, several elaborate methods are used to ensure that transmitters have the medium to themselves (see the Ethernet discussion). In Token Ring and star networks, any given cable has only two connected devices following rigid protocols; no conflict arises (see the Token Ring discussion).

As with any digital signal, the square waveforms degrade over distance somewhat more quickly than would be the case using analog transmission techniques. However, so long as baseband networks are engineered within their designed distance limitations, they experience a very low level of errors indeed.

In Figure 19.4, a representation of the degradation of the square wave is shown. Distances are limited in a coax cable system such as the Ethernet. Specification for a thick wire Ethernet is 500 meters. Yes, repeaters can be used to extend the distances, as seen in the chapter on Ethernet. Two repeaters can extend the linear distance to approximately 1500 meters (just under a mile of cable). But, this is still a single-user access at a time on the entire bandwidth available. The degradation of the signal is only one part of the equation.

A key characteristic of actual baseband network standards is that in general they have been designed to allow a "by the numbers" design approach. If one uses standardized components and stays within the very specific guidelines, the networks should work fine (barring failed components, of course). Also, the equipment, because it only must deal with that one signal at a time, is less costly than digital broadband components. Proponents of the baseband systems were die-hards in their assessments that this was the least expensive and the most robust operation to meet the needs of the LAN for the future. Realizing that these LANs were being built for a data-only transport, they felt that there was no need to do anything else. Xerox of Ethernet acclaim was one of the single largest proponents of using a baseband approach. This was natural because they were in that market. But several comments that they as an organization led one to believe that this was the technology that they would live and die on.

Several issues must be taken with the idea of the single bus topology or the Token Ring topology that only allows one user at a time to send in one direction. There has already been much said on the basis of how the network works from the chapters on Ethernet and Token Ring. Therefore, we concentrate on the differences of baseband and broadband. We are not trying to suggest that a "choose or not-choose" approach is covered here. We merely try to cover all of the issues in an unbiased fashion. Here are some of the counters to the arguments that always surface:

Baseband is cheaper than broadband and PBX solutions

Although this is a valid argument, the jury never sits for a global statement in this industry. Where the Ethernet cards are very inexpensive, and the ancillary equipment pieces such as hubs and twisted pair wiring are fairly inexpensive today, the PBX prices for 64 Kbits/s transport have also dropped through the floor. Second, the CATV has been in use for longer than we can remember. The parts are also available and mass produced. Thus, a per-tap arrangement makes this a somewhat moot point, because the cost per drop is about the same. If we look at the Token Ring services, the MAUs and the Token Ring cards are not as inexpensive. A 16-Mbits/s Token Ring card can cost upwards of $695, and the MAU can be as expensive as $600 to 700 for an 8-port tap into the baseband LAN.

Baseband is a very scaleable LAN service

In fact, it is not as scaleable as one would have believed. It has been some 15+ years since the initial baseband LANs were first contrived. Only recently has the ability to take the baseband from the 10 or 16-Mbits/s speeds to the 100-Mbits/s range. There is a fundamental need that is addressed with a baseband service. Now the application of trying to merge voice and video on the LAN, the speed constraints are not suitable for these applications. So, a full-duplex LAN either in Ethernet or Token Ring is now being worked on. The addition of voice and video will demand the duplex and higher speeds, else the network will come to a grinding halt, and quickly.

Baseband is highly efficient

Any protocol that we add on any topology and medium can be efficient. The use of baseband, Ethernet, and Token Ring in particular, have a certain amount of overhead. These protocols tend to make the baseband decision different than you might think. As you saw in the Ethernet discussion, there are minimum sizes that a frame must constitute. This is so the systems attached to the baseband cable can see the timing and listen to the information during specific time slots. All data transferred must abide by the rules or the system doesn't work. The overhead and the "wait until it's your turn to send" approach are what makes the 10 Mbits/s throughput really drop to approximately 3.3 to 4 Mbits/s. That happens to be 60% to 67% inefficient.

So the argument for the baseband systems is not as straightforward as you might believe. The heavier the load on a baseband system, the greater the efficiency loss. For example, a 10 Mbits/s Ethernet typically yields a net throughput of only 3 to 4 Mbits/s. The larger the population of users, or the density of the data, the greater the risk of collisions. See Figure 19.5 for the collision. This is such that if I send "peanut butter" and you send "jelly," halfway down the cable we collide. Now you have jelly in my peanut butter! This isn't what I sent, so it must be discarded and re-sent. In a baseband

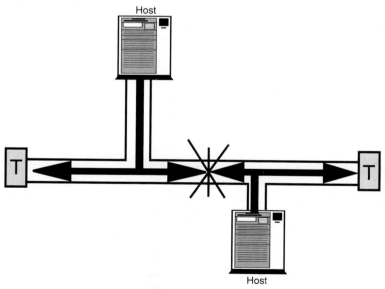

Figure 19.5 The more dense the data or the greater the number of users, more collisions are likely.

Token Ring, things are more deterministic, but the greater the population on the network, the greater the waiting time to finally transmit. See Figure 19.6 for a small scale look at the performances of CSMA/CD and Token-Passing Rings.

Broadband Cable Systems

The term "broadband" is used in several different ways in the communications industry. Among these various meanings are the following:

Any wide area communications channel of significant bandwidth

"Significant" in this case used to mean any circuit whose capacity equaled or exceeded 56 Kbps. Some analog facilities fit this description. Now, if a circuit is not at least a T-1 (1.544 Mbps), it is rarely referred to as a broadband facility. In fact, many of the T1 services and below are referred to as *narrowband*, T1 up to T3 are called *wideband*, and everything above the T3 level are referred to as *broadband*. Like anything else in this industry, it is all according to your own interpretation. Analog wide-area facilities, of whatever capacity, are rarely referred to as being broadband.

Any of a specific set of service offerings from common carriers

(e.g., SMDS, T-1 and T3, ATM, frame relay, SONET, etc.)

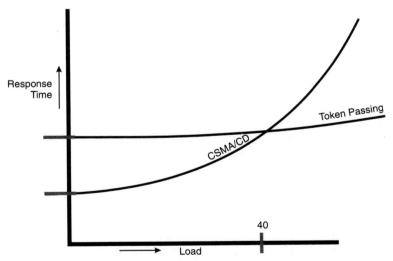

Figure 19.6 A comparison of load vs. response time for baseband networks. Token passing is a smoother degradation, whereas CSMA/CD works efficiently at lower loads.

The specific technique of using coaxial cable to carry multiple channels over LAN distances

This last meaning is the one discussed in this chapter. We have spent a lot of time throughout this book in defining the applications, uses, and definitions of baseband, yet very little time discussing broadband communications. Therefore, we finally delve into this as much as we dare. Although most people do not think in terms of baseband and broadband, there are significant differences and reasons that they exist. To fully understand the broadband communications, we use a CATV system as a comparison.

The LAN arena has always argued that baseband was the simplest and least expensive solution to installing a network. Fundamental differences exist with such a global statement. Further, the broadband communications on a cable system is perceived as an inferior product offering.

Broadband signaling on coaxial cable is one method of designing a local area network. It is unique in several respects:

- Multiple signals coexist simultaneously (really at the same time, not just looking as though they are happening at the same time because of speed).

- Broadband signaling is inherently analog—but digital signals can nonetheless be sent using it, just as digital signals can be sent over the public telephone network.

- It uses many components originally designed for a completely unrelated industry: cable television.

- It is not "plug-and-play," unlike for example Ethernet and Token Ring.

- It is targeted primarily at industrial environments because it runs only on coaxial cable, a medium more resistant than most (not counting fiber) to interference by industrial processes (e.g., arc welders).

Cables that use broadband signaling function like a superhighway with a few hundred lanes, all divided by solid double yellow lines (with no breaks). Each lane carries traffic more or less independently of traffic in other lanes. If traffic drifts across lanes, there will be accidents that disrupt traffic. But this doesn't normally happen as the lanes are separated by buffers (the double yellow lines provide some physical spacing, although not a great deal as to be wasteful).

Broadband LANs are based on the same underlying technology as cable television. To understand this, you only need to consider a typical CATV system in a hotel or large apartment complex. The building is "served" from some transmission system located outside the confines of the physical building. This serving end of the cable is called a *head end*. From the head end the television signals are distributed along the cable, and the cable forks as necessary to cover all locations. On the cable, there are many TV signals at the same time. The cable interface unit (or TV for short) selects a particular channel from the whole list of available channels on the cable. We do this by tuning into a specific frequency, what you do when you use the infrared remote control changer (Figure 19.7). The set is then tuned to the channel designator that you selected on your remote controller. The TV does this by tuning-in on a specific frequency range or "a frequency band." Because television channels need to carry a lot of information in the form of motion video, albeit one-way broadcast video, the bandwidth of the channel has been allocated a 6-MHz capacity (6 MHz or 6 million cycles per second). The lowest to the highest frequencies passed on this channel are 6 MHz wide. The channels are run next to each other similar to the way the telephone network operates. Channel 2 for example, on a CATV might operate on a frequency band of 48 MHz to 54 MHz. Channel 3 will operate at 54 MHz to 60 MHz. Standard cable systems have an immense amount of bandwidth available to them.

As the TV signals propagate down the cable, they will lose some of their strength. This loss of power is similar to the telephone network and the term is called *attenuation*. If the signal drops below a certain threshold, it becomes very difficult to separate the signal from the electrical noise always present on the cable. Of course, when this happens, the signal quality, or the output, to the TV drops and produces a sub-quality TV picture. To overcome this problem, an amplifier is placed at certain intervals along the cable. The amplifier boosts the signal strength so that it can continue to move along the wire in a usable manner.

Even though there is only a single physical cable, a number of channels can simultaneously be present on the cable. This is accomplished by sharing the frequency spectrum of the cable, and assigning different frequency

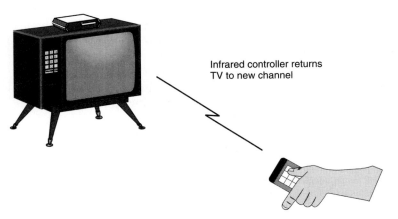

Infrared controller returns
TV to new channel

Figure 19.7 The remote controller causes the cable box and TV to re-tune to a new channel. All channels are on the cable at one time, only one can be viewed.

ranges to different channels. This was covered earlier in the data communications chapter, and is called *frequency division multiplexing (FDM)*. A broadband cable communications system operates exactly like this CATV system, and in fact uses the same types of cable and connectors. A counter to this FDM technique is where the cable is not divided into frequency bands, but in time slots. This is called *time-division multiplexing (TDM)* where each station can grab the entire bandwidth of the cable, but only for a very short period of time. The cable is currently capable of carrying up to 750 MHz of capacity. Early cable systems using the FDM techniques divided the channels into a number of dedicated logical channels as strictly an alternative to twisted pair wiring. This FDM technique did not address the interconnectivity solutions required by a LAN. Past innovations have allowed a time division multiple access (TDMA) within an FDM channel. A single cable can service a large amount of terminal communications. On this same cable, but on different FDM channels, data circuits, voice channel, and video signals can all coexist.

Modern systems allow multiplexing on two separate dimensions. The channels are separated by FDM and in a specific frequency band, the channel can then be shared via TDMA among multiple users. The most common of the TDMA access methods on a broadband cable has been the CSMA/CD developed by Xerox Corporation for Ethernet.

The considerable raw bandwidth of the cable is divided into many individual channels, each wide enough to carry one digital signal. However, the underlying structure of the network is different. On a broadband network, there is a single "head end" device that receives every broadcast signal on one of two major sets of channels, then retransmits the same signal on the corresponding outbound channel. Broadband bandwidth allocations are illustrated in Figure 19.8.

The actual cable runs might look something like those in Figure 19.9, a sample broadband cable layout. This diagram provides a clue as to why a small proportion of Ethernets are implemented on broadband cable plants; they are too complicated!

Being a truly multi-channel medium, broadband cable permits a surprising variety of applications to reside on a single cable plant. On one cable, a company can carry security video signals, terminal channels for mainframe access, one or two-way video signals for classroom use, and Ethernet. Ethernet? But Ethernet is baseband!

Well, yes, Ethernet is baseband. But baseband signaling, as stated, requires a dedicated channel on which one frame at a time is placed. But the Ethernet can use multiple media. Why not carve out part of a broadband cable's bandwidth for an Ethernet (or even several Ethernets—up to three at once are supported)?

Broadband cables in the CATV world have an inherent difference; they are designed around delivering the TV programming on a single cable in a

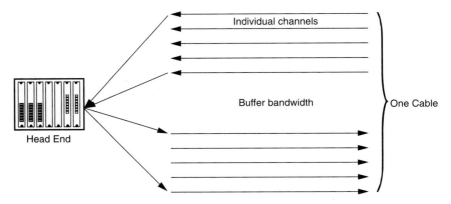

Figure 19.8 Broadband bandwidth allocation.

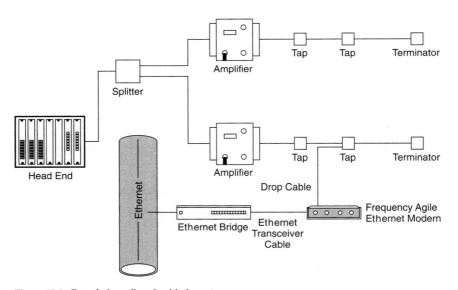

Figure 19.9 Sample broadband cable layout.

unidirectional flow. All signals come from the CATV head-end. A couple of variations are allowed to turn this cable into a bi-directional transmission. These are:

Using a single cable, a mid-split arrangement can accommodate two-way simultaneous transmission. A mid-split divides the cable into separate pieces. A portion of the cable is used to send signals to the head end from the transmitting devices called a *reverse direction*. The rest of the cable is used to send from the head end to the receiving devices, called the *forward direction*. See Figure 19.10 for this arrangement. A portion of the cable is allocated for a guard band to prevent the overlap between the channels. The head end's job is to receive the signals from the devices on the reverse direction, change the frequency and send the signal back out to the cable on the forward direction.

The second choice is to use a dual cable system. In this case, the entire spectrum of a single cable is used in the reverse direction to the head end. The second cable is used for the forward direction from the head end to the devices attached to this cable. See Figure 19.11 for a summary of how this will look. For those of us who remember the Wang net approach, Wang used a dual cable system. They split each of the frequencies up to provide services and utility bands for connecting other devices such as Ethernets to their broadband cable system.

Broadband is inherently an analog signaling method. Because, for example, video cameras are also analog devices, a signal from a video camera (or

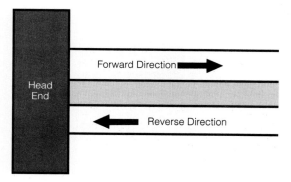

Figure 19.10 The mid-split cable.

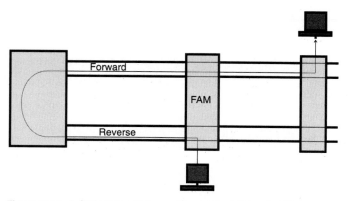

Figure 19.11 A dual cable can be used to get the full bandwidth in a system in each direction.

video recorder) can be directly transmitted onto a broadband cable channel in RGB format (red, green, blue). No conversions are required, as shown in Figure 19.12. Voice signals and conversations can also be modulated onto the broadband cable easily. In the case of a PBX on a site or campus, voice tie-lines can be used (Figure 19.13). Dedicated high-speed data links can also be set up on a broadband cabling system. Further, telemetry and paging type circuits can be modulated directly on and off the cable. Lastly, Ethernet and other digital technologies are indeed implemented on broadband cables.

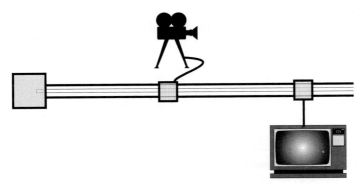

Figure 19.12 Video can be modulated directly onto the broadband cable without any conversion.

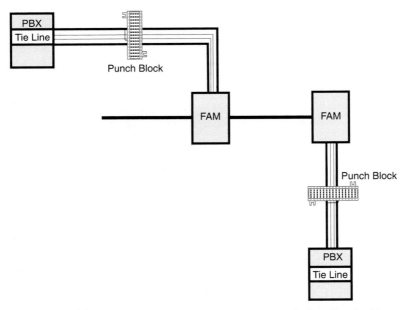

Figure 19.13 Voice grade tie-lines can be sent directly across the broadband cable.

TABLE 19.4 Summary of What Can Be Run on the Broadband Cable

Voice tie-lines such as 4 wire E&M

Logical connections of 3270 terminals to the cable and to the controller

CAD/CAM

Video conferencing

Closed circuit security systems

Building management systems such as energy management, security, alarms, etc.

LAN ala Ethernet

But additional hardware is needed. From earlier chapters you might remember another situation where digital signals were required to travel over an analog network, in that case the public telephone network. As in that case, to transmit digital signals over a broadband network we must use special modems in addition to all the normal components required to implement a cable television system. Broadband modems are called *FAMs, frequency agile modems.* Because frames are sent on one frequency and received on another frequency, they must be agile indeed.

However, all of the typical voice, data, video, and LAN services can be applied in some form onto the broadband cable. This was one of the motivating factors in making the decision to go to a LAN, wasn't it? The ability to create an integrated cabling solution for all of our communications needs has quite a bit of merit. Yet, unfortunately or fortunately, baseband cabling persevered and became the norm. Some of the reasons for this twist are because of the complexities of managing and maintaining the system. Table 19.4 summarizes the capabilities of using a broadband cabling system.

What more can we ask a single cabling system to do? Everything that we need to accomplish on a single site can be performed and modulated onto a single or dual cable system. Yet, the opponents of this cabling scheme have a tendency to knock it and recommend that it is foolish to even consider this as an alternative.

One significant disadvantage to broadband (vis-a-vis baseband) is that it must be installed by technicians qualified in radio technology. Moreover, as with all radio technology, it must:

- Be periodically checked to ensure compliance with FCC regulations.

- Be retuned when significant changes are made to the configuration particularly when additional taps are placed on the cable that would cause added loss. The use of more amplifiers might also be required.

In short, it is more of a hassle than baseband technologies. Therefore, the baseband war was won. There have been some very strong arguments for both sides of this equation. In a large campus environment, broadband might carry a lot of benefits. Whereas in a department-by-department basis, the baseband was less of a problem with installation and maintenance so this was the preferred route to follow.

Baseband vs. Broadband

Why use one over the other? Baseband should be your default choice because of its simplicity and lower cost. It is far more widespread; parts and service are much more readily available.

As with any default, there are a number of factors that might push one in the direction of other solutions. Those needs that might drive selection of a broadband rather than a baseband solution include:

- A cable television plant (with available bandwidth) is already in place in the facility to be wired (thus helping to limit costs).

- Video signaling is also required for a significant portion of the locations to be wired.

- Several separated Ethernets (but no more than three) are required, and which Ethernet a drop is to be on might change. (This reason is weaker today than in the past; such a configuration can now easily be accommodated with Ethernet hubs and appropriate 10BASE-T wiring.)

- The distances to be covered are longer than those that other Ethernet cabling can support, but within those supported by broadband.

- The network is to be installed in a highly electrically noisy environment, for example, on a factory floor. (Fiber is an alternative solution to broadband that should also be considered in such cases.)

Fiber-Distributed Data Interface (FDDI)

When we discussed the speed of the LANs as we looked at the Ethernet and Token Ring topologies, we were specific in terms of what can be expected. The older Ethernet operates at 10 Mbits/s, whereas the initial version of the Token Ring operates at 4 Mbits/s, with an updated version operating at 16 Mbits/s. These speeds were fine when the LANs were first introduced, but as we proceed toward the end of the century, the LAN throughput is becoming a potential bottleneck. The speeds were initially introduced and shared among 20 to 30 users to perform word processing, small spreadsheets and the like. However, as the LANs ballooned in the industry with devices appearing on nearly every desk, and the applications became more intense, the 4, 10, and 16 Mbits/s capacities clearly were signaling that there was a problem looming in the background.

More users are sharing more data on a single cable. We could use a subnetworking technique to bide our time, but this introduces smaller networks competing for a shared bandwidth. Further, the more we subnetwork, the more pieces on the network can fail. Consequently, the industry set out to develop a much higher speed network arrangement. The result was a technique known as *fiber distributed data interface (FDDI)*. FDDI was designed under the auspices of ANSI as a very high-speed, highly reliable data transport. The significance of the FDDI standard is that it is a national standard supported by the entire data communications industry. As a national standard, FDDI workstations, computers, and other peripheral devices work together regardless of the

TABLE 20.1 Summary of Existing Topology Problems and Limitations

Topology/item	Limitation
Bus speed	10 Mbits/s with effective throughput of 3.3–4 Mbits/s
Ring speed	4 Mbits/s with effective throughput of 3.3 Mbits/s
	16 Mbits/s with effective throughput of 12 Mbits/s
Bus attachments	Up to 1024 addressable nodes, but limited by cable lengths of 1500 meters
Ring attachments	72 addressable nodes on unshielded twisted pairs and 260 addressable nodes on shielded twisted pairs depending on distances
Unshielded twisted pair	Error prone due to wire characteristics, electrical, and mechanical interference
Shielded twisted pair	More expensive, bulky, and subject to ground loops
Bus and ring topologies	Subject to disruptions from cable breaks and other cable related problems
Bus topology	Collisions when the network gets 40% busy, resulting in less throughput

manufacturer. Moving away from the limitations of a 10 Mbits/s bus and the high-end 16 Mbits/s ring, FDDI was designed to have an effective throughput of 100 Mbits/s. Using a fiberoptic backbone, and a deterministic token-passing approach, the technology overcomes the constraints of the older systems. This is not a panacea for all LANs, but a stepping stone into higher-speed digital communications. There are certain strengths and there are limitations when considering the FDDI. It is not for everyone, yet for those organizations who have "hit the wall" with the throughput needs on the LAN, FDDI might offer some breathing room.

FDDI Design

When ANSI first started this long and arduous design back in 1988, the goal was to overcome more than just the throughput needs. Surely, a higher-speed capability was an earnest desire, and at 100 Mbits/s, no one would argue that a six to tenfold increase in throughput was not warranted. But there were other constraints that were also concentrated on. These constraints are summarized on Table 20.1. This table looks at the limitations of the bus and the rings of old combined. These were the targets for the standards committees and should be presented right up front.

To overcome these and many other limitations, the FDDI standard was written and technology developed to address each of these areas. As a result, FDDI is designed to:

- Support higher data rates at up to 100 Mbits/s.

- Work in a deterministic token-passing environment, preventing collisions.

- Support up to 1000 addressable nodes attached.

- Extend the distance of the ring up to 200 km (approximately 124 miles) with a single ring and 100 km (approximately 62 miles) with a dual ring.

- Provide robustness in that a dual ring can be used as a backup or the ability to wraparound equipment and cable failures.

- Use optical fiber as the preferred medium to overcome the physical limitations of electrical and mechanical interference.

- Allow the spacing at up to 2 km (approximately 1.25 miles) between repeaters.

Using the parameters listed, FDDI was born. You might expect that the industry and LAN administrators jumped for joy. Unfortunately, this did not happen. FDDI received a cool reception. The single largest reason (one we have stated over and over) was money! FDDI is just too expensive for most LAN administrators to implement. As it was introduced, a card operating in a high-end server could be purchased for $15,000 to $23,000. The individual cards that were designed to work in a PC or PC-based server sold for $1500 to $2000. Can you imagine what the cost for a 100-user LAN might be? For the sake of keeping it simple, we have summarized the costs in Table 20.2, than having to add a card. At the time of FDDI to the desktop, the PC hardware costs were rapidly dropping. The table summarizes the cost of a typical PC.

TABLE 20.2 Summary of Costs for a PC Equipped for the LAN Connection

Item	Cost
Change to PC, Intel-based 486D × 4/100 or Pentum 90 MHz w/8 mbytes RAM, 1.0Gbyte Hard Drive, single Floppy & SVGA monitor	$2500.00
LAN interface card Ethernet	$ 99.00[1]
LAN interface card ring (4/16)	$ 295.00
Cables and connectors	$ 25.00
Total	$2624.00–2820

[1] This is an either or decision with the Ethernet or the Token Ring card. Thus, the range of costs for the PC are included with either option.

From the cost of a $2600 to $2800—for the PC, one would reduce the incremental charge of the NIC card ($99 to $295), but add the cost of an FDDI card ($1500 to $2000) and you would double the cost per device on the network very quickly. Now using this example, a price comparison is shown for different sized networks, with the resultant differences as shown in Table 20.3. This is just so a true picture of the financial implications can be achieved. The table brings the reluctance factor to home really nicely.

Can you imagine the way management would feel if we were to request a LAN using FDDI and show the cost of a 100-user network? The PC-based solution with a standard NIC card is $200,000 less than the FDDI solution. What would be the benefit achieved for an added $200,000 in investment? But wait! There's more. The prices listed are examples only, so we should make sure that all the costs are associated with this comparison. We didn't add the differences for the server connection, the wiring differential, and the concentrators (MAUs) that work in the closet. So, taking this one step further these costs must be considered. To do so, we summarize the cost per connection as follows in Table 20.4 (by now you see we are getting cost crazy).

We are not trying to dissuade the reader from looking at the FDDI solution, but are trying to make sure that a true picture is achieved. Thus, if we take that 100-user LAN we dealt with in the previous example, and add to the cost of that LAN the variation (Table 20.4), a comparison of the LAN to that of the FDDI now totals up differently (Table 20.5). This is for comparative purposes only so that a set of numbers can be used. The actual prices will be completely different when you look at the building and workgroup specifics in your own network. But that aside, we continue our comparison.

TABLE 20.3 Summary of Cost Difference in a LAN

Number users	Cost/PC std NIC 2600	Cost/PC FDDI 4600	Difference
10	26000	46000	20000
20	52000	92000	40000
30	78000	138000	60000
50	130000	230000	100000
100	**260000**	**460000**	**200000**
250	650000	1150000	500000
500	1300000	2300000	1000000
1000	2600000	4600000	2000000

TABLE 20.4 Additional Costs of Installing FDDI to the Desktop Above Regular LAN Connections

Item	Cost per port/user FDDI	Totals
Servers for 100 users, 2 used	2 FDDI cards @$15,000	$ 30,000
Wiring for fiber to desk, with connectors, etc.	$50,000 for station wiring, $14,000 for backbone wiring $ 4,000 for connectors	$ 68,000
MAUs (called concentrators) for FDDI backbone	4 @$500	$ 2,000
Total difference		$100,000

TABLE 20.5 A Truer Comparison for the 100-User LAN with FDDI

Original cost of connectivity	Added costs for FDDI	Cumulative	Total
100-user LAN, hardware, etc., at $2600 per station	0	$260,000	
Cost of wiring for normal LAN at $350 per port/station	0	$ 35,000	$295,000
100-user LAN, hardware, etc., with FDDI to the desk	$200,000	$260,000 plus $200,000	
Cost of wiring for fiber to the desk and fiber backbone, connectors, etc., from Table 20.4 above	$100,000	$100,000	$560,000
Difference for 100 users			$265,000

This is a shocking result, because the actual cost to use the FDDI is double that of a regular twisted pair LAN. So why use it? Well let's back up a little. We have been using a 100-user LAN with Ethernet operating at 10 Mbits/s. For only twice the amount of investment, you get a tenfold increase in the throughput! Regardless of what management has to say, for this type of effective throughput, there are some benefits to be had. This is especially true when mission-critical applications reside on the LAN, when CAD/CAM or graphics applications are running, or when heavy database and spreadsheet activity is resident. A bit more on the graphics applications: When you are running a network operating system from any manufacturer and you combine a graphical user interface (GUI; such as Windows), you have added to the density of the data through the GUI. To further compound the issue, when users are using Windows on the network and they use a screen saver, they have all just added an immense amount of graphics on the LAN. This is proven from many an administrator who saw their 4, 10, and 16

Mbits/s LAN grind to a snail's pace, only to find that the GUI and screen savers were the culprits. Meanwhile, back to FDDI which would help to significantly reduce this problem.

FDDI Configuration

As we mentioned, FDDI is designed to run on a token-passing ring operating at 100 Mbits/s. The fiber used is a 62.5 μm center core with a 125 μm outer cladding. This constitutes a multimode fiber system. Other multimode fiber sizes can be used, but the specification states 62.5/125 as the transmission medium. The fiber is configured as a single ring (Figure 20.1). The single ring uses two strands of fiber to provide the resiliency of the ring to heal itself in the event of a cable problem. An optional second ring (Figure 20.2) can be installed. The second ring will be used as a backup to the first in the event of a major failure. When two rings are used, they operate in a dual counter-rotating manner. This means that they transmit the data in different directions. If dual rings are used, then four strands (or fibers) are run to each station. An alternative to a full dual-ring architecture can be used for critical applications and for critical user groups. This alternative is shown in Figure 20.3, where a single ring and a dual ring are intertwined. The single ring users are noncritical, whereas the dual ring users are the more critical and downtime cannot be tolerated.

Stations on the fiber can be connected differently as a result of these options. They can be:

- *Singly attached stations* These are devices that are noncritical and are attached only to a single ring.

- *Dual attached stations* These are the devices that have the circuitry and componentry to be attached to dual rings.

A definition is in order here because we have a tendency to intermingle words in this industry. We talk about critical and noncritical users. Then, we introduce the concept of a station. Other times we talk about nodes. Are they the same? The answer is no! There are some subtle and some logical differences in defining these terms. So let's try it this way:

A *user* is any workstation, PC, server, or any other device used to perform the work. A station is an addressable device on the network. This is the closest proximity to a user there is. The user is human, the station is the mechanical device. However, being an addressable device the station can either input or extract data from the FDDI.

A *node* is an active device on the network. The node can be a repeater or a device that completes the passing of information around the FDDI network. A node is nonaddressable. All stations are also nodes, but all nodes are not stations.

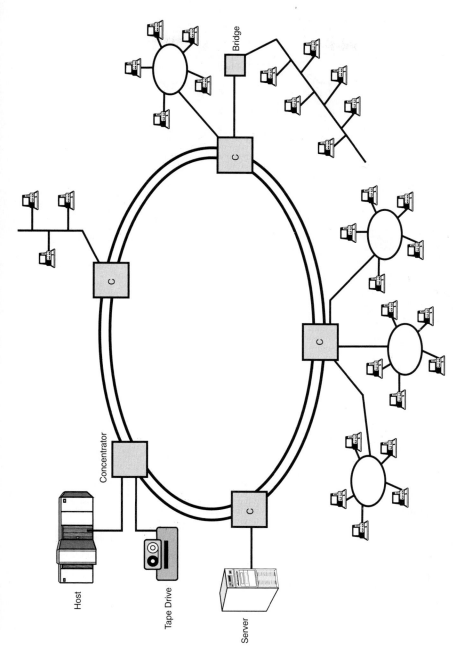

Figure 20.1 FDDI on a single ring topology.

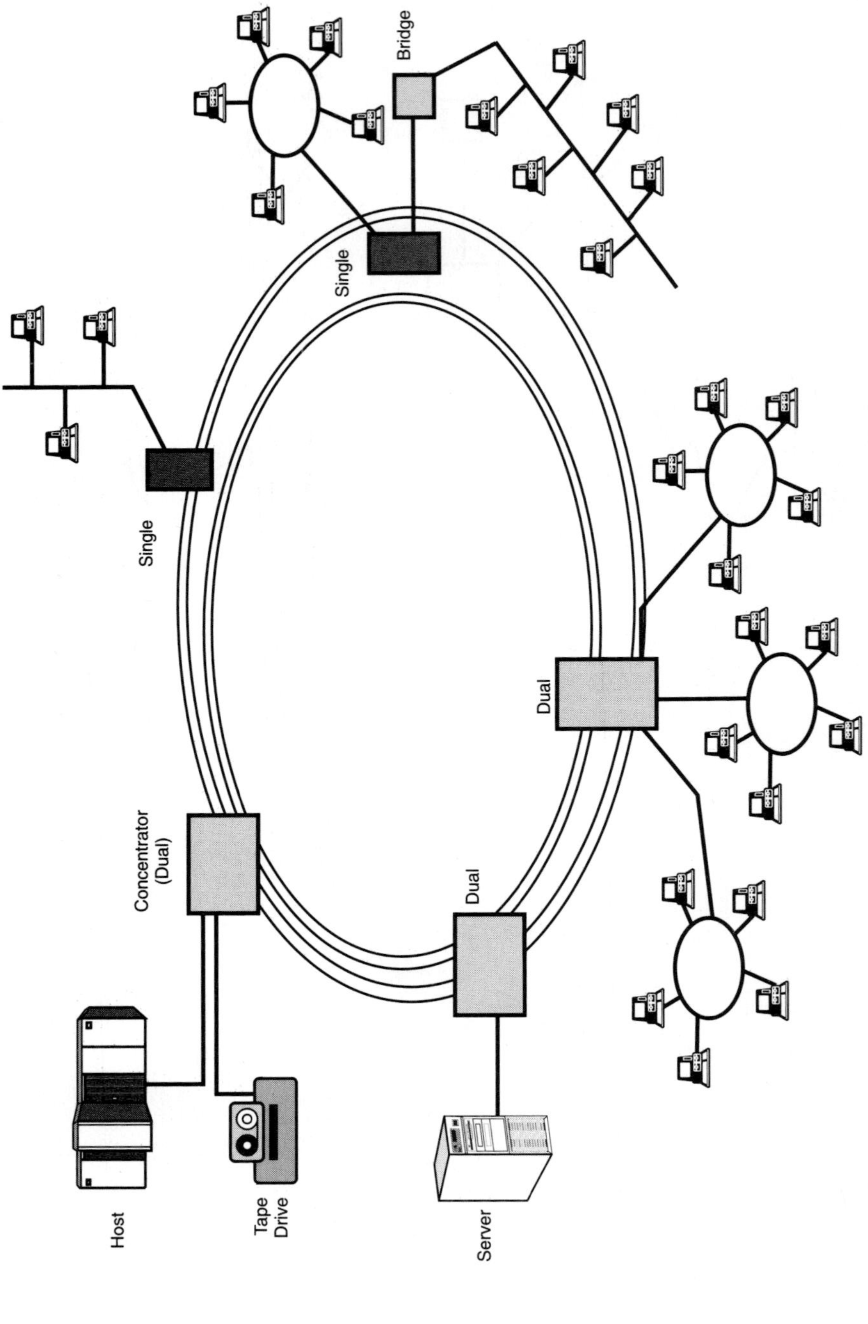

Figure 20.2 FDDI dual ring is operational.

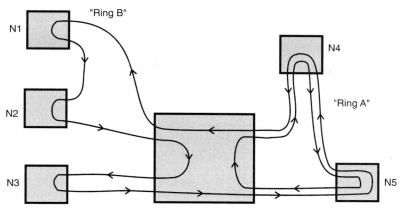

Figure 20.3 A mix of single ring and dual ring can be linked together.

Isn't that crystal clear? A node on the network can be a concentrator, the equivalent of a MAU or a multi-port repeater. When we refer to attaching users either singly or dually attached to the FDDI, it is the station that we are referring to. But, the physical attachment is likely through a node (a concentrator).

FDDI on the OSI Model

You knew this was coming! Where does all of this fit in our model? When we looked at the LAN architecture, we stated that the LAN typically works at the bottom two layers of the OSI model (physical and data link). Therefore, because we are really dealing with a LAN that is on "steroids" to get it up to a faster 100 Mbits/s, it should be obvious that the FDDI standards deal with the lower layers of the OSI also. The reference here is shown in Figure 20.4.

At the physical layer, the FDDI is broken into two separate sub-layers. There is the:

- *Physical media dependent (PMD)* sub-layer at the bottom. This PMD is responsible for such things as the specifications of sending and receiving the signals; making sure that the proper power levels for the light are applied; and specifying the physical cables and connectors used.

- *Physical layer protocol (PHY)* is designed to be media independent. This means that this upper portion of layer 1 doesn't care what the cables are. It does define the coding of the information to be sent, the decoding of the signals that arrive, the timing on the network, the status of the lines (wires or fibers) and the framing conventions.

OSI

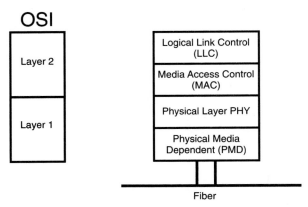

Figure 20.4 The FDDI stacks up against the bottom two layers of the OSI model.

The data link layer is again a sub-layered architecture. The two sub-layers are responsible for the following:

- *Media access control (MAC)* sub-layer protocols define the rules for formatting the frame of information to be sent/received, error checking, handling the token on the FDDI, and managing the data link addressing. Some of this is contained in what is called *station management (SMT)*. The SMT is responsible for managing the station attached to the FDDI. It provides the rules for node configuration, statistics gathering on errors, recovery in the event of an error and managing the connections.

- *Logical link control* (LLC) is as always the common control function of the LAN. It defines the protocols to be used in the upper layers of the model, because most LANs these days use multiple protocols. An example of this is a LAN running Novell Netware 3.X or 4.X. On this LAN Novell's SPX/IPX protocols are prevalent, but depending on the network, TCP/IP might also be running on the network. LLC manages which of these two stacks to use in delivering the information to the upper layer devices.

In this chapter, we stated that the PHY layer is responsible for the coding of the information, the timing on the network, and the framing conventions to be used. This is a fairly sophisticated set of responsibilities. When you think about it, this is heavy stuff. The coding is a classic example. FDDI uses symbols in the transmission of data onto the cable. To send the pulses down to the fiber, the PHY is responsible for performing a 4B/5B coding conversion.

Four bits (4B) of information are delivered to the PHY. The PHY converts these four bits into a five-bit symbol (5B). This allows for some synchronization (timing) on the FDDI and makes use of the extra bits inserted into the symbol by ensuring that enough pulses (ones) are used to keep the power of the light on the fiber. If there are too many zeros (0), then the equipment on the fiber will get amnesia and lose its place. If that happens the equipment will have to resynchronize (time) between ends. No data will be transmitted while the systems are trying to re-time, so a disruption will occur. Along with this creation of symbols, the FDDI uses electrical characteristics that are called *non-return to zero inverted* (NRZI). The use of ones and zeros will occur with a transition from the ones. The provision of the symbols allows that the transition will occur on a regular basis. This is all taken care of through the standards and specifications. See Figure 20.5 for the process as the actual data goes through the 4B/5B and the encoding being used in the NRZI format. For the actual data throughput, the 4B/5B encoding allows for a transmission rate of 100 Mbits/s on a 125 Mbaud link. A binary one (1) is equal to a pulse of light; the light is on. A binary zero (0) is equal to no light; the light is off.

Following on this discussion, the MAC layer is responsible for the construction of the frames at the sending side and the interpretation of the frames at the receiving side of the link. It is also responsible for sending and receiving the frames to/from the LLC, repeating frames similar to the Token Ring process. A number of other functions occur at the MAC layer, but a

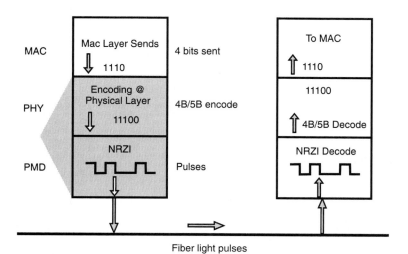

Fiber light pulses

Figure 20.5 The encoding and decoding process at the physical layer.

crucial one to be aware of is that it provides access to the ring using a timed-token protocol. A station on an FDDI network can send information after it detects an idle token. To do this the station must:

- Capture the token
- Remove the token from the ring
- Send the data (this can be a single or multiple frames)
- Upon completion of its transmission, the station then puts the token back onto the network so another station can send.

After transmitting information onto the network, it is the responsibility of the sending station to perform the clean-up operation by removing its own information from the ring. As the data makes a full circle and comes by the sender, it will remove the data from the ring. If something is corrupted or changed, the MAC layer will not try to solve the problem. The LLC or above layers are responsible for error recovery, not the MAC layer.

The timing of the token passing on FDDI is different from a typical ring topology. A timed token-passing concept is used. What this means is that as a station is ready to transmit its information, synchronous slots are made available to the station. When the station sends all of its data frames, if it does not have any more reserved synchronous traffic to send. (The synchronous traffic is the guaranteed bandwidth necessary for the particular station.) It can use the extra time left over to send asynchronous traffic. Asynchronous traffic is referred to as *random arrival* or a *bandwidth of demand capability for excess services*, not the start/stop communications process commonly used in data communications. Each station has a small clocking mechanism (a stop watch) to determine how long it will take the token to go around the ring (rotation) and pass by again. Based on this timing sequence, the station can determine how long it can hold the token and transmit its information. Once the station determines the timing, it sends all of its synchronous data on the network. If any spare time is still left over, it can send its asynchronous traffic.

The FDDI Frame

You knew that we would get into this piece, so here it is. The FDDI uses a frame size of 4500 bytes, which is different than the size of the other LAN frames we have already discussed. A comparison of the frame sizes based on the LAN type used is shown in Table 20.6. This is for comparative purposes so that the size of the frames can be seen to work in a suitable size for the FDDI and other LANs.

TABLE 20.6 A Comparison of the Frame Sizes Based on the LAN Type Used

LAN	Frame size
Ethernet 10 Mbits/s	1,518 bytes
Token Ring 4 Mbits/s	4,048 bytes
Token Ring 16 Mbits/s	16,192 bytes
FDDI 100 Mbits/s	4,500 bytes

The actual frame consists of similar size and structures of the existing LANs, but there are some subtle differences. This is a mechanism to transmit the data around the network. The frame will be broken down into several components (Figure 20.6). The token for FDDI prior to data is as shown in Figure 20.7. This is not unlike the token used in a token passing ring. After all, FDDI is a token-passing concept just operating at the higher speed. In Table 20.7, the actual frame codes are listed for reference. It would be wise to refer to this table when reviewing the figure so that the abbreviations will be easier to understand.

The frame definitions resemble that of a standard token-passing ring frame with the exception of the preamble. The preamble is designed to set the timing and synchronization of the frame for each of the devices.

Synchronization

Because there are timing arrangements used on an FDDI, it is critical that all devices keep the appropriate clocking mechanism. The PHY defines the clocking arrangement for the network. Every FDDI station on the network has its own independent clock. The receiver is designed to synchronize its clock to the incoming data and use its clock to decode the data taken off the

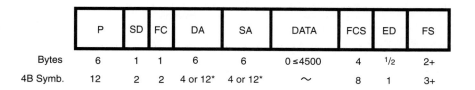

	P	SD	FC	DA	SA	DATA	FCS	ED	FS
Bytes	6	1	1	6	6	0 ≤4500	4	1/2	2+
4B Symb.	12	2	2	4 or 12*	4 or 12*	~	8	1	3+

*Some variation here based on whether 16 or 48 bit address

Figure 20.6 The FDDI frame.

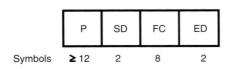

P	SD	FC	ED

Symbols ≥ 12   2   8   2

Figure 20.7 The FDDI token.

TABLE 20.7 Summary of Abbreviations Used in the FDDI Frame

Abbreviation	Description
P	Preamble, this is a pre-synchronization pattern that is made up of at least 6 bytes consisting of all ones. The 6 bytes (48 bits) minimum will be broken down into 12 each 4-bit patterns to create the 5-bit symbols (4B/5B).
SD	Starting Delimiter is comprised of 1 byte (8 bits) to indicate the start of a frame. When using the 4B/5B symbols the starting delimiter is created into 2 each 4-bit patterns and converted into 2 each 5-bit symbols.
FC	Frame Control is a 1-byte sequence broken down into 2 each 4-bit sequences and encoded into 2 each 5-bit symbols.
DA	Destination Address is a 6-byte address that will become 12 each 4-bit patterns and encoded into 12 each 5-bit symbols.
SA	Source Address is a 6-byte address that will again be broken down into 12 each 4-bit patterns and then encoded into 12 each 5-bit symbols.
DATA	The data that is being carried variable amount from 0 up to 4500 bytes. This will also be broken down into the 4-bit patterns and encoded into the 5-bit symbols.
FCS	Frame Check Sequence is a CRC-32 computed from the FC through the FCS. This is a 4-byte sequence (32 bits) or 8 symbols.
ED	Ending Delimiter is the signal to indicate the end of the frame. This is a 1-byte pattern that will go through the same 4B/5B sequence.
FS	Frame Status will be at least 12 bits or 3 each of 4-bit patterns consisting of three pieces of information. The first is the E bit that will be used to denote if an error occurred in the frame. Next the A bit signifies that the address is recognized or not. Third is the C bit to indicate if the information was copied. The EAC will be repeated at least twice as a check for errors since there is no CRD performed on the frame status.

network. It then retransmits the data with a new timing (clocking) that is generated by the local station. Sounds complex, but it is really quite simple. This keeps the network operating efficiently. The preamble in the beginning of the frame is designed to have pulses to create the clock synchronization. This again keeps everything in good working order. It should be noted that the bulk of the networks using the FDDI systems work extremely error free and the components work with exceptional performance statistics. This is a highly reliable networking service.

FDDI Applications

The primary application that FDDI is used for is in the backbone for many LANs. This high-speed backbone is used to connect a group of lower-speed LANs, such as Token Rings and Ethernets. Although there are other ways of using this FDDI network, users saw it as a means of extending the LAN into a campus area network (CAN). Using a high rise as one example, the backbone service is achieved. Figure 20.8 shows a group of individual LANs operating at various speeds and topologies connected to the FDDI backbone. Here is where the FDDI standards shine. FDDI is topology and protocol independent when connecting these other services. The high rise lends itself nicely in many older buildings where the cables might be in elevator shafts, running close to electrical closets or in proximity to other forms of interference (electrical, mechanical, or radio) that would corrupt the data on a copper-based backbone. Further, where riser space is at a premium, the fiber is less bulky and offers a very reliable connection between floors without taking up very large cable space or wall space. Moving out from the backbone in the high rise, the FDDI can also be used in the CAN. In this particular case, the fiber is immune from electrical hazards, such as lightning spikes and the like. This means that the typical protection normally used on copper cabling between buildings and the grounding and bonding requirements required by local fire and electric codes are nonissues. FDDI stands to provide the higher throughput between these buildings on a campus because the extended distances allow greater flexibility. Working on the 62/125 μm fiber and allowances for up to 2 km between repeaters extends the distances over conventional LANs.

Figure 20.9 is a representation of the campus connection. The cables require far less conduit space so the positive is that added ducts will likely not be required. What is also a positive is that if the outer conduit runs are jammed with existing copper wiring, the fiber can be laid in the same conduits with electrical that might be far less packed. If need be, they can be run through the same right of way with water pipes, something we could never do with copper. The water pipes can act as grounds and attract electrical spikes. Too many applications have been hindered because of the lack of bandwidth and connectivity in the past. FDDI offers the opportunity to overcome these limitations.

Figure 20.10 shows another variation of the application for FDDI. This case uses the FDDI as a collapsed backbone arrangement where multiple closets in a high rise or a campus can be linked with the fiber and an FDDI interconnection can support the higher-speed throughputs. The ability to bridge and route traffic across the FDDI is another positive. Because FDDI is protocol independent, one or two different flavors of equipment can be attached to the fiber ring. Bridges or routers can be either:

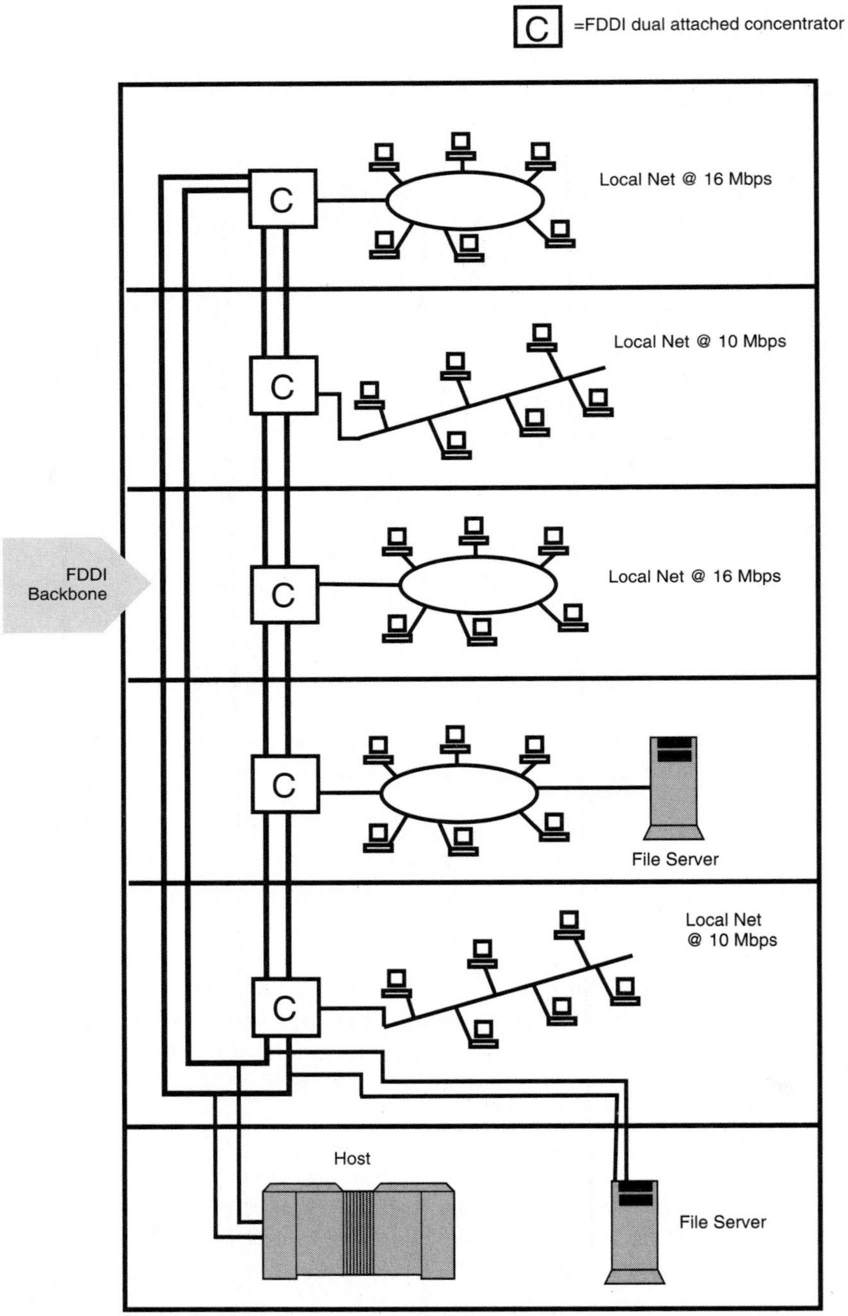

Figure 20.8 Various speeds on topologies can connect to the FDDI ring.

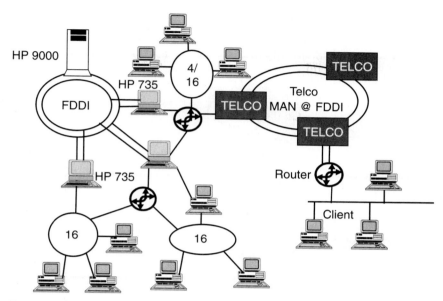

Figure 20.9 In a campus environment with FDDI as the host connector.

Encapsulating bridges

The data from the bridge (frame) is inserted directly into an FDDI frame and remains transparent to the network. As long as the addressing mechanisms are addressable, there shall be little concern for the use of an encapsulation technique. Although the FDDI nodes will not be concerned with the frame format, the stations attached to individual LANs will be. If encapsulation on FDDI is performed, the frame must be delivered to a station that can read the frame format.

Translating bridges

The data from a bridge (a frame) is stripped of the overhead that is associated with the original network topology and reformatted with new overhead to make the frame 100% FDDI compatible. This allows various equipment and topologies to coexist on the network harmoniously. Translation is more expensive than encapsulation, but the differences are becoming minuscule.

FDDI in the MAN

Recent developments in the local exchange carrier's areas have indicated that FDDI with the robustness and reliability can be delivered in the metropolitan area. The LECs are quickly moving ahead with plans to offer 100

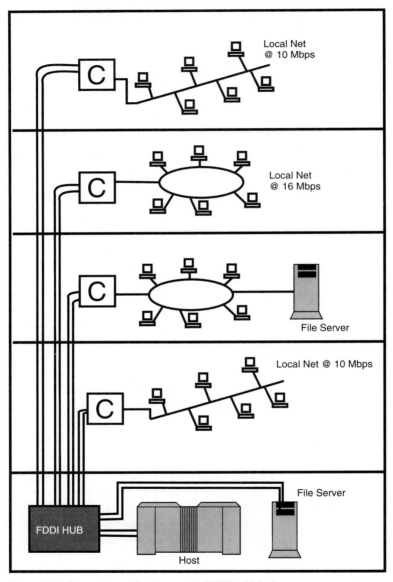

Figure 20.10 The collapsed backbone with FDDI in high rise.

Mbits/s throughput across their fiber rings in the major metropolitan areas. A particular scenario includes what is going on in the nation's capitol. Bell is providing an FDDI ring throughout Capitol Hill to interconnect LANs in various government agencies and buildings across the town. This form of a MAN is shown in Figure 20.11 with the MAN in place. The LECs have been using some conservative approaches in their deployment so that they do not overstate their offerings and so that they do not create a bottleneck. Table 20.8 summarizes the differences of the FDDI standard and the way that many of the LECs will offer this service.

FDDI Recovery

Clearly, the LECs and users alike will be careful to overcome risk where possible. Yet when placing an FDDI into service, the risk is exponential; all of the traffic on the network becomes super critical. The FDDI provisions call for the ability to select how the network will be used and recovered. FDDI allows for a bypass option so that a failed component on the network will be bypassed preventing downtime on the network. Figure 20.12 shows

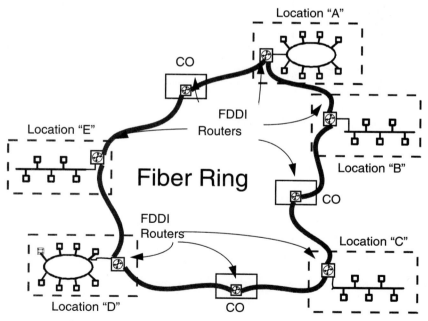

Figure 20.11 FDDI in the MAN.

TABLE 20.8 Comparison of FDDI Standards and LEC Implementations

LEC	Standards
Multimode 62/125 μm cabling, optional single mode fiber 9/125 μm	Multimode fiber 62/125 μm. Specs are completed for single mode fiber.
500 stations	1000 stations
100 Mbits/s using single or dual ring	100 Mbits/s with a single ring, optioned at 200 Mbits/s with dual ring
100 km maximum distance either ring	200 km maximum distance with a single ring, 100 km with a dual ring
Repeaters spaced at up to 2 km apart	Repeaters spaced at up to 2 km apart

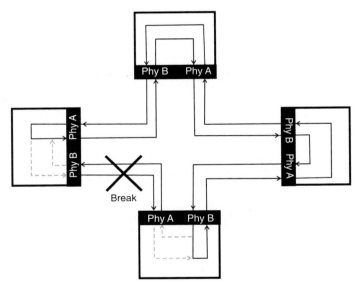

Figure 20.12 A dual counter rotating ring in healing wrap bypasses problem equipment or links.

the ability to perform the bypass of a failed component. Using the afore-mentioned dual counter-rotating ring topology allows an optical bypass ca-pability in the node electronics. If a station is not powered on, as was the case in the Token Ring problem lists, then the optical bypass allows the net-work to continue to operate. This is one solid way of guaranteeing less prob-lems on such a network.

Secondly, stations attached to the ring can be single attached stations or dual attached stations. The single attached stations are not perceived as mission critical. If a problem occurs with a cable break, the ring can close itself off and go around a problem. However, if the ring has multiple complications of failures, disaster will strike. With deeper pockets in the FDDI world, a user can take advantage of dual-attached stations that are connected to both the primary and secondary FDDI cable.

FDDI-II

Where FDDI is designed for the bursty nature of our data needs, there was no other alternative for other packet forms of transfers. Voice and video were considered constant bit rate, where data transmission is considered to be a variable bit rate service. Therefore, you could not attempt to put these other telecommunications services on the original specification FDDI. Newer specifications are being written to accommodate the voice and video needs of an organization. This is designed to give native throughput on a LAN to LAN connection, as well as packetized isochronous type services, such as video and voice.

Clearly, the FDDI-II was destined for greatness. Or so we thought! The evolution of FDDI-II has been slow in coming, partly because of the relative slowness of the standards bodies. However, the issue now is will it ever come about? With the evolution of SMDS and ATM, there are other alternatives to the FDDI-II standard. Many customers are looking at the ATM world as a better way to handle the integration of voice, video, data, and LAN traffic. Where FDDI will only give the user a total of 100 Mbits/s, ATM starts at 50 and 155 Mbits/s and aggregates the bandwidth up into the 2.4+ Gbits/s rates today. As you need added bandwidth, the ATM allows you to add another card that adds more cumulative bandwidth up to the new rate added. FDDI will remain a shared resource for the future so the cumulative bandwidth without any form of subnetworking will remain at 100 Mbits/s. You need only wait and see over a few short years to see if these two transport systems are complementary or competing service offerings.

Fast Ethernet

Another risk in the FDDI acceptance is the migration to the 100 Mbits/s Ethernet standards. In this particular case, these two separate approaches are rivaling the FDDI standard. The 100BASE-T standard will ramp up the Ethernet over twisted pair wiring in place (at level 5 wire) to a shared resource at 100 Mbits/s. This will also allow for the integration of the existing 10 Mbits/s Ethernet standard frame format into the new fast Ethernet standard.

Another approach is the 100 VG AnyLAN. It is a different concept, but will offer 100 Mbits/s throughput regardless of the topology and access method used. This technique will use the ring or bus capacities and allow the high-speed throughput at up to 100 Mbits/s. Sponsored and endorsed by Hewlett Packard, the 100 VG AnyLAN will be an inexpensive alternative to the FDDI world.

Part of the acceptance will depend on:

- The price of the interface cards which will work in a standard AT bus PC. Currently the price of the 100 VG AnyLAN cards are rapidly dropping. This might be a sign of acceptance and production cost decreases, or it might be an attempt by HP to drive the demand based on the cost of the cards.

- The cabling in place is already suitable on a twisted pair environment and should be easily migrateable to the higher-speed LAN capability.

- The ability to support any topology of a bus or a ring without a major change. Too many organizations want to provide some stability and not have to do mass migrations to newer high priced technologies.

The jury is still out on this acceptance and roll-out. However, as we mentioned in the beginning of this chapter, FDDI is still expensive. The cards for the individual workstation have not dropped to the reasonable level yet. The least expensive seen on the market has been a card priced at $1295. Whereas the fast Ethernet and VG AnyLAN cards are competing in the $895 range. The ATM to the desktop concept is now appearing with cards priced at the $895 range, but industry experts feel that by the end of 1995, the price of an ATM card in a PC will be $400. This will put significant pressure on the FDDI manufacturers to come up with lower cost chipsets if they plan to be competitive and stay in the market.

Ethernet Switching

Just when we thought it was safe to think about the options for faster LAN service with the myriad of capacities, such as FDDI, FDDI-II, ATM, Fast Ethernet and the 100 VG AnyLAN technologies and transport systems, a new twist is occurring. Using a high-end hub in the office environment that will have an aggregated bandwidth of multiple Gbits/s in the backplane, vendors are now introducing Ethernet switching systems. Using this arrangement the users will have multiple sub-nets that are all connected to the high-end hub. The high-end hub will support multiple ports (8-12-16), each operating at the 10 Mbits/s speeds today. Future boxes like this will more than likely double up on the number of inputs. Using a switching matrix inside this hub, users can switch from port to port inside the matrix, a cross-connect switch exists here. Using the 16-port model, an organization can have 8 simultaneous LAN to LAN connections, each operating at true 10 Mbits/s across the connection. In effect, it gives the entire 10 Mbits/s to each network (departmental) of a smaller group of users without the risk of congestion as would happen through a bridge or router.

This 10 Mbits/s of untouched bandwidth is a switched resource. As a switched resource, the constraints of meshing the networks together go away. Further, the congestion management states that if a problem occurs, a new port can be added and a new subnetwork can be incorporated into the LAN. This is unproved at the time of this writing because it is still in the definition and development stages. However, the opportunities to let user organizations stick with the installed base of Ethernet cards at 10 Mbits/s, which are also reasonably priced (under $100) and providing the aggregated bandwidth in a collapsed backbone hub is very attractive to some. No one really knows just how much bandwidth is required at the desktop. But the industry is pointing to this massive traffic jam looming in the future. How the individual LAN administrator will handle this problem will vary exponentially with the price, technology and comfort levels applicable to any specific vendor product.

Again, this is all in a "nether" zone and the winners and losers are not yet defined. It is possible that any one or all of these services will be around for years to come. Still, it is possible that only one or two survive.

21

Switched Multi-Megabit Data Services (SMDS)

The following section of information deals with a service rather than a technology, but we will probably bounce in between the two fields in such a definition. The use of high-speed data transfer has been around in various forms, and the last few chapters have been discussing ways of moving data across a network more efficiently and faster. These are goals in the industry that never get totally satisfied. As soon as an enhancement is developed, the industry says "That was good, but I want it faster, better, cheaper now!" We have mentioned catch-22 environments throughout this book, and this is one of them. We ask for it faster, and when it is delivered, we want it faster yet! SMDS is a similar situation, in that it is a service that delivers the data throughput at a very high rate of speed by today's standards. Yet, everyone has looked at it and said, "too little."

So, back to the drawing board. One quick word of caution for the reader at this point: the SMDS and some of the other high-speed communications services and technologies we are discussing go beyond a voice/data primer. We need to discuss these because the techniques are already here and many of the industry gurus will banter around the acronyms. Therefore, it is important for a user to understand a generic overview of the SMDS concept. Unfortunately, it can get quite detailed and technical. There is not a lot we can do about this, so bear with it and someday it will all make sense.

What Is SMDS?

As already mentioned, SMDS is a service provided on a high-speed communications channel to support what is called a *metropolitan area network (MAN)*. MAN is defined under the ANSI and IEEE specifications for an 802.6 network arrangement. In many cases, the user needs to provide connections between LANs in a major metropolitan area. This had been done in the past with leased lines, such as 56 Kbits/s and T1s. The use of a leased line was limited in a couple of ways, as follows:

The leased line might not be enough speed to connect various LAN-to-LAN services. If a user is running a 4, 10, or 16 Mbits/s LAN, the T1 will limit the throughput to 1.536 Mbits/s, leaving the possibility of severe congestion at the network interface. This has been the single largest concern in the industry. The 56/64 Kbits/s channels were even worse as far as throughput. A 4, 10, or 16 Mbits/s LAN trying to send data frames from LAN to LAN would certainly become congested and produce significant delays in the transfer of the data.

The costs of the leased lines are not insignificant. If the network shown in Figure 21.1 is connected with a full-meshed arrangement and (as in this case), four locations require 12 total leased access lines, the cost of a T1 at a representative cost of $700 per link per month would yield $8400 monthly, or slightly over $100,000 per year. Other ways of providing this connection

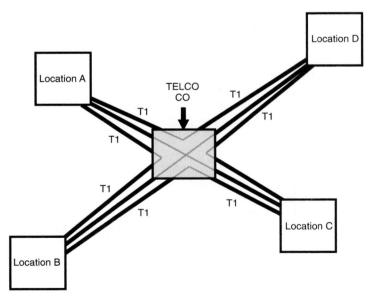

Figure 21.1 A meshed T1 network requires too many access lines, which creates a very expensive network.

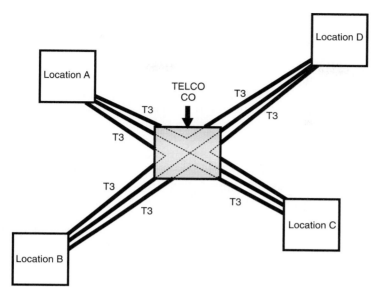

Figure 21.2 T3 installations to get native LAN throughput are an order of magnitude more expensive.

could be used, but, in each case, if the network is configured with less links, the congestion would become exponentially more severe. Using a 56- or 64-Kbits/s link to each of the locations does not satisfy the end result, so we will not go into the math on this type of linkage.

What if congestion is really bad? The user was faced with a dilemma. T1 was not all that attractive price-wise, but the only other choice to get more bandwidth between the sites was to step into a T3. This would be an astronomical fee to pay to provide the connectivity. In many parts of the country, the cost of a T3 access to a central office is in excess of $7000 per month. The cost now to connect these four locations in a meshed network (Figure 21.2) would approximate $84,000 per month or, annualized at $1,000,000. Few organizations can afford this form of connectivity.

Adding to the problems of too much or too little bandwidth, the excess capacity of a T3 would be wasted in most cases because this is a localized form of communications. Although you could multiplex voice and other data transfers onto this network (at the T3 level) the cost of the multiplexers would likely be exorbitant. Additionally, the cost to place a call or a data transfer to these local sites will be either free or very inexpensive. It would not be prudent to try to force this traffic onto the T3 when a less expensive technique already exists.

The results left users and LAN administrators in a bind. There was no reasonably priced service that would provide native LAN connectivity and throughput, yet the demands of the networks and resources were escalating. To conquer this problem, several of the standards bodies began to look at alternatives. This includes ANSI, Bellcore, and IEEE in the U.S.

SMDS is another form of high-speed public, connectionless, packet switching. In this regard, the SMDS services will handle the transfer of data across a MAN at rates of 4, 10, 16, or 34 Mbits/s. Users can commit to a burst rate that guarantees them the throughput they require. As a service and not a technology, the data delivery is at whatever rate the customer is willing to subscribe. This service is designed to assist users build network connections that require high-speed linkages without the added overhead and costs associated with fixed capacities. It is, therefore, a shared resource between and among various organizations in a metropolitan area. The connectivity can be intra-company or inter-company. As a LAN-to-LAN service, it uses a connectionless orientation. The users do not have to dial up or be connected directly to each other. Similar to the LAN, the frames or packets get sent out across the network with a destination address. All units on the network read the frame, but only those devices that are addressed will copy the data.

The Local Environment

With all of the higher-speed services discussed in the past chapters and those still to be discussed, the LECs have been noticeably absent from the discussions. Most of the higher-speed connections have been through the links provided by the IECs. Even in the case of leased lines, where the IEC must use LEC facilities (Figure 21.3), the IEC is still in control of the service. Most of our networks are wide area network (WAN) based. Therefore as a result of the divestiture in 1984, the LECs have been restrained from participating in this arena. Where the LEC did provide the local loop to the customer premises for the IEC, the LEC prices were considered too expensive. The tariffs did not allow for the negotiation of better prices because LEC services are supposed to be universally applied to any customer. The IECs in turn began to suggest ways to bypass the LEC connection, such as through the CAPs and CATV companies. These competitors are covered in earlier chapters. Consequently, the local environment for MAN services was limited and options were not readily available.

Bellcore developed the specifications of SMDS in the local arena for the Regional Bell Operating Companies (RBOCs) who fund the Bellcore research and development efforts. Just about all of the RBOCs have endorsed the use of SMDS (Table 21.1).

Using SMDS as a shared service among many users, the LECs now have an alternative to the competitive arrangements. Further, as the SMDS services are being developed, the LECs are offering these as value added ser-

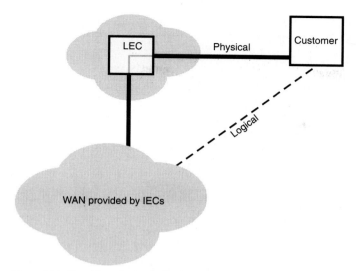

Figure 21.3 Even though physically the WAN passes through the LEC office, the connection is logically transparent to the user/IEC.

TABLE 21.1 Status of RBOCs Supporting SMDS

RBOC	Status
Ameritech	Adopted and available
Bell Atlantic	Adopted and available
Bell South	Adopted and available
NYNEX	Not adopted, no plans
PacTel	Adopted and available
SouthWestern Bell	Adopted, available in future
US West	Adopted and available

vices, rather than having the costs rolled into the embedded base pricing structure. This means that they are not tariffed the same, and that competitive pricing and special deals can be had. SMDS is designed for applications that are bursty in nature, requiring a burst of bandwidth for short periods of time. These applications do not need committed capacities and bandwidth all the time, but when they need it, they grab it and use it. Unfortunately, it is designed for the data traffic carried on LANs, not for the higher-density, quality graphics and interactive video applications that are emerging. These applications will be addressed in the future. It does support interactive sessions between users, so all is not lost by the restrictions and constraints of an SMDS network. Thus, the interactive applications that can be supported in a localized environment include:

- Computer assisted design (CAD)
- Computer assisted manufacturing (CAM)
- Source code transfer from host to host
- LAN-to-LAN native transfers
- Imaging services (magnetic resonance imaging, document imaging, etc.)
- Publishing services (compound documents)

Technology Used on SMDS

We have stated that SMDS is a service and not a technology, so we should describe that SMDS uses a technology called *distributed queue on a dual bus (DQDB)* to handle the transfer of the information. As a technology, DQDB is straightforward, it distributes the ability of users on the network into queues, and it uses dual bus architectures that are uni-directional (one-way) each. In LANs, most of what we discussed is a uni-directional data transfer. This architecture uses a controlled sequencing on a fiberoptic cable. However, the primary implementations have been single-direction transfers (one cable transmits one-way, while the second cable transmits the opposite direction). Using a distributed queue, a group of time slots are allocated on the network. Each device that wishes to transmit data across the cable must wait its turn for a queue slot. When the queue slot becomes available the transmitter can send data.

The SMDS Goal

The goal of SMDS then is to provide the high-speed data transfer for customer systems, and yet preserve the customer equipment investments. Too many systems and services in the past have required the use of fork-lift technology. When you need more, you brought in the fork-lift and removed the old equipment. Then, the lift brought in and plunked down the new equipment. This did not meet the user needs, nor did it encourage the migration to the higher-speed communications services until the equipment had been fully depreciated under IRS and financial rules. The second goal of SMDS is to roll it out as quickly as possible. If there is no incentive, then everyone will take a "wait and see" attitude. If users are not encouraged to use the service, they will seek other avenues of providing the services, which means at competitors. The LECs don't want this to happen, they want in as fast as possible. If they are the first providers and serve the needs quickly, users will be less likely to go out and seek other alternatives.

Further, if the SMDS proliferation can be rolled out quickly, users can migrate to higher rates of speed in the future as they are developed. The future of SMDS is to carry up to 150 Mbits/s in the MAN.

Access Rates to SMDS

As mentioned, this is a primary service offering from the LECs. As such, the access rates that they deliver today deal with the transport services of the past. This means that there are two basic platforms for accessing an SMDS network: T1 (1.544 Mbits/s) or T3 (44.736 Mbits/s). Unfortunately, there is little in between. However, the LECs have been offering this access at very attractive rates. Many of the LECs around the U.S., and the PTTs around the world (18 countries have already adopted the SMDS standards), are rolling out the service and the products. In the European marketplace, the PTTs are delivering speeds of 2 Mbits/s (called *E1*), 34 Mbits/s (E3), and 140 (E4) or on what they call *synchronous digital hierarchy (SDH)* at 155 and 622 Mbits/s. These higher speeds are still in beta, but will be the thrust of the future.

The LECs that have been aggressively pushing this service have come up with some very attractive rates to encourage its use. In the East, the LECs offer unlimited usage of the network with T1 access for around $500 per month/per site. This eliminates the need for the meshed network and allows a single connection into the network for on-demand bandwidth up to 1.5 Mbits/s (Figure 21.4).

Accessing the MAN

Next, look at the access method shown in Figure 21.5. The LEC will install a twisted pair (four-wire circuit) T1 access for the lower end user, or a fiber loop for T3 access. Variations are shown in this figure, where the LEC can run a fiber to the door and will deliver any rate on that fiber. Still another variation is the use of a fiber ring into the customer premises where a T1 or T3 access can be provided. If fiber is not available to the door, the LECs might bring the fiber to the curb and deliver a metallic link (coax) into the customer location at T3 speeds. These are all just different flavors of the same service provisioning, dependent on what the LEC has available to use. Future uses of SMDS might well wind up on microwave radio if that alternative is already in place. However, the primary method will work toward the deployment across a fiber-based network.

Once the local loop is installed, the customer will use a high-speed router at T1 or T3 speeds. An interface called the *SMDS network interface (SNI)* will be used as the connection at the customer location. Just a quick note here—many of the LECs have been delivering the SMDS service to the customer premises and providing the router. It seems that they have no means

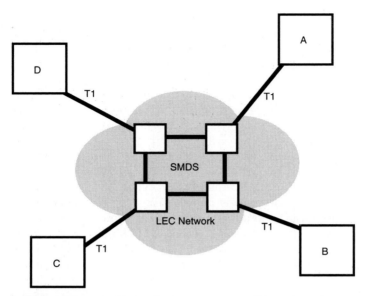

Figure 21.4 The revised network allows access up to T1 speeds on demand. This is much less expensive but it can be very robust.

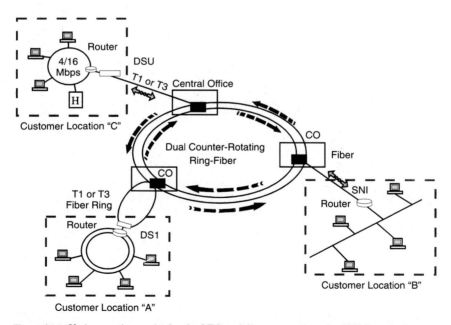

Figure 21.5 Various options exist for the LEC to deliver access into the SMDS network.

of controlling the data that is sent across the fiber and if the customer provides the router, there is no means of stopping the customer from adding cards in the router—thereby grabbing additional bandwidth for free. Because this is a value added service and not covered under the normal tariffs, this can be accomplished. The demarcation point on the fiber has now moved beyond the entrance of the customer's building into the equipment. The "demarc" will now be at the high-speed interface on the router card. With the SNI the customer now attaches to the router (Figure 21.6). The link might be on a single thread or a ring using a dual thread, it doesn't matter. The customer equipment, such as a LAN or a host computer, is then connected to this network. The access to the network is using a DQDB technique, as shown. The access using DQDB through the SNI then proceeds out to a switching system in the network. Protocols are written to provide the inter-switching systems services to make this all transparent to the end user. MCI has announced plans to provide connections on a long distance basis with Bell Atlantic and Pacific Bell creating a true cross-country SMDS network. This is shown in Figure 21.7 with the inter-carrier interfaces in place. This is a pilot service presently, but we can expect to see more of this in the future. In the event the LECs choose to connect to another service, called *ATM* (covered in Chapter 24), the overlay of SMDS will be directly onto an ATM network. Once again, this will be transparent to the end user.

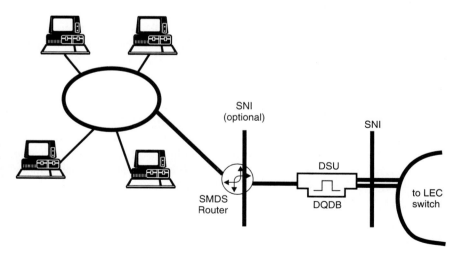

Figure 21.6 Customer connection at the DEMARC may change depending on who provides the router. The SNI is the interface between customer and LEC.

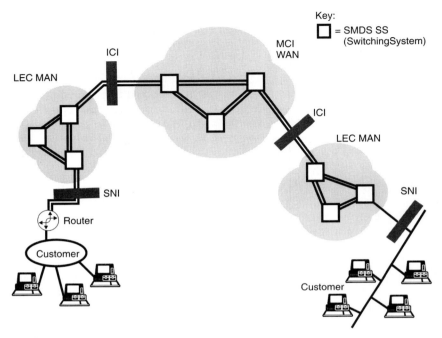

Figure 21.7 IEC to LEC arrangements are now falling in place, taking SMDS across the WAN. The inter-carrier interfaces have been defined.

The IEEE 802.6 Architecture

As with this entire book, we attempt to cover all the bases from a brief overview of the techniques, systems, and services; on to the more detailed discussion. It is only fair to explain just how the SMDS and DQDB work together in an architecture of protocols and services, and how they stack up against the IEEE 802.6 standards and protocols. The architecture is shown in Figure 21.8 with the dual bus at the physical layer, and DQDB at the link layer as they compare to the OSI model. Note that sitting on top of the DQDB (data link layer) there are four different services. The first is a medium access control to logical link control service, still working at the data link layer. The second of these services is a connection-oriented service. Third is a provision for isochronous service (such as voice, video, etc.). The fourth service is termed "other," and is under study for future use. The MAC layer contains several services.

The DQDB Architecture

Logically, the DQDB is structured as a dual bus, although in many cases it might also be in the form of a ring. In Figure 21.9, the dual bus architecture is laid out. There are two separate unidirectional buses. Each transmits in one

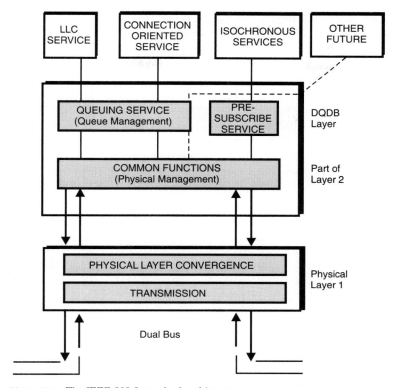

Figure 21.8 The IEEE 802.6 standard architecture.

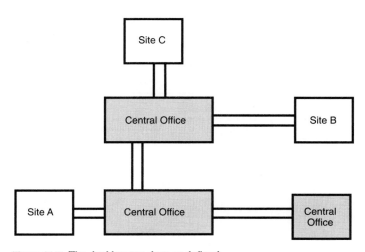

Figure 21.9 The dual bus topology as defined.

direction, but combined, they allow for a full-duplex operation. The buses are labeled as bus "A" and "B" to provide the necessary full-duplex operation between any two nodes on the network. An LEC lays out the cables in one of two ways: the cables can be laid out with a head-end in two different nodes (Figure 21.10). In this case, the head end of each of the buses appears at different locations, signifying the direction of the bus. In Figure 21.11, the two head ends of the dual buses are located in the same endpoint. Therefore, a looped arrangement is established. This resembles a ring, but in reality, it does not create the ring because both ends of the fiber are open-ended, meaning they are not closed off into a physical ring. Using either physical installation, the provision exists to recover the network in the event of an equipment failure or a problem on the fiber, such as "backhoe fade" where the cable gets cut. A break in the cable on the open bus architecture is healed as

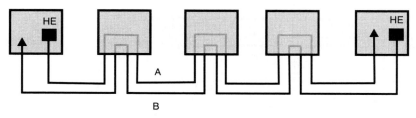

Figure 21.10 The open bus is one way to lay out the MAN.

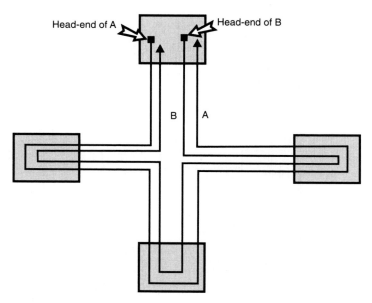

Figure 21.11 The looped bus has both head ends in the same location.

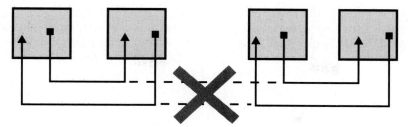

Figure 21.12 In an open bus, the result of a cable break is subnetworks formed into two networks.

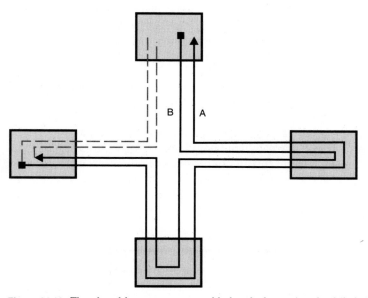

Figure 21.13 The closed loop overcomes cable breaks bypassing the failed cable. It looks like an open bus.

shown in Figure 21.12. The network will automatically reconfigure itself as two separate networks. Whereas the system in Figure 21.13 uses a looped architecture; if a failure occurs, the network will bypass the failed component or cable and reconfigure it into what constitutes an open bus. These robust arrangements will ensure more availability and more user satisfaction.

The MAN Access Unit

Similar to a multi-station access unit (MAU) on a LAN, each node on a MAN has an access unit called a *MAN access unit*. At the risk of confusing the user, we will refer to this as the *access unit* only, instead of a MAU. Each access unit is attached to both buses. The access unit can support a single node or act as a concentrator handling multiple inputs from a cluster. The access unit is used to read and write into the DQDB slots at the rate of 8000 samples per second. The buses have a slotted queue arrangement. As the access unit reads or writes into these slots, it can deliver information, read information, or request time on the network by reservation. The 8000 samples per second means that the network is timed to all of the synchronous time-division multiplexing systems in use for voice and data transmission, as well as the SONET standards. Figure 21.14 is a graphical representation of the functionality of the MAN access unit. The nodes read and copy data from the slots created in the queue. They gain access by writing to the slots.

The DQDB protocol is straightforward, based on a counter system. The counter system is used on each bus to determine if a slot is available to be reserved or to be written into. What happens is that the DQDB node reserves slots on one bus in order to use the slots on the second bus. Every time a slot comes by with a reservation tagged to it, the node increments a counter. It also watches the other bus and with every slot that goes by empty, it decrements its counter. Essentially what happens is that the node reserves slots on one bus to be used on the other. It sends a reservation request so that all of its upstream and downstream neighbors know it wishes to send information. Then, it begins to count the number of empty and reserved slots as they go by. When the counter equals 0, the node knows that it can now send its data. This is a form of carrier sense multiple access with collision avoidance (CSMA/CA) that prevents a problem from occurring with nodes trying to

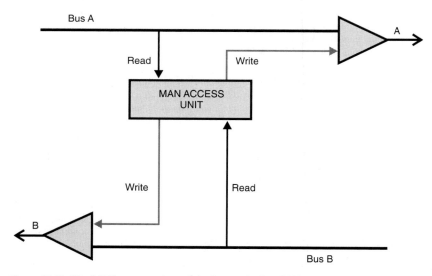

Figure 21.14 The MAN access unit reads/writes to the bus 8000 times/second.

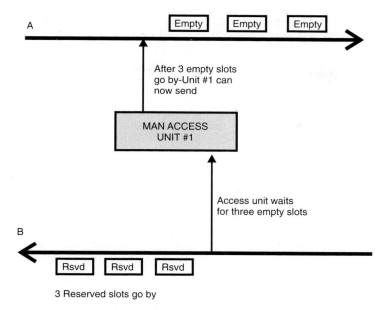

A

Empty   Empty   Empty

After 3 empty slots
go by-Unit #1 can
now send

MAN ACCESS
UNIT #1

Access unit waits
for three empty slots

B

Rsvd   Rsvd   Rsvd

3 Reserved slots go by

Figure 21.15 When unit 1 wants to send, it counts the reserved slots on bus B, then it decrements the counter on bus A until a slot is available.

send data at the same time. See Figure 21.15 for a sample of what happens on the bus from the access unit.

Because this is called *distributed queue* on a dual bus, it should stand to reason that each node maintains a list of queues for each bus. Although these two queues operate identically, they are kept separate from each other. This handles a deterministic approach to the dual bus, much similar to a token passing arrangement. To prevent any one node from saturating the network, provisions are in place to periodically send empty queue slots onto the network so that this deterministic MAN can be achieved. No one node shall be allowed to reserve all the slots on the network; they are negotiated on the basis of a quality of service (QOS), which equates somewhat to a throughput rate. You buy the throughput that you need, and you pay for the QOS that you want.

The Data Handling

A fair way to start into the data transfer on a MAN is to understand the formulation of the data units. The 802.6 can accept data up to 9188 bytes (octets) long. Then, a three-step process is used to break the data down into more manageable sizes. The end result is that the data will be transferred across the network in the slots (queues) at 53 bytes of information at a time. The process is described, and we will take it slowly to try to keep it going smoothly. This is analogous to the ATM, which is covered later, but is not the same. To prepare the data for transfer, the equipment is used to define the protocol data units of up to 9188 bytes. Several convergence layers (processes) act on the data to make it more manageable. These convergence layers are defined by the layer they are handled in. Think of this as slicing and dicing the data into smaller chunks, similar to the X.25 transport.

Segmentation and reassembly

Within the DQDB layer, the media access control convergence produces a segmentation and reassembly process (SAR). This process takes the very large frame of information and breaks it down into smaller pieces that can be put back together at the receiving end. If we start at the top (Figure 21.16), a large frame (MAC layer) of information is delivered from the logical link control unit. In this case, it might be a bridge, router, or LAN interface. At this stage, the header and trailer information is added to the MAC layer information. This is called an *initial MAC protocol data unit (IMPDU)*. The IMPDU is then chopped up into fixed length segments called a *derived MAC protocol data unit (DMPDU)*, with added header and trailer information attached to it. The numbering of these DMPDUs is handled by the relative position in

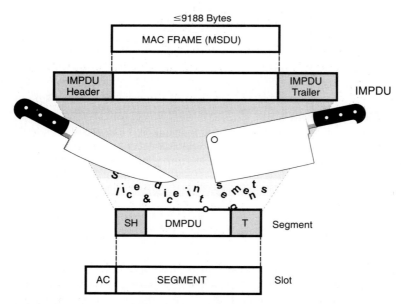

Figure 21.16 As the MAC layer frame (9188 bytes) is delivered, a header and trailer are added to create the IMPDU, then it is sliced and diced into segments, then header info is added, creating a slot (53 bytes).

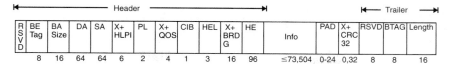

R S V D	BE Tag	BA Size	DA	SA	X+ HLPI	PL	X+ QOS	CIB	HEL	X+ BRD G	HE	Info	PAD	X+ CRC 32	RSVD	BTAG	Length
8	16	64	64	6	2	4	1	3	16	96		≤73,504	0-24	0,32	8	8	16

Figure 21.17 The SMDS frame (9188 bytes) is similarly converged. The overhead is shown in this graphic for the SIP level 3 PDU.

the slicing and dicing process. The DMPDU can be a beginning, a continuing, or an ending protocol data unit. The numbering is just that—beginning, continuing, or ending unit. As you can see from the figure, the data units prepared for the slots in the DQDB format will be a 53-byte segment. This will include 44 bytes of payload (the data), plus 7 bytes of header and two bytes of trailer information. This fixed-length segment is then sent across the DQDB in a time slot.

The SMDS data unit (SDU)

The encapsulation of an SMDS data unit is very similar in the way it is transferred across the network. First, the data of up to 9188 bytes is delivered to the interface equipment. Then, the process begins to slice and dice the data into smaller pieces that will wind up in the 53-byte slots using an SMDS interface protocol (SIP). Figure 21.17 shows the process from the very beginning. The 9188-byte data unit, plus the associated overhead, is added to the data unit. This is a layer 3 SIP-PDU. Table 21.2 describes the abbreviations of the information contained in the layer 3 SIP-PDU. You might want to keep this available as you review the diagram.

The layer 3 SIP-PDU then gets broken down into smaller pieces by going through another process that breaks the data into layer 2 SIP-PDUs. The layer 2 SIP-PDUs will be broken down into 44 bytes of data, and an additional 7 bytes of header and two bytes of trailer information. This is shown in Figure 21.18. Table 21.3 is a summary of the abbreviated definitions for this figure. Again, you might wish to keep this available as you review the framed format in the graphic.

This slot or segment is then delivered to the physical medium for transport across the network (layer 1 SIP-PDU). The physical layer is divided into two pieces or sub-layers called the *physical layer convergence process* and the *transmission system.*

SMDS in a LAN environment

Enough of the bits and bytes for now. When using a LAN-to-LAN in a metropolitan area, the concern is how do we transmit the information transparently. Most LANs use some protocols that provide for the reliable delivery of the actual data. In many LANs today, the protocol of choice is TCP/IP. If we

TABLE 21.2 **Summary of Fields in the Layer 3 SIP-PDU**

Abbreviation	Length (in bits)	Description
RSVD	8	Reserved
BEtag	8	Beginning and ending tag
BAsize	16	Buffer allocation size (length)
DA	64	Destination address (to)
SA	64	Source address (from)
X+		Ignored by the network
X+HLPI	6	Higher level protocol ID
PL	2	PAD length (if used)
X+QOS	4	Quality of service
CIB	1	CRC presence indicator bit, indicates the presence or absence of CRC
HEL	3	Header extension length (describes the header extension)
X+BRDG	16	Bridging
HE	96	Header information
PAD	0–24	Padding or filler
X+CRC32	0–32	Cyclic redundancy check (if used)
L	16	Length (must agree with the BAsize)

SOURCE: ©1991, Bell Communications Research, Inc.

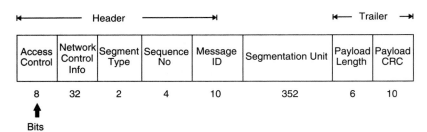

Figure 21.18 The layer 3 PDU is segmented into layer 2 PDUs at 53 bytes.

stack this up on an architecture, we need to see where the pieces all fit together. Figure 21.19 shows a LAN architecture running TCP/IP. This stack uses the TCP portion of the stack to guarantee reliable data transfer at the layer 4 of the OSI model. The IP is a packet-switching protocol working at layer 3 of the OSI model. IP breaks the data down into smaller chunks for delivery across a network. Then, IP will hand the packets down to layer 2 of the OSI, which is the data link layer. Here is where we can come into the SMDS world by using the equipment to transfer the data across the network. At the data link layer, the stack deals with the three SMDS interface

TABLE 21.3 Summary of Abbreviations for the Layer 2 SIP-PDU

Abbreviation	Length (in bits)	Description
AC	8	Access control
VCI	20	Virtual channel identifier
PT	2	Payload type
PP	2	Payload priority
HCS	8	Header check sequence (a CRC on the header)
ST	2	Segment type (beginning, continuing, or ending)
SN	4	Sequence number
MID	10	Message identification
PL	6	Payload length
PCRC	10	Payload CRC

SOURCE: ©1991, Bell Communications Research, Inc.

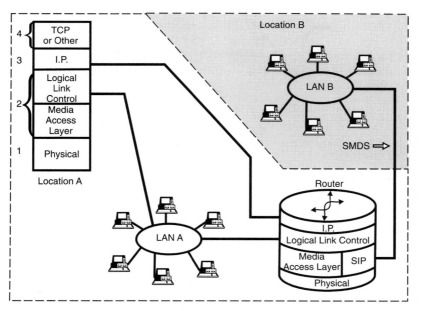

Figure 21.19 Typical LAN running TCP/IP into SMDS.

protocols (SIP). The data will be segmented into the slots, or 53-byte segments, and be prepared to be sent across the network. At the center of all this is the stack. This is where the convergence takes place, using the original TCP/IP data and sending it across the network in SMDS format. At the receiving end, the data is reassembled and sent back up the protocol stack to be delivered in a usable and understandable format.

Information Throughput

As mentioned earlier, the user will subscribe for an element of throughput. The word used in the SMDS world is a *sustained information rate (SIR)*. This is very similar to the frame relay term called the *committed information rate (CIR)*. The SIR is based on a rate of service called *classes*. There are five classes of service, defined for use depending on the application being served. The classes of service are shown in Table 21.4. These SIRs match very closely to the LAN throughput speeds that we are familiar with today. In several cases, we have discussed the use of 4, 10, or 16 Mbits/s. Two others exist. One is 25 Mbits/s, which is a LAN speed you will be seeing from several vendors in the future, supporting IBM's efforts to get LANs at ATM speeds. Although ATM is defined as a 50 or 155 Mbits/s rate today, IBM has conjured up the notion that a full-duplex LAN operating at 25 Mbits/s will yield an effective LAN speed of 50 Mbits/s, and that's ATM. This is not yet a supported speed by the standards bodies and the ATM Forum, but IBM and 32 other companies are planning on the introduction of this service and the chipsets necessary to deliver this speed. It will likely become a de facto standard in the future. The last SMDS class of service supports 34 Mbits/s throughput. The intent of the class of service is to provide control over how much information any node can place on the network as well as the ability to prevent or control congestion on the network.

TABLE 21.4 A Summary of the SMDS Classes of Service

Access class of service	SIR (Mbits/s)
1	4
2	10
3	16
4	25
5	34

The use of SMDS and the DQDB formats and protocols can deliver a myriad of services across the MAN. In the future, new capacities will be added. Where today the MAN is defined to provide LAN extension services, the ability to add other services will be introduced. The current use of SMDS in the industry includes the following:

- LAN-to-LAN in a metropolitan area
- Intra- and inter-company document transfer and sharing
- Collaborative processing and development
- Host-to-host transfers
- DASD mirroring
- Disaster recovery planning for remote hot site

Future Services on SMDS

Once the added speeds and service levels are provided, work will be done to provide several other services in the MAN, as well as a connection to the WAN. MCI is already positioned to offer nationwide SMDS services at up to 34 Mbits/s but will be looking for the 100, 155, and higher throughput speeds. These are definitely under study for future use at this time. New services will include the following:

- Video conferencing
- Interactive TV services
- Multimedia
- B-ISDN services and interfaces
- Voice
- Medical imaging
- Tele-radiology
- Tele-learning
- Others

You can only guess when all of these services and applications will be ready to roll out across the MAN and WAN. However, it would be a safe bet to suggest that by 1997–1998, many of them will be deployed by the LECs. They are moving in this direction and will want to be the first in the game. As added services are deployed, they will try to preserve their customer base and offer these services as reasonably priced as possible. The pilots that have been introduced and the rollout that should ensue indicate that customers will be looking for SMDS quickly. One publisher in Canada has already deployed a 14-site SMDS connection to Canadian and U.S.-based locations. A major University in the east has already integrated over 30 LANs into a MAN and plans to roll out more connectivity. The case histories of these developments are all positive. Delays across the network are insignificant, pricing is reasonable, and access is fairly straightforward. The protocols and rules are well defined. Look for a lot more on SMDS in the future as you journey into the communications world.

22

Frame Relay

One of the newer forms of data transport in the industry is a higher-speed packet-switching technique called *frame relay*. Frames of information are generated by most all of the data communications processes today. Although we call them by different names, such as packets, frames, or cells, they all are really just a means of transmitting a specific amount of information across a network in some logical order that is understandable to the device at the other end. The use of frame relay is new as stated, and is just now becoming an acceptable means of transporting information across a designated network. Although frame relay has been specified as a transport system for several years, the introduction and acceptance has been met with mild enthusiasm. This is not to say that it is not an efficient means of transferring the data, merely that the industry was confused about what the intended goals would be through the use of frame relay.

What Is Frame Relay?

Frame relay is a high-performance, cost efficient means of connecting an organization's multiple LANs and systems network architecture (SNA) services through the use of various techniques. Like the older X.25 packet-switching services, frame relay uses the transmission links only when they are needed. For all intents and purposes, the virtual circuit concept applies here as much as any other network service.

Virtual means almost or not quite. A virtual circuit then is almost a circuit, but not-quite a circuit. There is a physical connection into the cloud (Figure 22.1), where the customer rents or leases a circuit into the cloud.

Customer

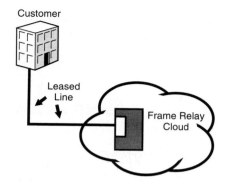

Leased
Line

Frame Relay
Cloud

Figure 22.1 The customer leases/rents a dedicated circuit into the "cloud" and terminates it on a port.

This circuit into the cloud is then terminated onto a port in a computer system. The computer system is trained to recognize where the connecting ends of the wires are located and grabs the use of the full set of wires when the customer has traffic to pass to the end location. Thus, a virtual network, or a virtual circuit connection is established into the cloud for future use as the customer needs dictate. Because the connection is not always "nailed up," other customers connected to the same network supplier can also generate and transmit traffic across the same physical pairs of wires within the cloud. Additionally, like private line connections, frame relay transports data very quickly, with only a limited amount of delay for network processing to take place.

When comparing frame relay to the X.25 services, much less processing is required inside the switches (network processors), therefore the reduced amount of processing allows the transport of data much quicker. By eliminating the overhead in each of the processors, the network merely looks at an address in the frame and passes the frame along to the next node in the network.

Why Was Frame Relay Developed?

Major trends in the industry led to the development of frame relay services. These can be categorized into four major trends as follows:

- The increased need for speed across the network platforms within the end user and the carrier networks. The need for higher speeds is driven by the move away from the original text-based services to the current graphics-oriented services and the bursty time-sensitive data needs of the user through new applications. The proliferation of LANs and now the client/server architectures that are being deployed have shifted the paradigm of computing platforms. The demands of these services will exceed hundreds to thousands of times the data transport needs of the older text-based services. Users demand more readily available connectivity and the speed to assure quick and reliable communications between systems or services. Fortunately, the bursty nature of the way we conduct our business allows the sharing of resources among many users, thereby sharing the bandwidth available. To accommodate this connectivity in a quick manner, some changes had to be made and the protocol dependency and processing of the networks had to be minimized. One way to accommodate the reduced overhead associated with the network is to eliminate some of the processing, mainly in the error detection and correction schemes.

- Increasing intelligence of the devices attached to the network. The use of data transfer between and among devices on the network has moved many of the processing functions to the desk top. Because the processing is now being conducted at the local device as opposed to using dumb terminals and a single host processor, the capacity to move the information around the network must meet the demands of each attached device. Increased functionality must be met with increases in the bandwidth allocation for these devices.

- Improved transmission facilities. The days of "dirty" or poor quality transmission lines required the use of over-correcting protocols, such as X.25 and SNA. Because the network performs better, a newer transmission capability was needed.

- The need to connect LANs to WANs and the inter-networking capabilities. Today's user wants to connect LANs across the boundaries of the wide area, unshackling themselves from the bounds of the LAN. The user demands and expects the same speed and accuracy across the WAN as they get on the local networks. Therefore, a newer transport system to support the higher-speed connections across a wider area was needed. Yet, the LAN to WAN inter-networking works fine in a simple point-to-point arrangement, except that the network is dynamic and the need to connect to multiple sites concurrently needed to be robust enough to support this newer need.

The Significance of Frame Relay

The network was originally brought up through the older analog transmission techniques that have been addressed several times throughout this book. As an analog transmission system, the network was extremely noisy and produced a significant amount of network errors and data corruption. It was this element of the network that was most frustrating to the data processing departments of the past. When data errors were introduced, a retransmission was required. The more retransmissions were necessary, the less effective throughput on the network. As a matter of fact several years ago, the use of a 4800-bps transmission service on the analog dial-up network might well produce an effective throughput of only 400 bps after all the errors and retransmissions were taken care of. This was intolerable and had to be corrected. To solve this problem, the network introduced the X.25 services, also called packet switching as addressed fully in Chapter 15. Where the X.25 was originally designed to handle the customer's asynchronous traffic, frame relay was designed to take advantage of the network's ability to transport data on a low-error, high-performance digital network, and to serve the needs of the intelligent synchronous applications of the newer and more sophisticated user applications.

When compared to private leased lines, frame relay makes the design of a network much simpler. The private line network shown in Figure 22.2 requires a detailed analysis to set all the right connections in place, which further accentuates the traffic-sensitive needs of the user network. The meshed network uses a series of connections that are N-1 points with links running from site to site. Therefore, if 10 sites exist in the network, 9 links will run from each site to every other site. This allows for the speed of connectivity, but the network costs are much higher. Further, depending on the nature of the data traffic, the bursty data needs of a LAN-to-LAN or LAN-to-WAN connection are not required full time. We, therefore, spend significantly more of the organization's money to support the meshed leased line network. In Figure 22.3, a frame relay access from each site is provided into the network cloud, requiring only a single connection point, rather than the nine of the earlier network. Data transported across the

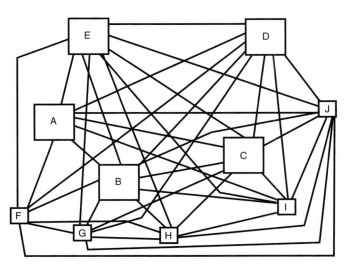

Figure 22.2 A leased line network requires more detailed analysis depending on the communities of interest.

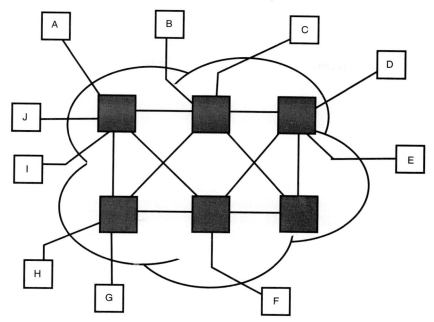

Figure 22.3 A single access is all that's needed into the cloud.

network will be interleaved on a frame-by-frame basis. Multiple sessions can be running on the same link concurrently. Communications from a single site to any of the other sites can be easily accommodated using the pre-defined network connections of the virtual circuits. In frame relay, these connections use *permanent logical links (PLL)*, more commonly referred to as *permanent virtual circuits (PVC)*. Each of the PVCs connects two sites just as a private line would, but in this case the bandwidth is shared among multiple users, rather than being dedicated to the one site for access to a single site. Using this multiple-site connectivity on a single link reduces the costs associated with customer premises equipment, such as CPU ports, router ports, or other connectivity arrangements. Because fewer ports are required, fewer connection devices are required; therefore, the customer saves money.

Because the PVCs are pre-defined for each in a pair of end-to-end connections, a network path is always available for the customer's application to run and transport data across the network. This eliminates the call set-up time associated with the dial-up lines and the X.25 packet arrangements. The connection is always ready for the devices to ship their data in a framed format, as the need arises. This takes away the need for the constant fine tuning of a private line or a dial-up network link arrangement.

Comparing Frame Relay to Other Services

When the network suppliers and the standards bodies were attempting to define the benefits of frame relay services, they looked at a comparison of the time-division multiplexed (TDM) switched circuit and the packet-switched network services.

TDM circuit switching

TDM circuit switching creates a full-time connection or a dedicated circuit for the duration of the connection, between any two attached devices. TDM divides the bandwidth down into fixed time slots in which there can be multiple time slots available, each with their own fixed capacity. This is shown in Figure 22.4, where each attached device on the network is assigned a fixed portion of the bandwidth using one or more time slots depending on the need for speed. When the device is in the transmit mode, the data are merely placed on this time slot without any extra overhead, such as processing or translations. Therefore, TDM is protocol-transparent to the traffic being carried. But unfortunately, when this attached device is not sending data the time slots remain empty, thereby wasting the use of the bandwidth. A higher-speed device on the network can be slowed down or bottled up waiting to transmit data, but the capacity that sits idle cannot be allocated to this higher device for the duration of the transmission. TDM is

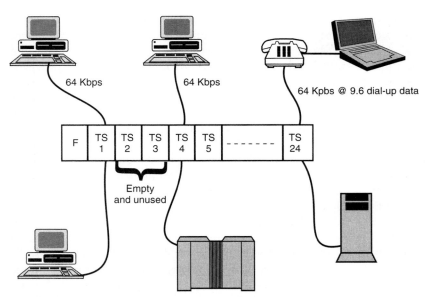

Figure 22.4 TDM switching forces the transmission into a fixed-capacity time slot, and idle channels are virtually wasted.

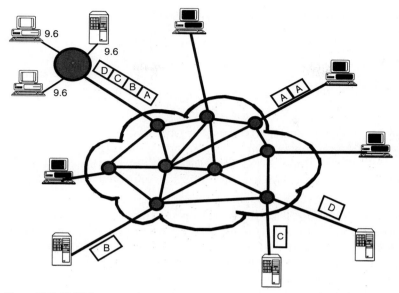

Figure 22.5 Multiple connections on virtual circuits are active on the line simultaneously.

not well suited for the bursts of data that are becoming the norm for the data needs in today's organization.

X.25 packet switching

Where the TDM world had its limitations of the fixed bandwidth allocation, a newer service was created that allowed the bandwidth to be allocated on the fly. Instead of simply putting the data into a fixed time slot, the user data is broken down into smaller pieces called *packets*, each containing both the source and the destination addressing information, as well as other control functional information. When a user sends data in a burst, multiple packets will be generated and routed across the network based on the addresses contained in the packets. The network creates a virtual circuit from each source to each destination to keep track of the packets on each connection. This is shown in Figure 22.5, where multiple virtual circuits can be active on a line at the same time. This is a different form of multiplexing, more of a statistical time-division multiplexing scheme. STDM uses the analyses of the past users to allocate more interleaved packet slots to the heavier users and less interleaved slots to the lighter users. The major drawback to this scheme is the penalty paid in "speed" of delivery. Where guaranteed data delivery and integrity was a prerequisite for the development of the X.25 networks, the delay in processing the data across the network was the price we paid. Clearly, something had to be done to overcome these limitations.

What the standards bodies attempted to accomplish was to use the best of all worlds. Taking the features of the switched network and the packet arrangements, the network arrived at a frame relay service that should meet the needs of the user. This comparison took into account the following service connections:

- *Speed* The speeds available to the user are important. As the capabilities of the dial-up digital network are now supporting the high-speed connections of up to 2.048 Mbps, the X.25 network only provides speeds of up to 64 Kbps. Thus, the decision is to use a digital connection of up to 2.048 Mbps.

- *Network set-up delay* The call set-up time on a digital switched network is relatively low. This is also true for the X.25 network, for general terms. However, as network congestion builds up the switched network circuits of the X.25, the network can introduce extensive delays. Thus, the network suppliers were looking to implement a low-delay call set-up. The PVC allows for this because the call set-up time is eliminated.

- *Routing* The routing of the switched network is fairly static. When one uses a switched call set-up, the routing is used once to establish the link. From the perspective of a network delay or a failed link, the user will have to reestablish the call through a new dial sequence. This is not as robust as expected. The X.25 network is robust, in that if a problem occurs on the link, the packet nodes in the network will immediately reestablish the connection with the next packet that runs across the network. Therefore, the suppliers and standards bodies opted for the robust and dynamic call set-up. The delay in this is very low.

- *Signaling* In the switched network design, the signaling is also static, in that it is used only in the initial set-up or tear-down of the call. Therefore, if the call is interrupted, the signaling disappears. In the packet-switched networks, the signaling is dynamic. The signaling is contained in every packet, thus it is easily reestablished. The standards bodies opted for the dynamic signaling arrangements of the packet-switched networks.

- *Bandwidth* Once again, the bandwidth consideration is important. In the circuit-switched networks the bandwidth is fixed. If you dial up a 56-Kbps circuit, you get a full 56 Kbps whether you need it or not. However, the variable length of a packet allows for the maximum of 56 or 64 Kbps, but you can use much less. The disadvantage here is that the speed cannot exceed the range of 56 or 64 Kbps. The robustness rests in the packet-switched service as opposed to the circuit-switched world. Therefore the standards bodies opted to arrive at a dynamically allocated bandwidth on demand concept.

- *Costs* Circuit-switched services are relatively low priced. The costs per minute for 56, 64, 128, 384, or up to 2.048 Mbps are very reasonably priced. Pricing approximates $0.50 per minute for a switched service, and a T1 switched service is dropping well below the costs that you might expect, putting pressure on the costs of a dial-up analog service. In the packet-switched services, the costs are very low, priced at a cost per kilopacket. There are numbers that approximate $0.23 per kilopacket. This is extremely low in terms of the dial-up network. The only issue here is what the best method of call processing will be if frames are routed across a PVC.

Unfortunately, this one issue was the stumbling block to wide acceptance and use of the frame relay networks. Pricing was introduced in various trends that went from low to very high. The standards bodies do not get involved with pricing issues. The network suppliers have changed their pricing plans, and are getting very aggressive. As a matter of fact, because frame relay falls under the auspices of a value-added service, it is not governed as a tariffed service. The network suppliers are now offering frame relay access and use on an individual case basis. They might require a nondisclosure agreement, before they will submit a proposal for pricing a frame relay network. This was initially a turn-off for customers who were interested in pilot programs introducing frame relay for the WAN connections. The problem was solved through the creative pricing arrangements, and now frame relay is becoming more readily accepted.

In Table 22.1, the issues and comparisons are summarized from this information. This is how the standards bodies arrived at the service offerings of frame relay.

TABLE 22.1 Comparing Frame Relay Services to Switched and Packet Services

Offering	Circuit-switched	Packet switched	Frame relay or fast packet
Speed	64 Kbps 2.048 Mbps	64 Kbps max.	0–2.048 Mbps
Delay	Low	Low to high	Low
Routing	Static (call by call)	Dynamic (packet by packet)	Dynamic (frame by frame)
Signaling	Static (session)	Dynamic (packet)	Dynamic (frame)
Bandwidth	Fixed	Variable, dynamic	Dynamic
Costs	Low	Very low	Variable (low to very high)

Service levels with frame relay are also a concern before critical data is placed on the network. The level of service is something that must be considered heavily. Response time across the network should equal that of a dedicated line. However, the use of a frame-based service must be carefully designed with the rates and that will deliver the desired response time. Table 22.2 shows a comparison of what can be expected from the use of a frame relay network. This assumes that the appropriate rates and that the number of hops allowed in the network are kept at a minimum. The rates are defined when the network is set up and the end points are established. The number of hops is the number of nodes within the frame relay network that the frame must pass through to get to the end point. The table below shows what can be typically expected from the use of this service.

Frame Relay Speeds

When designing a frame relay service, the speed of access is important to manage, both prior to and after installation. The customer must select and be aware of the need for a specified delivery rate. There are various ways of assigning the speed, from both an access and from a pricing perspective. For small locations, such as branch offices with little predictable traffic, the customer might consider the lowest possible access speed possible. The frame relay suppliers offer speeds that are flat rate, usage sensitive, and flat/usage sensitive combined. The flat rate service offers the speed of service at a fixed rate of speed. Whereas the usage-based service might include no flat rate service, but a pay-as-you-go rate for all usage. The combined service is a mix of both offerings. The customer selects a committed infor-

TABLE 22.2 Expected Service Delivery with Frame Relay

U.S. domestic network service levels	
Percentage of availability	Node to Node: 99.99% End to End: 99.85%
Bit error rates expected	Approximately 10^{-8}
Block error rates expected	Approximately 10^{-5}
Burst rates	To access channel rates or CIR
Burst timing	≤ 2 seconds if available
Delay on network	≤ 20 milliseconds average
Average round trip network delay	≤ 400 milliseconds maximum depending on CIR
Data rates	0–2.048 Mbps varies by vendor
Maximum number of hops	4–5 depending on network

mation rate (CIR) at a certain speed. The committed information rate is a guaranteed rate of throughput when using frame relay. The CIR is assigned to each of the permanent virtual circuits selected by the user. Each PVC should be assigned a CIR that is consistent with the average expected volumes of traffic to the destination port. Because frame relay is a duplex service (data can be transmitted in each direction simultaneously), a different CIR can be assigned in each direction. For example, a customer in Boston might use a 64-Kbps service between Boston and San Francisco for this connection, yet for the San Francisco to Boston PVC, a rate of 192 Kbps can be used. This allows added flexibility to the customer's needs for transport. However, because the nature of LANs is that of bursty traffic, the CIR can be burst over and above the fixed rate for a period of less than two seconds at a time in some carriers' networks. This burst rate (Br) is up to the access channel rate, but many of the carriers limit the burst rate to twice the speed of the CIR. When the network is not very busy, the customer could still burst data onto the network at a higher rate yet. The burst excess rate (Be) can be an additional speed of up to the channel capacity, or in some carrier's networks it can be a 50% rate above the burst rate. Combining these rates, an example can be drawn as follows:

$$CIR + Br + Be = \ Total \ throughput$$

$$128 \ Kbps + 128 \ Kbps + 64 \ Kbps = 312 \ Kbps \ total$$

Remember that the burst and the burst excess rates are for two seconds or less, depending on the carrier used. There are some carriers that will not allow any bursting across the network. Rather, they require that the maximum throughput be limited to the committed information rate. What this is basically emphasizing is that no standard offerings exist. Remember to ask before the assumption can be made that all frame relay networks are the same.

Another variation of this arrangement is the flat rate service offering, where a fixed price arrangement is provided regardless of the amount of traffic generated. This arrangement also allows the customer to select various CIRs at the home site, or any site.

The variable PVC arrangement with differing CIRs is a flexible arrangement. However, care must be taken when looking at this arrangement because there is a fee associated with the port on the switching node within the network that can be substantial. An example of this might be to compare a quick rate structure that was once a tariff. The vendor in question used a published rate at $275 per month, per site for a 64-Kbps CIR, and a $5895 per month charge for a CIR at 1.544 Mbps. This is exclusive of the leased line charges from the local telephone company for the access into the closest POP that handled frame relay services. Still other costs were included in the tariff. In some cases, the cost per account was charged at an added $10.00 per month plus $0.50 per kiloframe. This might suggest that the costs are variable—and they are.

In the initial example used above with the small branch office, a 0 bps CIR was selected, therefore every single frame generated on the network is considered a burst rate. The burst is a little more expensive than the CIR, but when dealing with the unknown, it sometimes makes sense to hedge your bets until a few months traffic is handled and the adjustment to a committed rate can be made. The variable rates of CIRs can be in 4, 8, or 16 Kbps increments, so there is a lot of flexibility in the selection process. Again, the variable rates are based on a compilation of the various vendor products. Some might have these and others will not. The use of frame relay was significantly delayed when customers attempted to figure the pricing arrangements out and make commitments to the vendor. The wrong choice can be expensive in either direction; too much capacity is expensive, too little might delay the traffic.

Guaranteed Delivery

Another issue that comes into play with the use of frame relay is the guarantee of delivery. When using a committed information rate, the guarantee is that the network will deliver traffic (frames) at the CIR, but bursts are another situation. When using the burst rate or the burst excess rate, the network will make its best attempt to deliver the frames, but no guarantees are made. As each frame is burst out through the network, it is marked within its overhead with a discard eligibility bit. This means that as the network nodes attempt to serve higher rates of throughput, the frames are given a designator by the end-user equipment. This designator lets the other hops on the network know that if the network suddenly begins to get congested, then the frames that are riding the network beyond the CIR can be discarded, and that other customer frames that are within the CIR will have priority. In essence, the network is providing some breathability for users by expanding and contracting based on how busy it is. The less busy, the higher a customer can burst without risk. But because bursts are for two seconds or less, the customer must be aware that the network can bog down quickly.

Because the network will only make its best attempt to deliver the data, the end-user equipment must be intelligent enough to recognize that frames have been discarded. In Figure 22.6, the frame format is shown with the set-up of the bytes in the header information, creating the discard eligibility setting. In this framed overhead, other pieces of information are also contained. Designators in each frame that are set by the network nodes along the path alert the nodes and customer equipment when congestion is occurring. The use of a *forward explicit congestion notification (FECN)* and a *backward explicit congestion notification (BECN)* are used for this purpose. These are shown in the framing overhead.

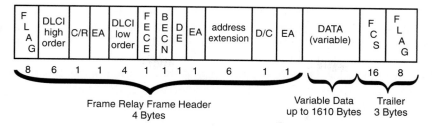

Figure 22.6 The frame format used in frame relay.

Another piece of the frame layout is the *data link control identification (DLCI)* that marks the PVC addressing scheme. The FECN and BECN are part of the 2-byte sequence along with the DLCI. Encapsulated in the frame after the header is the user data, a variable length frame of user information that can carry a LAN packet or frame of up to 1610 bytes. The bytes are not important at this point, but the variable length is. Because this is a variable, there are some inherent delays in the processing of the data across the network. For this reason, the packet-switching networks have received a negative image. The X.25 and the frame relay networks have timers and delimiters that the nodes must use to process the data across the network. These are set to certain parameters to allow for frames or packets to be transmitted at differing sizes. Along with the variable length data frames, the trailer in this figure represents the overhead associated with the error checking (frame check sequence), which is a form of CRC.

However, the frame relay nodes across the network only perform the error checking of the addressing; they don't check the user data contained inside. Actually, this is how some of the speed increases were achieved. Each node on the network only checks the address and passes the frame along the predefined path (the PVC) to the next node. Nowhere is the data being verified. This is no longer the responsibility of the network, as it was in the X.25 world. It is the responsibility of the end devices (customer equipment) to do the error checking and retransmission requests. The data integrity is assumed to be an end-user responsibility. This is a major digression from the packet-switching world, where each node along the way was responsible to perform a full CRC on every packet and accept or reject the packets based on the result of the CRC. Where the packets were guaranteed delivery and integrity, the frames are neither. How can this be? When the packet-switching techniques were first established, data integrity was all important, due to the nature of the analog network. The data errors were significant in that older environment.

To overcome the problem, the sequencing and the integrity checks were necessary at each step along the way to ensure that the customer received the correct data. This was at a price, the time and the delays inherent in this delivery mechanism. Trade-offs had to be made, either reliable but delayed data or unreliable but faster data. When the frame relay system emerged in the network, the digital world was already well in place. Therefore, through the use of digital transport and the fiberoptic networks, errors on the network are much less likely to occur. Not that they will not occur, but there will be less of them because of the characteristics of the fiber-based systems. As we have already seen, the fiberoptic networks are not prone to the same kind of errors that were specific to the radio and copper-based networks. These radio- and copper-based networks are prone to disruptions and errors from noise that will not be a problem for the fiber-based frame relay services. Secondly, the copper-based networks are prone to errors from electro-magnetic interference (EMI) which again, the fiber networks are impervious to this type of disruption. Thus, the frame relay services can afford the customer a choice as mentioned. Because you can expect fewer errors from the transmission medium, then the intense error checking at each step along the network can be eliminated and save valuable transport time.

Advantages of Frame Relay Services

The benefits and advantages of frame relay services are many. These will at a minimum include the following:

Increased utilization and efficiency

The network uses a combination of frame relay and cell relay services to allow the user network to "breathe" when necessary and dynamically allocate the bandwidth to a logical pool of capacities real time. The support of multiple connections simultaneously allow multiple sessions to be on line concurrently. The nature of the bursty data transmissions on our applications does not require the full time allocation of bandwidth until it is needed. This allows shared resources on corporate LANs in remote sites.

Savings through network consolidations

Data from various sources can be applied to the network from various sources, such as SNA traffic from the host, LAN bursty traffic from the desktop devices, and other application specific devices. The use of a single connection to support all of these connections at various speeds and at variable times allows the end user to consolidate services and therefore, save money. The frame relay service will typically be used to connect three or more sites; therefore, it provides cost-effective full logical mesh connectivity. Frame relay enables the user to save on transmission and switching costs.

Improved network uptime

Network downtime is a phrase that will make even the most staunch data processing person shudder. The risk of downtime was always a major consideration in the design efforts of the past. Redundant links were required to prevent a single event from bringing the entire network to a halt. This redundancy consumed a significant portion of the data processing budgets. The use of a single link to any point across the network was at risk of "backhoe fade" or other network disruption. Automatic rerouting of the links was not easily accommodated. The use of a frame relay service into the virtual cloud allows the reestablishment of the network connections within the cloud automatically. The only single point of failure that really remains is in the local loop or last mile. This can be redundantly protected, if necessary, at a greatly reduced rate. Because the network supplier is responsible for the robustness of the recovery and delivery mechanism, the customer can save a considerable amount of funds that are better spent on other areas. Uptime can achieve 99.9% availability on a network of this sort. Further, because only a single connection into the cloud is required, the customer can reduce the equipment costs for the interface equipment (i.e., DSU/CSU, routers, MUX ports, etc.) because the number and type of interfaces can be limited to a single one.

Improvements in response time

With direct logical connectivity through the use of the PVC to the multiple locations on the network, a single interface improves the response times that we are accustomed to. By limiting the number of hops that the frames will traverse through, along with the elimination of the call set-up time, response times are vastly improved. With the allocation of bandwidth to support the bursty data needs, the response times are also improved, especially when a burst rate can be accommodated. Allowing the link to support higher throughputs than originally committed, the user can achieve greater throughput in a shorter period of time, by 2-second bursts across the network. Allowing the network the breathable concept to support the initial bursts of data as opposed to the fixed bandwidth of the leased line concept, users will achieve higher throughput. Frame relay is the best transport protocol for high performance and fast response interconnection of intelligent end-user devices, providing the highest ubiquitous speed, lowest overhead protocol standard that is available for the WAN.

Easily modifiable and fast growth

The logically connected PVCs can be adjusted fairly easily through the network administration group of the carrier. New PVCs can be assigned on a single link quickly, or existing PVCs can be increased in capacity up to the access channel capacity needed to support the dynamics of the organization's transport needs. With the older private line meshed networks, when a new site needed to be included into the network, a whole new configuration was required. Special time frames were required to rework the entire network, or a regional portion of an existing network. With the PVC concept, a virtual connection is created quickly, and without major modifications to the network. One merely orders the physical link from the local exchange carrier into the point of presence (POP) of the frame relay supplier. Then the frame relay supplier configures the logical connections through the use of PVC arrangements. The speed of service can be cost effective and much less tedious than the networks of the past. The frame relay service supports interconnection of LANs running multiple protocols, and transports the data over the WAN protocol transparent. Some of the LAN protocols supported across the frame relay networks include Appletalk,[1] TCP/IP,[2] IPX/SPX,[3] XNS,[4] OSI,[5] DecNet,[6] SNA,[7] and NetBIOS.[8] This is a fairly robust set of protocols in any environment today.

Standards based

The use of frame relay falls into the CCITT (now the ITU-TSS) and the ANSI standards infrastructure. The interoperability between various switching platforms has been proven, and it is a logical progression toward the future switching services for broader bandwidth services, such as B-ISDN and ATM, allowing a smooth transition or migration path to these newer, emerg-

---

[1] Appletalk is the Apple Computer protocol.

---

[2] TCP/IP is the Internet Protocol (Transmission Control Protocol/with Internet Protocol).

---

[3] IPX/SPX is the Novell Netware Protocol (Internet Packet Exchange/Sequenced Packet Exchange).

---

[4] XNS is the Xerox Networking Protocol (Xerox Networking System).

---

[5] OSI is the Open Systems Interface reference model (Open Systems Interconnect Protocol).

---

[6] DecNet is the Digital Equipment Corporation protocol (Digital Equipment Corp. Networking Protocol).

---

[7] SNA is IBM's standard protocol (Systems Network Architecture).

---

[8] NetBIOS is the IBM standard protocol (Network Basic Input/Output Systems Protocol).

ing services. Future services that can be possibly interfaced with the system of frame relay might include video conferencing (using compressed motion at 384 Kbps and packetized transfer using the H.261 standard), facsimile services (CCITT groups III or IV), X.25 services, and integration or gateway functions onto a packet-switching network.

Services Available

The following services and protocols are usually readily available from the carriers, allowing the transparent flow of data across the network. Those that are not will be added as the network services are proliferated and the interfaces are developed to support the various service offerings:

TCP/IP and Novell IPX/SPX

Using the packet-level routing capabilities of the frame relay network through a certified or type accepted router (the word internationally is homologated), any type of data can be transported—although each respective router must be checked for compliance. Some routers might not support all of the protocols that are needed within an enterprise's network. The particular router to support the specific protocols are normally tailored by the carrier in their recommendations for the network connections. In the event that a customer does not have a router capable of inter-networking with the suggested network or other pieces, the carriers are in a position to rent, lease, or sell the specific routers that will work on their network. The routers support a variety of LAN topologies such as Ethernet, Token Ring at 4 or 16 Mbps, and FDDI. These routers will also support a wide range of media interfaces from a coaxial, twisted pair wiring (shielded or unshielded), and fiberoptics. Newer versions are currently being explored to support the wireless connections also.

CCITT X.25 protocol

Convenient interfaces to the X.25 networks are being introduced to allow a customer to connect and transport data to sites that might not have the frame relay services available (an example is in Mexico where the only packet-switching service available is X.25) or cannot justify this connection financially. The X.25 interface can connect the host computing environment to the network interface across a public packet-switching service, then into an X.25 gateway on the public network (Figure 22.7). Or packets of data can be rerouted across the frame relay network through the router at the customer location (Figure 22.8). In either case, the connections are the important issue, not the location of the connection.

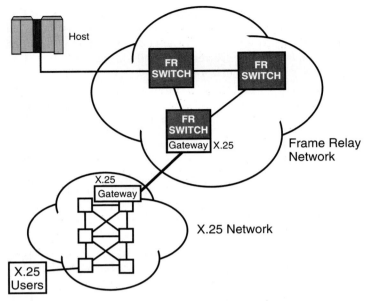

Figure 22.7 Access to frame relay can be through the public packet switching service to an X.25 gateway.

Facsimile (CCITT Group III or IV) traffic

The use of an X.400 protocol electronic messaging service for domestic and international messaging services can be accommodated by the carrier or through the interfaces at the customer location. Facsimile traffic is extremely time-sensitive and that packetized fax has not been truly developed. The use of X.400 is a "store and forward service" to accommodate this form of communications. Future developments might well change this into a frame relay capability.

Frame Relay Standards

Clearly then, the standard form of frame relay has some merit. However, at the time of this writing, there are problems with the standard form of introduction and use. The standards bodies have completed a significant amount of work, but there still remains a lot of work to be done because of the varied implementations by the carriers, each with some proprietary or modified standard in use. Many of the network suppliers (although not all, by any stretch of the imagination) use a standard frame relay switching system. The system of choice is the StrataCom IPX switching system. This is a choice that the vendors had to arrive at to support the high speed architecture of the frame relay networks. This is a system that uses a dual-bus 32-Mbps mid-plane, using redundancy at every module, bus, power, etc. The choice of the IPX is because of the flexibility of the services offered in either circuit mode, packet mode, or frame-mode service capabilities. The IPX is a typically proprietary hardware/software implementation of the standards, yet it works efficiently and was the first frame relay switch used. It, therefore, became the de facto standard that all other switches were based on. The benefit of the IPX is that it offers the single platform that most vendors have accepted and, therefore, the greatest chance of compatibility of systems in the network.

The Major Players

The two major standards players in the specification and support for frame relay services are the CCITT and ANSI. Together, they have defined a standard data (packet mode) interface for ISDN networks and frame relay services. The basics of frame relay were actually developed in the CCITT blue book I.122 (Packet Mode Bearer Services) in 1988. Because this specification has been around for awhile, you can see that the implementation was slow to start. The speed at which frame relay was developed from that point on has been directly attributable to the demand for a simple, easy to use, high-performance service for LAN-to-LAN connectivity. Additional standards are covered in the following paragraphs.

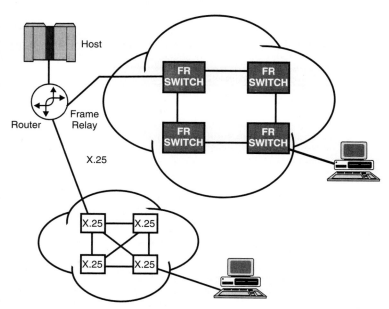

Figure 22.8 An alternative is to use X.25 to a customer's router that redirects the packet onto frame relay.

Others

In 1989, StrataCom, Digital Equipment Corporation, Northern Telecom Inc., and Cisco Systems began a joint development effort to specify added frame relay services, under the auspices of the Frame Relay Forum.[9] Their efforts were specifically geared toward addressing the needs of LAN-to-LAN communications, including the need for a local management interface (LMI). This collaboration of these groups led to the publication of a joint specification for the first implementations of frame relay by numerous vendors in 1990. Portions of that specification were adopted by the CCITT and ANSI standards for inclusion in the standards. The standards that are associated with the frame relay services are shown in Table 22.3.

In 1988, CCITT approved the Recommendation I.122 "Framework for additional packet mode bearer services," which is part of a series of ISDN-related specifications. ISDN developers had been using a protocol called *LAP-D*—link access protocol with D channel, to carry signaling information on the D channel of ISDN (LAPD) as defined by CCITT Recommendation Q.921. LAPD has the characteristics that could be useful in other applications, such as provisioning for multiplexing of virtual circuits at level 2 in the frame level (as opposed to level 3 in the X.25 networks). Therefore, I.122 was written to provide a framework outlining how such a protocol might be used in applications other than ISDN signaling.

TABLE 22.3 The CCITT and ANSI Standards Involved with Frame Relay Services

ANSI standards:	
T1.606	The Frame Relay Architectural Specification
T1.618	Data Transfer Protocol
T1.617	Signaling Specifications (control protocol)
CITT Standards:	
I.122	Framework for Providing Additional Packet Mode Bearer Services
Q.922	ISDN Data Layer Specification for Frame Mode Bearer Services (LAP-F)
Q.931	ISDN user-network interface layer 3 specification (Blue Book)
Q.933	ISDN DSS1 Signaling specification for frame mode bearer services

---

[9] StrataCom was one of the founding members of the frame relay forum, a vendor interest group that accelerates the acceptance of frame relay in the vendor community. After the joint publication, StrataCom wanted to form a vendor interest group to accelerate implementation of frame relay within the structure of the standards.

LMI Specification

In September 1990, the joint group of DEC, StrataCom, NTI and Cisco Systems wrote the added specification known as the *link management interface (LMI)*, or what was also referred to as the *interim link management interface (ILMI)*. For its basic operation within frame relay, it simply refers to the ANSI standard. LMI defines an added interface function, the local management interface, as a proposed method of exchanging status information between the user device (e.g., router or switch) and the network. The LMI defines a number of other useful specifications and options to simplify the implementation of frame relay and enhance its operation. More than 50 vendors accepted support of the functions of the LMI, so ANSI included portions of the LMI specification (the ANSI standard T1.617) with some minor modifications.

What the Standards State

To fully understand how frame relay works, the following three areas must be incorporated in the networking technology:

- The basic data flow of frame relay
- The interface signaling used at the network/user interface
- The internal operation of a frame relay network

The frame relay standards define an interface and the flow of the data, not the whole network. The internal operation of the network is up to the vendor's implementation of goods and services to support their network.

The basic data flow

In most popular synchronous protocols, data is carried across a communications line based on very similar structures. The standard HDLC frame format is used in a myriad of these protocols and services. Frame relay makes a very slight change to the basic frame structure, redefining the header at the beginning of the frame (two bytes long). The basic frame structure for other synchronous protocols is shown in Figure 22.9. In this case, the HDLC frame header consists of an address and control information. For frame relay, the header is changed as shown in Figure 22.10, which uses the 2 bytes (octets) to define the following pieces.

- The data link connection identifier (DLCI). This identifier uses up to 1024 LCNs.
- C/R—command/response bit. Is not used in frame relay.
- EA—extension bit. When set to 0, it extends the DLCI address.

- Forward explicit congestion notification (FECN). This is set in the frames going out into the network toward the destination address.

- Backward explicit congestion notification (BECN). This is set in frames that are returning from the network to the source address.

- Discard eligibility (DE) bit. This bit is used by the source equipment to denote if the frame is eligible to be discarded by the network if the network gets congested. When set to a 1, it indicates that the frame is eligible to be discarded during congestion period.

- Extension bit (EA). When set to a 1, it is used to end the DLCI.

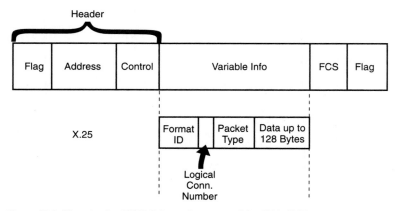

Figure 22.9 The standard HDLC frame format used for SNA, X.25, etc.

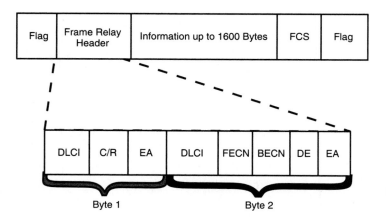

Figure 22.10 The change in the HDLC frame for frame relay.

The FECN, BECN, and DE parts of the addresses are designed to alert the end-user equipment on the status of the network's ability to process frames. Higher-level protocols use a window-size control procedure or other forms of control to handle traffic flow. These are flow control mechanisms that are designed to alert the network components and the end-user equipment to slow down the delivery of frames to the network. If discarding is taking place while a router or other piece of equipment is in burst mode, the router or other equipment will begin to buffer the frames until the congestion is resolved.

Interface signaling for control

When frame relay was first proposed to the standards bodies, the concept was fairly straightforward: keep the network protocols simple and let the higher layer protocols at the other end of the link deal with any problems. To this end, the proposal was very practical. However, it soon became apparent that the deployment of frame relay in the industry where real world situations will arise, the networks would need to have a signaling mechanism to address several areas:

- Allow the network to notify devices when congestion exists. Becoming aware of the status of the permanent virtual circuits (PVCs).

- Guaranteeing throughput and equality for all users on the network, so that a single user would not occupy the network and deny others their just access.

- Creating the opportunity to expand the services and features of the network in the future.

Therefore, the standards players had to incorporate the use of a signaling system to solve these issues. Some of them make use of certain bits within the 2-byte header. Others are addressed by using certain DLCIs as being the interface control mechanism. Unfortunately, these mechanisms add to the complexity of the frame relay network, so the actual use of the signaling mechanisms are optional. If a vendor decides to implement frame relay services, but chooses not to implement the signaling standards, the result will be a compliant network, so the data will still flow. However, if a vendor chooses not to implement the signaling, other complications can arise. The effective throughput of the network, the expected vs. actual response times on the network, and the efficiencies gained from frame relay might all be lost. You must be aware of how the carrier or vendor has implemented the frame relay service. This could be a critical portion of an evaluation to use frame relay services.

Internal networking

All of the standards and modifications define what is supposed to happen at the frame relay interface, normally called the *user-to-network interface (UNI)*. How well the network performs is also determined by what happens inside the cloud, which unfortunately the standards do not attempt to specify how to accomplish this. Some of the functions that occur within the cloud include:

- Determining the path for the PVCs in use.
- Estimating when congestion is building up, and what to do to stop it.
- Responding quickly to mitigate the congestion when it does occur.
- Deciding which frames to discard when this becomes necessary.
- Providing guaranteed rates or quality of service levels to users.
- Establishing or permitting multiple levels of priority traffic.
- Providing the correct throughput levels of service.
- Effectively monitoring performance and producing reports on statistics, routes, etc.

Integrated Services
Digital Network (ISDN)

What Is ISDN?

Integrated services digital network (ISDN) is one of the latest carrier product offerings to support what the carriers perceive as the customers' need for the transport of information. Simply stated, ISDN is a mechanism that allows the on-demand transport of voice and nonvoice services on a call-by-call basis. It is based on all-digital transmission, using state-of-the-art technology to allow access to the raw bandwidth that the carrier believes is needed by the customer. In Figure 23.1, the basis of ISDN is shown as the combination of present and future services.

The industry as a whole is somewhat in a state of turmoil. Many users feel that their needs for ISDN have not yet materialized. Some feel that it is too little, too late. Still others feel that the use of ISDN is the wave of the future, allowing them the flexibility and the capability to address their present and future needs for telecommunications services.

ISDN defined

ISDN as defined by the CCITT (or ITU-TSS as it is now called), is the following:

ISDN is a network service evolving from the telephony networks, that provides end-to-end digital connectivity to support a wide range of services, in-

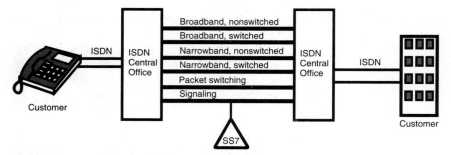

Figure 23.1 ISDN provides a mix of current and future services to the customer.

cluding voice and nonvoice services, that users will have access to through a limited set of standard multi-purpose interfaces.

Proponents of ISDN state the services will include the ability to:

- Transmit the name and number of the calling party to your telephone, prior to you answering the call. Although, this would be of some interest to the residential user as a means of screening calls and allowing that user to refrain from answering a call from an unknown caller. However, the issues that have surfaced from this concept are numerous and have worked their way into the court system. This leaves a lot of room for interpretation and for the courts to get involved with our telephone network once again. Think of the advantage to a bank (or other telephone intensive business) if an incoming call can deliver the customer's information (phone number, contact person, account status, etc.) as the agent answers the call. This is being done on some ACD systems, but it requires special arrangements. Estimates of the use of this caller ID are that 20 to 30 seconds could be shaved off each call. This would, of course, mean that improved service and less labor costs would be benefits of this service.

- Allow access to multipoint, and dial-up voice and data conference calls from the desktop. The use of voice and data calls through digital technology will be at the rate of 64 Kbits/s. You can get a customer on the phone, access a database, and display your marketing information on his terminal, PC, or other device all at the same time. Further, as the discussion rolls along, a conference call to a major supplier or distributor can be accomplished, without disrupting the initial call.

- Pick and choose the services to be provided on a line-by-line basis, with access to various services as needed. The use of voice, data, and video conferencing can be allowed on a call-by-call basis. It can transmit time and billing information for the current call to a telephone or PBX system for accountability purposes instead of keeping separate time slips.

- Use the same line to transmit telemetry information without interfering with the ongoing voice and data calls in progress. The utility companies see the benefit of this to remotely read meters in a residential application. The use of these techniques would also be invaluable to large corporations to dial access to remote sensing systems, industrial control systems, or in the case of medical applications, to remotely monitor vital signs of patients from a distance.

You can imagine the impact theses features would have on the industry as a whole if the need for the labor-intensive resources that are becoming scarce at an alarming rate were to be reduced further.

So, ISDN is a series of products and services designed by the carriers, for the users, to increase the productivity and contain the costs associated with doing business. Wrong!

No doubt these enhancements offer a wide variety of applications that will only be limited by the user's imagination. However, the truest concept of ISDN is that it is a transport vehicle to deliver the digital data (voice and data services). It is not a product and it is not a package of hardware offerings from the suppliers.

ISDN will also be an evolutionary grouping of transport bandwidths. The implementation of the network services will not be 100% from the start. Therefore, users have come up with a series of comical statements regarding the use of ISDN and what the acronym means to them, such as:

- I still don't need
- I still don't know
- I smell dollars now
- Innovations subscribers don't need

The list is endless, but the message is the same. How do users see or perceive the benefit of a network topology and the ensuing benefits that can be derived? For many, the offerings of ISDN today match what can be done today without ISDN. For example, the caller ID display, which equates to automatic number identification (ANI) can be accomplished today using the central office digital switching technology that's already there. Another example would be the use of ISDN for telemetry purposes, but by using a dial-up modem and some mechanical interfaces, the same functionality can be accomplished.

So why all the hype? If everything can be done using existing technology, why do we even need the ISDN? The answer is obvious, once you look at the possibilities of using the same lines and interfaces that you have today, integrated across the network on call-by-call basis, we can eliminate the need for special lines dedicated to specific applications. Moreover, because the total functionality is evolving, the additional features to be derived from the network in the future will far surpass the technologies we know today. Think of the possibilities, let your imagination be your only limiting factor.

Who Is Making the Rules?

The Consultative Committee for International Telephony and Telegraphy (CCITT) is the responsible body for the development of ISDN standards on a global basis. CCITT's emphasis will be to ensure that ISDN will be just as ubiquitous as the present day public-switched telephone network when fully deployed. Because the ISDN is an enhancement to the PSTN, this should pose no great threat. However, the CCITT has also expressed the need for portability in the deployment of ISDN in its universal state. This means that an ISDN terminal should be capable of being plugged into an ISDN anywhere in the world and still provide full features and functions. This creates a dilemma for a CCITT because the PSTN really hasn't achieved the same portability on a global basis. The interfaces are different, the line terminating impedances aren't the same, and the signaling is different. So how can the CCITT expect the evolution of the universal access to a network, from anywhere in the world? In fact, this hasn't been accomplished in our old analog telephony network which has been evolving over 100 years. Look, for example, at the potpourri of services and carriers available here in the U.S. and the special needs this multitude of carriers presents. The rest of the world typically has one or two networks at the most.

The CCITT has created an ISDN reference model or a picture of what an ISDN should look like. This model identifies a series of functional groupings of equipment and reference points where the critical functions must be performed to ensure communication between the groupings. Don't confuse the groupings and functions as different pieces of equipment, rather they are functions that must be performed. These recommendations were contained in the findings, known as the *Red Book*, dated in 1984. Subsequent work has been done on the definition and implementation of ISDN services. The additional information is contained in the *Blue Book* dated 1988. Figure 23.2 is a representation of the ISDN reference model.

The implementation of these recommendations is the hot topic in the industry. Many vendors and carriers purport to have a fully ISDN-compatible terminal or network, when in fact the recommendations aren't completed yet. Thus, these implementations are at some risk of final acceptance, when the final standards are set. Changes might be required when finalization is complete.

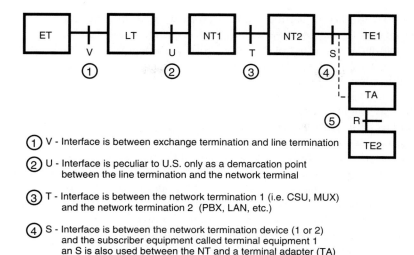

① V - Interface is between exchange termination and line termination

② U - Interface is peculiar to U.S. only as a demarcation point
   between the line termination and the network terminal

③ T - Interface is between the network termination 1 (i.e. CSU, MUX)
   and the network termination 2 (PBX, LAN, etc.)

④ S - Interface is between the network termination device (1 or 2)
   and the subscriber equipment called terminal equipment 1
   an S is also used between the NT and a terminal adapter (TA)
   which allows non-ISDN equipment to be used on the network

⑤ R - Interface is between the non-ISDN equipment and TA

Figure 23.2 The CCITT ISDN reference model.

Why Do We Need ISDN?

The carriers and vendors will have us believe that we need ISDN now, and the only way we can successfully use the all digital end-to-end capabilities of the network is through the implementation of ISDN. They might be correct, but they also have a tendency to oversell the capabilities and possibilities. The most overused and abused term these days is the concept of ISDN. Everyone will have us believe that the system and network is there today for our unmitigated use of all of the features and capabilities. Unfortunately, nothing can be further from the truth. Sure pieces exist, and trials have been underway for some time to evaluate the worthiness of the ISDN. But, the entire network will take quite some time for the full implementation. The main reason is one of money. In order to implement a network digitally at the local loop level, the investment to the local exchange carriers is substantial (one estimate is that the installed base of local loop is worth $275 billion, whereas to change the local loop to fiber-based systems would be in excess of $482 billion).

So who will pay for all these upgrades? You and I will! Unfortunately, the way this works is the LEC would build the prices into its rate base, which must be approved by the regulatory bodies governing the setting of prices (such as the public utilities commission). The regulators looking at the increase in the rate base will surely have a problem with the LEC adding to the monthly cost for dial tone, especially for the casual residential user who stands to gain little from the ISDN concept up front. Because this is a technology that will most benefit the business user, the regulators will frown on and possibly disapprove any rate hike to support the upgraded facilities from the LEC. Because this is a business service, the installation will have to be borne by the business user. That means higher costs for the local loop to our offices. Many of the LECs have tariffs in place to offer the service. Unfortunately, the tariffs are as the service offerings.

But back to the question of why we need ISDN. As the need for higher speed transport increases, the limitations of the available analog interface to the LEC and the IEC will be quickly surpassed. More and more users are demanding added call-carrying capacity to move information around the world. The primary areas that will consume bandwidth are additional speeds for data communications, transfer of files from host to host, host to PC, PC to host, LAN to LAN, LAN to WAN, etc. Video conferencing will also stress our telecommunications systems to the maximum.

Another need is the demand to transfer graphical information. As the desktop applications communicating complex documents from workstation to workstation are enhanced, even greater carrying capacities of bandwidth are necessary to move these files. The typical multi-page, text-based document will be fine, but the complex document that incorporates the use of graphics, annotation with voice, data, text, video, and any other format, will require the transmission of files in the multi-megabit rates. We will not be using ISDN to start at the multi-megabit per second speeds, but files transmitted at switched 64 Kbits/s up to 1.5 Mbits/s will provide greater throughput than the old analog dial-up data communications at 9.6–19.2 Kbits/s that we are more accustomed to.

Thus, as more complex information is entered into the systems, and the demand for greater bandwidth increases, the use of ISDN clear-channel capacity (which was covered in Chapter 13) will increase productivity and reduce costs. This isn't portending that you must have ISDN to send documents at higher speeds. Specific needs will dictate the use of higher throughput, and the technology to be used.

The main thrust is that ISDN is supposed to reduce call lengths because of the higher throughput on a dial-up basis. The shorter the call, the less expensive to support it. From the carrier's standpoint (both the LEC and the IEC), the shorter call duration will aid them in providing services to a greater number of users without adding circuit capacity and overbuilding their networks. Therefore, the carriers also stand to improve efficiency in the use of their networks, which should reduce their operating costs. This sounds too good to be true, both the supplier and the user will have lower costs! In theory, it should work, the rest remains to be seen.

Now, you can probably appreciate why the push is on for the implementation, acceptance, and standardization of the ground rules for this capability. The other obvious benefits to the use of ISDN include:

- More flexibility for the user on a call-by-call basis.
- Better control over the network by the carriers.
- All-digital transmission is more reliable and less expensive than analog.
- Quicker set-up time on calls.
- Additional services from the carriers, which equates to more revenue from new sources.
- Bypass defense by the LECs in providing service to the customer.
- Less new cable facilities needed into the office and the residence, because multiple simultaneous services can be used over a single 2-wire connection at the local loop.
- Analog to digital/digital to analog conversions become the customers' responsibility.
- Signaling out of band becomes the customers' responsibility.
- Carriers can provide improved network management and maintenance control on ISDN.
- High-speed facsimile, dial-up video, slow-scan video, and packet switching will become more cost advantageous, thereby proliferating more use of these services.

The Overall ISDN Concept

The objective of ISDN was originally introduced in 1979. It is the evolution of telephony applications that provide digital end-to-end connectivity to support a wide range of services to include voice and nonvoice, through a set of limited multi-purpose user interfaces. The reason for this concept was the obsolescence and cost-prohibitive nature of the older analog technology. Since the days when a dial-up voice connection was riddled with crackle and popping on the line, users have been complaining to the local and long haul carriers to improve service. The evolution of analog-to-digital transmission techniques significantly improved the transmission of voice and the ultimate introduction of higher-speed data communications. It would be appropriate to summarize some of the telecommunications requirements placed on the networks and carriers. These are the candidates for integrated digital communications through the evolved networking techniques:

Telegraph

The telegraph was the first electrical communications system to be introduced. Using direct current pulse signals on single wire earth-return lines, and later on 2-wire lines, it was possible to detect signals over long distances by using sensitized galvanometers as the detectors. Transatlantic cables were in use in the mid-1800s. Pulse codes such as Morse Code were devised, and operators who were highly skilled could achieve speeds of 30 words per minute. The use of telegraph and the ensuing delays over the transmission path distorted the actual human side of communications. Unnecessary words were eliminated as an expedient in getting the transmission across the cable. The telegraph introduced the first forms of digital coding and used a protocol to ensure accuracy and the reception of messages.

Telephone

The invention of the telephone by A. G. Bell in the late 1800s, and the subsequent proliferation of telephones in the U.S. transformed telecommunications. Users could now communicate directly with each other, without a protocol and intervention of the skilled operator to interpret the message. After the skilled operators were removed from the picture, telephone communications needed improvements to allow a novice user to use the system comfortably over great distances. Automatic switching, an invention necessitated in the network, and invented by an undertaker (Strowger) increased the ease of use and the ability of unskilled users to access called parties without human intervention.

Telex

The development and manufacture of complex machines prepared the way for the production of delicate mechanisms for telephone switching. At about the same time A. G. Bell was promoting the telephone, the typewriter (invented in 1867 by Christopher Sholes) was finding its way into the office. It took only a small adaptation to use an electric typewriter to send telegraph code for each letter and number on the keyboard. It wasn't until the 1930s, however, that interconnection using Telex actually expanded without the need for the skilled telegraph operator. One of the reasons that this process was slow in acceptance was the protocol needed for the machines to call and answer back with a discrete identity. Telex machines that used a punched tape were the first attempt in introducing direct machine to machine communications without the intervention of the operator. Many people in the industry thought this was a dead technology, especially because of the proliferation of facsimile and view data terminals—however telex is still alive and strong on an international spectrum.

DataComm

Telex and telegraph are actually the original data networks introduced into the industry. The term *data communications* became more commonplace after the introduction of digital computers, which encouraged the concept of time-sharing tasks requested by many terminals and devices. From the original dedicated star network in the office or factory, data networks have become true worldwide transport networks interconnecting mainframe computers with compatible terminals, and with other compatible devices. Today, the use of data communications over the public-switched telephone network is on the continual rise, using a binary code converted into tones (or analog signals) operating within the speech bandwidth spectrum (300 to 3300 Hz). Speech bandwidth as provided by the local exchange carriers limits the speed of information flow between communications devices recently (up to 28.8 Kbits/s on dial-up analog lines). Therefore, dedicated data lines and data networks have expanded in use to allow more data throughput in the communications world. The need to provide private line service impedes the user's overall flexibility in connecting to any device during peak loading periods. Many circuits have been justified on the basis of part-time use, while the line sits idle the rest of the time. This does not constitute effective use of our communications expenditures.

Packet switching

Using a dedicated or a circuit-switched facility to communicate information has been one of our primary means of communicating data. An evolutionary product, public packet-switching techniques were introduced. These networks were modeled after the bursty nature of data communications and the concept that all data calls do not necessarily need simultaneous two-way communications. Data can be broken down into smaller pieces, called *packets*, and sent from device A to device B. The messages or packet can be transmitted across the network at the convenience of the network rather than the convenience of the user. This method allows for the efficient use of facilities and lines within the network. Several users' packets might be traversing the network on the same virtual link, without any degradation of the response or loss of data integrity. This method allows the carriers to gain more utilization of their facilities and allows users to communicate at decent rates of speed without the use of dedicated or private lines. Refer back to the chapter on X.25 packet switching as a refresher.

Other services

Other services might also require separate networks, or can be shared across one of the existing networks. Some of these include:

- Radio telephone services/cellular communications
- Telemetry services
- Facsimile
- Video teleconferencing
- Radio paging
- Viewdata
- Slow scan video
- Teletext
- Home news services
- Electronic funds transfer
- Electronic mail
- Wideband services

Integration

The use and existence of separate networks to provide the above services has in some ways been an advantage in gaining access to the services. However, using separate lines and access methods to these services has also been a disadvantage. To gain access to the various services, the user had to subscribe to that network. This represents an overhead and management problem for the user to administer the effective use of each of the facilities provided. Moreover, in periods of peak loading, many facilities were overburdened, while others sat idle. To eliminate this problem, users and suppliers alike began to explore and demand an integrated plan. This plan calls for access through gateways or other access methodologies to any service by any user through a common set of lines. The existence of integrated networks will allow for the user management of their entire telecommunications function through the single thread to the outside world. This is what the overall objective of the ISDN is really all about: to provide the user with easy access to multiple services over a single connection to the network. The least important element of the title (ISDN) is the word digital, even though it is the digital network that allows for this universal and simple access to the wide range of services to be provided. To meet this objective the following comparisons must be implemented as shown in Table 23.1.

These are no simple tasks. The concept is complex to define, and even more difficult to perform. This is particularly true in the regulatory environment here in the U.S., where clear lines of demarcation have been established with the divestiture of AT&T from the telephone company subsidiaries.

TABLE 23.1   Comparing the Objectives of ISDN

Locally	Publicly
Duplex operation of simultaneous voice, data, telemetry, and signaling on a single link.	Capacity must exist to independently switch voice, data, telemetry, and signaling information from a subscriber location over a single line.
Transparency must exist, on the message content used in the various services.	The ability must be put in place to extract the message address information, check information for validity, and transmit the message information to the desired address.
Operation must take place over the 2-wire connection (twisted pair) within the distance limitations and other characteristics of the copper wire.	Provide the ability to access specialized networks (packet for example) and pass message information to and from these networks, in a protocol transparent way.
Extended areas must be serviced through the use of digital regenerators. The old analog amplification equipment must be removed.	Charges and billing information must be identified and billed across the various services between the providers.
Rapid access and set-up times must be available. Power consumption across the line (span power) must be kept at a minimum during idle stages. Set-up times must be provided end to end within 2 seconds, replacing the older methods of call set-up requiring 6 to 30 seconds.	Provide the capability to address or subaddress the customer premises equipment and ensure proper operation when addressing and connection takes place.
Transmission and timing systems to detect framing and bit timing must be provided in both directions.	Provide some form of local power to the customer premises equipment from the network switching exchange. This must be able to switch on local power to "power up/power on" the customer's equipment.
Good error-free performance (a typical error rate of 10^{-7} is expected).	
Compatibility in interfaces to other forms of digital transmission media such as radio, fiberoptics, cellular, coax must exist.	
Protection must be provided against power cross connect faults, lightning, etc.	
Remote equipment must be suitable for mounting independently on premises for the integration with customer provided terminal equipment.	

The ISDN Architecture

ISDN is based on the open systems interconnection (OSI) layered communications model adopted by the CCITT. The architecture generalizes the

OSI model and applies it to information and signaling transfer, and system management. The inclusion of the transmission media in the model high-lights the need to evolve ISDN from today's metallic environment in the local loop plant to a fiberoptical environment and on the premises wiring to higher grades of twisted pairs or fiberoptics, while the system management reflects the market's need for the customer controllable and easily maintainable systems and services.

Physical View

Two CCITT ISDN user network interfaces are used for connection to end-user devices. They are the *basic rate interface (BRI)*, 2B+D, and the *primary rate interface (PRI)* 23B+D/30B+2D.[1] These are covered later. How-ever, see Figure 23.3 for the basic rate concept and Figure 23.4 for the primary rate concept. While the most common PRI configuration is a single D channel per physical facility, multiple links (up to 20) can share a single D channel. This will lead to increased information-carrying capacity of the interface (AT&T method of sharing a single D channel for multiple PRI). The D channel on one 23B+D can be used to control a number of 24B PRI connections.

The PRI offers an economic alternative for connecting digital PBXs, LANs, host computers, and other devices to the network. The BRI brings Centrex and PBX customers integrated voice and data as well as advanced

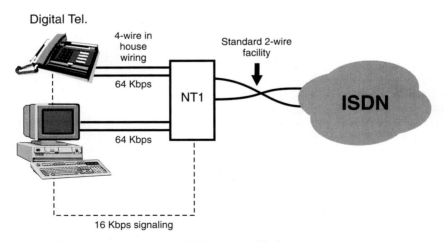

Figure 23.3 The basic rate interface (BRI) supports 2B+D.

---

[1] 30B+2D is the international version of ISDN PRI, using an E1 with the framing and signal-ing channels.

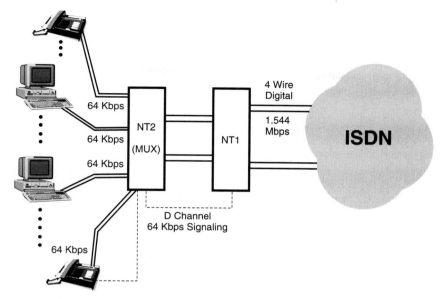

Figure 23.4 The primary rate interface (PRI) supports 23B+D in the U.S. and 30B+2D internationally.

voice features. Contrary to many of the proprietary digital interfaces available in many of the PBXs on the market, the BRI is a standard interface, and offers the user the benefit of multi-vendor compatibility. This statement assumes that the PBX makers implement a standard. This is questionable today. An example of this point lies in the realization that every PBX manufacturer advertises that their system is ISDN compatible. If this was a true statement, then a user should be able to buy one manufacturer's system (AT&T for example) and plug in a different manufacturer's digital ISDN telephone set (Rolm/Siemens, for example) and they would work transparently. This is not the case today because every manufacturer builds a proprietary means of using their equipment. The future offers some hope, but the standards must be cast in concrete first.

To maximize terminal portability and ensure easy change from one network to another, the same BRI and PRI interfaces are used for interconnecting terminals, hosts, digital PBX, and other devices in the premises network. The ISDN architecture includes premises applications processors that provide information movement and management, and end customer control features. Examples of some of these features are:

- Voice messaging
- Text messaging

- Message center attendants
- Message detail recording
- Electronic directory
- Traffic data
- Customer station rearrangement

At the network ISDN node, the integrated channelized access is separated into components, either physically or logically, depending on the implementation, and diverted to the appropriate functions. The network ISDN node also provides interconnections to four types of networks:

- Channel-switched networks
- Circuit-switched networks
- Packet-switched networks
- Common channel signaling networks

Another important ISDN interface unique to the U.S. is the internetwork interface between local exchange carriers (LECs) and interexchange carriers (IECs). The court system and the FCC require that equal access be provided between the networks of the LECs and the IECs. Today, network application nodes are working hand-in-hand with network switches to provide sophisticated services. With progress in distributed processing, the use of application nodes will expand.

Logical View

Circuit mode services

The 1984 CCITT Red Book defines in Q.931 (I.451), messages and procedures for basic call set-up and disconnect, and includes a potential framework for supplementary voice services. AT&T uses LAPD as the link layer protocol and the extended Q.931 as the network layer protocol for the D channel. The extension furnishes signaling for implementing all of today's Centrex and advanced PBX features in ISDN. New functional messages are added as well as messages conveying activation and deactivation of feature buttons and dial feature/access codes. The extended version of Q.931 protocol is being expanded to include functional messages for advanced signaling in the premises network and for features unique to ISDN. For an overview of the message format, see Figure 23.5. This is the HDLC frame format for LAP-D messages that will traverse the data channel.

ISDN provides transparent circuit-mode services for the transfer of information. Protocols for layers 2 to 7 are supplied by end-user devices.

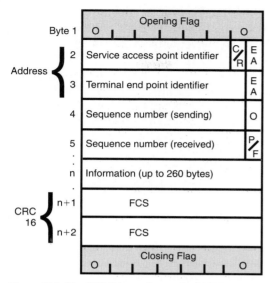

Figure 23.5 The HDLC frame format for LAP-D signaling.

The out-of-band signaling infrastructure for the premises network is provided by the D channel in the BRI and PRI. It can also be provided by the public common channel signaling (CCS) network that transports the user-to-user signaling transparently from one premises switch to another. This extends the benefit of today's CCS network to premises network users, and facilitates the interconnection of geographically scattered premises switches.

In the public network, the ISDN user part (ISUP) and the signaling connection control part (SCCP) of the CCITT signaling system 7 (SS7) supports end-to-end ISDN features, while the transaction capability (TCAP) supports communications between the network switches and network application nodes. The interworking between Q.931 and ISUP is being defined. In particular, mapping of messages and procedures for supporting network-wide features, such as call waiting, originating, and automatic callback are being addressed. The automatic callback feature should work from any user to any user, no matter where they initiate a call or what carrier service they use.

Packet mode services

The ISDN architecture supports packet-switched data services on the 16 Kbits/s D channel and the 64 Kbits/s B channel, using X.25, X.31 (an enhanced ISDN version of X.25) and the LAPD-based packet mode protocol.

X.31 extends X.25 to provide dial-up packet-switched services in an ISDN. It uses Q.931 signaling to establish the physical channel connection to the packet handler, after which the conventional X.25 in-band call control is used to establish the logical connection (Figure 23.6).

The LAPD-based packet mode protocol is the next step toward integrated signaling. It uses Q.931 signaling to establish the physical and logical connections. The LAPD-based packet mode protocol also decomposes bearer capabilities into building blocks to allow the freedom to distribute functions between the network and the premises system. This allows tailoring of individual services to application needs.

LAPD layer 2 supports multiple logical links on a single physical link. The minimum functionality packet-switched service uses a portion of LAPD, consisting of:

- Multiplexing
- Error detection
- Frame delimiting

Other LAPD functions like error retransmission, and packet sequencing and acknowledgment, can be performed on an end-to-end basis by end-user devices. The LAPD-based packet mode protocol initially supports X.25-like services. Unlike an X.25 service, full X.25 is terminated only by the end-user devices, the intermediate switch provides the minimum functionality packet switching. As a result, the X.25 services supported by the LAPD-based packet mode protocol have lower trans-network delay and higher through-

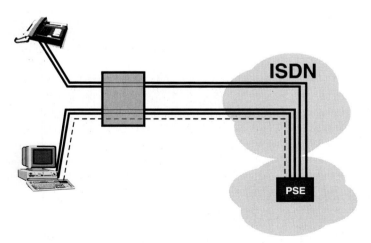

Figure 23.6 The Q.931 signaling establishes a physical connection to a packet-switching (X.25) handler.

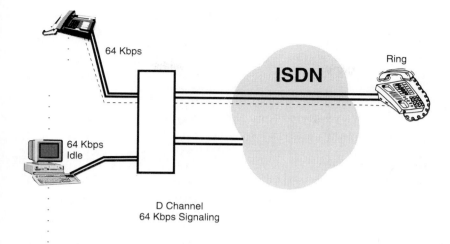

64 Kbps

ISDN

Ring

64 Kbps
Idle

D Channel
64 Kbps Signaling

Figure 23.7 The user can establish a 64 Kbits/s voice connection on ISDN in sub-second times. The call will be physically connected until the parties hang up.

put. The LAPD-based packet-mode protocol, together with its Q.931 signaling procedure, provides a mechanism for the customer to request services based on their application needs on a call-by-call basis. This means that a user can use the signaling channel to establish a voice call on the link at 64 Kbits/s as an independent action (Figure 23.7). The call can proceed as normal until the two parties agree to end the conversation and mutually hang up, ending the connection between them.

Immediately following this call, the user can now use the signaling system to link the 2 B channels together and form a 128-Kbits/s data connection to a remote host (Figure 23.8) to conduct a file transfer to the host system.

Application services

The marriage of the terminal and telephone expands the horizon of services that the network must provide. More and more required functions go beyond information movement into information management and its applications. Voice messages are left on recording systems that can be controlled by the end user for playback and editing. Electronic directory services can request the switch to dial a number directly when a button on the phone is pushed. These applications require the higher layer protocols available in ISDN. CCITT's recommendations for the various layers are:

- X.224 as the transport layer protocol
- X.225 as the session layer protocol
- X.226 as the presentation layer protocol

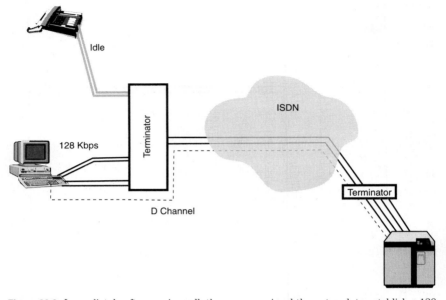

Figure 23.8 Immediately after a voice call, the user can signal the network to establish a 128-Kbits/s connection to a remote host computer.

Application (being deferred)
Presentation   X.226
Session   X.225
Transport   X.224
Network   Q.931
Data Link   Q.921
Physical

Figure 23.9 The ISDN protocols as they compare to the OSI reference model.

These are shown in the context of a comparison in the OSI model in Figure 23.9. At the application layer, CCITT and ISO are still working on defining the protocols.

Architecture Reality

The roll-out of products and services to implement the ISDN concept has been underway for some time now. In the network, AT&T has approached the implementation in three categories:

- Exchange carrier networks (Bell and PTT networks)
- Interexchange networks (interexchange carriers)
- Premises networks (customer premises equipment)

Their flagship vehicle for these services is the 5E-ESS, which offers the features and capabilities in ISDN. Their switches have had ISDN-like interfaces for years, at the BRI and PRI interface levels. Evolving from that base, they have continued development of interfaces to the ISDN standards, beginning with the PRI, to take full advantage of ISDN services in the public network.

The exchange carrier networks

The key element in realizing the ISDN exchange carrier networks is the local switch. As long as the carrier (LEC) has a switch that offers the basic interfaces to ISDN, the switch acts as the ISDN node. The switch (typically the end office) will provide simultaneous voice and data services to Centrex and local dial-tone users over a basic rate interface. This switch will support the full range of Centrex voice features using the extended version of Q.931, and circuit-switched and packet-switched data. In addition, the switch will more than likely support modem pooling for inter-networking with non-ISDN data terminals through the public network. See Figure 23.10 for this, as shown in a modem pooling arrangement. This moves the services (ala data circuit-switched) onto the network. Users will access theses services on an as-needed basis. The network now becomes the service provider, instead of just a dial-tone provider.

The switch will also support the end-user applications processor environment. This application processor will provide message services, electronic directory, message detail recording, and traffic data statistics. Using SS7 transaction capability (TCAP) and switching services point (SSP), the switch will interact with network databases to process calls (Figure 23.11). This will allow the exchange carrier to implement an intelligent network architecture and offer such services as virtual private networks, enhanced services, and centralized network control as part of the ISDN capabilities. The use of the primary rate interface connected to the switch will allow the transparent end-to-end transport of user-to-user information for applications, such as caller name, security check, and feature transparency.

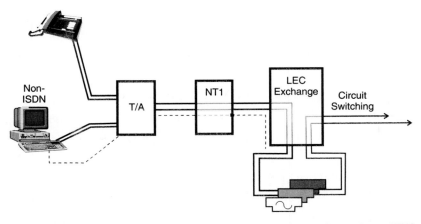

Figure 23.10 The exchange office ISDN node can support modem pooling with non-ISDN terminals. The modem can be added to the link as needed from the pool.

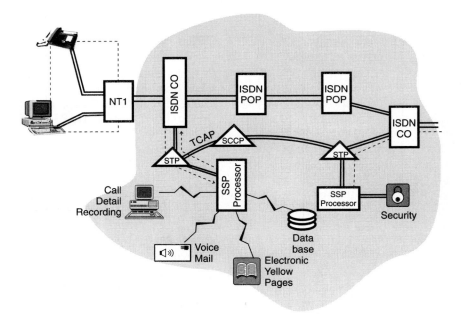

Figure 23.11 The ISDN carrier office allows access to a full range of services on the network. This allows call processing and features in an intelligent network architecture.

Additional flexibility can be derived through the use of remote switching modules (RSMs) and subscriber loop carrier (SLC) carrier systems, shown in Figure 23.12. Using these devices, ISDN services can be provided to remote locations. On the network side (IEC), using signaling system 7 (SS7) in conjunction with the ISDN user part, access can be provided. The combination of channel, circuit switching, X.25 packet switching and common channel signaling, is derived through a series of ISDN gateways for the inter/intra-LATA services.

The interexchange network

The interexchange carrier network services can be accessed through a major ISDN node either through switched services through the local exchange carrier or directly to the IXC using private line services. The same four supporting transit components are accessible through the IXC network services. Service functionality can be attained through a digital cross-connect system (DCS) supporting speeds of 64 Kbits/s, 384 Kbits/s, and 1536 Kbits/s circuit mode service that is shown in Figure 23.13. In the AT&T environment, the use of a 4 ESS with DACS or a 5 ESS supports these services. The use of a packet-switching system (1PSS) supports the X.25 packet data network. DACS-based private line services and customer controlled reconfiguration (CCR) are also provided. AT&T ISDN services also add capabilities through network applications processors that operate off the common channel signaling network to provide advanced 800 services,

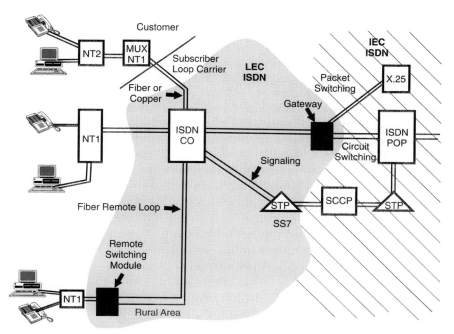

Figure 23.12 The LEC to IEC hand-off will function through gateways to combine long distance circuit/packet/signal switch. Remote areas will be served from the LEC via remote partitions.

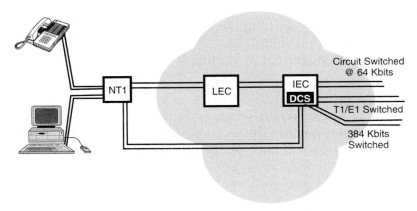

Figure 23.13 Access to the interexchange carrier network can be through the LEC, or directly to IEC bypassing local carrier. At the IEC network 384-1536 Kbits/s can be accessed through a digital cross-connect system (DCS).

software-defined networks, etc. Further the use of CCR provides the customer dynamic reconfiguration of their private line networks.

The premises network

The premises line of services and products is designed around the full integration of voice and nonvoice applications, as shown in Figure 23.14. Intelligence is also provided through a network gateway or node in an ISDN environment. ISDN-capable PBXs acting as the gateways to the public network, as provided by the exchange carriers, are accessed through the PRI. Using this arrangement, a variety of transport services and facilities can be accessed, taking full advantage of the call-by-call service selection capability of Q.931. Add-on capabilities will be available through these ISDN-capable PBXs, such as:

- Message center services
- Voice messaging (voice mail)
- Office telecommunications services (OTS)
- Electronic mail (E-mail)
- Centralized management systems
- Least cost routing (LCR)
- Address translations
- Security
- Feature transparency
- Station message detail recording (SMDR)

Locally, the PBX will interface to a wide range of equipment for connectivity on the premises. This will include:

- Analog telephones
- Digital telephones
- Switched data services for async and sync terminals
- PCs and integrated workstations
- Cluster controllers (327X) and terminals that are typically hardwired
- Integration of LANs
- Integration to WANs
- Integration of services of ETN/ESNs

Basic Operating Characteristics

The basic structure for ISDN operating characteristics is digitally encoded data using a standard of 64 Kbits/s at the basic rate, and using 1536 Kbits/s at the primary rate. This is a derived channel using pulse code modulation techniques, the North American standard for digital transmission.

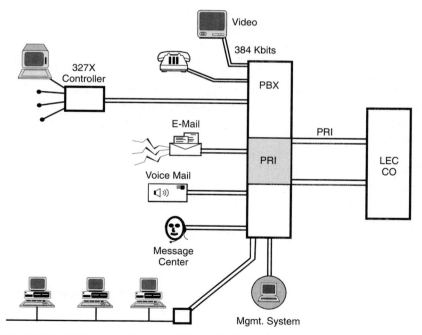

Figure 23.14 A full range of services will be available through the PBX for voice and non-voice applications. The PBX will use the primary rate interface.

However, the concept of ISDN is based on a global perspective. Therefore, the use of North American standards for the concept of network transport has to accommodate interfaces to an international standard. In order to fully appreciate the magnitude of the digital hierarchy, perhaps a quick review of digital technology is in order.

First, the analog signal is converted to digital signals using pulse code modulation (PCM) techniques. For a refresher on the pulse code modulation technique, refer back to Chapter 13 in the discussion of T1.

North American T1 differs from the international, or European standard of 30 channels at 64 Kbits/s, plus overhead of 2 channels at 64 Kbits/s, yielding a digital stream of 2.048 Mbits/s, which is the E1. Other differences exist in the encoding and companding techniques which require the translation from North American to European standards, and vice versa.

When sub-rate channels are needed or desired, a different encoding and modulation technique is used. The most common is adaptive differential pulse code modulation (ADPCM) that produces 44 channels at 32 Kbits/s each. The need for extra overhead for signaling and control dictates that only 44 instead of 48 channels are derived. ADPCM is a standard rate accepted by CCITT for use in ISDN.

Other rates of bandwidth are used in the ISDN concept. They use fractional portions of the DS1/E1 for speeds higher than the 64 Kbits/s rate. This is also referred to as *multi-rate ISDN*. The use of fractional rates works under a nonchannelized format today, but will be defined as a standard in ISDN. See Figure 23.15 for a representation of this.

Although the channel speeds are defined at 24 or 30 channels at 64 Kbits/s, the requirement to provide signaling for status (dial tone, ringing, busy, off-hook and on-hook) leaves an effective data throughput or usable bandwidth of 56 Kbits/s. Again, the refresher on T1 (Chapter 13) will help to bring this back in focus.

CCITT and the ISDN study groups are studying the use of the 8 Kbits/s of overhead for framing channel on a T1, and the format to be used. North American standards call for ESF under the AT&T guidelines and the ANSI recommendations. Significant differences exist in the use of these standards here in the U.S. These differences deal with the format and protocol used in error detection, with AT&T's recommendation being a passive arrangement and the ANSI recommendation being an active approach in support of the LEC environment. Both recommendations are being implemented in the U.S. An announcement by AT&T to conform to both the AT&T and the ANSI recommendations has relieved some of the pressure at the LEC and the user's end.

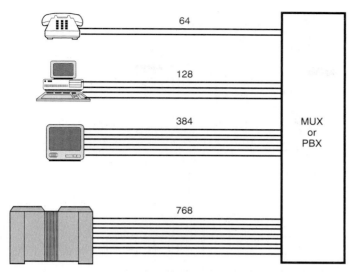

Figure 23.15 The ability to use various bandwidth requirements through multi-rate ISDN in a PBX or MUX.

Bearer Services

ISDN works on the principle of transport services known as *bearer services*. The basic operation of the bearer service is the 64-Kbits/s channel capacity. The bearer service offers the ability to transport digital voice or nonvoice services across this standard. Packet services can also be transported across the bearer service at higher rates of speed (up to the 64 Kbits/s speed). Bearer services, as mentioned before, are divided into two categories: the basic rate interface (BRI) that uses two bearer services at 64 Kbits/s, and one data channel operating at 16 Kbits/s, with an additional 48 Kbits/s of overhead, yielding a 192-Kbits/s data stream over the typical two twisted wires (Figure 23.16), from the business or residence to the local exchange.

The second category is the primary rate interface (PRI) that uses 23 bearer services and one data channel, all at the 64-Kbits/s per channel speed, yielding a 1.536-Mbits/s data stream from the business to the local exchange. As mentioned before, this is the North American standard (used in U.S., Canada, and Japan), whereas the international standard will use 30 bearer services, plus 2 data services all at the 64 Kbits/s rate, yielding a 2.048-Mbits/s data stream. These framing formats are shown in Figure 23.17.

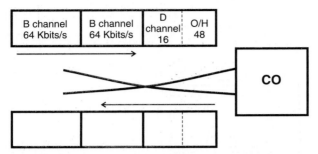

Figure 23.16 The actual throughput on the ISDN BRI is 192 Kbits in each direction. Only 144 Kbits/s in each direction is available to the BRI user in a duplex mode.

T1

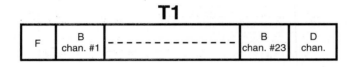

E1

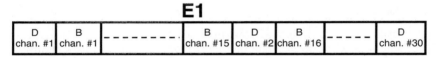

Figure 23.17 The PRI format will deliver T1 @23 B+D and 8 Kbps framing, or the E1@30 B+2D.

The primary rate under either the North American or international standard will use existing T1/ E1 technology and be the main emphasis for the business world. Applications will be based on the use of WATS, INWATS (800), FX, tie line, and data lines, all being combined onto a single primary rate. Because T1/E1 technology has proliferated so much in the U.S. and the rest of the world over the past few years, because of the economics involved, the evolution to a PRI should be fairly simple. The connection through a digital PBX, multiplexer, network node, DCS, or other T1/E1-compatible equipment will require little change in the network, and simple card changes in the customer premises network.

Initially, the basic rate interface posed some additional problems. These have since been resolved. For example, the two-wire connection to the small business, small branch office of a larger company, or residential service, is typically an analog service. The problem to be addressed entailed how to digitize the customer information at the customer premises, and carry the digital information across the 2-wire circuit. Digital circuitry has a

4-wire, full-duplex capability. The North American approach to this is to use a technique known as *2B1Q* (Figure 23.18).

In all line coding techniques there is a need to provide for:

- Continuously alternating signals with low bandwidth and good facilities for timing recovery.

- Low attenuation and therefore long line lengths (> 2.5 miles).

- Protection from noise interference, in particular near-end crosstalk, the transmitted signal interfering with the received signal at the near end terminal from which it was transmitted.

- Low complexity to the echo-cancellation function. CCITT has specified the use of echo-cancellation in the transmission mode of ISDN.

- Fast convergence of echo-cancellation. On switch-on, there is a perceptible time delay before the echo canceling routine gets into sync. On the subscribers' systems that discontinue transmission during idle time, the delay must be kept at a minimum.

- High signal-to-noise ratios. This is affected by the number of decision levels in the signal. Pure binary signals with just two decisions will, all things being equal, have a higher signal-to-noise ratio than multi-level signals.

Line coding methods can be divided into two basic categories: those that involve watching and converting data as it passes (called *linear coding methods*); and those that require recoding the bit stream through a look-up table (called *block coding methods*). Alternate mark inversion (AMI) and HDB3 (high-density binary modulo 3) are examples of linear coding methods. 4B3T (4 binary bits to 3 ternary) is an example of block coding. Early line coding systems used bi-phase techniques, while the two main contenders under CCITT guidelines have been AMI and 4B3T. North American standards have introduced the newer contender, the 2B1Q (2 binary bits to 1 quaternary).

The linear codes with two decision levels tend to send information to the line at the same rate as they are received in binary form (160 Kbits/s for the

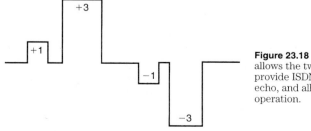

Figure 23.18 The 2 B1Q technique allows the two-wire facility to provide ISDN, reduce noise, cancel echo, and allows full-duplex operation.

CCITT basic access). The multi-level codes tend to reduce the sending rate because more information redundancy is available in the code. Therefore, such codes are more attractive, even though they require more complexity to combat crosstalk and inter-symbol interference. 4B3T, for example, uses a 120-Kbits/s line rate and 2B1Q uses 80 Kbits/s, but at the expense of lower signal-to-noise ratios, and more ISI (lower baud rate). Many of the LECs have chosen the use of 2B1Q as the method of delivering ISDN to the business and household, for explicit cost reasons.

The original concept of ISDN in the user and the exchange carriers' perceptions was that fiberoptics or 4-wire circuits would be required to introduce ISDN to the basic rate access. This included business or the residence service. However, to bring fiber to the door, or to reinforce the copper facilities to every door in North America, would have been cost prohibitive. 2B1Q allows the introduction of BRI services over the standard 2-wire interface, with no appreciable plant facilities costs. This will allow the introduction and proliferation of ISDN services. One estimate was to bring fiber to every door costing $482 billion, whereas the installed plant was only valued at $275 billion. Thus, an investment of this magnitude would have impeded, if not killed, the ISDN concept of services here in the states.

Clear-Channel Capability

The use of the bandwidth in an ISDN world requires clear-channel capability. *Clear channel* means that the full 64 Kbits/s is available for the data stream. As mentioned earlier in Chapter 13, there is a problem with clear-channel capability in the PCM techniques used in the North American standards. Many of the LECs have held back in deploying the use of clear channel services because of the financial burden and the indecision of the standards committees here in the states. As the LECs were evaluating clear-channel capacities, the costs of upgrading central office technology to support this were quite high.

In North America, the technique used is B8ZS. Recall the discussion of B8ZS from our discussion in Chapter 13. Data is now transparently transmitted through the network without conforming to the imposition of the 1s density rule. A good deal of activity has taken place to provide a migration to B8ZS to provide the customer a clear channel 64 Kbits/s transport.

Applications for ISDN

A myriad of potential applications exist for ISDN in the marketplace. Many of the vendors have announced products and services that support attachment of other devices to an ISDN network capability. The name of the game is transparent, digital, switched connections, from any service to any feature or service. In itself, this definition covers a broad range of applications that might have many users skeptical of the real ability of ISDN to perform all things for all people.

Skepticism is probably a good way to approach any new technology. But, is ISDN really a new technology? The carriers who have begun to roll out the network services believe this is nothing new, only a modification of the evolutionary network services. The newness really comes in the form of the reference points and the interface standardization, which users have been after for years. The carrier network has been all digital for quite some time now, the analog network remaining at the local loop level. Even the local loop has been evolving to a digital access point over the past decade, with the introduction of T1/T3 as a customer available service.

The applications for voice and nonvoice service are limitless. The only limitation is the ability to introduce the technology and the user's creativity. Some of the applications have been demonstrated in this list:

- Desktop conferencing
- Async protocol conversion to SNA/SDLC
- VTAM INN link replacements
- Async protocol conversion to SNA/SDLC with muxing
- 327X emulation
- Async access to private packet-switch network
- Async access to Unix hosts
- Async access to Unix hosts via Pad and X.25 Muxing
- 327X coax elimination
- Extended LAN
- Async access to Ethernet terminal servers
- Replacement of dedicated lines (64 Kbits/s)
- Async modem pooling (in or out)
- ISDN wide area networking (WAN)
- Access to Apple Talk Network and Gateway
- Ethernet LAN bridging and routing
- Customer service call screening/handling
- Incoming call management

Obviously, more applications and demonstrated capabilities exist, but it would be too difficult to generate an all-inclusive list in this format.

ISDN Centrex vs. PBX Service

ISDN isn't really a product or a service, but a transport mechanism to carry raw bandwidth of user information. It will however, be implemented through a series of products and services from the carriers and vendors of hardware solutions for the end user. These products and services as we have seen, will handle a variety of applications based on the users' needs.

In many cases, the concept of ISDN has been introduced as a feature of Centrex services. Users are being lulled into the false impression that if they want ISDN, they must use a LEC Centrex offering. Nothing could be further from the truth.

Centrex services make ISDN implementations a lot easier and less risky today. The reason is the carriers, and in particular, the LECs have provided the digital switching capability in their exchanges to provide the transport across the network. Because Centrex is also a CO-based service, that is, the Centrex is a software partition of the CO, it would make sense that the ISDN capabilities are merely a logical extension of what the LEC can readily provide to their Centrex users.

However, the capability is also a transport mechanism from an end user's PBX, located at the customer's premises. We are dealing with the ability to deliver digital connectivity to the PBX and interface at the BRI or PRI level. No magic, just connectivity. So, for now let's dispel the rumor that it is a Centrex service.

In the U.S., telephony service falls into different categories (Table 23.2).

TABLE 23.2 A Summary of the Telephony Services in the U.S.

Basic Dial Tone: Standard telephone service from the LEC to the customer locations (business or residential), terminated on a single line set.

Key Systems: The ability to bring multiple lines to the customer premises and terminate in a key service unit (KSU) for access to multiple lines by multiple lines. Usually, in a small office environment, a one-for-one line to user fits.

Centrex: Comes in a variety of names but still does the same thing. Centrex is a C.O.-based technology which uses a line to every user in the location over a twisted pair of wires. Additional features are available, over the basic and key system. All service and maintenance of the Centrex is provided by the LEC.

PBX: Also called by many names, but extends the dial tone from the C.O. to a demarc point at the customer premises. The LEC provides trunks which are competed for by multiple users. The mix here is far greater users competing for fewer trunks, an economic issue. Fully featured similar to Centrex, but the end user provides all of the administration and maintenance over the PBX.

Hybrids: Some mix of features and capabilities which match PBX, Centrex and/or key systems.

Centrex

Centrex is typically a service you'll find in North America, but is now finding its way into the international marketplace. The international market has no equivalent service at this time. The two main suppliers of Centrex capabilities through their switches and offered by the LECs are Northern Telecom and AT&T. Each has a slightly different approach to offering ISDN through their products, however they are coming closer and closer together as the standards near completion.

Northern Telecom markets their product known as *Meridian Digital Centrex*, offering a full range of features in the Centrex world using a proprietary digital technology. In 1986, NTI first offered BRI ISDN on a limited basis, as part of the Meridian Digital Centrex products.

AT&T markets the 5 ESS and 5E-ESS that offer Digital Centrex services through a full array of features and capabilities. However, AT&T has gone to a full ISDN environment. Therefore, AT&T only offers ISDN Centrex in 5 ESS; no other type of digital Centrex is available.

Both vendors have caused a lot of confusion in the marketplace as they compete for market share in the ISDN arena. Each has published various reports on their implementation of ISDN. Yet, looking closer at the real world, many of the customers were actually installing Centrex service. The manufacturer, depending on philosophy, stated it was ISDN. This is especially true in AT&T's case because ISDN Centrex is all you can get.

This is where the user gets confused. Is ISDN really proliferating, or are the vendors just forcing the numbers to show greater acceptance?

To use ISDN Centrex, the process is straightforward. Simply order digital Centrex from a LEC running a 5 ESS or order Centrex in a DMS world (NTI product) and get the BRI capability. This would provide 2B+D to every extension.

ISDN is a reality today, but it is not limited to Centrex services. The 2B+D capability to the extension and the 23B+D at the access through the CO as a gateway to the Interexchange network, using the B channel to access packet services, provide automatic identification of outward dialed calls (AIOD), and user-to-user information through SS7 connections.

ISDN PBX Capabilities

The PBX hasn't been neglected during the evolution of the central office product lines. Many manufacturers have been introducing support of the PRI (23B+D) interface at the PBX level. Every PBX manufacturer has some plan to introduce products and services used in an ISDN, depending on the maker.

The CCITT standards never really addressed ISDN at the PBX level, so many of the manufacturers have created their own "steak and sizzle" marketing campaign to compete for the market share of ISDN PBX capabilities. Primarily, they are providing interfaces for the PRI, but they can also access the BRI at the PBX level. Many have introduced BRI services on a station-to-station basis and support for signaling and control within the PBX for use across public or private (switch-to-switch) networks. Most of these products are proprietary inside the PBX, so the buyer must beware of potential incompatibilities.

Telephone terminal equipment manufactured by one manufacturer will not work in a PBX by another manufacturer. Typically, when a PBX is selected and installed, the buyer is making a sole source purchase for new or add-on equipment. This is also true in the Centrex environment, where a digital Centrex telephone set from AT&T will not work on an NTI ISDN Centrex. So much for standards!

Each of the PBX manufacturers are now expounding their ISDN-compatible PBX's capabilities to deal with private networking arrangements. Once again, they will have some proprietary software in dealing with PBX-to-PBX connectivity through a private network. The users must be aware of how the system will integrate into their networks before making a purchase/lease decision.

Applications for ISDN in Vertical Markets

The use of ISDN still remains an issue of not if, but when, each of us has to decide that the evolution of the network has sufficiently taken place. At that point, we must decide when is the best time to support and implement ISDN in our organizations. No one best answer exists, this is a decision that must be handled on a case-by-case basis. Remember the discussion about the management style in your organization? Each company must choose the time to experiment with the technology, once that company feels the technology has advanced enough to support and not hinder their operations. Through a detailed study of the individual requirements, the implementation scope and timing should point the way.

Some of the vertical markets being addressed by the purveyors of the technology (ISDN) are:

- Telemarketing industry
- Legal industry
- Financial industry
- Insurance industry
- Manufacturing industry

- Retail marketplace
- Medical profession
- Petroleum industry

Messages and Frame Formats

D Channel protocols

The D Channel adds a new dimension to digital networking. It provides for an out-of-band common-channel signaling facility for ISDN. Common-channel signaling uses a single channel to convey the necessary signaling information, through special messages. This information is used to identify called and calling stations, set up the connections, maintain the connections, release connections, identify line and network status, provide billing and other operational data, and provide input for network management.

The D Channel for the basic rate interface (BRI) is a 16-Kbits/s derived channel that is shared for signaling, low-level packet switching, and telemetry. The D Channel for the primary rate interface (PRI) is a 64-Kbits/s derived channel that is used exclusively for signaling.

The D Channel uses a specific bit-level time slot for the BRI and a specific byte-level slot for the PRI. The network termination function (NT) builds the message-oriented signaling procedure from these bits. The NT uses layer 2 and 3 functions that will use the services of the layer 1 to create a standard signaling channel.

CCITT recommendation I.400 describes terms as Link Access Procedure on the D Channel (LAPD). LAPD is a protocol that operates at the data link layer of the OSI model. LAPD is independent of transmission rate and requires a duplex, bit-transparent D Channel. The D Channel protocols are shown in Table 23.3.

The LAPD layer 2 functions include:

- The provision of one or more data link connections on a D Channel.
- Discrimination between the data link connection via a data link connection identifier in each frame.
- Frame delimiting, alignment and transparency, allowing recognition of a sequence of bits transmitted over a D Channel as a frame.
- Sequence control, to maintain the sequential order of the frames.
- Detection of transmission, format, and operational errors on the data link.
- Recovery from detected transmission, format, and operational errors with notification to the management facility of unrecoverable errors.
- Flow control.

TABLE 23.3 A Summary of the Protocols Used on the D Channel

	D Channel protocols	
Layer three	Q.931/I.451	
Layer two	LAPD/I.441	
Layer one	BRI/I.430	PRI/I.431

The LAPD layer 3 functions are:

- Provide a means to establish, maintain and terminate network connections across an ISDN between communicating application entities.
- Routing and relaying.
- Network connections.
- Conveying user-to-user information.
- Network connection multiplexing, segmenting, and blocking.
- Error detection, error recovery.
- Sequencing, flow control, and reset.

D Channel message-oriented signaling

The D Channel handles both the transfer of user data and signaling information. The signaling information facilitates the establishment, maintenance, and clearing of ISDN channels. ISDN uses out-of-band signaling over the D Channel for a more efficient transfer of data and the overall efficiency of the bearer channels (B channels). The D Channel is presently defined as a 3-layer protocol set. These layers are the bottom three layers; physical, data link, and network layers.

The network layer protocol, CCITT Q.931/I.451, uses messages to convey signaling information between ISDN layer 3 entities. The components of the messages are referred to as *information elements*. Many types of messages exist at layer 3, depending upon the network connection. ISDN will support circuit mode, packet mode, and frame relay mode connections. At layer 3, the information field from the LAPD frame is processed. The LAPD frame was constructed at layer 2 from the D Channel layer 1 time slots. I.451 defines the message-oriented signaling application for the information field. The messages transported on the D Channel are:

Call establishment messages:

- Alerting
- Call proceeding

- Connect
- Connect acknowledge
- Setup
- Setup acknowledge

  Call information phase messages:

- Resume
- Resume acknowledge
- Resume reject
- Suspend
- Suspend acknowledge
- Suspend reject
- User information

  Call disestablishment messages:

- Detach
- Detach acknowledge
- Disconnect
- Release
- Release complete

  Miscellaneous messages:

- Cancel
- Cancel acknowledge
- Cancel reject
- Congestion control
- Facility
- Facility acknowledge
- Facility reject
- Information
- Register
- Register acknowledge
- Register reject
- Status

Asynchronous Transfer
Mode (ATM)

ATM Capabilities

Asynchronous transfer mode (ATM) is one of a class of packet-switching technologies that relays traffic via an address contained within the packet. Packet-switching techniques are not new; they have been around since the mid-1970s (such as X.25 and the original Arpanet). However, when packet switching was first developed, the packets used variable lengths of information. This variable nature of each packet caused some latency within a network because the processing equipment used special timers and delimiters to ensure that all of the data was enclosed in the packet. The X.25 packet-switching techniques were covered in Chapter 15. As a next step to create a faster packet-switching service, the industry introduced the concept of frame relay (covered in Chapter 22). Both of these packet-switching concepts (one a layer 3 and the other a layer 2 protocol) used variable length packets. To overcome this overhead and latency, a fixed cell size was introduced. In early 1992, the industry adopted a fast packet or "cell relay" concept that uses a short (53 byte) fixed-length cell to transmit information across both private and public networks. This cell relay technique was introduced as ATM. ATM represents a specific type of cell relay that is defined in the general category of the overall broadband ISDN (B-ISDN) standard.

ATM is defined as a transport and switching method in which information does not occur periodically with some reference, such as a frame pattern.

Where all other techniques used a fixed timing reference, ATM does not. Hence the name "asynchronous."

What Is ATM?

ATM is a telecommunications concept defined by ANSI and CCITT standards committees for the transport of a broad range of user information, to include voice, data, and video communication on any user to network interface (UNI). Because the ATM concept covers these services, it might well be positioned as the high-speed networking tool of the 1990s and beyond. ATM can be used to aggregate user traffic from multiple existing applications onto a single user-to-network interface (UNI). As shown in Figure 24.1, the ATM concept aggregates a myriad of services onto a single access arrangement. Some of these applications include:

- PBX-to-PBX tie line/trunks
- Host-to-host computer data links
- Video conferencing circuits
- LAN-to-LAN bridged or routed traffic
- Multi-media networking services between high-speed devices
- Workstations
- Supercomputers
- Routers
- Bridges
- Gateways

All of these services can be combined at aggregate rates up to 155 Mbits/s today. However, the future speeds will be at rates of 622 Mbits/s, and go into the 1.2 Gbits/s class by the turn of the century.

Broadband Communications

The use of fixed statistical (and synchronous) time-division multiplexing services on the telecommunications networks of old had certain limitations. Some of these were the restrictive bandwidth available to perform functional transmission. Fixed time slots were used, requiring that amount of bandwidth to be allocated to a specific service at a specified transmission rate. For the evolution and migration of these network services, remember that these came from the telephony world. In telephony, a 4-kHz bandwidth was used, and a digital time-division multiplexing scheme was applied to this bandwidth. The result was fixed time slots of 64 Kbits/s baseband ser-

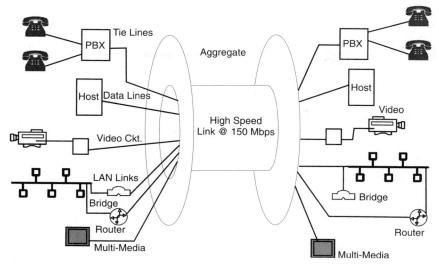

Figure 24.1 Multiple inputs are aggregated onto a single UNI.

vices delivered to a user network. Where larger amounts of throughput were required, these fixed slots were concatenated together to form higher rates of speed (i.e., 128 Kbits/s, 256 Kbits/s, 384 Kbits/s, etc.).

However, the fixed slot was just that: fixed. If a user only needed 128 Kbits/s for 1 hour, the time slots would go unused for the rest of the day. Newer multiplexers called *inverse multiplexers* or *I muxes* are available to allocate time slots together by groupings of channels (i.e., 128 Kbits/s, 256 Kbits/s, 384 Kbits/s, 512 Kbits/s, and 768 Kbits/s). However, some of these are not intelligent enough to pull noncontiguous channels together. For example, if a user wanted to pull together six channels and conduct a video conference at 384 Kbits/s, the I mux would look at channels 1 through 6. If one of these six channels was already in use (for a voice or data transfer), then the I mux would have two choices: either knock the existing connection down, or wait until the channel became free. This means that the I mux also has to reserve the other five so that another connection does not set up. The more expensive I muxes can allocate noncontiguous channels for this video call, but the price for this type service is prohibitive. Choices have to be made. The choices are less cost and less convenience, or more cost for more flexibility. Either way, this is not consistent with the way we run our everyday business. Broadband communications service was introduced as a breathable demand network concept. This allows an organization to use as much as required, when required. Upon release of 384 Kbits/s capacity, the networking equipment would reset the values and create a resource that was available. Users, therefore, could select and use the required throughput rate as needed. This "breathing" concept makes the service far more usable.

In ATM, a contrast exists with the synchronous transfer mode (STM). Synchronous transfer mode, which sounds as though it has something to do with synchronous optical network (SONET), actually describes the way the digital telephone networks have worked since the inception of time-division multiplexing. Most synchronous transfer mode signals run on the standard asynchronous DS1 and DS3 used in today's telephony networks.

ATM on the other hand is an outgrowth of BISDN standards that is intended to be carried on SONET. So, the asynchronous transfer mode is designed to be carried on synchronous facilities, whereas synchronous transfer mode deals with asynchronous traffic. Talk about confusion! However, STM is based on the fixed time slot that can carry asynchronous traffic, such as voice and video signals. The transmission is based on fixed time slots to prevent delays and latency on these time-sensitive communications. Whoever thought these terms up had to have a perverse sense of humor.

Thus the need for broader communications channel capacities emerged. *Broadband* is defined as a higher throughput rate of bursty traffic than what is traditionally available on the telephone companies' channel capacities. Instead of a 64 Kbits/s (DS0) channel, the broadband communications will handle multi-megabit transport starting at the very low end at 1.544 Mbits/s (this is not currently approved but is being worked on by the standards committees because the bulk of the business users are at the T1 and lower rate) progressing upwards in chunks or rates up to 155 Mbits/s today. Tomorrow's needs will deal with rates up to 622 Mbits/s and 1+ Gbits/s.

The characteristics of broadband communications will be for the transport of higher-speed communications of a continuous or bursty nature from a variety of inputs to a mix of outputs. All of these techniques and characteristics are based on the strict TDM/PCM and SONET. Before going any further, here's a quick review of both TDM and SONET so that we are all on the same wavelength.

Time-Division Multiplexing

Time-division multiplexing and pulse code modulation were designed as the first steps of a digital carrier and transport system. In 1958, the Bell system had created this concept to migrate from the older analog system (Chapter 13).

A review of the digital multiplexing function

First to create a TDM/PCM scheme, the techniques were evolved from the 4 kHz of bandwidth typically allocated to a telephone circuit. Remember that this 4 kHz has already been referred to earlier in this book. See Figure 24.2 for the graphical representation of the 4-kHz channel.

Using the 4-kHz channel, a sampling rate of the introduced signal will occur at twice the highest range of frequencies on the line. This means that a sampling rate of 8000 times per second is achieved as follows:

$$Sampling\ rate = Bandwidth \times 2$$

$$8000 = 4000 \times 2$$

8000 samples per second yields a standard 125-μs sample time

$$1\ second/8000\ samples = 0.000125\ second$$

Time-division multiplexing/pulse-code modulation

Once the sampling rate (8000 samples/second) is established, the value of the sample is then established. Under normal conventions, a signal is converted from the analog wave into a digital pulse stream. The pulses, therefore, must relate to some value along the curve of the wave. This is done by considering the coarseness of the analog wave. We use a set of values based on 256 combinations, which is derived by placing sensitivities and using binary coding based on an 8-bit sample. This is derived by:

$$2\ states\ in\ binary = 0/\ 1$$

$$2^8\text{-bit sample} = 256\ combinations$$

Using an analog wave with amplitudes peaking on both the positive and negative sides of the zero line results in 128 possible values on the positive side of the wave, and 128 values on the negative side of the wave. This is pulse amplitude modulation.

The final step is to digitally encode and transmit the signal in pulse code modulation. Using the typical sine wave if we select a value of one sample at a value of 5, the value of the sample of this wave is digitally encoded into a data stream of:

$$0000\ 0101$$

This is then transmitted in digital form. Refer to Figure 24.3 for the graphical representation of this consolidated onto a single drawing for ease

4 kHz Bandwidth

Figure 24.2 The 4-kHz channel is used to create a digital data stream.

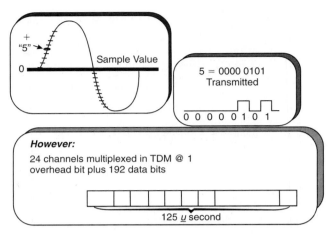

Figure 24.3 The PCM technique transmits a binary data stream.

of viewing. TDM/PCM using the 8-bit value and 8000 samples per second therefore operates at a channel speed of 64 Kbps/s. This is the basis for the digital carrier system.

Using the 4-wire circuit for transmitting the digital signal, the time-division multiplexer can take 24 inputs and multiplex them onto a single-carrier system called the *T1 carrier*. Therefore, a circuit carries 24 simultaneous channels multiplexed together with an extra overhead of 8 Kbits/s (1 bit per sample of 24 channels × 8000 samples/ second). The resultant yield is 1.544 Mbits/s. This was covered in much greater detail in Chapter 13, so refer back to this chapter if you need more detail on how this works.

$$[24 \times (8000 \times 8)] + [1 \times (8000)] = 1.544 \text{ Mbits/s}$$

$$(24 \times 64,000) + (8000) = 1.536 \text{ Mbits} + 8 \text{ Kbits} = 1.544 \text{ Mbits/s}$$

The Digital Hierarchy

Using these TDM/PCM values, a North American Digital Hierarchy was developed. Everything is based on the standard 8000 samples/second and a digital multiplexing scheme. The hierarchy used below set the rules for future levels of multiplexing. This North American digital hierarchy is shown in Table 24.1.

Each of the previous rates was established on the same rules. A frame of information is transmitted in 125 microseconds that carries an 8-bit sample from each of the channels within the frame. Thus, a DS1 frame consists of 193 bits (192 bits of data and 1 framing overhead bit), whereas a DS3 frame (called an M13 frame) consists of 4760 data and overhead bits. In a DS3

TABLE 24.1   A Summary of the North American Digital Hierarchy Values

Name	Capacities	Equiv.	Yield
DS 1	24 channels @64 Kbit/s	One T1	1.544 Mbit/s
DS 1C	48 channels @64 Kbit/s	Two T1s	3.152 Mbit/s
DS 2	96 channels @64 Kbit/s	Four T1s	6.312 Mbit/s
DS 3	672 channels @64 Kbit/s	Twenty-eight T1s	44.736 Mbit/s

multiplexing scheme, the timing is adjusted by adding (called justification) bits to make up the 125-microsecond time slots. The M13 frame is shown in Figure 24.4, showing the 4760 time slots. A problem exists with the M13 asynchronous protocol used to transmit this information. The frame is too small to multiplex 28 T1 frames into a single M13 frame. The thought behind this is that 28 T1 frames will require 5404-bit slots. The M13 only has 4760 (only 4704 of these are available because of the overhead added, as shown in Figure 24.4). Thus, the M13 frame must be transmitted 9366 or 9367 times per second. This makes it asynchronous because the arrival rate of the frames will be different. Further, the 125-microsecond clock will be lost. The bit sequencing within this frame is also very ambiguous. You cannot trace a channel through the M13 asynchronous protocol because the bits are moving from frame to frame into different bit slots. In order to perform any diagnostics or troubleshooting, you must demultiplex the entire DS3, a process that is inconvenient and disruptive.

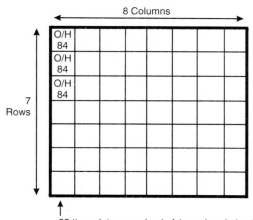

85 time slots comprised of 1 overhead plus 84 data bits yield 4760 per frame.

Figure 24.4 The DS3-M13 frame uses a 4660-bit framing sequence.

SONET

To rectify the problem with the M13 frame format, SONET was created. All of the digital transmission speeds were initially set to be carried on the existing wired systems in place at the telephone company and interexchange carrier levels. These systems included microwave, coaxial, and twisted pair copper wires. As newer fiberoptic systems were introduced, a new synchronous digital hierarchy (SDH) evolved. Throughout the world, SDH was accepted, yet in the U.S. the term *synchronous optical networks (SONET)* was adopted. Although the international and U.S. versions of SDH/SONET are very close, they are not identical.

SONET specifies a synchronous transport signal (level 1) at 51.84 Mbits/s, which is a DS3 with extra overhead. The overhead allows for diagnostic and maintenance capabilities on each Synchronous Transport Signal (STS). The optical equivalent of the STS-1 is an optical carrier level 1 (OC-1). This is the basic building block for chunks of bandwidth as they are multiplexed together to form much higher capacities. Increments of SONET include the capacities shown in Table 24.2. These transport systems and carrier levels work from the base of a T3 plus overhead creating an STS-1.

SONET goes further than just defining the multiplexed values of speed. It breaks the architecture of the link into three separate steps for purposes of defining the interfaces and defining responsibility. These layers include (a fourth is added here to describe the layer-by-layer responsibilities):

- *Photonic layer* Deals with the transport of bits and the conversion from an STS electrical pulse into an OC signal in light pulses.

- *Section layer* Deals with the transport of the STS-N frame across a physical link. The functions of the section layer include scrambling, framing, and error monitoring.

- *Line layer* Deals with the reliable transport path of overhead and payload information across a physical system.

- *Path layer* Deals with the transport of network services such as DS1 and DS3s between path terminating equipment.

SONET defines each of these into an architecture such that overhead and responsibilities are clearly defined. This layered architecture is shown in Figure 24.5. Note that the four layers are shown in this graphic. Actually, the photonics and the section layers are one, but have been subdivided for clarity in showing the structure. Photonic and section deal with the physical medium, or the OSI layer 1 protocols.

The SONET architecture actually consists of three pieces:

- The *section* is defined as the transport between two repeating functions, or between line-terminating equipment and a repeater.

- The *line* is defined as the transport of the payload between two pieces of line-terminating equipment.

- The *path* is defined as the transport of the payload between two path terminating equipment (multiplexers etc.) or the end-to-end circuit.

This is shown in Figure 24.6 to lay the pieces out on a plane. The graphic shows how the pieces all work together in a fiber link. The fiber specification is what was originally used in SONET (and SDH internationally), however newer microwave and satellite termination equipment is being introduced to provide the same capacities and formats. SONET defines the responsibility of the carriers in multiplexing the signal onto the physical medium, as well as the points of demarcation for the various carriers involved with the circuit.

SONET name	Speed
OC-1*	51.84 Mbit/s
OC-3*	155.52 Mbit/s
OC-9	466.56 Mbit/s
OC-12*	622.08 Mbits/
OC-24**	1.244 Gbit/s
OC-36	1.866 Mbit/s
OC-48**	2.488 Gbit/s
OC-96	4.976 Gbit/s
OC-192	9.953 Gbit/s
OC-255	13.92 Gbit/s

TABLE 24.2 Typical SONET Specified Speeds. The Asterisked Ones Are the Common Speeds for ATM, the Double Asterisked Are the Future Speeds for the Broadband ISDN Networks

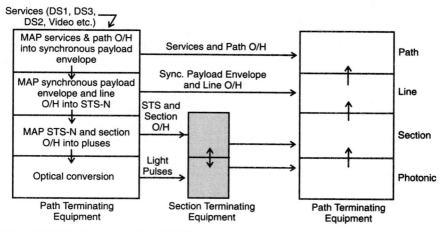

Figure 24.5 The four layers of the SONET protocols.

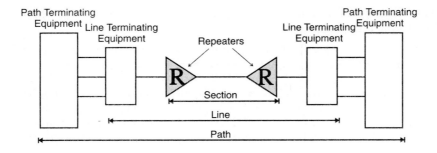

Section, line and path definitions

Figure 24.6 The SONET architecture is laid out to define responsibilities for the signal.

SONET frame format

In each of the layers listed, overhead is added to the data that is being transported to assure that everything will be kept in order, and that errors can be detected. This overhead allows for testing and diagnostics along the entire route between the functions specified in the SONET layers. A new frame size is also used in the SONET architecture. The older M13 asynchronous format just would not cut the mustard. In order to get the network back to a synchronous timing element, a frame is created that consists of 90 octets (8-bit byte) across and 9 rows down (810 bytes). This is the OC-1 frame, or the equivalent of the DS3 frame plus the extra overhead. However, with the change in size the frequency of frames generated across the network is now brought back to 8000 per second. The 125-second clocking and timing for the digital network can be reestablished. The frame is shown in Figure 24.7 that is 90 bytes (columns) wide and 9 rows high, yielding the 810-byte frame format.

Of the 90 columns, the first three columns ($3 \times 9 = 27$ bytes) are allocated for transport overhead. The transport overhead is divided into two pieces; 9 bytes (3 columns by 3 rows) for section overhead. The remaining 18 bytes (3 columns by 6 rows) are allocated for line overhead. This is for maintenance and diagnostics on the circuit.

783 remaining octets are called the *synchronous payload enveloper* or *SPE* (87 columns by 9 rows). From the SPE, an additional 9 bytes (1 column by 9 rows) are set aside for path overhead. The path overhead accommodates the maintenance and diagnostics at each end of the circuit; typically, the customer equipment. Thus, the leftover 774 bytes are reserved for the actual data transport.

$$(774 \times 8 \text{ bits} = 6192 \text{ bits} \times 8000 \text{ frames/second } 49.536 \text{ Mbits/s})$$

ATM uses the STS payload as the transport of information by inserting 53 cells into the STS payload. This allows for the fluid and dynamic allocation of the bandwidth available. As it is being advertised, ATM starts at 50 Mbits/s and 155 Mbits/s. To achieve the 155 Mbits/s rate, three OC-1s are concatenated (kept together as a single data stream) to produce an OC-3C. This works out rather nicely.

The Cell Concept

In the past, using the traditional time-division multiplexing arrangement, bandwidth was provided on a fixed time slot basis. If a user wanted a high data-rate transfer, this fixed time slot limitation got in the way. Using the standard 64-Kbits/s transmission speeds, you would be able to derive rates based on standards of 64 Kbits/s chunks. However, this worked for a 64 Kbits/s or 1.536 Mbits/s channel capacity only.

If a user wanted 384 Kbits/s (6 channels multiplexed together), the service was not ubiquitous. In many cases, the user had to derive either a 64 Kbits/s or use a full T1/E1 service. There just wasn't much in between. Secondly, even if these services were available, the price might have been too expensive. The network did not lend itself to user needs, so the user had to fit the application to the capacities offered by the network. Obviously, this impeded the ability to use bandwidth efficiently and cost the user in either money or throughput performance. Something had to be done!

Network suppliers began to look at the inefficient use of capacity. They found that they were consistently overbuilding their infrastructure of links or trunks only to lease it off at lower rates than they should have. These low

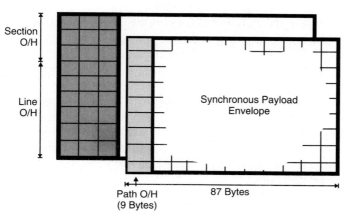

Figure 24.7 The SONET frame is 810 bytes long; it will be represented 8000 times/sec. Overhead is allowed for maintenance.

rates were justified to make the service more attractive to the end user. However, this overbuild is expensive because they were providing service that was only minimally used and they could not reclaim the bandwidth that sat idle for 8, 12, or 16 hours a day.

Looking at user demands and their own financial picture, a new concept emerged. Rather than forcing the user to adapt to the constraints of the network, why not let the network adapt to the needs of the user? Thus, ATM was born. ATM gets around the inefficiency of the fixed time slot, rates, and formats of the TDM world by allocating whatever is necessary to the user whenever the user wants it. To do this, the network suppliers looked to the packet-switching world and settled on a service that mimics the use of packet switching. Differences exist in this concept. Instead of using the processor-intensive slow speed services of X.25 packet switching (that tops out at 64 Kbits/s), or the speedier frame relay (at 2 Mbits/s), a mix of packet/frame technology evolved using a fixed-size cell that offers higher throughput due to efficient use of bandwidth without the overhead of X.25 and frame relay.

The Importance of Cells

These cells can get around the waste of frame relay or other frame concepts. In frame relay, the frames are larger, but variable in length. Therefore, if a frame size is set to accommodate LAN traffic, such as an Ethernet frame of 1500 bytes, the network deals with a frame of the same size. Consequently, the 1500-byte frame is used, even if the frame only has 150 bytes in it. A pad function (filler) is used to fill the frame for transmission. The network has been used for a much longer period of time to send an insignificant amount of data. The use of this frame was only 10% efficient. This was done through the fills and the buffering of data until the network switch was sure that all of the data was received before sending it on to the next location in the network. While the network was handling this partially empty frame, other users who wanted to transmit were delayed. All because the network had to recover timing, framing, and formatting on the fixed time slots.

On the other hand, a much smaller cell allows the transmitter to break down large blocks of information into much more manageable pieces of data. If the frame used (in this example), is only partially filled, the network need not be concerned. It will only be required to send the 150 bytes of information plus any associated overhead in a couple of cells. Thus, the network performs more efficiently.

Deriving bandwidth

As these fixed cells are being used, another benefit is achieved. A user who needed to transmit 8 Mbits/s was stuck. Either they would have to settle for a T1 (1.544 Mbits/s) transmission, which will ultimately cause congestion, or they would have to lease a T3 (44.736 Mbits/s), which would be too expensive and only a small portion would be used. Once again, the fixed time slot arrangement of the TDM world got in the way.

Using a cell concept, the user would transparently send interleaved cells across the network regardless of the amount of bandwidth needed, and consequently get the effective throughput of 8 Mbits/s. This is based on the availability of the fiber distribution to the user's door because it is the fiber (using SONET) that will deliver the capacity necessary to derive the bandwidth being discussed.

As cells are transmitted, they will be stacked up in the SONET frame (STS-n frame) to derive the necessary bandwidth that the user needs. Once the transmission of 8 Mbits/s is completed, a new user might only need 384 Kbits/s so the cells will be used to derive that amount, making the rest available to others.

Cell Sizes and Formats

The standard fixed cell size for ATM will be a 53-byte (OCTET) cell. This will be comprised of a 5-byte overhead for addressing etc., and 48 bytes of payload. The cell is shown in Figure 24.8. The 48-byte payload will be a variable, depending on the information and control necessary. There are several types of cells used that could take 4 bytes of the 48 away for control, leaving the user 44 bytes of effective data; or another option that gives the user all 48 bytes of payload for data. It depends on the implementation and the service run on ATM. Either way though, the cell is still fixed at 53 bytes long.

The cell is broken down as follows. Five bytes of overhead are shown in Figure 24.9.

GFC

The first 4 bits of byte number 1 contain what is known as a *general flow control identifier*. This will be used to control the flow of traffic across the user-to-network interface (UNI) out to the network. Presently the standards bodies are deciding how to best use the GFC, and are not explicit in their definition of its use. Remember that this is only used at the user inter-

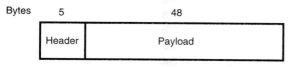

Figure 24.8 The ATM cell consists of 5 bytes of header and 48 bytes of payload.

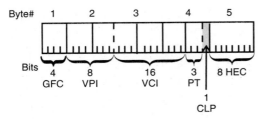

Figure 24.9 The overhead uses 5 bytes to keep addressing in order.

face because once the cell goes onto the network, the network-to-network interface (NNI); that is between network nodes; these extra 4 bits will be reassigned for network addressing.

VPI

The next 8 bits of the header are called the virtual path identifier. A *virtual path identifier (VPI)* is part of the network address. A virtual path is a grouping of channels between network nodes.

VCI

The *virtual channel identifier* is a pointer on which channel (virtual) the system is using on the path. The combination of the virtual path and virtual channel make up the data link running between two network nodes. The VCI is 16 bits long. The VCI uses the second 4 bits of byte 2, all 8 bits of byte 3, and first 4 bits of byte 4.

PT

Three bits in byte 4 are allocated to define the payload type which indicates the type of information contained in the cell. Because these cells will be used for transporting different types of information, the network equipment might have to handle it differently, several types of payload type indicators have been defined:

- 0 to 3 are for the data types inside the cell.
- 4 to 5 are for network maintenance and control.
- 6 to 7 are being defined.

CLP

Bit number 8 of byte number 4 is a cell-loss priority bit. The user can define whether or not to discard the cell if congestion occurs on the network. If congestion occurs and the bit it set to "1" by the user, the network can discard the cell. If it is set to "0," then the cell might not be discarded.

HEC

The 5th byte of the overhead is used as a header error control. This is an error-correcting byte that is conducted on the first 4 bytes of the header. It is used to correct single-bit errors and detect multiple errors in the header information. If a single-bit error occurs, the HEC will correct it. However, if multiple errors occur in the header, it will discard the cell so that cells will not be routed to the wrong address because of errors occurring on the network. The HEC only looks at the header information, it does not concern itself with the user data contained in the next 48 bytes.

The Cell Format for User Data

Once the header is completed, the user information is then inserted (Figure 24.10). As already mentioned, the user field is either 44 or 48 bytes of information, depending on the process used. Here's how this works:

The Adaptation Layer

Called the *ATM adaptation layer (AAL)*, this is probably the most significant part of ATM. The adaptation layer provides the flexibility of a single communications process to carry multiple types of traffic such as data, voice, video, and multimedia.

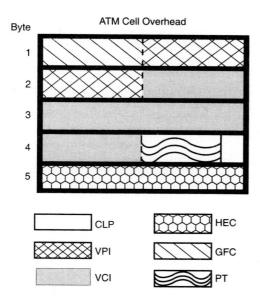

Figure 24.10 Once the header is computed, the payload is attached.

Each type of traffic has varying needs. Voice needs constant data traffic, LAN is bursty in nature, video is time sensitive, yet data transfers from host-to-host can be delayed without problem. It is in this adaptation process that the network can deal freely with varying types of information and only route cells based on the routing information in the header.

Just about any type of transmission will require more than a single cell of information (48 bytes), so the ATM adaptation layer divides the information into smaller segments that are capable of being inserted into cells for transport between two end nodes. Depending on the type of traffic, the adaptation layer functions in one of five ways. These are shown in Table 24.3, with various types of information and adaptation layers involved.

The adaptation layer process

The AAL is broken down into two sublayers; first is the convergence sublayer, the second is the *segmentation and reassembly sublayer (SAR)*. This SAR is similar to X.25 PADS. The purpose of the process is to break the data down into the 48-byte payloads, yet maintain data integrity and pointers for ID purposes. This process of two sublayers produces a protocol data unit (PDU). The convergence sublayer PDU is a variable length that is de-

TABLE 24.3 A Summary of the Five Different Adaptation Types

Type	Name	Description
1.	Constant bit rate (CBR) services	Allows ATM to handle voice services @ DS0, DS1, and DS3 levels. Recovers timing & clocking for voice services.
2.	Variable bit rate (VBR) time-sensitive services	Not finalized but reserved for data transmissions that are synchronized. Also will address packet-mode video, in a compressed mode using bursty data transmission.
3.	Connection-oriented VBR data transfer	Bursty data generated across the network between two users on prearranged connection. Large file transfers fall into this category.
4.	Connectionless VBR transfer	Transmission of data without prearranged connection. Suitable for LAN traffic that is bursty and short. Same reasoning as X.25 where dial-up connection and set-up is longer than data transfer.
5.	Simple and efficient adaptation layer (SEAL)	Improved type 3 for data transfer where higher level protocols can handle data and error recovery. Uses all 48 bytes as data transmission and handles message transfer as sequenced packets.

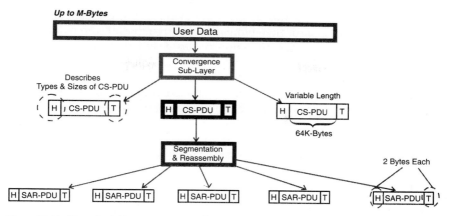

Figure 24.11 The adaptation process goes through two steps to produce the ATM payload.

termined by the AAL type and the length of the higher layer data passed to it. The SAR-PDU is always kept at 48 bytes to fill an ATM cell data stream. This is shown in Figure 24.11 as it proceeds through the process.

As the user data, that can be multi-megabyte files, is passed down to the convergence sublayer (CS) process, the data is broken down into variable block lengths. A maximum of 64 Kbytes is used in this process. The large user file is broken down into the 64-Kbyte segments. A header and trailer are added to the CS-PDU, that describes the type and size of the CS-PDU. This is then passed down to the next sublayer process.

The SAR then receives the CS-PDU and breaks it down into 44-byte cells (if less than 44, the rest is padded). To the SAR-PDU is added additional overhead 2 bytes of header and 2 bytes of trailer. The SEAL instead uses all 48 bytes for user information, and therefore uses the bandwidth more efficiently.

Lastly, the PDU is then inserted into an ATM cell with the 5-byte header. Following this, the cell is handed down to the physical cable system. These cells are then delivered across the physical media into SONET, or other transmission. Thus, the speeds used involve continuous or variable bit rates, depending on the application. Being carried in a SONET frame, the cells can be strung out to get the highest bandwidth possible. Unlike DS1 or DS3 services, ATM cells are slotted into the SONET frame horizontally, rather than vertically.

ATM Standards Protocols

The development of communications standards takes on a global perspective and coordination. There are many standards bodies involved with the often-lengthy task of generating a consensus view on implementing a broadband communications technology. Standards bodies are continually refining and extending the standards, adding more advanced features and capabilities. At each revisit of a standard, the cumulative experience with the older implementation is used to update and improve the standard. The older standard implementations remain useful; they are simply built upon. The standards process has always been a fine balance between technology and commercial interests. A recent phenomenon among standards groups is the emergence of the forum from various commercial groups. The purpose of the forum is to impact positively the standards process for a particular communications technology. The forum consists of a group of interested carriers and vendors supporting education and implementation activities.

CCITT

The International Telegraph and Telephone Consultative Committee Communications Standards Organization (CCITT) provides communications standards for the International Telecommunications Union (ITU), which is supported by the United Nations. The 15 study groups of the CCITT have specific technical responsibility for developing international standards. For example, ATM and SONET/SDH standards are being developed in study group XVIII. Every four years, the study groups submit recommendations to the CCITT plenary for approval. The CCITT standards are law in some European countries, but in most countries they are treated as recommendations. The clout of the CCITT in setting standards comes from coordination with national standards groups, such as ANSI, the thoroughness of its standards process and its worldwide stature.

CCITT has recommended that ATM be used in worldwide broadband networks. The first standards were produced in the mid-1980s and provided a basic outline of the service. ATM was chosen as the transfer mode for BISDN in 1988; and an initial set of recommendations were agreed upon in 1990. These recommendations specify ATM among the protocol suites at the lower layers of the OSI reference mode. Most ATM standards are already specified by the international CCITT committee. However, work continues in the area of signaling, call set-up definitions and network management functions.

The B-ISDN protocol reference model shown in Figure 24.12 is defined in CCITT recommendation I.121 into multiple planes:

- *U-plane* The user plane provides for the transfer of user application information. It contains physical layer, ATM layer, and multiple ATM adaptation layers required for different service users (e.g., continues bit rate and variable bit rate service).

- *C-plane* The control plane protocols deal with call establishment and release and other connection control functions necessary to provide switched service. The C-plane shares the physical and ATM layers with the U-plane. It also includes the ATM adaptation layer procedures and higher layer signaling protocols.

- *M-plane* The management plane provides management functions and the capability to exchange information between the U- and C-planes. The M-plane contains two sections; layer management that performs layer specific management functions, and the plane management that performs management and coordination functions related to the complete system.

Each of these protocols are contained in the *user network interface protocol architecture (UNI)*. The UNI specification involves the protocols that are either terminated or manipulated at the user-network interface. ATM bearer services, defined by both ANSI and CCITT, provide a sequence of preserving connection-oriented cell transfer service between the source and the destination, with an agreed-upon quality of service (QOS) and throughput. ATM bearer services involve two lower protocols, the ATM and physical, of the B-ISDN protocol stack. These two layers are service-independent and contain functions applicable to all upper layer protocols. The ATM bearer service at the public UNI is defined as the point-to-point, bi-directional virtual connections at either a virtual path (VP) level and/or a virtual channel level (VC). The UNI (private and public) are limited to the physical and ATM layers and higher-level protocols required for UNI management.

Physical layers can include DS3 and SONET for both public and private UNIs. Additional physical layers can be specified for the private UNI.

Rates

Supported U.S. access rates to a public wide-area network supporting cell relay service (ATM) include those shown in Table 24.4.

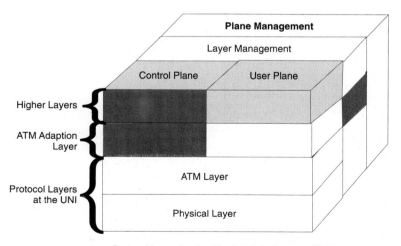

Protocol layers involved for initial deployment (PVC).
Extra layers necessary for switched service (SVC)

Figure 24.12 The B-ISDN protocol reference model (*CCITT*).

Supported rates
DS3
STS-1
STS-3c
STS-12c

TABLE 24.4 The Basic Rates Supported on ATM Under U.S. Standards

The physical access channel can be a T3/DS3 or SONET-based (STS-3c or STS-12c) facility. ATM cells must conform to the ITU-T recommendation (CCITT) I.361. Cells are transported to their destination with a level of assurance consistent with a pre-specified quality of service (QOS). The actual services provided vary, and service is delivered in a transfer rate of cells per second. Supported cells per second transfer rates in the U.S. range from 150 to 1,470,000 cells per second. The cell/channel control protocols are based on CCITT Q.93B, which is further defined by the ATM Forum.

ATM is a standard based on the overall B-ISDN reference model, as already stated. It is based on a layered architecture concept similar to that used by the International Standards Organization (ISO) seven-layer open systems interconnection (OSI) model. Basically, these models divide any communications process into sub-processes, called layers, arranged in a stack. Each layer provides services to the layer above to aid in communications between the top layer process or applications at the end of the connection.

The ATM relevant portions of the BISDN model comprise the bottom three layers of the BISDN protocol stack; the ATM adoption layer (AAL), divided into two sublayers; the ATM layer and the physical layer, also divided into two sublayers. There is not a one-to-one relationship between the BISDN and OSI stacks. However, the three layers (the ATM portion of the BISDN protocol stack) are roughly equivalent to the ISO layers one (physical) and two (data link), as shown in Figure 24.13. Be careful to keep these separate because they are not identical.

ANSI

The American National Standards Institute (ANSI) was created in 1918 to coordinate private sector standards development in the U.S. ANSI is the U.S. representative to international standards groups and includes 300 standards committees, as well as associated groups, such as the Exchange Carriers Standards Association (ECSA). The ANSI subcommittees active in broadband network standards are T1X1 and T1M1. T1X1 plays a role in SONET rates and format specification, and T1M1 guides the effort to define the standard for operations, administration and maintenance, and provisioning (OAM&P).

The physical media-dependent sublayer deals with aspects that are dependent on the transmission medium selected. The PMD sublayer specifies physical medium and transmission (e.g., timing, line coding) characteristics, and does not include framing and overhead information. Physical characteristics of the UNI at the UB TB and SB reference points are defined in ANSI T1E1 2/92-020, and shown in Figure 24.14. Other SONET physical medium specifications are defined later. This is a moving and emerging target as the standards bodies move toward the broadband specifications.

The transmission convergence sublayer specification deals with physical layer aspects that are independent of the transmission medium specifications. BISDN independent TC sublayer functions and procedures involved at the UNI are defined in ANSI T1.105-1991 and T1E1-2/92-020.

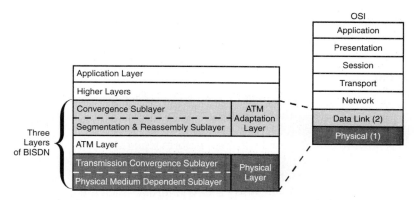

Figure 24.13 BISDN 3 layers closely resembles the bottom two layers of the OSI.

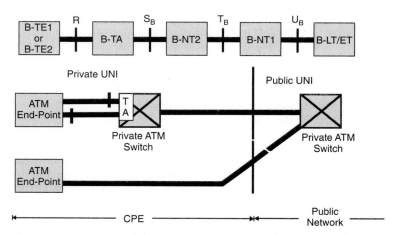

Figure 24.14 The physical access reference points of the UNI developed by ANSI and CCITT (*ANSI*).

ATM Forum

The ATM Forum is an international consortium of members that are chartered to accelerate the acceptance of ATM products and services in local, metropolitan, and wide area networks. Although it is not an official standards body, the forum works with the official ATM standards groups (ANSI and CCITT) to ensure interoperability among ATM systems. The consortium accelerates ATM adoption through the development of common implementation specifications. The forum's first specification on the ATM user network interface (UNI) provides an important platform upon which vendors can design and build equipment. It defines the interface between a router or a workstation and a private ATM switch, or between the private ATM switch and a public ATM switch.

The ATM Forum has created interest in private uses of ATM in LANs. LAN management standards such as SNMP are being promoted for ATM network management and miniature switched ATM networks as LANs.

Equipment

At this point in your path to understanding ATM, the logical point will be to consider how we'll access these services. Several products have been introduced, others announced for future delivery. The real gist of the connection process is how we can migrate from one platform to another without scrapping everything and starting over. Many of the early products were quite expensive and only attractive to the very large corporations.

Consequently, the 500+ members of the ATM forum which is comprised of carriers, equipment suppliers, consultants, and end users, have all been active in trying to specify the best approaches to allow ATM evolve in the workplace. They are discussing the ability to interconnect equipment, such as:

- Bridges
- CSU/DSU
- Gateways
- Multiplexers
- Routers
- Servers

Industry experts foresee a time when ATM technology will be used from the desktop, across a LAN, and through a WAN to serve an entire enterprise. Today, ATM vendors expect the ATM switching technology to be obsolete and overtake conventional switching techniques over the next 10 years. By 1997/1998, many of the Fortune 500 companies will likely be deploying ATM switching in the LAN/WAN arena. The real issue is where ATM will take off first; at the local LAN level, the LEC network, or the public IEC network. Surprisingly, the vendors of ATM services already have announced or have plans to announce products in the near future. At that point voice, data, video, and images will be transparently transported across both private and public networks.

Bridges

As we know them today, bridges link similar LANs together across a short (departmental) or a long (WAN) boundary. Overall bridges will likely continue to be used as an interconnection device for LANs. However, newer bridges will emerge that perform bridging and routing services into a high-speed hub or concentrator. The bridge will act as the filter or buffer device onto the higher-speed hubs.

As each of the high-speed networking technologies continues its evolution, the ability to link services together will become more crucial. The bridge will be required as more networks are broken into smaller pieces to improve throughput performance. Thus, as network segmentation continues, the bridge becomes one of the major interfaces to the hub or controller (Figure 24.15). The hub is an ATM format service, whereas the controller could be an intelligent switching concentrator at raw or native speeds. The

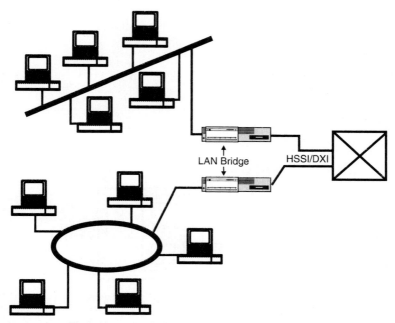

Figure 24.15 The bridge uses a high-speed interface into an ATM hub. The hub created the ATM interface.

bridge's significance will be in the form of providing a high-speed access capability into the hub. This will use a high-speed serial interface (HSSI) or digital exchange interface (DXI).

Routers

Routers are becoming more sophisticated every day. They can be used to interconnect LAN segments for high-speed access, and provide enhancements to network security. Firewall routers allow for the screening, access, and denial of data transfer between segments.

Upgraded port speeds, increased numbers of LAN and WAN ports, and software will all add to the router's capabilities. The routers will also be tied to the high-speed hubs and bridges to access high-speed ATM services. However, the router user will most likely be looking for interim steps to access ATM, as shown in Figure 24.16. Not every user can access ATM via T3 services because of the cost for this service. Therefore routers will be available to connect to ATM via a simple T1 interface. Although this can only be viewed as a near-term solution, it is the first step toward accessing ATM at a reasonable price. Router hardware and software changes will allow this device to create the cells at the local level to attach and forward the cells to a switching system. As a connectionless device, the router will deliver data packets to the ATM device, which will forward them to a cell processor.

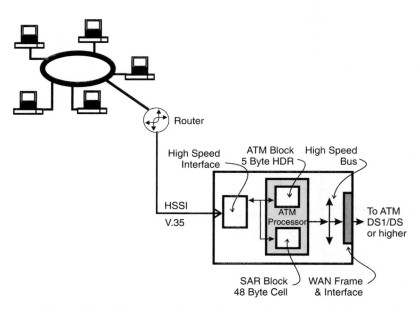

Figure 24.16 Initially, the router will use a high-speed interface to high-speed ATM hubs. Later, the router will become fully ATM-compatible.

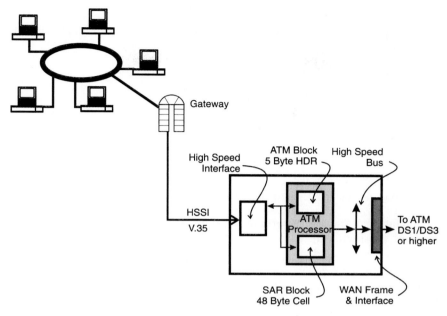

Figure 24.17 Even gateways will interface to a high-speed ATM hub, providing protocol independent service.

Gateways

These devices are designed to interconnect a potpourri of dissimilar connections or networks. Working up the OSI model, the gateway performs the full seven-layer services. As a connection to the network, the gateway will use ATM as the interface at layers 1 and 2 of the OSI model.

Because ATM will be protocol independent, the gateway will provide high-speed connections to an ATM hub or switch (Figure 24.17). This will require upgrades of hardware and software, but the impact will be somewhat minimal. Such traffic services as TCP/IP, IPX/SPX, LU6.2 APPN/APPC, and SNA tunneling will still be services provided via a gateway. High-speed inter-connectivity across these services at 45, 100, and 155 Mbits/s will be achieved at the hub or switch. But the gateway will function at native speeds of 4, 10, or 16 Mbits/s today; or a subset at 64 Kbits/s–1.544 Mbits/s transport. Several LAN and WAN ports will be available through the gateway offering aggregates of bandwidth at much higher speeds.

Servers

Servers in this regard will be the access capabilities to the services offered on a network via ATM transport. The public-switched and the private network arena will have a series of server functions (Figure 24.18) to provide the end-user functionality and features. These high-speed multi-tasking servers will offer:

- Remote video and cable services
- Plain old telephone services (POTS)
- Pretty amazing new stuff (PANS)
- Multimedia
- Video services
- Remote database access
- Dense graphics and CAD transfer and access

The fabric of the inner working will require dual-bus parallel/serial interfaces at the high end. High end would include the 100-/155-/622-Mbits/s throughput range and lend itself to a series of channel attachments to dual fiberoptic links for server access. These dual fiber links will provide a level of fault tolerance and an aggregate throughput in the Gbits/s spectrum. The end user needs only to request the high-speed service and the server does the rest.

Switches

ATM switches will come in a variety of flavors, speeds, and access ports. At the heart of the switch will be a dual-bus architecture with throughput starting at 622 Mbits/s on up to 2.4 Gbits/s and more (Figure 24.19). Several inputs (starting at 4 and currently up to 16 in the user environment) will

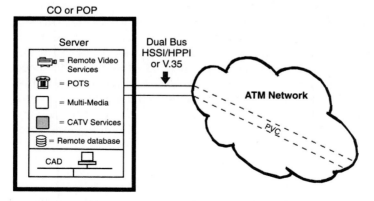

Figure 24.18 Servers will be accessible for a variety of applications and multitasking.

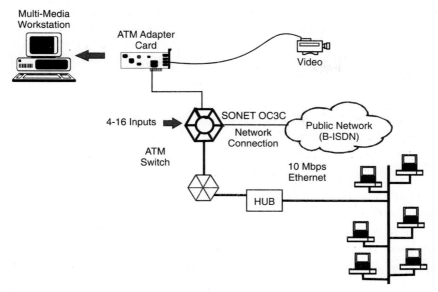

Figure 24.19 ATM switches at the user side will have 4 to 16 ports and support various data streams.

match across a matrix of outputs (ranging from 4, 10, 16 Mbits/s and more). The ATM switch will process the AAL 3-, 4- and 5-level services in the enterprise network.

The local switch will handle the switching and routing of cells in a building or department. This local switch will connect to a network backbone switch via an OC-3 (155 Mbits/s) or less service to start. Faster, higher throughput will become available through upgrades at the backplane level and the number and speed of ports attached. This requires that the switch is scaleable and migratable to the Gbits/s ranges. Traditional switches support 4–90 ports at speeds of 45 Mbits/s to 155 Mbits/s. Parallel processing of an ATM switch will allow concurrent switching among many parallel paths, providing each port full access to the allocated bandwidth. Cell switching breaks up data streams into very small units that are independently routed through the switch. The routing occurs mostly in hardware through the switching fabric. The combination of cell switching and scaleable switching fabrics are key ingredients of ATM.

Public switches

The public network ATM switch is a larger, more intelligent version of the customer premises switch. A public switch is capable of handling hundreds of thousands of cells per second and has thousands of switch ports, each operating at rates of up to 622 Mbits/s. All cell-processing functions are performed by the input controllers, the switch fabric, and the output controllers. In an ATM switch, cell arrivals are not scheduled. The control processor resolves contention when it occurs, as well as call set-up and tear-down, bandwidth reservation, maintenance, and management. The input controllers are synchronized so that cells arrive at the switch fabric with their headers aligned.

The resulting traffic is said to be *slotted*, and the time to transmit a cell across the switch fabric is called a *time slot*. All VCIs are translated in the input controllers. Each incoming VCI is funneled into the proper output port, as defined in a routing table. At the output controllers, the cells are formatted in the proper transmission format. For example, a broadband ISDN output controller provides an interface that consists of a line terminator to handle the physical transmission and exchange terminator for cell processing.

Multiplexers

The slowest defined rate for ATM today, 45 Mbits/s, is faster than most routers, bridges, and video codecs operate. This leaves the network manager with the choice of dedicating a 45-Mbits/s circuit to each lower-speed device, possibly wasting bandwidth, or multiplexing several lower-speed signals onto an ATM circuit. An ATM multiplexer might be considered similar to a switch, but it does more than just provide connectivity to multiple users. It handles isochronous data, such as voice and video. The multiplexer accepts data from routers or bridges through a standard interface (V.35 or HSSI), segments the data into cells, addresses each cell, and maps the cells into a WAN framing structure (T3/E3 or SONET/SDH).

CSU/DSU

The purpose of the CSU/DSU is to encapsulate information into the proper framing before it enters the WAN. The CSU/DSU converts between one communications technology and another by providing the interface between the on-premises HUB, switch or router formats, and the broadband network. The CSU/DSU regenerates the signals received from the network as well as the sent data. Thus it can also serve as a way to troubleshoot the transmission line. The CSU/DSU automatically monitors the signal to detect violation and signal loss. When problems are detected, the CSU/DSU allows remote network testing from the central office, including loopback testing of the transmission line.

ATM in the LAN Environment

LAN bandwidth continues to increase. Although fiber distributed data interface (FDDI) pushed the bandwidth of the LAN to 100 Mbits/s, it really just delayed the bandwidth congestion problems that are inherent in a shared LAN. The capacity of shared-media networks is only as large as the speed of the common line (or bus). Bandwidth cannot be added simply when needed. If added capacity is needed, we either have to segment the network into smaller sub-networks or a change of the network interface cards is necessary. As added speed is needed, an upgrade from 4 to 16 Mbits/s Token Ring cards can be used, or as an alternative 100 Mbits/s FDDI/CDDI cards are added. This can be both expensive and difficult to manage. In the Ethernet world, the expansion from 10 Mbits/s is not as simple because the 100 Mbits/s fast Ethernet standard is still to be resolved and completed. Now, emerging standards and products for the 100 Mbits/s VGAnyLAN are appearing, but the jury is still out on the acceptance and transportability of these products.

ATM-based LANs allow users to connect to the network at their required speed, adding bandwidth as needed. ATM removes the limitations of shared LAN backbones by collapsing them into the ATM switching fabric. The typical extended ATM LAN will employ an ATM switch to link slower LANs (Figure 24.20). To interconnect a series of LANs, multiple switches can be concatenated together.

The switch will be more powerful than today's intelligent LAN hubs in terms of processing power, routing capability, and network management. Also, the ATM switch is distinguished from today's hubs in that its total bandwidth is the sum of the bandwidth from its input ports, rather than a fixed bandwidth. Thus, ATM switches allow information to travel from any port to any other port without the risk of blockage. In contrast, traffic on a bus-based LAN or hub can be blocked, if there are too many sources on the bus.

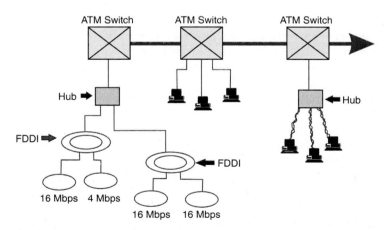

Figure 24.20 ATM switches will provide connections to multiple LANs in the office.

Evaluating the Need for ATM

As any system goes, the need for ATM will be based on the traffic loads and future demand for bandwidth. In general, if congestion is occurring on a LAN today, the obvious solution is either to segment the LAN into smaller pieces, or add the necessary capacity to support the traffic.

As new applications emerge the drain on available transport will be greater. Particular attention must be exercised in the areas of heavy traffic loads. The applications will include:

- CAD
- Magnetic resonance imaging (MRI)/electronic media imaging (EMI)
- Multimedia
- Voice on LANs (future)
- Video on LANs (future)

As these applications begin to emerge, the need for ATM might well become evident more quickly. The use of the ATM switch will relieve the congestion problem, but at a cost. What is your motivation to grow the network? If service and function are important, then ATM can serve the demand.

What to Expect at the Desktop

As ATM adapter cards become more available, so to will the higher-speed services move to the desktop. The intelligence of the devices at the customer's location has increased dramatically. As the intelligence of the desktop computer increased, so has the reliance on them. In just a couple of years, we've migrated from the 286, to 396, and 486 processors. Now the P5 Pentium has emerged and brought higher processing capacity and bandwidth demands to the desktop. The future of the P6, a 64-bit processor, will add more demand and complexity at the desktop again. Traditional LAN and WAN technology and products have not kept pace with the power required of these machines.

It is not unrealistic to expect 100 to 155 Mbits/s at the desktop in the near future (Figure 24.21). Applications and computers will continue to be bandwidth intensive and capacity hungry. In the next generation devices, a 622-Mbits/s to 1.2-Gbits/s adapter will likely become available. Not every desktop will need this degree of speed, but ATM is flexible, allowing a mix of services to be driven on the LAN/ATM platform. Therefore, traditional low-end users might stay with 4 to 10 or 16 Mbits/s, intermediate users will be served with 100 to 155 Mbits/s, and the heaviest of users will require the higher range of services at 622 Mbits/s up to the Gbits/s range.

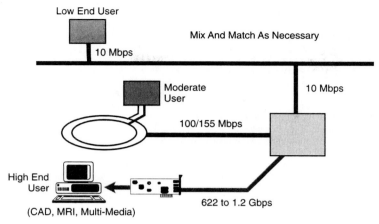

Figure 24.21 ATM switches will support low-end to high-end applications to the desktop.

LAN-to-LAN via ATM

If all goes according to plan from vendors, carriers, standards bodies, and users alike, then ATM should have plug-and-play capability. The use of shared-wire, single-bus technology with all stations tapped in might become a thing of the past. As the cable originally weaved from office to office, distances became the bottleneck. Therefore, when the cable had to be made longer, two pieces were joined together with repeaters. However, a single break in the cable could bring down the entire LAN. The idea of a hub started with the application of a star wiring topology for Ethernet LANs. Now hubs are becoming the central switching point for the entire enterprise. In the near future, hubs will embrace internetworking switching and multimedia.

The next generation ATM-based hub will provide an integrated framework (a star-shaped backbone with a high-capacity switch) that provides support for multimedia applications (Figure 24.22). This design eases the task of routing video streams, as well as simplifying network management. As the capacity of a single switch is exceeded, switches would be added to the backbone, the high-speed intelligent hub will provide a structure to interlink networks.

LAN-to-MAN via ATM

As the services of ATM roll out into the telco arena, the use of LAN connectivity within a metropolitan area network (MAN) will become more accessible. Many of the telcos have announced some implementation plans like Bell

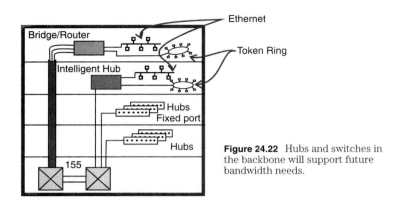

Figure 24.22 Hubs and switches in the backbone will support future bandwidth needs.

South, who linked a supercomputer in North Carolina to medical imaging equipment in other parts of the state. NYNEX and US West have trials underway also. Currently, GTE is building SPANet to provide real-time, gigabit-rate fusion of voice, video, and image traffic.

Internationally, several European network operators (PTTs) are testing ATM across their boundaries. British Telecom International, Germany's Bundespost Telekom, France Telecom, Italy's ASST/STET, and Spain's Compania Telefonica Nacional de Espana will implement local and cross-border ATM links. Their networks will conform to CCITT recommendations, ETSI and Eurescom specifications. These carriers will be interconnected through plesiochronous 34-Mbits/s links. Regardless of the carrier approach, the need to interconnect across a MAN from multiple LANs is a driving force. This will include a plethora of devices to provide the service.

Local Exchange Carriers

Just about all of the local exchange carriers have begun discussing the use of ATM in their backbone networks. Their plan is to migrate the existing technologies they have now into the broader bandwidth transport services. However, this will likely be a slow migration path, unless they see the use of ATM as a quick ramp-up at a reasonable cost.

Further, Bell South and other RBOCs have already announced plans to support ATM. As a means of economizing, their approach currently calls for a limited deployment with the possibility of a single major ATM switch per LATA. This will have some consequences as users try to draw bandwidth on demand. Equally important, on a LATA-wide basis, all traffic would then have to be backhauled from various points within the LATA.

Currently, the LECs have been jumping into just about everything hoping to find the one standard bandwidth application that will carry them into the future. Their hopes are to limit the constant changes of the past. In other

LEC arenas, the use of ATM will overlay the existing time-division multiplexing infrastructure with the single or a few switches within a LATA. Early ATM users who attempt to deploy their applications over ATM, or who choose to carry all traffic (i.e., voice, data, video, LAN, and multimedia) on a single platform, might well be frustrated and seek out other opportunities.

It would probably be safe to say that many of the early adopters will be looking to pilot the service before jumping headfirst into something new. Their requirements will probably be low, something in the T1 range. In order to provide such a limited channel capacity over a SONET or SDH backbone, with all the bandwidth that will be available, the LECs might see a lot of capacity sitting idle. This, of course, won't support a favorable return on investment; therefore, it will likely hinder the acceptance at the LEC level. Secondly, if everything is backhauled to a single ATM switch, how the LECs address remote and dispersed sites for a single organization will also have a "make or break" approach to contend with.

Because ATM is really based on the network being shaped to fit the customer's needs rather than the age-old philosophy of the customer fitting the network needs, we will likely see the LECs offer their early services as connection-oriented service for virtual private data networks (VPDN) with the intent of placing fiber to the door and displacing copper T1 circuits. Literally, they will have to start moving the capacity closer to the customer's premises. A combination of services will have to emerge to draw the customer's data out to these virtual private data networks. Much the same way that the LECs offered VPN/SDN voice networks, the application and bursty data will have to be fully and transparently supported. The LECs might have to initially offer a meshed private line network to get the customer to migrate out of the older leased line services. Such a network might be expensive at first, but the longer range strategy is what's at stake here. As the LECs gain a larger customer base, they can begin to move ATM switches closer to the customer and offer the high-speed switching at a more reasonable price.

As this scenario moves out into the LEC environment, there will be a separate data networking service and a separate voice switching service. For several years to come, there will be no driving force or incentive to merge voice and data into a single homogenous network. The current installed base of TDM multiplexers, digital cross-connect systems (DCS), and switches cannot support the high-speed data networking needs, yet that base is far too expensive to scrap and start all over again. Many of the TDM services are new (particularly those supporting the ISDN arena), so a period of 10 to 15 years might still exist in the depreciation cycle for the equipment.

Is this an epitaph for high-speed voice services? Not really, just a statement of fact and a reality check. The LECs have struggled long and hard to bring ISDN into their networks, waited for standards to be set, and fought for user acceptance. Unfortunately, it is too little, too late because our data needs have far outstripped the services to be offered on ISDN. The LECs have been burned once, they will be very careful to not let this reoccur. From the LEC standpoint, things can be complicated. What might look like a private network is in actuality a shared resource. The LEC must be very cautious of congestion and traffic management. When things first start out, usage will be limited and will appear as though no problems exist.

ATM represents a major leap for the LEC, but user's needs might not jump as much. Users are running at the capacity of existing T1, but the existing network doesn't offer the scaleability or granularity needed. Unfortunately, as the user goes beyond the T1, the only logical next step is the T3. This is overkill because the demands aren't that high for a full 44.736 Mbits/s, but there's nothing else. Pricewise, this is a strain on the user. Therefore, the LEC will have an initial line of defense in the use of ATM to deliver the bandwidth as needed, without breaking the bank. Hubs might well be used to concentrate all of the aggregated services from the customer to the central office (Figure 24.23). These will help to quickly provision the necessary data stream for customer locations. This all implies that fiber be readily deployed to the local loop (or curb).

IEC

Just as the LECs are going to be cautious in the provisioning of ATM, we can equally expect the IECs to be equally careful. Major vendors (three to five) will likely begin offering ATM in their backbones in major metropolitan areas only. Once again, this means that the IECs will be backhauling the service from the central office to points of presence (POP). WILTEL (now LDDS/WorldCom) was the first of the IECs to announce plans to support ATM in

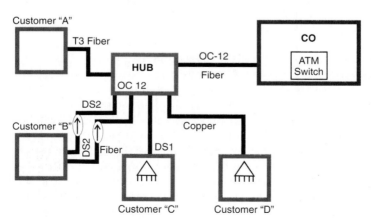

Figure 24.23 Aggregating services in fiber to the curb will help to bring ATM quicker.

1993, but further announcements have been sparse. It would be expected that Sprint will also come quickly with ATM services because they were one of the original founders of the ATM forum. MCI (now partnering with BT North America) has announced plans to support ATM, but has not yet given firm timetables. Incidentally, MCI was also strongly leaning toward a major nationwide deployment of SMDS, but indicated that it might well back off from that position. Lastly, AT&T has been developing its own services under the name of Software Defined Broadband Networks (SDBN), which is geared toward very specific service offerings of voice only, data only, or multimedia applications. It appears that the timetable within the IEC arena will be in the 1995/1996 range to roll-out services, but reality says that 1997/1998 is when we can expect to see availability of switched services instead of the private network implementations that are currently in place.

Each of the major players has been quickly trying to deploy the range of services that they feel the customer is looking for. Internetworking has been an issue with many of the past services, because different technologies manufactured by different vendors have been used. In many of these situations, it was the proprietary handling of the information that restricted ubiquitous access across carrier platforms. We should hope that this won't occur with ATM.

Each of the IECs has a software or virtual defined network (SDN/VPN) for voice. Additionally, the same deployment of ISDN has been used for the IEC as for the LEC. Consequently separate networks might still exist for the short term (1995 to 2000). As each of the carriers is waiting for standards to be developed that address critical issues (billing, internetworking, operations, and maintenance, etc.) the hesitancy to jump too soon still reigns true. For now, we can expect to see service offerings that will allow customers to migrate from SMDS, ISDN, FDDI, and frame relay to the preferred services, such as ATM, and ultimately, the B-ISDN world.

CAPs and CABLE TV Suppliers

With all the competition at the local loop, and the caution at the LEC level, a new set of players might emerge to bring ATM to the forefront of their networks. The two major competitors are the alternate access suppliers or competitive access providers (CAPs) as they are now known, and CATV suppliers. These providers have an edge on the LECs because they are not stuck with the standard copper cable plant. As a matter of fact, their fiber to the major users in metropolitan areas could quickly become the winning force for them. The CAPs are co-locating in the central offices to deliver high-speed services at T1/T3 rates today. They could carry the voice dial tone needs, and they are in a better position to upgrade their equipment and roll-out services quickly.

These providers can selectively choose who they will deliver service to, and have far less equipment to update. Their investments are relatively new, so they could be easily migratable. The fiber is multi-mode or single-mode, depending on the provider, so the installed base is readily migratable. Interestingly, they have been extremely quiet!

CATV Companies

The battle over bandwidth is never over until the last round is fired. CATV companies have also been quiet about the ATM world. Yet, most of the major providers are in the process of upgrading their coaxial-based networks with fiber to the door. Could it be that these suppliers have a larger vision than just delivering TV signals across the fiber? You might think that this is not a good application for delivering bandwidth. But looking at the cost to upgrade the cable and the limitations on customer services, this could prove to be a logical extension of the metropolitan area network (MAN) and access to the WAN arena. Many of the CATV suppliers have been acquiring ATM switching equipment and voice switches for their networks. Do they know something we do not?

After all, the CATV companies go to every door. ATM to the door is what this is all about. Further, ATM will support voice, data, video, and multimedia, so the CATV suppliers might well be in a position to transport door-to-door applications. Then, if needed, they can access the long-distance WAN arena through a simple connection to the IECs.

Application Needs Driving

ATM is designed to meet the requirements of next generation networks on a more global scale for LAN and WAN; for data, voice, video, and image for both public and private networks, and for various countries worldwide.

As enterprise networks continue to grow in complexity and diversity, newer demands are being put on the existing network infrastructure. Crucial issues that must be addressed include such areas as network management and ease of use. Users continually seek out higher-speed communications, cost efficiency, the ability to truly scale bandwidth as needed, the capacity to aggregate services in a super or subrated format, and independent protocols on the network. Further, as needs change, users desire the ability to reconfigure the network on the fly.

Consequently, the TDM schemes of the past do not meet these needs of the future bandwidth-hungry applications that are emerging. Because voice and video networks grew apart from data applications, something had to be done. The industry sought out and found the transport system that should support such applications as:

- Enterprise networking on a wide area or global area. As new techniques are developed, the applications can be migrated onto the WAN or GAN (global area network). Local area and campus area networks can be quickly added and tied to the wide area network. Dynamics will foster, rather than inhibit, the interconnectivity of corporate computing and human resources. It is possible that by the turn of the century such mundane terms as *local*, *metropolitan*, and *wide area networks* will be history. Instead, the network manager will manage and control all resources on a single domain; that being the enterprise. Regardless of where a resource exists, the addressing scheme will likely be available to route requests and subsequently receive responses from and to the desktop of any authorized user.

- Virtual/dynamic workgroups, regardless of their location geographically (whether next door or around the globe), will have instantaneous access to the corporate data and to each other. Simple forms of electronic mail, irrespective of complex addressing schemes, will be defined as entities encased within the network. As more organizations continue to downsize, skill sets in various locations can easily be assembled and linked together for the culmination of organizational projects and goals. Long response times associated with the traditional networking techniques will become passé. The speeds to be offered can allow for response times to improve with latency across the network valued at less than 80 m/sec.

- Through this dynamically assembled workgroup, the physical boundaries of distance become obsolete. The users are virtually connected as though they were in the same room.

- Joint development on various R+D or engineering project work through the ability to conference engineers from various parts of the country, or the world. As complex documents, such as those required for design and engineering, are used, the work force can physically be anywhere. Simple text manuscripts or complex documents, such as CAD files, video clips, voice notation, and database or spreadsheets, can all be amassed together as a single document. These can easily be ported from site to site across the ATM network for revisions, comments, or new input as an original; then ported right back into their original host system. Thus, duplication of effort can be reduced, if not totally eliminated. Through this ability, these enormous files ranging in the 200-plus Mbyte sizes can easily be accommodated by the network.

- This is also the beginning stages of multimedia application deployment. Today's transport systems make these files unmanageable because they take too long to transport and cause severe congestion on the network.

- Distributed computing, using the future client/server architecture will surely cause network congestion and stress the current levels of bandwidth availability. As networks (local, metropolitan, and wide area) continue to grow because of the proliferation of servers and super-servers, the need for added bandwidth grows almost exponentially.

- Databases that are truly distributed will require large throughput capacities because the actual program might reside in one site, while a shell interface resides at another. To update these distributed databases, huge amounts of data will be competing for bandwidth.

- Further, as massive file transfers are required for journaling or vaulting electronically to a remote site, the network will be called upon to deliver more capacity. Using ATM in any of these scenarios, the network could be very flexible. When 155 Mbits/s is not sufficient, users can grow to 622 Mbits/s by increasing ports and speeds without any undue hardship or impact to the rest of the network.

- Multiwindow desktop conferencing via video or computer conferencing can be supported on ATM. Cost-efficient services with reliable transport will be the key operatives here. Teams of business administrators or senior staff can review plans and strategies for new product introduction or development, while a team of researchers can be reviewing current results of testing and diagnostics through electron-microscopes, or a group of physicians can be conferring on the real-time output of a medical resonance imaging system where the patient is in the MRI lab across town, and the physicians are in various departments or buildings around town.

- Files can be created that include video conferencing on one screen (or window), an MRI process on the other, the patient's historical files on a third, and other diagnostics (i.e., EKG, EEG) machines can be on the next. These pulled together in a window environment can give a total picture of the patient's current and past medical condition. Better-quality health care can result without disrupting a physician's schedule via electronic viewing.

- Tele-learning at colleges, universities, and even K-12 locations can support multiple sessions of learning material. These of course will be handled through multimedia applications. Many schools are currently downsizing on staff, and resultant decreases in course offering have occurred. Rather than transporting staff from school to school, which is wasteful and time-consuming, or transporting students to remote schools that bear insurance and safety penalties, the schools can be linked together for real-time compressed video on an interactive basis. Many of the past video training experiments have been reviewed as too limiting because students really are in a delay mode and one-way mode from remote campuses. Many of the limitations were financially restricted because dedicated lines of sufficient capacity were too expensive and could not be justified for limited (one to two hours per day) scheduling and use.

With the ability to connect as much bandwidth as necessary on-demand, these limitations should soon be past history.

- Full-service networking—the reality where voice, data, video, text, and imaging can all be truly integrated onto a single network platform. This totally integrated solution requiring hundreds of Mbits/s to the Gbits/s range will have to be accessible and price-sensitive. Thus, the network will finally be used to match the customer needs, rather than the other way around.

Transparency across the globe will exist from every desktop to any other desktop or service, at any location and at any time. The network will have delivered the real image of communication; that being the transparent, native (not compressed or throttled) throughput across the network at native speeds (4, 10, 16, 100 Mbits/s) and linking diverse speeds to the single-network platform that is capable of handling multiple inputs to single threads at on-the-fly speeds.

Timing and availability

Fundamental changes are taking place in the telecommunications, LAN, MAN, and WAN arenas as they relate to technology and business applications. The telco and PTTs are installing sophisticated broadband switching systems and services for their customers. Meanwhile, the enterprise network is reeling under the impact and demands of high bandwidth telecommunications terminals that will become as much a part of the desktop over the next decade as the PC did during the 1980s. Only broadband networks provide a viable solution. Now is the time to devise ways to gain competitive advantage through the logical and methodical application of their telecommunications network.

Accommodating the bandwidth needs of the organization, which are constantly increasing, is crucial to taking full advantage of the communications network services.

The following outlines some of the steps toward deciding on what to use and when. These are not all inclusive, but should help to set the pace for the process.

Assess the Situation

The infrastructure decisions for both public and private networks are tied to the bandwidth needs of everyday business applications. The simple applications used at the desktop in the past included:

- Word processing
- Small databases

- Simple spreadsheets
- Simple CAD drawings
- Line art graphics

However, these applications have migrated and evolved into far more complex systems, with object linking and embedding graphics into them. The requirements at the desktop are rapidly increasing and demanding higher throughput. Where a simple network using a single shared cable satisfied the original needs, today these don't fit. Managers are, therefore, faced with the task of calculating expected demands and sub-networking in order to finesse the network by transitional approaches. Then, the following step is to take a quantum leap. For example, FDDI and frame relay are making serious progress but ATM is also developing rapidly. Should an investment be made in FDDI or frame relay when ATM offers to deliver true voice and data integration? The progression of ATM from LAN switching, followed closely by desktop applications then finally in the local and long-distance network will change the equation.

Match the network to the needs of the business

Managers have to keep a critical eye on costs because this is always a requirement, but they should not lose sight of preparing for the future. Today, telecommunications is a critical link in the strategy of business development and growth. Cutting costs at the expense of performance or access could be catastrophic. The promise of broadband applications has prompted a re-evaluation of network design away from the traditional mind-set. This has to begin with a complete analysis of needs and an evaluation of vendor abilities to deliver products. The best place to start with a pilot is in a controlled environment at the LAN level. Experiment with products that offer high-speed transport that will meet business needs.

As products become available for the construction of an ATM backbone, organizations will install ATM-based LANs to gain the economics and efficiencies of increased backbone bandwidth. High-speed routers and bridges will be used to interconnect these LANs. These organizations will then take the next step to second generation routers, multiprocessor RISC-based machine, supporting very high-speed routing. These will be connected to an ATM infrastructure, providing the path to migrate to the desktop.

Link communications into a single platform

One possibility is to consider the use of a hybrid arrangement where current data communications can coexist with ATM. This hybrid will protect the investments already installed, allowing continued use of a Token Ring,

Ethernet, and FDDI LANs, while adapters are being developed to add ATM into hubs, MAUs, or workstations.

Consider what the LECs and IECs will offer

When evaluating an enterprise broadband network, compare it to alternative networks to determine what broadband applications can and cannot deliver. The real challenge is to incorporate these applications into powerful, cost-effective on-demand solutions.

Unconnected pockets of ATM might already exist

Most major manufacturers have some products announced for ATM. ATM-based LAN switches are already available, some installations already exist. For the first time ever, a logical way to integrate local and wide-area solutions exists. ATM defines an open, interoperable interface between networks, allowing users to take advantage of a high-performance set of features and services. Initially, ATM will advance in large organizations with bandwidth-hungry application in a LAN or CAN. Campuses (college, hospital, business) will spur growth in the interconnection market.

Interconnect pockets of ATM

In two years, a much wider variety of ATM products will be available that will interoperate cleanly and transparently. Some of these products will be available from public carriers. Frame relay and SMDS will be fully supported over ATM; SONET will be more widely installed and evolved. The LAN-to-MAN and LAN-to-WAN barriers will disappear as a single transparent protocol—independent network emerges.

Ubiquity follows

Expansion of fiberoptic installations and improvements in the reliability of high-speed transmission over existing copper facilities will lead to the eventual widespread installation of ATM. Already telcos and PTTs are deploying ATM in their networks. This is being simplified by the cooperative development of standards in the U.S., Europe, and Asia. These carriers will offer transport services that deliver multiple channels of high-speed data, voice, video, image, multimedia, entertainment, etc., to the general public. Simple investments can be made today that will be protected and will evolve with these deployments.

25

Cellular and Personal Communications Systems

Radio Transmission

Cellular transmission is the latest and greatest front in the industry. In 1946, Illinois Bell Telephone Company introduced a mobile phone service that would allow users driving in and around the Chicago area to communicate directly to and from vehicles. Using a radio-to-telephone interface, the phone companies had a lock on this service. Users lined up by the hundreds, if not thousands to request mobile phone service.

Unfortunately this was not as simple as driving up to the telephone company and getting the service requested. There was a 2-year wait in many areas of the country as the service rolled out. The reason was simple, radio waves are not intelligent! The telephone companies mounted high-gain antennae on top of high-rise office buildings in the major downtown areas. From there, they would boom out a signal at approximately 250 plus watts. This is a very strong signal in any transmission. The antenna was on top of the highest building, and the power output was so great because the telephone companies wanted to get the best coverage possible. The FCC had only issued a limited number of frequencies for use in the AMPS and IMTS networks. Twelve channels were typically allocated in each of the metropolitan areas. Therefore, the channels were limited, and the telephone companies needed to provide the greatest area of coverage possible. All calls within the city had to be routed through this centralized tower because of the limited amount of channels. In this particular network, 12 channels were available. Each channel was used on a high-powered radio transmitter to provide the required coverage. Further, the system operated on a one-way (simplex) basis. Only one side of the call could speak at a time. This was analogous to the older two-way "push-to-talk" radio systems. They were trying to provide coverage in approximately a 25-mile radius from the downtown area. Back in the late 1940s that was sufficient because the work force was less mobile.

Problems with the AMPS/IMTS

This arrangement worked for short periods, but as newer demands were placed on the system, channel capacity was very limited. Only 12 users could be on the network at one time. The limited number of channels (frequencies assigned by the FCC for this service) led to the frequency reuse planning in the telephone companies.

However, because they were also trying to get the greatest coverage, the output signal did not stop at the 25-mile marker. Radio-based communications will travel greater distances as a function of the frequency used and the power output. If you want greater coverage, you need to use a higher power output system and antennae, or you have to use more equipment. Even though the signal pattern was designed to provide coverage only in a 25-mile radius, it kept on going beyond the distance planned. This meant that if another system was set up in the nearby area, using the same frequencies, the telephone company system would cause interference. As clearly shown in Figure 25.1, the overlap areas were bounded by another 50 to 75 miles. This means that the radio-based system could only use the assigned channels every 100 miles apart. The areas around the circle in this figure, called *buffer zones* exist to prevent interference. You might recall in the bandwidth discussion earlier in this book, sidebands and frequency bandpass filters are used in the telephone wired world. This is essentially the same thing in the wireless world, where the telephone company allocated the buffer zones so that two conversations would not occur on the same channel. The figure shows that the signal was used in the center (thicker) core, but the overlap areas are shown in the lighter core.

Cellular Communications

Prior to the introduction of an improved telephony service through radio-based systems, the industry standards bodies faced a dilemma. They wanted to meet the demand for more services, but they just did not have the available frequencies. Therefore, the engineers went back to the drawing board and created a new technique called *cellular communications*. Using a frequency pattern from radio transmitters, the concept will allow a honeycomb pattern of overlapping "cells" of communication. Because these

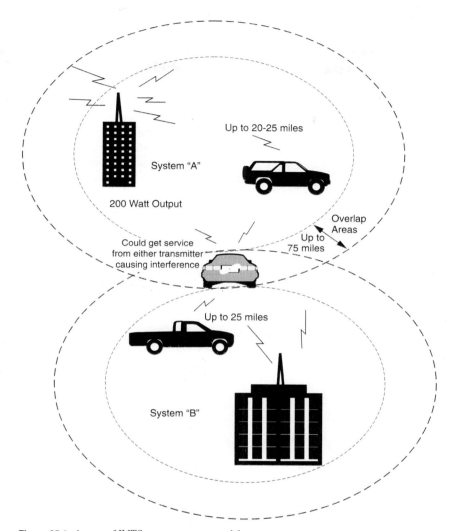

Figure 25.1 A map of IMTS coverage areas and frequency reuse.

cells can be minimal in size, a reuse of the frequencies, over and over, is possible. Thus, cellular communications was born. The goal of cellular was to make more service available to vehicular users.

In 1981, the FCC finally set aside 666 radio channels for cellular use. These frequencies were assigned or set aside for two separate carriers. The lower frequencies were reserved for "wireline companies." *Wireline companies* are the regulated providers (the local telcos). The higher frequencies were assigned or reserved for nonwireline carriers, which are the competitors to the telephone carriers. Both operating carriers (the wireline and nonwireline) are licensed to operate in a specific geographical area. The areas are classified as the metropolitan service areas (MSA) and rural service areas (RSA). Each carrier uses approximately 312 frequencies for voice/data communication and 21 frequencies for control channels.

Another major difference in cellular communications is the use of control supervision and switching of calls to adequately serve the cellular user. This is particularly true in a mobile environment when the vehicle moves from one cell to another. The dynamic switching and control necessary to facilitate a smooth and seamless hand-off from one cell to another is paramount. If this did not work properly, all communication could be terminated.

Conventional mobile radio telephones would degrade as the vehicle moved farther away from the base station. This degradation frustrated both the caller and the called party. As the user went beyond an area of coverage, the call would be cut off further, frustrating both calling and called parties.

With cellular communications, when a call is in progress and the caller moves away from the cell site toward a new cell, the call gets "handed off" from one cell to another. In Figure 25.2, this process begins to take place. Here's how the process works. Once a call is in progress and the hand-off of from cell-to-cell becomes necessary to keep the call in progress, a hand-off takes place. As the cellular phone approaches an imaginary line, the signal strength transmitted back to the cell site will start to fall. The cell site equipment will send a form of distress message to the mobile telephone switching office (MTSO), indicating that the signal is getting weaker. The MTSO then orchestrates the passing of the call from one cell site to another.

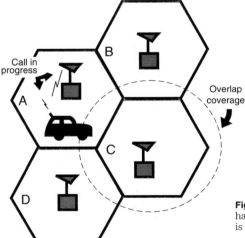

Figure 25.2 A call in progress will be handled by the cell in which the user is located.

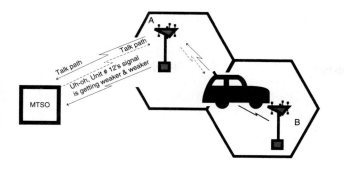

Figure 25.3 The cell site senses the drop in signal strength and alerts the MTSO.

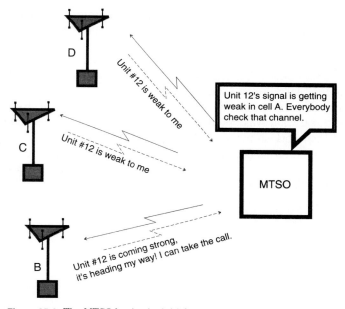

Figure 25.4 The MTSO begins its initial sequence to prepare for the hand-off.

In Figure 25.3, the initial sequence begins where a cell site notifies the MTSO that something is going on.

Immediately after receiving the message from the cell site, the MTSO sends out a broadcast to the other cell sites in the area. Its request is to determine who is receiving the cellular user's signal the strongest, called a *quality of service measurement*. Each site responds accordingly. In Figure 25.4, the MTSO's initial broadcast goes out across the network.

As the responses come back to the MTSO, the particular cell receiving the signal strongest is then selected to accept the call. The MTSO directs the targeted cell site to set up a voice-path in parallel to the site losing the signal. When this is ready, the MTSO sends a command to the cellular set to go to the new frequency (resynthesize). The cellular set then retunes itself to the new frequency assigned to the receiving cell site, and the hand-off takes place. This takes approximately 100 milliseconds to enact. Figure 25.5 shows the hand-off.

This might sound a little more complicated than it really is. The designers recognized that the original radio systems put out way too much power, therefore, frequency reuse was not possible in a 100-mile radius. Using a lower-power output device, the radius of the radio transmission system was smaller. As a matter of fact, the cellular system was designed to operate in a range of approximately 3 to 5 miles. If the power output is reduced to 10 watts, the radio transmission will travel less distance. Therefore, the user must be closer to the equipment to receive the call. To accommodate this, the cellular "cells" include equipment placed (typically) in the center of each of these overlapping cells. The distance from the radio equipment to the user will be approximately 1 to 2 miles. Therefore, the system should work more efficiently. Using only a 10-watt output, the frequencies used in each of the cells can be reused over and over again. A separation of at least two cells must exist, but that was taken into account. Figure 25.6 is a representation of a cell network showing the honeycomb pattern.

The carriers have been getting better at providing this coverage as the years have gone by. Many of the cellular suppliers now have provided for a seamless transition from cell to cell, anywhere in the country. To do this, they have developed an interface into another telephone (landline) based technique called

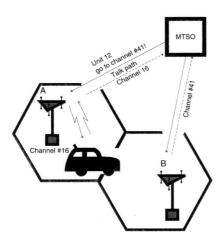

Figure 25.5 The alert goes out to the cellular site, advising it to go to a new channel.

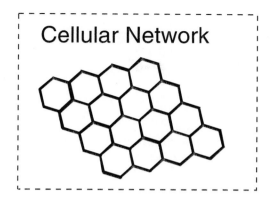

Figure 25.6 The cellular honeycomb pattern is a series of overlapping cells.

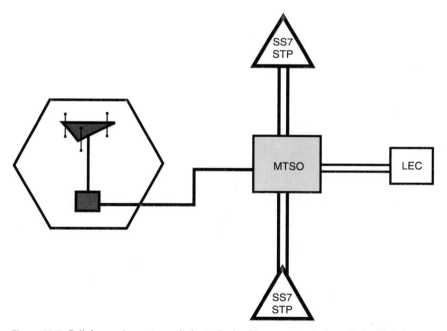

Figure 25.7 Cellular equipment now links to the backbone network through SS7 linkage.

signaling system seven (SS7). Through the use of SS7, the carriers can now hand a call off across the country without the user involvement. It just happens transparently. The SS7 linkage is shown in Figure 25.7, where the MTSO equipment is tied into a computer system called a *signal transfer point (STP)*. In this figure, the MTSOs are tied to duplicate STPs as a means of providing redundancy in the network. Through the use of this interconnection, additional features were also provided in the cellular networks similar to the telephone networks.

Meeting the Demand

The capacity to provide communications from a car phone was easily met, but was expensive to start. A vehicle-mounted cellular phone in 1984 retailed for $3000. But since then, the cost has declined dramatically.

The Telephone (Mobile) Set

The set houses a transceiver capable of tuning to all channels within an area. These are frequency-agile units capable of receiving/transmitting on all 666 frequencies, as opposed to the fixed-frequency units of old. The mobile set is shown in Figure 25.8. The major components of the unit are:

- The handset
- The number assignment module (NAM) an electronic fingerprint (a 32-bit binary sequence)
- Logic unit
- Transmitter
- Receiver
- Frequency synthesizer (generation of frequencies under control of the logic unit)
- Diplexer—separates transmitter/receiver functions
- Antenna

Cellular's Success and Loss

In 1984 cellular communications became the hot button in the industry. This was an original attempt to add capacity to several systems around the country. In the preceding years, the Advanced Mobile Phone Services (AMPS) and the Improved Mobile Telephone Services (IMTS) were very limited. Part of the problem stemmed from the fact that all systems were analog. The use of frequency-division multiple access (FDMA) is an analog technique designed to support multiple users in an area with a limited number of frequencies. An analog radio system will use analog input, such as voice communications. Because these systems were designed around voice applications, no one had any thought of the future transmission of data, fax, or packetized (X.25) data from a vehicle. When cellular was first introduced, the industry experts were predicting that by the year 2000, approximately 900,000 users would be on the network conducting voice phone calls. After all, it was a voice network designed to carry dial-up telephone calls.

No one was sure what the acceptance rate would be of cellular radio telephone services either. Currently, there are approximately 27 million cellular users in the U.S. This was a surprise to the industry experts who had no

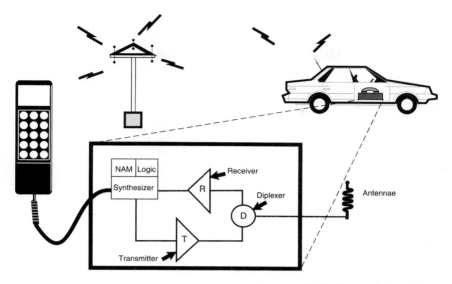

NAM = Numeric Assigned Mode

Figure 25.8 The cellular unit sets up the calls to the cell site, using a frequency-agile capacity to synthesize the necessary frequency.

idea of the pent-up demand. Approximately 100,000 to 150,000 new users sign up monthly. The problem is not one of acceptance and signing up new users, but retention of users and encouraging them to use the service more. The churn ratio has been as high as 15 to 30 percent. This is a costly process and is one that the providers are trying to overcome. But, frustration exists on three counts:

- *The cost of the service can be quite high* Depending on the location the basic monthly service will range from $19 to $39, and the per minute charge for incoming or outgoing calls is $0.35 to $0.45.

- *The possibility of getting poor transmission or reception on the network* This is a design problem but must be overcome before total acceptance is achieved.

- *The lack of available service* The network can get congested very quickly and if you are paying for the service, you obviously want it available all the time. Users typically don't want to hear about the providers' problems. To overcome this problem, many carriers have split their cells into half and more, just to be able to use the frequencies again and again. Figure 25.9 shows where cell splitting has been performed. The graphic shows multiple smaller cells inside a normal cell. This might be over-accentuated, but it is primarily a means of showing how this can be accomplished.

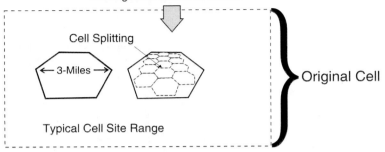

Figure 25.9 Cell splitting is required when constant congestion occurs. This helps to use more frequencies in smaller areas.

Many users sign up for the service, but rarely use it. Users might buy a cellular phone and keep it for emergency purposes only. This is fine for the user because the costs are minimal, but for the supplier, it means they have to prepare to serve more users than will actually use service. This causes concern for the carriers because their revenue is based on usage. Without usage, there will be little to no revenue. Yet, the capital costs of building the network are fixed and must be recuperated through the usage. The carriers are trying to figure out how they can:

- Accommodate more users
- Encourage subscribers to use more price-sensitive services
- Generate new revenue

We face a quandary in the industry:

- The carriers need users (more) to generate higher revenues to pay off the investment.
- The carriers must begin an evolution from analog to digital systems that will allow more efficient use of bandwidth (frequency spectrum).
- The need for security and protection against theft is putting pressure on the carriers and users alike.

Analog systems do nothing for these three needs, per se. Using either amplitude modulation or frequency modulation techniques, which were discussed in the data communications section of this book, to transmit voice on the radio signal uses all of the available bandwidth (or most of it). Distance, noise, and other interference demand power, amplification, and bandwidth to deliver a quality service. This means that the cellular carriers can support a single call today on a single frequency (specific frequency). As such, the limitations of the systems are in channel availability on a given set of technologies. If you consider the noisy nature of analog signals, you can also determine that quality is a matter of multiple factors; not the least of which will be congestion and atmospheric conditions, as covered in other sections of this book.

Other concerns with analog systems rise from both a security and fraud issue. The systems using a single frequency are subject to monitoring, but this is no different from any other airborne technique. Further, when the analog network can be penetrated from a fraudulent perspective, we must be concerned. This is not strictly an analog systems problem.

As mentioned earlier in this chapter, the analog system was designed to provide the benefit of quick communications while on the road. Because this was considered a service that would meet the needs of users on the go, for telephone, the thoughts of heavy penetration were only minimally addressed. However, as the major metropolitan service areas (MSA) began expanding, the carriers realized that the analog systems were limited. With only a single user on a frequency, congestion in the MSA began to be a tremendous problem.

When cell splitting takes place, other issues become a problem for the carriers. The more cells in use, as a result of splitting, the more critical the arrangement of frequencies available to use. Further, the smaller the cell created through splitting, other issues (Table 25.1) become problematic.

This put an added financial burden on the carriers as they attempted to match need with return on investment. These problems have plagued the industry for quite some time now. As the industry begins to search for new solutions, other uses of cellular communications began to crop up putting an added load on the service.

TABLE 25.1 Some of the Problems Carriers Will Experience with Cell Splitting

Issue	Problem/symptom
More equipment	Cost issues associated with buying the equipment. A cell site costs in the range of $650–800K.
Real estate	Getting the space to mount added quipment plus local ordinances in the area. Consumer pressures leading to the "not in my back yard" syndrome.
Power	Remote generation equipment required that is a cost and a security problem.
Logistical problems	Managing the new sites, new frequencies and other ancillary services for a much larger equipment base.

As mentioned, the cellular systems were designed around the "road warrior" driving around using a vehicular phone. Initial installations were handled in cars and trucks. However, as the technology improved, transportable telephones became nouveau for cellular users. Why should they be required to remain in the vehicle to use the service? Units could be mounted in the vehicle, but as the user departed the vehicle, the set could be removed and used for several hours of "standby operation" and one hour of on-line operation. This opened a whole new avenue for service opportunities. The transportables were initially expensive, but prices have dropped to a far more reasonable rate (in the range of $500 to $600). Mobility became more flexible for the cellular user. No longer were users relegated to the vehicle. The transportable unit is shown in Figure 25.10.

Portable handheld units began the next wave for cellular users. If a set could be removed from the vehicle, why then should it even be tied to the vehicle? If a set can be designed with a rechargeable battery pack, a briefcase model could be made available. Better yet, why not come up with a flip-type phone or a shirt-pocket phone? As the prices were dropping for the vehicle phone, so too were those of the handheld phones. Recently, ads in the trade magazines and local newspapers around the U.S. have been offering the handheld devices for a mere $0 to $15 with a one-year subscription to the carrier's network. The retail value of these phones is as low as $360 to $1500. This has come a long way from the initial days of the cellular networks in 1984.

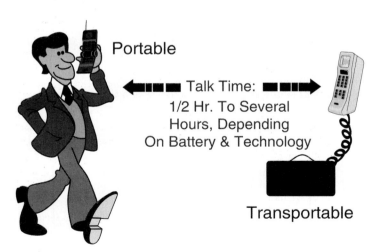

Portable

◄■■■ Talk Time: ■■■►
1/2 Hr. To Several
Hours, Depending
On Battery & Technology

Transportable

Figure 25.10 The transportable set took the set out of the vehicle. Several hours of standby operation are available.

Features exploded also, with the introduction of:

- Voice messaging
- Redial
- Display
- Call forward
- Memory
- Hands-free operation

Sub-compact or shirt pocket flip-phones caught on. Users became enamored with the technology, and the price was almost too good to pass up. These latest and greatest models have between 0.3 and 0.6 watts of power output; totally portable and accessible. Professionals who had meetings (i.e., doctors, lawyers, etc.) outside their normal office environment liked the idea of using a flip phone (cellular), giving them the ability to be reached at a moment's notice, but freeing them from the space requirements in the briefcase (which is valuable), and freeing them from the vehicle. These became the talk of many social events.

As mentioned earlier, these evolutions in the analog arena were all driving forces in the use and acceptance of cellular communications. However, as the congestion problem continued to go unsolved, the carriers entered a new era. The whole world has been evolving to digital transmission systems on the local and long-distance scene. This evolution to digital has many benefits for the standard wire based carriers. These include:

- Higher usage
- Better quality
- Multiple techniques
- Digital speeds

Digital Transmission

The cellular carriers are looking to step into the digital world. This stems from a compatibility and a frequency utilization perspective. If users can share a frequency or a range of frequencies, then more users can be accommodated on less bandwidth. This has some definite financial impact on the carrier investment strategies. They could support more users on less frequencies and require less equipment.

Digital

Digital transmission will introduce better multiplexing schemes so that the carriers can get more users on an already strained radio frequency spec-

**TABLE 25.2   A Summary of the Key Benefits for the Cellular Carriers'
Migration to Digital Communications**

Key Benefits of Digital
Less costs overall. Initially the migration will be expensive, but longer range will be less
More users on the same or less frequencies and spectrum
Better security than offered by the analog transport systems
Less risks of theft and fraud

trum. Additional possibilities for enhancing security and reducing fraud
also can be addressed with digital cellular. Again this appears to be a win-
win for the carriers. A summary of the key benefits for the cellular carriers'
migration to digital communications is shown in Table 25.2.

Once the decision was made to consider digital transmission, the major
problem was how, what flavor to use, and how to seamlessly migrate the existing customer base to digital. The digital techniques available to the carriers are:

- Time-division multiple access (TDMA)
- Extended time-division multiple access (ETDMA)
- Code-division multiple access (CDMA)
- Narrowband advanced mobile phone service (N-AMPS)

Carriers and manufacturers are testing and evaluating each of these
systems. There are several pilots and installations ongoing. Many have
been successful, others have been extremely disappointing. This creates
a split in the industry, depending on the technique chosen and the carrier
support. You must wait and see what happens here. But users who have
recently purchased a portable phone are also a factor to keep in mind.
When the network goes digital, those new phones just purchased or subscribed to will have to be changed. Can you imagine the hundreds of
thousands of users who are going to rebel against the changeout of
equipment?

Voice Technology and Applications

Because cellular was designed for voice communications originally, all of
the development was in the ability to service voice transmissions and applications. Many of the technological discussions have already been addressed. However, the applications for voice are just as many. Some of these
include the following:

- *Dial-up calls from and to the user on the road* This is what the cellular network was designed to do. The primary goal of cellular addresses the voice dial-up call without wires.

- *Dispatch services for maintenance and repair crews* Many municipalities are now using cellular communications to notify the crews on the road. Very specific two-way communications can be handled. This certainly beats the disruptive two-way radio systems that were used in the past.

- *Trucker availability is a prime value* Because truck drivers need to be reached anywhere and anytime, the use of a portable phone keeps them in touch with the dispatchers. This saves valuable time and money for the trucker and the company. Hopefully, this savings also gets passed along to the customer.

- *Safety and security (accident reporting and 911 services)* The books are full of cases where lives have been saved and emergency response services have been dispatched quicker. The use of a telephone in a vehicle to 911 (999 to some parts of the world) services can get ambulances, fire trucks, and police agencies on the roll much quicker than ever before. No single benefit has surpassed the value of cellular more than this one issue.

- *Secondary services for the Coast Guard* Where two-way radio-based services were limited and subject to extensive monitoring, the Coast Guard now has a more reliable means of communication, rather than having to go ashore and find a landline.

- *Voice service for police vehicles* Using cellular phones, the police agencies can eliminate the radio patch system that was predominant in the day-to-day operation. Motor vehicle registrations, discussing problems with other officials, and private communications, as opposed to talking on an open radio channel, all have their unique benefits.

- *Disaster recovery for office or manufacturing locations* The environment for quick reestablishment of voice communications can be handled on a limited basis with cellular phones. Call forwarding arrangements to a portable phone after a building is destroyed works well. Many of the key players in an organization will be far more mobile and will, therefore, benefit from the phone on the run, as opposed to having to locate in one spot.

- *Replacement for two-way radio applications (such as taxi, ambulance, and other on-demand services) can benefit from cellular* Two-way radio is difficult to use for the novice, noisy, sporadic, and congested in most metropolitan areas around the country.

- *Air-phone type services* As the nomadic user moves from the office to the vehicle to the airplane, being reachable is still the name of the game. Therefore, a special form of cellular communications will make the traveling executive or salesperson more available.

- *Vehicle scheduling/diversion for dispatchers in a storm, major traffic jam, etc.* The vehicle can be notified to use alternate routes.

- *Fire departments as a back-up technology* While out at a fire scene, things can get quite hectic. Fire fighting personnel must always compete for resources and get second opinions from experts. Safety ties in closely here. The ability to reach a chemical expert regarding the mixing of water with certain materials to extinguish flames, makes critical decisions more reliable. Many cases are documented that indicate if a mixture of water with a specific chemical occurred, a reaction that could be explosive could occur. This is not easily discussed on a radio, but it makes sense on a portable phone.

- *Any wired system (PBX, key system) can benefit by supplementing the wires for the nonwired user in an organization* Within the walls of an organization, catching anyone on the phone these days is a hit-or-miss proposition. Meetings, social gatherings, and group informal discussions all have a propensity of drawing us out of our office areas. By taking the phone with us, there should be no missed calls in the future.

Facsimile Technology

It was only a matter of time before the users of cellular communications started looking for and finding new applications. A voice application is fine, but supplementing that capability with fax was natural.

Facsimile transmission has been used in trucking industries for some time. A bill of lading or customs form can be shipped right into the cab of the truck. The advantages of this technique are varied. Other applications, such as sales literature, quotations, medical records, etc., could have strategic and saving benefits. An example of a fax transmission across the network to a trucker is shown in Figure 25.11. This particular case involves the arrival of a trucker at a border crossing without all of the appropriate paperwork. Assuming that perishable goods are on the truck, the dispatcher needs to get the truck rolling as soon as possible. However, from a governmental control arrangement, they need to see that the appropriate permits, fees, and inspections have been performed. Rather than delaying the movement of the vehicle, the trucker can immediately call the dispatcher and request the appropriate forms. When these forms are ready, the dispatcher can immediately fax them into the cab of the truck. What used to be an overnight and very expensive proposition has been reduced to a few minutes of time and far less expense.

Cellular systems transmit information from sender to receiver. As a mobile unit is transmitting or receiving fax, hand-offs from cell to cell might take place. This hand-off might cause a loss of information during the cycle.

Figure 25.11 The post-fax scenario allowed information to be shipped directly into the vehicle.

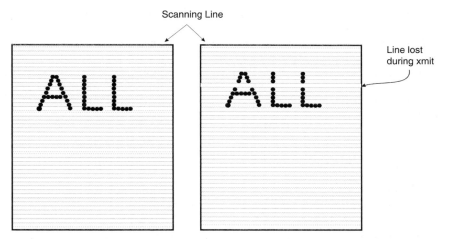

Figure 25.12 Interruptions on the circuit do not cause disconnects. A line of dots may be lost during hand-off.

But because pixels (or dots) are being transmitted, the loss of information will be minimal. See the chapter on facsimile systems later in this book for a more detailed discussion of this service.

As facsimile is being scanned, a Huffman or Read (pronounced REED) code set is being used. Actually, the facsimile machine is scanning a page of information on a line-by-line basis. During hand-off, a 100-ms delay could occur while the cellular user is handed from one cell to another. For data communications, this is a long time, but fax is far more forgiving. The facsimile machine waits for carrier detect, then continues to send. However, any data that was enroute during that hand-off could be lost. Because the line-by-line scan and transmit is used, all that will be lost is a line of dots. This shouldn't impact the information very much, yielding a high-quality facsimile transmission. This is shown in Figure 25.12, where a line of dots, or one scanned line might be lost. This is not a major problem because even though the line is lost, the rest of the document is still legible. So, using a fax modem allows for a wait time of up to 400 ms, as opposed to a data communications modem that will drop the line if the carrier is lost for greater than 80 ms. A hand-off can take from 100 to 200 ms in an average cell. The information should, therefore, be usable and understandable. Further, as the fax modems communicate to each other, they have the ability to monitor line quality and fall back to slower speeds, if necessary, to get the transmission through.

Data Transmission

Many users say that data transmission does not work on cellular communications. Although this was limited in the early stages, things have gotten much better. The first problems occurring were the hand-off problems. As a vehicle using data services rolled from cell to cell, the hand-off would disrupt the call for 100 to 200 ms. This was just enough to disrupt the carrier detect (CD) cycle, so the modem assumed that one of the callers had hung up. Thus, the modem would hang up, too! Obviously, this problem could be overcome by having data transmission take place while the vehicle is stationary. However, that becomes a limitation of communications. As Figure 25.13 represents, when data is disrupted, the modem will fall off and disconnect. This graphic is designed to show the early problems with the wireless transmission services.

Newer data modems for cellular communication use a technique similar to the fax modem. The modem will delay 400 ms before hanging up. Because the hand-off only takes 100 to 200 ms, the modem has time to spare without losing the call. Some data might be affected, but error detection, ACK/NAK, and CRCs all will come into play in detecting and correct-

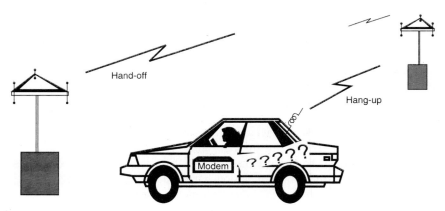

Figure 25.13 Early implementations of wireless data caused disconnects during cell-to-cell hand-off.

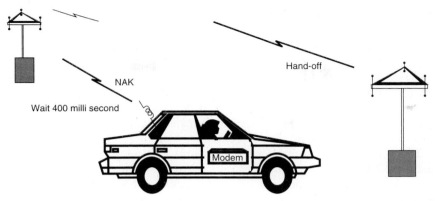

Figure 25.14 Improvements in the data communications prevent the disconnects as hand-off occurs.

ing the errors that occur on the cellular network. See Figure 25.14 for a representation of the data-correction problem. Conventional data was always regarded as suspicious on cellular airwaves. Because the modems had to hang onto the line for 400 ms waiting for carrier to return, everyone thought that data would only be effective at 2400 BPS. Motorola has introduced a 9600 bits/s cellular modem, and another company has a 19,200 bits/s cellular modem. As these techniques become more common, we can expect to see more users clamoring for the capabilities. Unfortunately, across a data link via cellular (analog technology), the bandwidth constraints will limit this service to casual, bursty data needs. If large amounts of data are to be transmitted, the costs and the available frequency spectrum might get in the way.

Digital Cellular Evolution

As the spectrum for radio cellular users becomes more congested, two primary approaches are in offering:

- Time-division multiple access (TDMA)
- Code-division multiple access (CDMA)

However, standards bodies and manufacturers are all coming up with variations of the use of the frequency spectrum. TDMA has been discussed, and was pretty well settled upon in the past. But, manufacturers have been developing an extended-TDMA (ETDMA) that will bring more use to the system. TDMA will derive a 3-fold increase in spectrum use, whereas ETDMA will produce 10- to 15-fold increases. The concept is to use a digital speech interpolation technique (DSI) that uses the quiet times in normal speech, thereby assigning more conversations to fewer channels, gaining up to 15 times over an analog channel. Each of the moves to digital requires newer equipment, which means more capital investments for the cellular carriers.

TDMA uses a time-division multiplexing scheme where time slices are allocated to multiple conversations. Although TDMA deals typically with an analog-to-digital conversion using a typical pulse code modulation technique, it performs differently in a radio transmission. PCM is translated into a Quadrature phase-shift keying technique, thereby producing a 4-phased shift, doubling the data rate for data transmission (actually driving a voice call at 8 Kbps).

Spread spectrum, which was first used back in the 1920s, has evolved from military security applications. Spread spectrum uses a technique of organizing the radio frequency energy over a range of frequencies, rather than a modulation technique. The system uses frequency hopping (similar to ETDMA) with time-division multiplexing. At one minute, the transmitter is operating at one frequency, and at the next instant, it is on another—according to some code. The receiver is synchronized to switch frequencies in the same pattern. This is effective in preventing detection (interception) and jamming. Thus, additional security is derived. These techniques should produce increased capacities in 10- to 20-fold over existing analog systems.

Another possible solution for bandwidth or user expansion is a technique developed by Motorola, Narrowband Advanced Mobile Service (N-AMPS). N-Amps is a channel-splitting technique where channel bandwidths are allocated in 10-kHz chunks, rather than the traditional 30 kHz. However, industry experts look at this technique as more complex and required stringent filtering, a problem not associated with the TDMA, ETDMA, and CDMA solutions.

Personal Communications Services

Why require a person to be tied to a pair of wires or to a fixed location with an associated number, when the networks can offer a tie to the individual through an intelligent network that is capable of finding a called party no matter where he or she is? This concept will appear on the mass market at a time when the population is becoming far more mobile. The use of a cordless telephone technology has already made its niche in the industry, particularly in the home where teens use a cordless phone as the norm. Security or convenience is not a concern. Service is an issue because the set must be kept directionalized with the base station. The limitation of distance in a 300-foot radius of the base station works fine in the residence. However, this does not work as well in the business environment because of all the concrete and steel around the set.

Therefore, it is only natural that the next logical extension of this service will be to provide unrestricted service and access regardless of where the recipient of the call might be at any time. Consider the use of a cordless telephony capability anywhere in the world, and the reality of personal communications becomes less myth, more fact. Throughout the evolution of wireless communications, users have always wanted the ability to be reachable at any time. Unfortunately, the use of earlier systems was always limited to very specific areas, such as a major metropolitan area and the surrounding areas (typically, a 40- to 50-mile radius from the user's home area). This distance limitation led to many complex gyrations in the use of paging and radio handheld systems. The industry finally responded with a more robust application through the use of satellite and radio repeaters. Now an individual could move out of the home area and be reached virtually anywhere around the country.

Pockets of "dead space" always exist, but these are limited. Even when the use of two-way improved mobile telephone services (IMTS) was deployed, several limitations were inherent. Limited frequencies, long waiting lists and expensive service all kept the use of this technology at a minimum. Cellular improved the situation, but the cost of the air time was considered very expensive. At a time when the cost of air time should have been dropping, carriers and providers of cellular service were raising their rates to compensate for losses because of the high churn (turnover) rates of customers, and through a new phenomenon of theft and fraud. The real user was penalized because of the losses being sustained by the providers.

Now the industry wants to deliver wireless communications in the form of telephony, data, paging, and e-mail applications to the mass market. To do this, a newer concept is needed that will allow for the rapid deployment of equipment and the more diverse coverage of all areas around the country, then the world. This will make major investments far less lucrative, so the proposed carriers and providers of personal communications services are looking for ways to introduce the technology to the masses with limited investments and large scale frequency reuse. The resultant discussions being held by these carriers and providers is to purchase and install microcells and pico-cells (Figure 25.15). First and foremost, the newer carriers are looking at providing much smaller systems and equipment and continue to reuse the frequency spectrum more and more. The obvious goal is to produce the equivalent of a wireless dial tone to the masses (both businesses and residences) for a reasonable, affordable price. When the technology and service is first introduced, the industry pricing might be as high or higher than the monthly costs for cellular communications and wired dial tone from the local telephone company. However, as the proliferation of the service takes place, the service costs will begin to fall dramatically. One could ultimately expect to see this form of service renting for $10.00 per

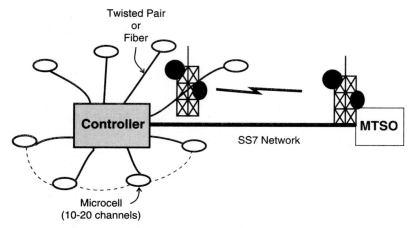

Figure 25.15 A single controller will be able to manage up to 96 microcells, with each cell handling 20–25 calls.

month and $0.10 per minute of usage. It will take some time to arrive at this point, but it should happen prior to the turn of the century. Spotty services and coverage might prevail during the beginning deployment of these services, which will cause some dissatisfaction by the pioneers who sign up for the service early in the implementation stages. This too will pass as the providers gain footholds in their market penetration and service areas.

All in all, this equates to a catch-22 problem, the carriers and providers need users to sign up and use the service so that they can deliver more benefits and coverage. Yet, the users will be frustrated as the deployment takes place and have a propensity to migrate back to the wired dial tone because this is an older by proven technology. As the definition of the personal communications is playing out, the world is wrestling with not only the service, but how to name the service. Where the FCC has dubbed this as *personal communications services* and the rest of the world has dubbed it *personal communications networks*, the vendor community has begun to use both terms. Thus, the dilemma. Variations of the service and the offerings have cropped up everywhere.

Technology

The technology decisions in this arena are equally perplexing. Where the use of typical frequency division services (an analog transmission system) in normal multi-user wireless communications is proven, it will not allow for the frequency reuse, as described. Further, any attempt to use a narrower band for transmission would only provide a limitation to the services anticipated for use. Consequently, the industry has been wrestling with what technology to deploy for the future. Again, the costs will be astronomical when a nationwide or worldwide network service is installed. Therefore, the use of time-division multiple access (TDMA), enhanced time-division multiple access (E-TDMA) or code-division multiple access (CDMA) are all being considered by the various future offerings. Each of the vendors or purveyors of service could theoretically take a different approach. The use of these various technologies are summarized in Table 25.3. In this table, the various options are shown with the respective advances in frequency reuse increases, the comparison of digital versus analog capabilities, the services that might well be provided and the potential number of suppliers that will exist are all shown in this table.

One further added comparison includes the use of narrow-band advanced mobile phone service (N-AMPS), which is being considered and deployed on a trial basis by some of the cellular carriers. This adds a degree of complexity because the variations will not be fully compatible. Therefore, a change in the transmitted signal or the format of the information will be required if an inter-networking application is used.

Cost penalties will exist with the various degrees of technology that are selected. The most expensive solution will be the use of CDMA, a technology that is still new and will require the total change of the equipment and techniques used.

The Concept of PCS/PCN

Visualize how this system might work! The local wired telephone services from the local exchange carriers will still exist. Therefore, the use of the wireless local loop in the dial-tone arena will still use a pair of twisted pair wires as always, as has been the case. However, the personal communicator will now act as a cordless telephone in the business or residence.

TABLE 25.3 A Comparison of the Technologies Being Considered for PCS/PCN

Technology capability	TDMA	E-TDMA	CDMA	N-AMPS
Frequency re-use gains	3X	10–15X	20X	3–5X
Digital	Yes	Yes	Yes	No
Services to be offered	Voice Data Paging	Voice Data Paging E-mail	Voice Data Paging E-mail Radio determination	Voice Data Paging
Costs	Low–Med	High	High	Low
Quality	Fair	Good–Excellent	Good–Excellent	Good

As shown in Figure 25.16, a cordless telephone arrangement will be set up with a smart terminal attachment. This smart box will allow the user to place a personal communicator in the charging unit at the location. While in the charging unit, the set will act as any cordless phone. A call coming into the telephone will travel across the local telephone company wires and ring into the network attached box. When the user picks up the phone and answers the call, the set will now be cordless. In Figure 25.17, the same scenario will be used, but the most significant difference will be the lack of a pair of telephone wires into the business or residence. In this scenario, the set is 100% wireless—with a smart interface in the location, a call coming into the business or residence will come in across the airwaves, rather than the wires, as defined in the wired scenario. The primary difference here is the absence of a telephone wire.

Taking this one step further, when the owner of the telephone set chooses to leave the primary location, the set will be taken from the charging unit and possibly clipped onto a belt loop or placed into a pocket (purse, jacket, or other). Newer sets might also appear as a wrist-mounted telephone set. AT&T has a wrist mount that is spring loaded. When the user opens the set, the hand is cupped around the ear to act as the receiver. Actually, the spring-loaded receiver slips into the palm of the hand and allows for a simulation of the normal telephone set. The transmitter is located on the wrist band. This is just one variation of the telephone set that might appear in the personal communications world.

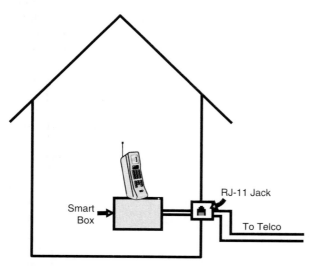

Figure 25.16 The personal communicator placed into a charger or smart box acts like a standard cordless phone when used with telco wires.

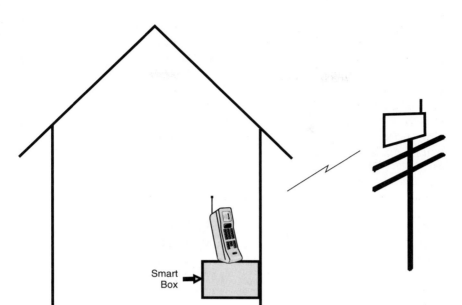

Figure 25.17 In the wireless arena, the communicator is connected to the telco via microcell technology.

A conceptual rendering of this is shown in Figure 25.18. As the user leaves the building and steps out into the open space of the neighborhood, the set will immediately begin communicating with a local cell. This local cell might be provided by the local telephone company or other supplier. However, the cell will be a microcell, possibly mounted atop a light pole or telephone pole. In Figure 25.19, the cells are located on top of a telephone pole line. These cells will be wireless to and from the personal communicator set, but the individual cells will be hard wired along the pole line via fiber, coax cable, or twisted pairs of wire. These wired facilities will be carried back to the telephone company central office. In this particular discussion, the telephone companies have an edge over the competitors because they already have the poles installed and own the rights of way to these poles.

Continuing along this discussion, the user's communicator will continue to communicate to the local microcell as the user proceeds down a path or walkway. As the user gets close to a boundary of the cell a hand-off will be required from one cell to another. The microcells will only cover a limited distance to begin with, possibly in the range of 80 to 200 feet apart. This implies that hand-offs will happen on a far more frequent basis. As the user continues down the path, the hand-offs continue along the way every 80 to 200 feet. As we enter into an area where the poles are no longer around, the

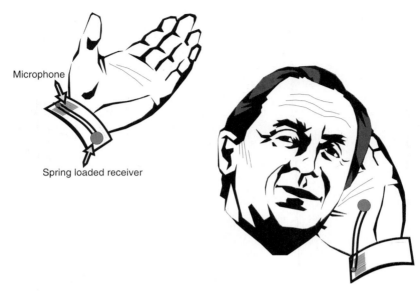

Figure 25.18 The spring-loaded wrist phone places the receiver in the cupped palm to accommodate full handset operation. The microphone stays at the wrist.

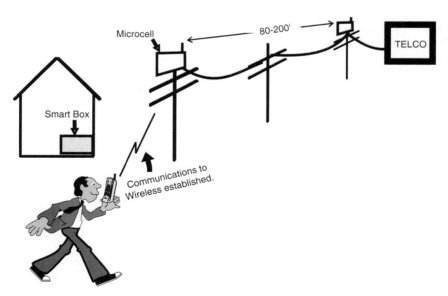

Figure 25.19 Once the set is removed from the smart box and the user leaves the locale, an instant communication is established via the microcell to alert the network that the user is mobile.

microcells are now mounted as telepoints along the way. A telepoint will be a wireless access mounted atop of a building or mounted on the side of a building. All that will be necessary will be a close proximity to the telepoint or microcell. In Figure 25.20, the telepoint is mounted on the side of a building, and in Figure 25.21, the microcells are mounted on top of the buildings serving a radius of 200 feet between cells. Each of these cells will be controlled by a central cell controller, with a cell serving 20 to 25 simultaneous calls and a controller managing between 32 to 64 cells. At some future point in time, these controllers will likely manage up to 96 cells, each controlling 20 to 25 simultaneous calls. This amounts to a single control device managing between 1900 to 2400 concurrent calls. Imagine the efficiency of this concept!

Why Personal Communications?

In the original concept of business communications, as well as the connection to personal communications, the use of cordless phones emerged. Serious efforts have always been made to eliminate the "telephone tag" problem within the business community. Instead of leaving messages and waiting for a return call, the industry has been attempting to provide the connection directly. However, whether in an office environment or a mobile user, the limitations of the telephony world still prevailed. Too often, the user was relegated to a pair of physical wires. To overcome some of these

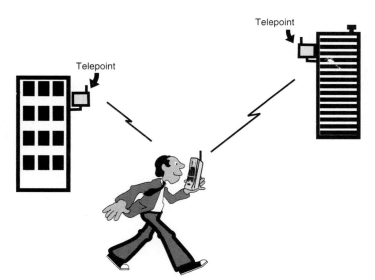

Figure 25.20 When no pole lines are available, telepoints can be mounted on the sides of buildings.

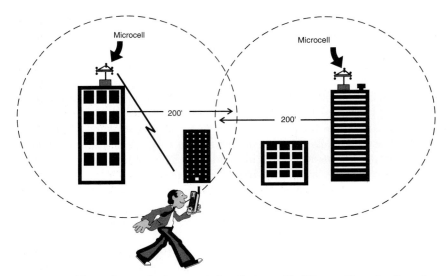

Figure 25.21 Microcells can also be mounted on the tops of buildings, with each cell serving a radius of 200 feet.

problems, cordless phones were introduced both in the business and residence to allow for added flexibility. Unfortunately, the same problems of distance limitations reigned. By using a wireless arrangement, the amount of wires needed in an office environment, coupled with paging systems (referred here as the use of loudspeaker paging within an office) and the use of personal pagers (the wireless notification services on a radio-based system), business felt that added flexibility and mobility could be achieved. Yet, human nature crept into this picture, where individuals could plead that they were out of range and did not receive a page, or that the batteries to the portable pager or telephone set were dead.

Hence, the problem was magnified. Frustrations by management, customers, and clients were building up. The real intent of a call is to connect two people who want to talk to each other, not to leave a paging message or a voice mail message. Using a voice mail arrangement only allowed us to leave a message, not to actually speak to the person desired. Nothing has really been gained through this technology. Further, many organizations have learned that by introducing voice mail as a tool to assist in the passing of information, it quickly eroded to the point where the voice mail system was being used as an excuse to never have to answer the telephone again (not in real time anyway). The message could be left and the individual could selectively decide whether to return a call or not. Pagers with alphanumeric displays allowed the caller to leave more information while trying to reach a party, but again, this only lent itself to the recipient using a valuation of the content of the message and when to return a call.

Clearly, the attempts to put two people together were only creating logical steps toward the ultimate goal. Therefore, the next logical step was to replace the pagers and the voice mail (tag) problems with a device that would be used for real-time connectivity. The use of the cordless phone, as mentioned, was the first attempt at accomplishing this replacement. Despite what many feel, this cordless phone was very successful, although limited. In particular, the cordless phone was extremely successful in the residential and small business market. Overcoming industry concerns of interference, lack of security (or eavesdropping) and lack of total privacy were the unabated desire to be untethered. Retail organizations, distribution and warehousing facilities, and small branch offices all found the use of a cordless phone an improvement over the hard-wired environment. This movement was occurring around the world, but legal issues of eavesdropping, smuggling, and interference had to be dealt with. The evolution from these single-line cordless telephones that operate in the shared unlicensed frequency ranges led to the need of a licensed (or controlled device) technology that would eliminate the fear of eavesdropping and interference. Hence, the evolution to the personal communications arena.

Personal communicators have been experimented with by various purveyors with mixed results, under experimental license agreements from the regulatory bodies around the world. The results have been so mixed and the offerings so varied that there had to be another round of discussion about the packaging and delivery of the services. Personal communications is the attempt to rectify the ills of all the past experiments and trials to deliver the call to the originally intended recipient.

It is not quite as clear cut as simply stating that a portable communicator will solve these ills. Personal means that the individual, not a company or department, will be reachable. Yet, the decision still remains with the individual of whether to accept a call, forward the call to another location, allow the call to route to an answering machine or voice mail system, or just ignore it. The fact that communications means that the recipient can accept or reject the call is only another part of the equation. This has to cover a much broader range of services, and a much wider area of coverage. The device should allow for the connection anywhere in a building, city, state, or country. Further, the services should include other connections, such as paging, e-mail, data transmissions, facsimile traffic, and other types of communications that would normally be provided in the wired office environment. Thus, the use of the personal communications concept must include any form of telecommunications.

The Evolution of Personal Communications

Since its inception, the wireless personal communications concept that started early in the 1980s has mystified even the industry experts. What services could be provided at a reasonable cost, be ubiquitous enough to serve the masses, and still produce a profit? Further, although voice telephony is the primary goal of providing wireless services, newer applications are constantly surfacing. New systems should include a provision for the following capabilities, based on user-selectable options:

- Voice dial-up services
- Data communications at various speeds, but minimally at 9.6 Kbps
- Paging and alphanumeric messaging on a pager
- E-mail access and file transfer capabilities
- Radio determination, vehicle location services
- Personal digital assistant type services for calendaring, mail, contact management, etc.
- Facsimile at Group III speeds or higher
- Future access to higher bandwidth applications such as multi-media
- Message center operations to leave or retrieve messages
- Telemetry services for process control and alerting services
- Dial-up video in a slow-scan mode (7 to 10 frames per second)

Comparing PCS to Cellular Networks

As this chapter flows along, you might begin to ask, "What are the differences between PCS/PCN networks and the already existing cellular networks discussed?" The answer is that there probably isn't going to be a lot a technological differences between the two networks, but the functionality and the population served will likely be the differentiating factors. PCS has been described (hopefully) as the replacement for the local loop, or the hardwired service.

Understand that this section is not purporting that all local wired facilities will disappear or be abandoned. Nothing like that will likely take place in our lifetimes. However, as a secondary means of providing the single number telephone concept, with the ability to be totally mobile yet still be reached, the PCS will fit the niche. However, on a higher quality, 64-Kbps standard, premium sustained service for the fast moving vehicular-based user, cellular will offer the appropriate connectivity. Therefore, we must take a position here that the two technologies and services should be viewed as complementary, rather than competitive. The use of a cellular network will continue to support the office in the car concept for voice, data, video, fax, and image services. The costs will continue to be higher for this service because it will be more robust and dynamic.

For the casual user, or the mobile person who needs to be near the residence or the office, the PCS will suffice in most cases. Therefore, two separate target markets will still be evident. This is not to say that the providers of the service will necessarily be different, the cellular provider of today might well be the cellular and PCS provider of the future. On an international plane, this has always been true in the past where the local Post Telephone and Telegraph (PTTs) organizations were the only authorized providers. However, on a global scale the shift to competition in this local loop bypass is already underway. What was once the sacred monopoly of the government around the world is now breaking down to a competitive service offered by many. At this point, the market will likely be two major cellular players and up to five PCS suppliers. But as time goes by, newer services are offered and newer technologies are used, the number of players in this market will be wide open.

Consider that in October 1993, the FCC has already announced its intentions to provide PCS services through a series of auctioned licenses to approximately 2500 licensees. There are likely to be 11 national PCS providers, followed by 49 regional, and the rest will be more localized in a specific metropolitan area, or rural area. The market penetration at this juncture will be wide open. Further, the smart phone, smart card phone (Figure 25.22) or the wrist phone shown earlier will emerge as the wave of the future to allow total connectivity anywhere and anytime. This is the vision of how the world will work in the 1998 (and beyond) time frames. To see this in motion, several experimental steps are being conducted at this time and will likely continue through the end of 1995.

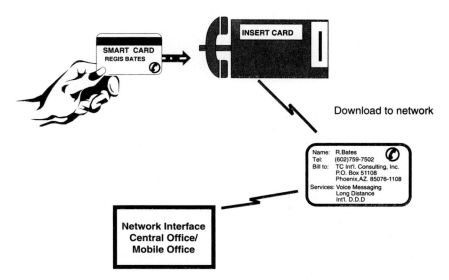

Figure 25.22 A smart card will be a newer evolution in the PCS/PCN arena.

The FCC has already started to auction off the licenses with some mixed results. FCC was disappointed with the responses and proposals from the perspective providers. Where the FCC expected every company to knock down the doors to get licensed, many of the larger players appear to be taking a wait-and-see attitude. This auction of licenses is only for players who can prove that they have the financial stability and backing to sustain large capital investments while the return revenue streams are still low. This is a departure from the lottery system that was used in the allocation of frequencies and licenses for the cellular industry. The government expected to raise approximately $10 to $12 billion through the auction of the spectrum, although that might not happen.

By the turn of the century, we will all likely look back and see that there was a plan to roll out these services. Further, everything that we know today about the telecommunications arena will likely change. Wireless telephony will probably be the basic service for business and residential users. Wireless PBXs will constitute a good portion of the market for new installations, and wireless LANs will be integrated into the wired backbone and infrastructure. Wireless tracking systems for just about anything we want will be easy and inexpensive. We'll be tracking our vehicles, people, and resources through a manufacturing environment and our staff throughout the world. Is this exciting? That depends on your perspective and how you view this implementation of the locator and tracking services.

26

Radio Systems

Radio-Based Systems

Satellite and microwave communications are radio-based systems that have been around for a number of years. At one point, much of the long-distance services in the network were served primarily via microwave, and the international long-distance services via satellite. A good deal of these systems are still in effect serving long-distance networks, rural areas, and international long-distance connectivity. Many organizations have built their own private networks by either launching their own capacity into a satellite orbit, buying into their own rights of way for microwave towers and radios across the country, or just renting the capacity from other suppliers. This is by far, not a dead service or technology.

Much has changed with the introduction of fiberoptics. Yet both of these techniques still have a use for the high and low bandwidth services for:

- Voice
- High-speed data
- Low-speed data
- Video
- Facsimile
- LAN-WAN connectivity

Satellite

Satellite uses a microwave transmission system, with radio connectivity being the primary means of broadcast. This technique has long been considered useful in very long-distance communications. Satellite transmissions were initially begun in the early 1960s, but several enhancements have been made over the years.

Typically, an earth station (uplink and downlink) is used to broadcast information between receivers. Current technology uses a geosynchronous orbiting satellite. *Geosynchronous orbit* means that a satellite is launched into an orbit above the equator at 22,300 miles. This orbit means that the satellite is orbiting the earth as fast as the earth is rotating. Therefore, it appears to earth stations that the satellite is stationary, thus making communications more reliable and predictable—and earth stations less expensive, because they can use fixed antennas.

This communications technique uses a frequency (uplink) broadcast up to the satellite, where a transponder receives the signal, orchestrates a conversion to a different frequency (downlink) and transmits back to the earth. As the communications is transmitted from the satellite toward the earth, a 17° beam is used. This produces a pattern of reception known as a *footprint*. The footprint is shown in Figure 26.1, with the entire U.S. covered by a single satellite. There are obviously many more than just three satellites in orbit, circling the earth, but only three are needed to provide global coverage. The satellites used for commercial applications are in what is called a *geosynchronous orbit*.

Frequencies

Satellite frequencies currently being used are categorized in bands. These bands are divided into the RF frequency spectrum that allow for different capacities and reception variations. The frequency bands are divided into two separate capacities, called *uplinks* and *downlinks*. The service provides for full-duplex operation, so a pair of frequencies is used. The uplink is used for transmission, the downlink is used for receiving (Table 26.1).

Satellites are good for broadcast communications; one to many locations. They are also termed distance insensitive, because once the signal is traveling up (22,300 mi) and back down (22,300 mi), the position of the receiver doesn't matter on a land line basis. We do not rent or lease the channel based on distance, but on the capacity of the channel. If two sites are using a satellite to communicate, the price is the same whether they are 10 miles or 3000 miles apart. The round trip from earth to the "bird" and back is essentially the same distance.

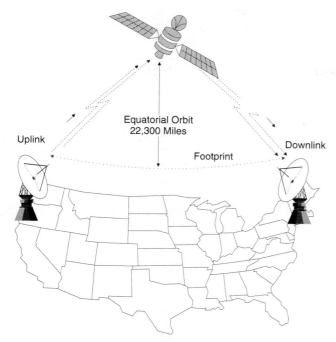

Figure 26.1 The satellite footprint covers ⅓ of the earth.

TABLE 26.1 Summary of the Frequency Bands Used for Satellite Transmission

Band	Uplink frequency	Downlink frequency
C	5.925–6.425 GHz	3.700–4.200 GHz
Ku	14.0–14.5 GHz	11.7–12.2 GHz
Ka	27.5–31.0 GHz	17.7–21.2 GHz

Advantages of satellites

Obviously, there are some distinct advantages to the use of satellite, or else no one would want to use the capacity. These advantages include some of the following:

- *Distance insensitive* As stated, the distance between the two end points is not a consideration in the pricing.

- *Single hop* With just a single shot up and back from the satellite, most communications coverage should be easily addressed. There is only one repeater function taking place, which helps to eliminate some of the errors in data transmission whenever a signal must be repeated.

- *Good for remote areas and maritime applications* Ships at sea have a problem in their communications needs. They cannot see a microwave dish based on their position on the horizon, thus no line of sight is available. Further, it is not practical to use any landline systems, because wires will not be available to run from the ship back to the land-based telephone networks. For remote areas where it is impractical to get wiring because no major development has occurred, a satellite link might be the only true solution. Good error performance for data: once a signal has penetrated the earth's atmosphere and gotten into space, little can corrupt a data stream. Therefore, after traversing approximately 5 to 20 miles in the atmosphere, the signal is relatively immune to any other problems. The bit error rate of a satellite transmission is approximately 10–11, which is very good compared to other transmission systems.

- *Broadcast technology* Because the footprint is so large, the transmission can be to many locations simultaneously. There is no need to mesh a private line network, or to transmit serially and sequentially when using satellites. The signal broadcast, once up to the satellite and once back down, can be sent to as many locations as are tuned to the specific frequencies. This can be a time savings factor and a cost savings opportunity.

- *Large amounts of bandwidth* An average satellite has 24 transponders. A transponder is a transceiver, a transmitter, and receiver. Each of these transponders can handle approximately 36 MHz of bandwidth, which as you will recall from earlier chapters, is significant. Using some various modulation techniques, such as QAM from Chapter 12, we could achieve approximately 3.4 Gbits/s of overall data throughput. Of course, this is a variable, depending on the modulation technique used and the channel spacing (separation). But, the idea of massive amounts of bandwidth is what counts. Many of the TV broadcasters use satellite communications to serve their distribution of broadcast-quality video communications at 45 or 90 Mbits/s.

Disadvantages

No system offers only great advantages, so the opposite side of the transmission system must be viewed to keep all things fair. Satellites have some distinct and peculiar disadvantages. Many can be overcome, but they do surface whenever a discussion is introduced.

- *One-way propagation delay* This delay is about ¼ to ½ second, and can be disruptive to voice and asynchronous mode data protocols or any protocol that requires an "ack" or a "nak" before transmitting its next block of data. The turnaround on a data stream would be double the ¼ second, or ½-second average turnaround. In using a bisynchronous protocol, a 4800-bits/s transmission speed could be relegated down to 400 bits/s because of the transit delays.

- *Multihops increase delay, detrimentally impacting voice* If one cannot transmit on a single satellite link, for example, around the earth from Boston to the Middle Eastern countries, then a double hop is required. This means that the signal will be transmitted up to one satellite and back down to the earth station, then retransmitted up to a second satellite, then finally back down to the earth station around the horizon. The delay will be at least one second from the time a message is originally sent until it is finally received. The disruptive nature of this for voice makes it virtually unusable. For data, however, protocols can be "spoofed" or faked into thinking that the data got there sooner, and the problem can be overcome.

- *High path loss in transmission to satellite* The transmission path will introduce more loss to the signal while it is making its way to the satellite. Therefore, more power is required and signal-to-noise ratios must be carefully administered or else the information could become unusable.

- *Rain absorption affects path loss* Any radio-based medium (especially in the microwave frequencies) is subject to a certain amount of absorption from water. Rain in a channel path can absorb much of the signal strength, leaving a far weaker signal to find its way to the receiver. Once received, the signal (now weaker and noisier) will have to be amplified. Remember the analog discussion in Chapter 6 about the impact of analog amplification.

- *Congestion building up* The satellites were originally designed in a spacing of 4 degrees apart, based on a 360-degree circular orbit. However, as more countries in emerging worlds have sought to use radio rather than wire-based network access, the satellites have been placed in closer orbital slots. They are now being positioned at 2-degree increments. This can lead to congestion on the transmission paths, cause occasional interference and (of course) limit future parking places.

Very Small Aperture Terminals (VSATs)

Very small aperture terminals came into being in the mid 80s, when the size of the satellite terminal shrunk. Since then hundreds of networks have cropped up. The pricing schemes have not yet made it totally feasible for more, but as technology continues to shrink, the use might expand. Figure 26.2 is a graphic representation of the VSATs located around the country.

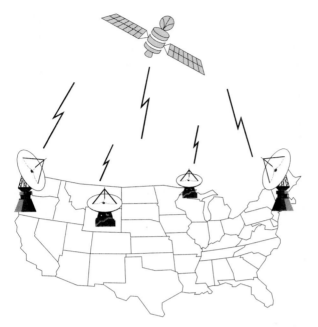

Figure 26.2 V-SATs are located around the country to provide coverage.

Costs are continually falling, and benefits derived from VSAT networks are rising, breathing new life into this technology. No longer do users have to own and operate their own hubs, where shared hubs are more realistic. Prices from suppliers have dropped; VSAT units can be as little as $250/month in quantities of 1000 units. Many specific data applications are now being migrated to satellite, via VSAT. In 1991, more than 35,000 VSATs were on record. Newer regional operators are now approaching smaller network users (10–65 nodes). The dishes (VSATs) can be as small as 1.0 meter to 2 meters, thus making it more realistic to place units in remote sites.

However, these smaller terminals are limited in their applications; voice and video (motion) applications aren't generally feasible. VSATs do, however, provide sufficient bandwidth capacities for data transmission (host/host, file transfer LAN/LAN, and some receive-only broadcast video).

Microwave

The local telephone companies, long-distance carriers, and users alike have all been successful in the use of microwave radio systems. Microwave was the predominant means of handling long-haul transmission here in the U.S. The advantage is that rights of way and natural barriers are more easily overcome. A single radio channel can carry as many as 6000 voice channels in 30 MHz of bandwidth.

Microwave, as any radio technology, is designed around space and frequency spectrum, which is a finite resource. Thus, coordination between paths is extremely important to prevent interference. Microwave radio can use either an analog or a digital modulation method. Each operates differently.

Analog radio uses either amplitude or frequency modulation, usually frequency modulation. In 30 MHz of bandwidth, 2400 voice channels can be carried, equating to a voice channel of 12.5 kHz.

Digital microwave

Digital microwave has been available since the mid 70s, with the direct modulation of an RF carrier. Because a modulation technique uses 1 bit per hertz, using a 64-Kbits/s channel would be far too consumptive of radio frequency spectrum. Therefore, a better modulation technique is needed. A microwave installation is shown in Figure 26.3. Phase shift keying or Quadrature amplitude modulation techniques are more common. Using a 16 QAM, a total of 1344 voice channels can be carried on 30 MHz of radio, yielding 90 Mbits/s (3 bits/Hz). Newer 64 QAM supports 2014 channels at 135 Mbits/s (4.5 bits/Hz). Using digital microwave allows for the direct interface of T1/T3 carrier circuits.

Service in the 18- to 23-Hz radio spectrum is called *short haul microwave*. Typically, this is a point-to-point service operating between 2 and 7 miles and supporting up to 4 T1s. Newer systems support up to 8 T1s.

Digital termination services that operate in the 10-GHz range began in the early 1980s. This short-haul service is designed for metropolitan area digital services, similar in concept to cellular radio. This TDMA technology has access to the system for 24 users at speeds up to T1. As when dealing

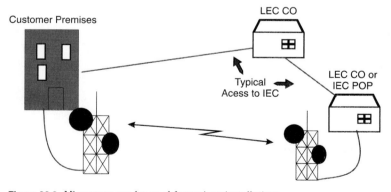

Figure 26.3 Microwave can be used for various installations.

with any radio frequency spectrum, the need for licensing exists. Satellite services are usually coordinated by the network suppliers. Microwave can be coordinated by either the network suppliers or user. You will need:

- Line of sight
- Path coordination
- Frequency/licensing
- FCC-trained personnel
- Tower/antenna location
- Power/backup power
- Time

Future Use of Microwave and Satellite Systems

Because the radio systems can carry such broadband capacities over radio frequency, we will see several opportunities to use these technologies.

- *CATV* Uses coaxial broadband cables to delivery compressed video to users. These systems use digital modulation at 90 Mbits/s. However, as high-definition TV becomes less expensive and more widespread, the 135 Mbits/s of bandwidth throughput can be achieved on microwave/satellite systems to a coax or fiber interface.

- *DBS* Direct broadcast satellite systems, a receive-only (one way) transmission for individual reception. Using a very small aperture antenna (1 meter or less), a viewer will receive TV signals through a personal earth station (or some small component). DBS will be available to CATV, MATV, and broadcast users alike. This is perceived as a supplementary service, rather than a replacement.

- *Interconnection already exists between cellular systems and microwave* However, as Iridium comes about in the late 90s, satellites in low polar orbit will be used for worldwide cellular connections. This will go further, into the personal communications networks (PCN), using a personal communications server (PCS) system to deliver calls to/from the cellular arena.

- *VSATs will continue to get smaller, and newer modulation/time sharing techniques will provide greater access to bandwidth* Prices will continue to drop as these services continue to compete with fiber and microwave alternatives.

Another form of wireless communications is provided by transmitting infrared light through the air. This can be an exciting opportunity to provide basic connectivity in a localized area for a very reasonable cost. Infrared light systems have been around for three decades, but have never really caught on until recently. Clearly, the need to provide connectivity is the issue in any telecommunications environment. Infrared or laser beams can be considered in the overall corporate strategy if the bandwidth needs are limited to a few channels, or if the communications system is being complemented or supplemented with this technique. Operating up in the frequency spectrum of laser beams in the terahertz (trillions of hertz per second), an invisible light beam is focused from a transmitter to a receiver over a very short distance, usually under one and a half miles. Similar to radio technologies, infrared light is subject to specific curtailments and impairments in transmission paths, such as:

- Very limited distances, typically at something under 1½ miles.

- It requires a clear line of sight in order to communicate with the other end.

- Limited bandwidth of approximately 16 Mbits/s throughput.

- Very prone to environmental disturbances and limitations, such as movement, electrical disruptions, and so on.

- Subject to disruption from fog, dust, heavy rain, and other objects in its path because the fine beam of light might not be able to penetrate these impairments.

With this in mind, the use of laser on infrared light still has its benefits. One would certainly have an opportunity to gain benefit under the right conditions. No single technique can be 100% effective; each has its own limitations and advantages. This is why so many options are available. We can pick and choose the best fit of transport systems to meet the needs of the organization.

Despite the technical bandwidth limitations of 16 Mbits/s, there are certain cost advantages and ease-of-use considerations. These factors keep this technique high on the acceptability scale of many users for many applications. The availability of low-powered solid-state laser diodes has opened up many possibilities. Producing cost-effective transmission capabilities through the air for voice, data, video, and now LAN traffic at native speeds will enhance a network immensely. Transmission through the air without a cable system infrastructure reduces the need for rights-of-way, as would be necessary with a cable system. Further, the use of infrared technology falls under the domain of nonlicensed technology. This removes many of the licensing requirements and associated delays and costs with other systems. For under the two-mile range, this technique holds a wide range of exciting possibilities. To use such a system, three criteria must be considered though:

- The basic geometry of the system
- Atmospheric conditions that will affect the transmission
- Site selection for the installation of the link

The System Geometry

As already mentioned, the use of an infrared system is generally used in a point-to-point, line of sight application. Some multipoint and repeated systems have been developed, but these are few and far between. There are three elements that make up the basic system which include:

- The transmitter, where the signal gets modulated from its original electrical form onto a light beam. Modulation was covered in great detail in Chapter 5.
- The medium used, in this case air.
- The receiver that demodulates the light back into its original electrical form.

The basic components of the laser system are shown in Figure 27.1, comprising the whole system. In this figure, the components look like guns; this is for representation only.

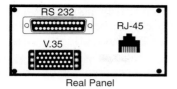

Real Panel

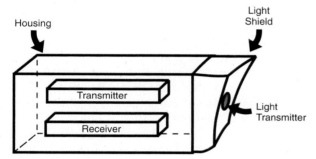

Figure 27.1 The basic components of an infrared system.

The data is input into the transmitter, where it is modulated onto a light beam. The raw laser beam is a very narrow beam of light (the actual light is referred to as monochromatic) that has very little divergence or spreading. Theoretically, the beam is useable to the receiver in its raw form, but from a reality standpoint, it is extremely difficult to aim and maintain this raw light beam. Therefore, the raw light beam must pass through a lens that will produce a divergent beam. This helps to solve the initial limitation of the beam itself. However, this also creates some of the distance limitations of the transmitted signal. Because the lens creates the divergence, it reduces the distances from the transmitter to the receiver by diffusing the strength of the beam. Safety requirements also limit the output power of the beam. The transmitter generally uses a constant diameter of light, called a *footprint*, regardless of the distance being used.

The incoming light is focused by a *collector*, or collecting lens, that sends the received beam to an optical detector. The receiver uses an angle of acceptance (RAA) of approximately 3 to 5 mR to provide a degree of selectivity. The transmit beam divergence angle (TBA) is always greater than the receiver angle of acceptance. The receiver needs a minimum level of incident light in order to demodulate the light back to the original electrical input. There are trade-offs in the size, complexity, and cost of the receiver lens that limit the range. The receiver angle of acceptance and the transmitter beam divergence angle are shown in Figure 27.2.

The setup in the previous paragraph is a one-way, point-to-point system (called a *simplex system*). However, most applications need two-way communications. A duplex system can be used. To create a duplex system, two simplex systems are coupled together. The creation of a duplex system is shown in Figure 27.3. Both of these systems will operate with lasers at the same wavelength (which is typically 830 nm). This is an advantage of a light-based system because the directionality of the lasers does not migrate from a transmitter to a receiver like a radio system would. Remember that in radio systems, two separate sets of frequencies are used to keep the transmit frequency apart from the receive frequency. A further feature of these focused light transmission systems is that multiple systems can be mounted close to each other without causing interference, so long as the angle that the systems make with each other is greater than the original angle of the receiver. The use of multiple systems is shown in Figure 27.4. Two light beams are transmitted across the same path, yet are kept separate because the angles of acceptance are different.

Atmospheric Conditions

The effects of atmosphere on a light transmission will affect the total overall performance of the system. The severity and the length of these conditions will affect the distance and performance. All electromagnetic radiation used in

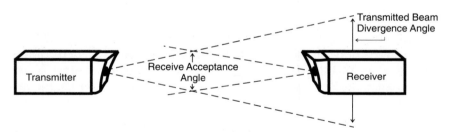

Figure 27.2 The transmitted beam divergence angle compared to the receiver acceptance angle.

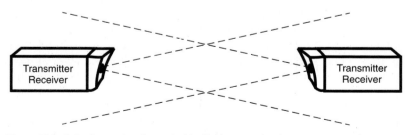

Figure 27.3 A duplex system is created by linking two simplex systems together.

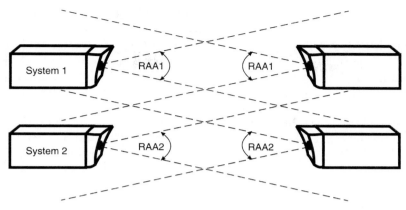

Figure 27.4 Multiple systems can be installed close together without conflict.

any communications system is affected by the atmosphere. However, each of the transmission systems is affected differently, whether it is low-high frequency radio, microwave radio, or laser transmission. The wavelengths of each transmission determines the effects on the actual transmission. The three most significant conditions that affect laser transmission are:

- Absorption
- Scattering
- Shimmer

All three conditions can reduce the energy at the receiver, which will then affect the reliability and the bit-error levels. These three conditions are covered in more detail.

Absorption

Absorption is caused primarily by the water vapor (H_2O) and carbon dioxide (CO_2) in the air. The density of water and carbon monoxide are conditioned by humidity and altitude. Gases that form in the atmosphere have many resonant bands, called *transmission windows*, that allow specific frequencies of light to pass through. These windows occur at various wavelengths. The window we are most familiar with is that of visible light. The near infrared wavelength of light (830 nm) used in laser transmission occurs at one of these windows; therefore, absorption is not generally a big concern in an infrared laser transmission system.

Scattering

Scattering has a greater effect than absorption. The atmospheric scattering of light is a function of its wavelength and the number and size of scattering elements in the air. The optical visibility along the path is directly related to the number and size of these particles. The three most common scattering elements in the air that will affect a laser beam transmission are:

Fog and smog. Fog appears when the relative humidity of air is brought to an appropriate saturation level. Some of the nuclei then grow through condensation into water droplets. Attenuation by fog is directly attributable to water droplets less than a few microns in radius. The result of the scattering is that a smaller percentage of the transmitted light reaches the receiver.

Visualize the situation when driving a vehicle in dense fog. The effects of the headlights on the car are somewhat useless. Try turning on the high beams and see what difference this action makes. The resultant scattering of the visible light from the headlights makes it almost impossible to see. The light is being reflected back at the vehicle, as well as being scattered in all directions. This is exactly how fog affects the infrared light beam being transmitted.

Smog's effects are similar to those of fog but to a lesser degree. This is because of the difference of the particulate matter radius. Consequently, fog is the limiting factor in laser transmission performance.

Rain. Although the liquid content of a heavy shower is 10 times the density of a dense fog, the rain drop radius is approximately 1000 times that of a fog droplet. This is the primary reason that attenuation via rain is 100 times less than that of fog. Think about this for a moment; when fog is created, the saturation level is very high in the air, but the droplets are much finer. In rain, although more liquid is falling, the coarseness of the rain doesn't affect the transmission as much. Visualize that there is more space between the drops of rain than between the droplets in fog; therefore, the light can get through rain much easier.

When driving in a heavy rainfall, the headlights still penetrate the raindrops and give greater visibility over a greater distance. Even by turning on the high beams, the distance is better than with the low beam. A certain amount of the light is still being reflected back at the vehicle and scattered in all directions, but more of the light is penetrating through the rain. So, the same results occur with the transmission of the laser beam through the rain. It would take an extremely heavy rain (greater than five inches per hour) to significantly reduce the transmission of the laser beam.

Snow. The effects of snow on a laser transmission fall somewhere in between those of fog and rain, depending on the degree of water particles in the snow. A very wet snow is close to rain, and an extremely dry snow is similar to fog. Once again, try to visualize the difference of driving through a dense snow storm. Driving through a heavy wet snow is like driving through a heavy rain. It is not all that pleasurable, but at least the light penetrates the snow and visibility is achieved. However, when a "white-out" occurs, the ability to get the light through the snow is significantly reduced. The light appears to be reflecting right back at the vehicle, rendering all visibility useless. The same effect occurs with a laser transmission.

Shimmer. Picture a hot day while driving down a long road. As you look toward the horizon, it appears that water is on the road. As you look closer, you can see what appears to be heat waves shimmering off the ground. This is the direct result of a combination of factors including:

- Atmospheric turbulence
- Air density
- Light refraction
- Cloud cover
- Wind
- Others

These factors combined will cause a similar disturbance to transmitting a laser beam through the atmosphere. As local conditions begin to combine, shimmer begins to affect the transmission. Shimmer will impose a low-frequency variation on the amount of light being received by the receiver. This variance could result in excessive data error rates or video distortion on a laser communication system. By using simple proven techniques such as frequency modulation (FM) and automatic gain control (AGC) in the system, the effects of shimmer can be minimized. Moving the equipment several meters above the heat source will also reduce the effects of shimmer.

Site Selection

An equally important factor includes proper site selection. To assure successful installation and use several factors must be considered, such as:

- Clear line of sight
- Mounting conditions
- Avoiding adverse conditions
- Mounting structures

Each of these factors is important. Diligence must be applied.

Clear line of sight

When setting up a transmission system, a clear line of sight is needed. This means that nothing can obstruct the light. The two ends must physically "see each other" or the system will not work, it is as simple as that. The laser beams should not be directed through or near wires (i.e., electric wires, telephone, cable TV, etc.). Trying to force the light through tree limbs is not a good idea. As Figure 27.5 shows, getting over cables and wires is important. Allowances must be made for the growth of tree branches over time (Figure 27.6).

The system can be mounted either indoors or out. Figure 27.7 shows the systems mounted indoors. The light beam passes through glass. The glass must not have any reflective coating or tinting on it. Reflective coatings might impair the ability of the beam to pass through it.

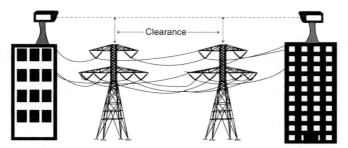

Figure 27.5 Towers may be necessary to keep the path above electric or telephone wires.

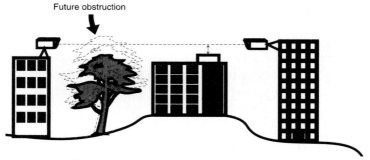

Figure 27.6 A clear path must be ensured for the transmission.

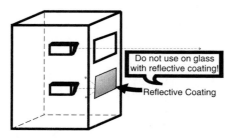

Figure 27.7 With proper care, the laser can be used indoors to pass through glass.

Adverse mounting conditions

Few precautions must be dealt with when mounting. It is important to ensure that the maximum signal strength is received and that the beam is not inadvertently misdirected. These considerations include avoiding heat sources, water obstructions, east-west orientations, and vibration.

Heat sources. Heat scintillation (or shimmer) occurs as a result of differences in the index of refraction of air with temperature. The air might act as a lens that incorrectly steers the laser beam. The beam should not be directed over exhaust vents, cooling towers, chimneys, smoke stacks, or large air conditioning units.

Water obstruction. Water droplets on glass can redirect the light. To prevent this when using the system indoors, use an awning above the window. If building codes preclude using an awning, mount the unit as high as possible on the glass. This will prevent the buildup of water on this section of the glass because the water will run down quickly.

East-west orientations. The angle of acceptance is designed to be quite narrow. Any extraneous infrared light passing directly into this angle of acceptance will be accepted, thus causing errors. East-west shots that have direct sunlight on the same axis will cause interference. The sun contains infrared light in its total light source, and this cannot be filtered out. Therefore, a system will be "blinded" for periods when the sun's axis is the same as the beam from the system.

Vibration. Both the transmitter and receiver deal with very narrow beams of light. The smallest movement of the transmitter or receiver can severely limit transmission or reception. The system could even lose its angle totally, resulting in an outage.

Licensing Requirements

Because the use of infrared and laser-beam technology is in the invisible light spectrum, there is no licensing required. The use of light luckily falls into the frequency spectrum where there are no regulatory guidelines. The U.S. Government has not figured out how to charge for the use of light, nor can they determine how much we use. So, they can't tax it on a usage-sensitive basis. Because the government cannot figure it out, neither can the telephone companies. This, of course, makes it much easier to deal with the technology itself, allowing the organization to place a system in operation quickly.

Unlike radio technology, where the bulk of the frequency spectrum requires licenses and path clearances (with limited exceptions) from the regulatory bodies around the world, the use of light theoretically causes no interference between systems. Thus, use of this spectrum is allowed with little to no coordination effort. So long as the receiver acceptance angle does not coexist on the exact same axis, there should be no problems. This is an attractive feature of light-based systems.

Bandwidth Capacities

The system's bandwidth can cover a myriad of capacities, depending on the configuration purchased. The specific application must be considered in terms of the bandwidth necessary as well. At the high end of the bandwidth, the systems today can deliver as much as 16 Mbps throughput for LAN-to-LAN connections on a 16-Mbps Token Passing Ring network. Thus, one could say that the bandwidth is 16 Mbps. However, other systems are designed to meet the needs of the user interface. An example of the capacities is summarized in Table 27.1. In this table, the needs are met by different systems at the capacities needed to meet the application. In several of these systems, the pricing is dependent upon the actual needs.

Applications

As noted in Table 27.1, several applications can be applied to the use of infrared and laser technologies. Specifically these are addressed below as they pertain to a need for connectivity from site to site. In the following scenarios, a typical customer with two sites (buildings) located at approximately 1000 meters apart, whether on a single campus or in two buildings in the same town, needed to provide various forms of connectivity between

TABLE 27.1   A Summary of Various Systems Available on the Market and the Application Used

Bandwidth	Application
10 Mbps	Ethernet data link
4 and 16 Mbps	Token Passing Ring
1.544 /6.312 Mbps	T1/T2 capacities data/video
4 × 1.544 Mbps	Quad T1 voice/data
RS-232 to 19.2 Kbps	Data transceivers
RS-422A to 2.048 Mbps	Data transceivers
Audio 20 to 20 kHz	Audio services

SOURCE: Laser Communications Inc. Lancaster, PA.

these sites. After looking at the alternatives, the choices were whether to use:

- A direct burial cable (copper, coax, or fiber)
- A microwave radio system
- An infrared laser system

Although more considerations and information would be needed to arrive at any conclusion, the following might play as an active comparison.

The cable decision

Upon looking at the cable scenario, the customer was faced with extraordinary costs to dig a trench, lay a conduit, and run a cable connection to the two buildings. Additionally, the necessary right-of-way might be extremely difficult to obtain. The customer decided that a cable-based solution was not appropriate in this situation. A note here, another alternative exists, that being the use of local leased line facilities from the local exchange carrier (LEC) or a competitive access provider (CAP). However, the long-term leased-line rental costs were not attractive to the customer.

The radio-based decision

Looking at the area and the potential ordinances and variances that would be required in his scenario, the microwave decision was also rejected as being a second choice only if another alternative would not work. Realizing that certain forms of licensing would be required and that a clear path is necessary, in a very heavily installed radio area, the costs and time constraints were greater than the customer wished to endure. They were not even sure that a license would be granted, even if they invested the time and effort into the process. Hence, the microwave choice was ruled out.

The infrared decision

Because the cable and the microwave choices were not that attractive, the logical choice was to use an infrared system. No licensing requirements were necessary, so the system could be put into action as soon as the acquisition was made. Because no right-of-way constraints were in the way, the process could be handled appropriately fast. An added benefit was that there would be no monthly recurring charges for the use of the facilities as would be the case with the leased lines. Lastly, the cost of implementation required no major tower, power equipment, cable entrances, or other construction needs.

The cost of buying the system was also very attractive because a one-time cost of less than $20,000 was all that was required. The customer also took into consideration that if a move from one of the buildings was ever necessary, the system could be packed up and moved along with the rest of the office equipment. Assuming that a clear line of sight could be obtained, and the distance did not exceed the specifications of the system, it could be put back into action in a matter of hours. This was not the case with the other options looked at. The clear winner in this particular scenario was the infrared laser system. But what of the application? Regardless of the application, the same rules would have applied. However, in this particular scenario the applications might have included the following:

The use of infrared technology to link two LANs together

The technology can link a Token Ring to Token Ring network together at native speeds. Either a 4- or 16-Mbps capacity can be applied here. In Figure 27.8, two LANs are linked together at native speeds. What is meant here is that the LANs are connected at the same speed as the LAN operates. This allows for transparent connectivity at the full rated speed of the network itself for a price of approximately $35,000 for a full 16 Mbps, or $26,000 at 4 Mbps, rather than slowing the transport down to some lower-level speed. If a bridging arrangement is used here, the total throughput would be limited to a 1.544 Mbps or lesser speed.

Infrared laser technology used to link two Ethernet LANs

The use of infrared laser technology to link two Ethernet LANs together at a native speed of 10 Mbps is shown in Figure 27.9. A different system is used here, with the native throughput operating at the full-rated 10 Mbps,

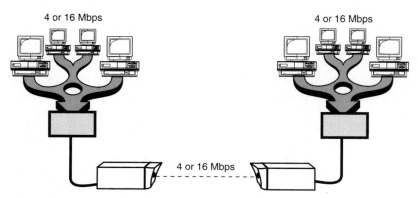

Figure 27.8 Linking two Token Ring LANs together at a native speed of 4 Mbps or 16 Mbps provides transparent connectivity.

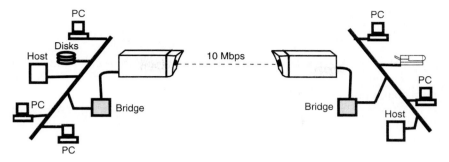

Figure 27.9 Infrared can link two Ethernet LANs together at a native speed of 10 Mbps.

rather than using a slower-speed bridging arrangement that runs at 1.544 Mbps or less. The system is priced around the $15,000 range.

In both of these cases, the alternatives to these connections could be leased T1/E1 speed lines at slower connections than the LAN operates. This could potentially cause network congestion because the throughput is buffered to accommodate the line speed. Further, the leased line of up to 1 mile from a local operating company (telco) could cost as much as $700 to $1000 per month. The payback on this arrangement would obviously be dramatic, gaining the additional speed and still paying for the system in less than 2 to 3 years, depending on the system selected. Further, if the native speeds were leased from the local telco the costs for the services could be as much as $1800 to $2400, depending on the speed selected thereby producing a return on the investment in less than 12 to 16 months. This is for the close-in services. A microwave system could cost as much as $80,000 to $100,000 for a complete installation, which is three- to four-fold the cost. As can be seen for this environment, the use of the infrared system has some definite advantages.

Connecting PBX to PBX using a digital transmission system

The connections of PBX to PBX using a digital transmission system for TIE line capabilities between systems using a quad T1 capability for voice connectivity is another application. Figure 27.10 shows this arrangement with a digital trunk interface connection at each end of the system. In this particular situation, 56/64 Kbps of voice connections are provided with 24 channels on each of the four T1s, resulting in 96 digital voice connections. This would be a prime case for connecting two PBXs in very close proximity to a single network, or a means of providing other voice networked services such as voice messaging, off-premises extension, etc. For a quad T1 service, the infrared system would price out at approximately $23,000, yielding a net payback of less than one year, given four T1s at $1000 per month rental from the local telco.

T1/T2 service between two sites

The provision of a T1/T2 service between the two sites to provide a 6-Mbps channel capacity or four 1.544-Mbps channels for video conferencing capabilities at compressed video standards are shown in Figure 27.11. However, these compression rates deliver superior-quality video compared to the more compressed speeds currently being offered in the industry (i.e., 256-, 384-, 512-Kbps video services). This service would cost approximately $15,000 for this type of service, and would certainly deliver a return on investment in less than one year.

Infrared high-speed digital dataphone services

The use of an infrared service to provide for high-speed digital dataphone services between a host computer is shown in Figure 27.12. In this scenario, a DDS service can be replaced with the system between these two buildings, whereby a statistical time-division multiplexer (Stat-mux) or a subrate multiplexer can be used to provide several 9.6- to 19.2-Kbps data channels at the remote site. This system would cost approximately $10,000, but a DDS service between these two sites at 56 Kbps could cost as much as $1800 per month. This would yield a payback in less than six months.

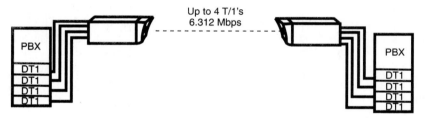

Figure 27.10 Linking PBXs together with up to four T1s for TIE lines. DTI cards are used in the PBX to yield up to a 96 PCM circuit at 64 Kbps.

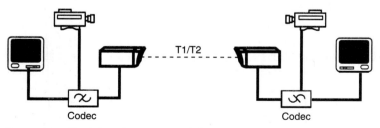

Figure 27.11 The infrared system can deliver speeds of up to 6 Mbps (T2) for video conferencing, or four links at 1.544 Mbps (T1).

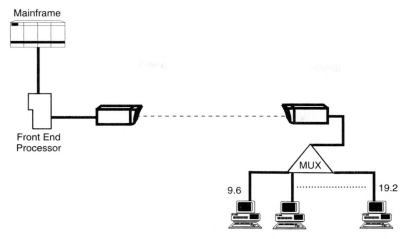

Figure 27.12 The infrared system can replace multiple DDS service for data connections to a host computer at 9.6–19.2 Kbps each.

These examples are merely representative of the types of connections that can be provided on a 1:1 ratio. Clearly, mixing and matching the services on these infrared systems would offer even greater returns on investment.

Can the System Carry LAN Traffic Transparently?

Obviously, the question always comes up with the reality of an infrared system being able to carry true LAN native speeds. As an example of providing this analogy, several conditions must be met:

First, the system must have the bandwidth capacity to carry the native LAN traffic. This has been evidenced in these scenarios, whereby the system throughput is at least 4, 10, or 16 Mbps.

Second, the system might not add significant latency to the network because the entire LAN is based on very specific timing for the delivery of the frames on the network. Should the latency become a problem, the frames will not be delivered in time, causing a discarded frame or a network timeout, which could cause the network to require a resynchronization to recover timing. In either case, the use of the infrared system across the air would approximate 3.3 ns per meter. A 1000-m distance between these systems would add only 3300-ns latency to the transport of the frame. Further, the system must modulate the signal onto the medium with some added delay based on the equipment. If a standard delay of approximately 30 ns is added for the modulation and the demodulation process, then an additional 60 ns of delay is added. These delays should fall well within the tolerance levels of the networks, as defined.

Third, the connection arrangement must fall within the scope of either an access to a bridge or repeater for the cable system in an Ethernet world, or else be connected as a station or at a multi-station access unit (MAU) for a Token Ring network. Once again, these connections are easily met with the use of the infrared interfaces and the network components.

Advantages of Infrared

The use of infrared technology brings advantages along with it. The benefits are quite numerous and should be continually considered whenever a connection is required, rather than immediately ruled out. Many organizations automatically discount this as an old technology, not worthy of consideration. The more astute telecommunications person keeps it in mind as an option, at least until some other technology proves out as either better suited for the job or more cost effective. Some of the major advantages are:

- Immune to radio-frequency and electromagnetic interference (RFI/EMI)
- Reasonably high bandwidth available
- Secure technology, more so than a radio-based technique
- Cost effective
- Easy to install
- Efficient when dealing with a potential move of the system
- No license required
- Systems can traverse similar paths without interference
- Can be up and running in a matter of hours
- Supports multiple standard interfaces
- Handles voice, data, video, and LAN traffic either separately or on the same system, depending on application

Disadvantages of Infrared

As with any technological invention and use, there are always the reverse sides of the positive. The primary disadvantages to the use of infrared are not as dramatic, but should be factored into the equation. Because the use of a light-based system, such as this, is primarily based on the actual need, these disadvantages exist:

- Very limited distances, up to 1½ miles or shorter distances, depending on the manufacturer and the rate of speed being used.
- Concern over the use of a laser in an office environment.

- Negative effects of vibration on the use of a system.

- Atmospheric disturbances that could impair the reception of the beam from the transmitter.

- Risk of exposure to the light (laser) beam from cleaners.

- Needs clear path in a line of sight, regardless of any other condition.

- Mounting must be stable because of the sway of buildings, mounts, poles or trees, where very little tolerance can be allowed as a result of the small footprint of the received signal.

Fiberoptic Compatible Systems

A newer product has emerged on the market that utilizes a fiberoptic system of wave-division multiplexing services with up to four clear channels of pulse frequency modulation video or digital data signals per wavelength window via one multimode fiber. Although this is not a wireless technique, the concept is similar to a development that has been touted as *free-space optics*. In the conceptual model of this newer service known as *free-space optics*, the vendors are creating the bandwidth of a fiberoptic channel capacity across free air waves. Hence the name "free space" as opposed to inhibited by a piece of glass cable. A standard fiber-based system uses different wavelengths to carry the channels of communication across a glass fiber. Whereas, the free-space optics system will use a different color of light in a different wavelength to carry various streams of voice, data, or video transmission over short distances. Obviously, because this concept will use laser beams in free space, the need exists to use low-power output devices. Distances of approximately three to five miles are the numbers being discussed at this time. Some of the applications for this free-space optic system will include, but not be limited to:

- Security and surveillance systems

- Multimedia learning over a short distance

- Video multiplexing at NTSC, PAL, and SECAM standards

- Digital data at native speeds

- Digital voice transmission

The low power will be a requirement to prevent any damages from the effects of the laser output while operating in close-in areas, such as major metropolitan downtown areas. The system will also have to meet these requirements for infrared systems, taking care not to be too low to the pedestrian traffic. However, because this is in the visible light spectrum, the signal will be less impeded by the various atmospheric drawbacks that affect the operation of an infrared system.

The systems will also be fully modular and easily transportable, as with an infrared system. This will allow the flexibility to set up and operate a transmission quickly. Because this will be light beams traveling across the airwaves, the licensing issue will not come into play. The bandwidth of these systems will range from approximately 2–40 MHz capacities, yielding data rates of up to 16 Mbps for native LAN traffic at the LAN speed. Wavelengths will operate at 750, 780, 810, and 840 nm respectively.

The use of this form of communications and medium will obviously offer some very attractive options to end users for access to the long-distance carrier's point of presence (POP) and the ultimate access to the long-distance networks. Local area networks, campus area networks, metropolitan area, and wide area networks will all have some use of the free-space optic systems. Consider the use of such a system as an alternative to access the long-distance network, or as a means of by-passing the local loop in the telephone company arena.

Chapter

28

Video Conferencing

Video Conferencing Systems

Up to now, this book has covered the use of the older analog dial-up voice network and the newer techniques used to produce digital transport systems to carry voice and data transfers. However, another use of this network is coming of age. The ability to conduct face-to-face communications between and among sites in the business organization. The age-old concept of jumping on a plane and traveling to a remote location is fast becoming passé. The reasoning behind this is the wear and tear on traveling executives, professionals, sales forces, and others.

Timing is also a critical concern. To be competitive in today's marketplace, decisions must be made quickly and decisively. Many times the information necessary to make these decisions requires the face-to-face meeting, or the ability to see a product. In the engineering world, the ability to see a component or piece of equipment that needs modification is also crucial. In the past, the only way to handle this was to go to the location and conduct the meeting or review. The options were limited and many times they were too expensive. Now this is all changing quickly. The use of cameras that can project an image of the remote site across the network is far more prevalent. Technologies have improved, the network has increased in reliability and performance, and the capacities needed to transmit such information have become far more reasonably priced. Therefore, more and more organizations are deploying video conferencing systems throughout the organization. It is becoming one of the tools of the late 1990s and will carry into the turn of the century.

The use of video conferencing systems in an organization is not new. These systems appeared in the industry back in the early 1960s. The concept is a simple one: conduct a face-to-face meeting with members of the organization anywhere in the world without the travel associated with such a meeting. This takes advantage of the leased-line or dial-up capabilities of the telecommunications industry techniques that use a device to convert the analog signal of a video conference into digital pulses. This device is called a *coder/decoder (Codec)*.

Video conferencing has been technically possible for several years, but at a steep price that made it a practical alternative for only very large organizations with specially designed facilities. In the late 1970s and early 1980s, an organization might have spent $1 million just to prepare a site for the video capability. This was particularly true in the boardroom conferencing. This was considered a senior-level management tool because of the cost. Now, the arrival of universal standards and improvements in equipment and signal transmission have led to much smaller systems with significantly lower costs. The costs of the equipment were $200,000 to $300,000 in the early days. New systems are available in the $20,000- to $60,000-range. These developments have made video conferencing an affordable alternative for businesses of all sizes.

Roll-about and tabletop video conferencing systems have rapidly replaced their larger and more expensive predecessors. A new market for video conferencing between groups of two to three users emerged. The video conferencing systems and their use is no longer relegated to the very expensive boardroom environment, it can now be easily accommodated at the desktop or in a regular conference room. Special facilities and lighting constraints that were prevalent in the larger systems have been displaced with the newer, more sophisticated systems that overcome the environmental situation. Small groups of engineers, designers, financial personnel, or managers can conduct ad-hoc meetings. These can be impromptu meetings because the equipment can virtually operate anywhere. Desktop units that cost as little as $10,000 to $15,000 make this a more viable solution and palatable solution to the cost and time associated with travel.

The logistics of transmitting a video signal across the leased line or dial-up telephone network are no longer as complex as in the past. For the average business, the bandwidth (or pipe) is already in place to support the video conferencing systems. Therefore, video equipment transmission needs only to be added and piggybacked on the existing voice or data communications links. Digital leased services, such as T1 or switched 56 Kbits/s are readily available in organizations. Additionally, the emergence of integrated services on a digital network (ISDN) has opened the door for the impromptu dial-up connection at speeds of 56 Kbits/s up to 1.544 Mbits/s. The most commonly used is in the range of 128 to 384 Kbits/s. This is easily accomplished through an inverse multiplexer (I mux). You might remember the discussion of the I mux in Chapter 24. The lower economic barriers to entry in video conferencing and the managerial efficiency sought in corporate America make video conferencing the opportunity of the 1990s.

What Is Video Conferencing?

Back in the early 1960s, AT&T introduced the concept of video conferencing at the New York World's Fair. Their idea at the time was a product known as the Picture Phone. In this early introduction, callers could see a motionless snapshot and hear the voice of the party on the other end of the connection, at the same time. Even though this exhibit generated a lot of excitement and opened the eyes of many organizations, it was really just a demonstration of how far the technology could go. The service and the technology would have to evolve and mature before it could be taken seriously as a business tool. Early attempts at video conferencing were stymied by several factors.

The systems had to be connected to similar devices on both ends, and only a handful of suppliers existed at the time. Standards for interconnectivity were nonexistent. The special circuits required at the time also led to delay because they were extremely expensive and not ubiquitously available. The operating costs for such a system were astronomical.

Eventually, AT&T's picture phone died a slow death, but the spirit and the concept survived in the minds of the engineers. Research continued and improvements were made. During the 1970s, full-motion, two-way, closed-circuit video emerged in the boardroom environment. Also, one-way noninteractive broadcast private TV networks that used satellite transmissions or very high bandwidth began to appear. Although firms made good use of their networks, others were not willing to make the investments in the build-out of special rooms and circuits at each of their sites. Despite the improvements that were evident, the financial burden slowed the development and installation of video systems.

During the 1980s, things began to change. Several major breakthroughs opened the door to reasonably priced video conferencing. The more significant accomplishments included the following:

- Improved video compression techniques made the bandwidth constraints less difficult to deal with because the systems could operate at less than full T1 speeds.

- The divestiture of the AT&T and Bell Companies opened the door for competition in the telecommunications industry, reducing transmission costs by 40% and more.

- The evolution and roll-out of digital transmission systems throughout the industry rather than just in the major metropolitan areas.

- Computer chip technology progressions that led to the decline of the cost of the electronic componentry.

Prior to 1986, a typical video conferencing system cost in excess of $200,000. Because these systems required special acoustic conditions, lighting, and other environmental conditions, they were installed in the special video rooms. The treatment of these rooms could be as much as $700,000. This was obviously too rich for the average organization.

A mere one year later, the cost of the equipment had dropped to $120,000. By the end of the decade, a standard video conferencing system bore a $60,000 price tag. Price competition increased significantly in the early 1990s, as shown in the graph in Figure 28.1. Although some companies are still spending over $100,000 to build and equip lavish video rooms, the market trend is toward a functional, accessible, modular system that can be used by all levels within the organization.

Video Conferencing Overview

When we think of video conferencing, the image of high-tech multimedia conference rooms looms in our minds. Although this is true in some environments, the reality of the situation is that more and more lower-end type services are what drives this application. Video conferencing has a significant potential as an efficient, productive means of conducting point-to-point or point-to-multipoint meetings. Identifying and achieving that goal depends on the organization's acceptance and knowledge of video conferencing systems. Teleconferencing in itself offers many benefits. Some of these can be summarized as follows:

Increasing managerial productivity

The time for travel to and from sites is normally nonproductive. Regardless of whether the individuals involved with travel are dedicated, the problem is putting that time to productive use. This is becoming increasingly more difficult when air travel is involved. Most of us have been crammed into an

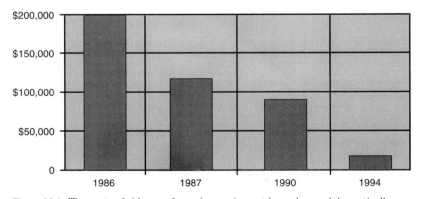

Figure 28.1 The costs of video conferencing equipment have dropped dramatically.

airplane at least once. Dread the thought that you wind up in the center seat on aircraft. You just don't have the space to move or spread documents out. Coupled with this is the fact that many of the airlines are now cramming every seat they can get onto a plane to produce more revenue. The lack of space is compounded when the plane is chock full. The quality of air travel has decreased over the years, despite what the airlines want us to believe. Constant interruptions, constant passenger fidgeting on the plane, and climbing over disrupts the work. Even the road warriors who have computers (laptops and notebooks) are pressured with lack of space to conduct much useful business on the plane. So, if this time is nonproductive, the counter to this is to replace the travel with a technique that allows the executive to attend the meeting for as long as necessary. Upon completion of the meeting, the executive walks down the hall back into his or her office and gets right back into the thick of things.

Quality of meetings

A funny thing happens when video conferencing systems are used: The quality of the meetings held usually increases. This is in part because of better planning and execution of the meetings. When someone needs to travel to a meeting the inevitable happens, they get to the meeting, but left something behind. Or, as an alternative to this, they still have to get back to the office to check with others. Clearly, an entire organization or department cannot travel to a meeting. When video is used and we need to check with another department, this is just down the hall. Decisions and discussions can be facilitated much more expeditiously.

Pre-planning a video conference usually makes the time more efficient. If everyone has an agenda and pre-meeting materials, the time is used better. With video we tend to do more of this pre-planning. Because the cost per minute used to be so much more expensive, everyone tried to be as quick to the point as possible. Otherwise, the conference was going to be exorbitant.

Time and money savings mount up. The savings in pre-planning mentioned above make better use of our time. Further, the cost of travel is escalating at a very rapid pace, despite what the travel industry tell us. Consequently, we can involve more people, bring in conferees only when needed, rather than have them sit around for hours waiting their turn to speak, etc. This all adds up to a benefit in more ways than money, but also valuable time.

Improvements in the quality of life

We travel a lot (Bud runs up 350,000 to 400,000 miles a year, and Don is no slouch either). The amount of time dedicated to road warrior events is constantly demanding five to six days of their time. This means that the traveling person similar to the authors, only gets one day a week at home. This puts a lot of pressure on the home life, as well as what happens in the office. We have to be more accessible, and we have to work much longer hours. There are no substitutes to getting person-to-person communications taken care of. Even though techniques such as e-mail, voice mail, network access from remote, etc., all exist, they still do not replace 100% of the need to speak with a human. As the travel gets eliminated or reduced with video conferencing, the amount of free time that the individual retrieves is phenomenal. This is not to imply that some limited amount of travel will not be required, but when one day per week is compared to six days per week, there is no contest. Unfortunately the senior levels in organizations wind up in this catch-22 race, where they are on the road five and six days per week. They will be prime candidates for video conferencing applications.

The Pieces to a Video System

The components of a video conferencing system vary depending on budgets, applications, and personal preferences. However, certain basic components are required. These include the following:

- A camera
- One display screen or monitor, at a minimum. Many of the systems come with two.
- Codec (the heart of the video system is the coder/decoder)
- Audio input/output capabilities (unless you are very good at reading lips, you should have an audio service integrated)
- Control, either manual or automatic
- The communications link

The camera

Crucial to a video system is the ability to see what's going on at the other end. The camera is the basic entry into this service. The camera is one element that should not be overlooked: the better the camera, the better the presentation will be facilitated. Hopefully, the camera system is something that is nonobtrusive in a conference environment. In the early days, the cameras were a constant source of confusion. As needs dictated, the camera had to be shifted from one speaker or object to the next, then back again. This required that a controller in the room was assigned to move the camera from position to position. Unfortunately, one of two scenarios constantly got in the way. First, the controller got so engulfed with the meeting that the camera never got moved. Therefore, people would see blank walls and not the body language of the speaker. The second might be that the controller got so involved with the positioning and movement of the camera that he or she lost all interest and input into the meeting. This should be taken into consideration right up front. Most systems now come with simple to use camera arrangements that allow fixed and variable movements on the camera so that positioning can be simplified. In many systems, the camera might be mounted inside the cabinet so it is not visible. The quality of the camera cannot be overstated. A transport system and the codecs cannot make up for poor-quality input from a cheap camera.

The monitors

Every video conferencing system will include at least one monitor. Larger systems will typically be offered with two or more, using standard TVs. One will be for the graphics camera, the other will display the participants. Some will use the two monitors to display the participants at each end. Still others might include picture in a picture (PIP) so that both ends of the transmission can be seen at once. The variables are many, depending on the manufacturer, the system, and the depth of the budgetary pockets. Monitors should also be a concern so that sharp images can be projected, but much of the quality of the sharpness on the monitor will be determined by the speed of the link. Poor-quality reception might not be a problem in the monitor. As we move the systems to the desktop, the variations of the monitor size increase exponentially. Typical monitors for conference room units are 35" screens, whereas the roll-about systems today use a 25" to 35" monitor. Desktop units are now using a 19" monitor. The size is not a function of the quality, but of the overall display that must be viewed by the number of participants.

Audio Capabilities

Most people take the audio portion of a video conference for granted. Unfortunately, this piece of the connection is equally if not more important than any other. Most of the actual dissatisfaction that comes from a video conference usually involves poor audio quality. There is a reason that the audio information is so critical: Research has proven that 60% to 70% of all information on a video conference is still audio transfer of information. Although we supplement the audio portion of the meeting with visual input, the bulk of what happens is verbal.

Other ideas state that the combination of the visual and the verbal communications enhances the retention level to approximately 85%. Therefore, the audio portion of the conference is as important as the video. Another way of looking at the situation, if a video conference is established and the video portion goes away, a good part of the meeting can still take place. Attendees might be able to refer to handouts or other materials and follow along with the audio presentation. However, if the audio goes away, the connection will be rendered useless. This points further to the total need for all pieces to work, but also underlines the need for the audio connection. Typical audio systems allow for up to 15 kHz of voice quality in a video system to give the best-quality voice communications possible.

The Control System

The actual brain of the overall video conferencing system will incorporate the control system and the software portion of the system. These components help to orchestrate the overall performance of the video conferencing capabilities. Every part of the system interacts with the control and software functions. The audio mixers use the control to allow for full-duplex communications, the cameras are controlled in the angle, pan and tilt, and zoom functionality. The control portion that is developed in the software makes the system transparent to the participants in a conference. Executing the preset functions for scanning the audience, or locking into a specific camera is done in this part of the system. Also, adjustments to focus and volume controls on the audio can all be accomplished here.

The Coder/Decoder

Using compression techniques to send a lot of information over the communications link is part of the function of the coder/decoder. This is the chipset that will take the analog picture of the participants and digitally encode it for the transmission link. At the receiving end, the decoding function will take the digitally encoded picture and convert it back to an analog form for delivery to the monitor. This acts similar to a modem; it is the change agent in the video conferencing system. The Codec will handle the digital compression of the information so that we can get away with using a much lower data rate on the actual physical link. Compression algorithms are becoming so sophisticated that the Codecs can do on a 128 Kbits/s or better what used to take 90 Mbits/s transmission. This is a big savings in terms of the availability and acceptability of the video conferencing deployment.

The Communications Link

The less visible but very important component of the video conferencing systems is the communications link. Improvements in video conferencing, both from a quality and cost standpoint, closely parallel that of advances in the transmission technology. The demands for speed have decreased over the years from a full T1 speed to today's technology that requires as little as 64 Kbits/s (Figure 28.2).

Transmission or data rates vary, but most common systems use either 56,000 bits per second (56 Kbits/s), 64 Kbits/s, 112 Kbits/s, 384 Kbits/s, or

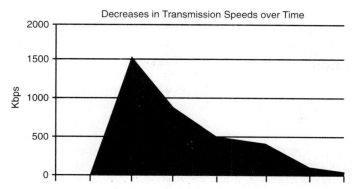

Figure 28.2 The overall improvements in technology have reduced the bandwidth for video over the past decade.

1.544 Mbits/s at the high end of the spectrum. In comparison, these data rates are a dramatic reduction from the 90 Mbits/s of bandwidth needed to support the original picture phone at the World's Fair, which actually produced a very hazy picture at best. That 90 Mbits/s transmission rate in 1960 was considered quite reasonable. This is also the speed and throughput used by the major networks to bring us one-way broadcast-quality transmission services on our TVs. Compare a still frame of video at that rate to that of what you see on TV, and you have the differences in the world of video.

The higher the data rate, the better the picture quality and the clarity of the motion. A data signal transmitted at 56 Kbits/s has difficulty handling a lot of motion and picture detail. Even though strides have been made to achieve tolerable video conferencing at 56 Kbits/s, the average participant would not be happy with the results. Signals sent at 128 Kbits/s (or multiples of 64 Kbits/s higher) produce details that increase proportionately with the increments of bandwidth added. From a personal perspective, the only starting point that really works in a dynamic meeting is the transmission at 384 Kbits/s. However, this is not a ubiquitous service, and it is very pricey. The average organization will start at very high expectations, but as the pricing starts to evolve, their expectations in picture quality start to drop. It is amazing what the financial picture will do for a person's tolerance level. There are basically two options for transmission:

Dedicated or private line services that include high-capacity T1 circuits between and among locations. This can be very expensive.

Switched-carrier services are also available from the long-distance companies. Switched services are often referred to as *bandwidth on demand* services because the user only pays for the usage when it is needed. These switched-carrier services can also fall into two different categories: Switched-digital services (such as switched 56 Kbits/s and 64 Kbits/s) or ISDN (which is a digital switched service that provides two 64 Kbits/s channels). Although this service is not available everywhere it might be needed and differences exist in the implementation of ISDN around the country and the world. ISDN is offered in two interface rates:

Primary rate interface is a 23-bearer service and a data link channel, all operating at 64 Kbits/s each. This is essentially the dial-up version of a T1, with a signaling system 7 channel (the D Channel) for the call set-up and tear-down.

Basic rate interface is a 2-bearer service operating at 64 Kbits/s each, plus a D Channel operating at 16 Kbits/s. One BRI can support a dial-up video conference at up to 128 Kbits/s. If more is needed, an I Mux can be used to pull together the increments of 64 Kbits/s needed, such as 256 or 384 Kbits/s. This is a very effective tool toward using the video conferencing systems today.

Many of the systems on the market today use the BRI as the interface for their video dial-up services. The use of a low-end I Mux costs approximately $10,000 to $20,000 and allows the user to pull together the capacity of the BRI or multiple BRIs together.

Standards in Video Conferencing

Although video conferencing has been around for so many years, it is only recently that it has taken the industry by force. Why now? What is different? Actually, the result of 30 years of pent-up demand might be one of the reasons that video is such a "hot button" in the industry. However, the major contributor to the most recent interest and hyperbole is the result of newer and better standards for interoperability in video systems.

The ITU made significant strides in the video conferencing standards arena. The standards fall under a blanket of the H.320 specification, which details the way video systems will communicate with each other. Recommendation H.320 developed by the ITU Study Group 15 still has some hurdles to overcome, especially in the LAN arena because many organizations are now considering the use of video on the LAN. However, the outlook is a positive one. The interoperability between differing systems marks a turnaround in many of the obstacles that kept a number of organizations away from the use of video. The P × 64 standard permits equipment from different manufacturers to speak the same language based on the bandwidth used. In these standards, the P stands for the multiples of 64 Kbits/s being used. The number can range from 1 to 30 (or 64 Kbits/s to 1920 Kbits/s) and conforms to international standards.

Prior to the adoption of the standards, choosing a video conferencing product was a risky proposition. To guarantee compatibility, users were restricted to the purchase from a sole provider for every point in their video network. As a result, the market was filled with a diverse product line that was generally unable to work with each other, unless a mirror image was installed on the other end of the line.

The most notable of the standards is the H.261, which specifies how the codecs (coder/decoder) decipher the digitally transmitted signals from dissimilar systems. It stipulates a minimum design to ensure all codecs will interoperate at the lowest level. H.261 covers transmission from the basic ISDN rate of 64 Kbits/s or 56 Kbits/s through the full T1 rate. In addition, H.261 defines two resolution standards:

- Quarter common intermediate format (QCIF), a format used mainly in desktop and videophone applications.
- Common intermediate format (CIF), a format that is optional under the H.261 standard used for room systems.

CIF requires a coder/decoder to transmit video at a rate of 30 frames per second with 288 lines of lumina pixels per frame, 352 lumina pixels per line, 144 chroma lines per frame, and 176 chroma pixels per line. This does produce a picture with lower resolution than broadcast TV. QCIF is even lower resolution because the pixel and line numbers are both halved, yielding 144 lines of lumina pixels per frame, 176 luminal pixels per line, 72 chroma lines per frame, and 88 chroma pixels per line.

Flexibility is built into this standard to allow each manufacturer to fine-tune the performance of their codec. Although most manufacturers have developed their own proprietary algorithms that translate the analog information into digital data, the standards ensure that divergent systems will be able to communicate on a generic level. In early 1992, the three largest manufacturers successfully completed interoperability using the P × 64 standard, proving that once noninteroperable systems could internetwork. This has led to a significant increase in the demand for the video conferencing systems, because one of the major barriers was removed.

Other standards are also included in the use of video conferencing systems. Some of the more common ones are shown in Table 28.1 with a representative summary of what it is designed to accomplish. These fall under the umbrella of the H.320 standards.

Most manufacturers now offer video conferencing equipment that is H.320 compliant. Therefore, if a room system is H.320 and so is a desktop that is H.320 compliant should interoperate. However, while virtually all group systems are H.320 compliant less than half of the desktop video conferencing systems on the market support the standard. The buyer must beware. As the H.320 continues to roll out, then it must be expanded to address the video on the LAN and the interactive sharing of documents. Study Group 15 is currently working on standards for screen and document sharing, known as *collaborative computing*.

The major emphasis on video systems today in the industry is the desktop and the roll-about system. However, sales of these systems are projected to skyrocket over the next few years. These projections are shown in

TABLE 28.1 A Summary of the Standards in Video Conferencing

Standard	Description	Status
H.261	Uniform coding of a signal. Outlines how to compress digital video information, the syntax and semantics of a video bit stream (P × 64).	Adopted
H.221	Communications framing, specifies what information is in a bit stream so each codec can keep track of video frames. Outlines whether a bit is audio, video or control as well as how to weave bits together.	Adopted
H.230	Specifies how commands between codecs are exchanged during a session for control and diagnostics.	Adopted
H.242	Transmission Protocols for Call set-up, transfer, and tear-down.	Adopted
H.320	Technical requirements for low-bandwidth audio-visual systems. The overall specification for video conferencing.	Adopted
H.331	Half-duplexing multiplexing. Tells the codec how to string together data for transmission over half-duplex channels such as satellite.	Adopted
G.711[1]	64 Kbits/s pulse code modulation for audio.	Adopted
G.722	48/56/64 Kbits/s variable adaptive differential pulse code modulation (VADPCM) audio.	Adopted
G.728	16 Kbits/s audio.	Adopted
H.233	Transmission confidentiality defines the method for identifying and negotiating Encrypted Data	Adopted
H.243	Multipoint handshake, defines communications between a multipoint control unit (MCU) and the codec. Multipoint video conferencing.	Adopted
Not named	Still-frame graphics coding	
	Data transmission	

[1] The G.711, G.722, and G.728 standards specify the audio compression parameters that range in bandwidth speeds of 3.5–7 kHz.

the graph on Figure 28.3. Clearly, this is going to be a big movement in the industry, if the sales skyrocket as they are projected. The use of video will become another tool in the conduct of our everyday business, regardless of where we reside in the chain of command.

Multipoint Control Units

The use of a multipoint control unit is beginning to climb as more organizations are bringing in video conferencing systems, and the cost is making it easier to use these systems in more sites. Once a video conferencing program begins, the next logical step is to begin the multipoint conferences. What was once just a point-to-point service now needs to adapt to the changing business needs. Ad-hoc meetings among sites is becoming more prevalent. The collaborative and dynamic workgroup formulation requires that three or more sites communicate with each other simultaneously. To perform this function, the user is faced with a decision to either bring the MCU functionality in-house, use a bridging function from an outside resource, which there are limited-service bureaus (such as Sprint and AT&T), which perform this function or use a combination of the two.

The MCUs are devices that will allow the bridging and concatenation of the necessary sites into a single conversation. These devices will range from $10,000 to $100,000, depending on the system configuration. Typical sites use a device that will allow up to 8 to 20 ports. These range from $10,000 to $20,000 on average.

Many of the organizations that use a mix of these services will start off slowly with a four to eight port MCU. When larger meetings are required, or when a boardroom type meeting is scheduled they use the services of the bridging companies. Unfortunately, the use of these systems results in mixed reaction and varying satisfaction levels. Normally the quality of the connection is not the issue. Rather, it is the attitude and scheduling conflicts that result in user dissatisfaction. Forgotten appointments, uncooperative bridging operators, and even the cost issues have created a demand for the MCU to be moved in-house. More organizations are looking at the use of these systems internally. This actually comes when most organizations were considering the use of outsourcing all of their communications to a service provider. This is yet to be a resolved issue, but a significant amount of activity will spur a new business opportunity. When there was little need for such a service, there were few players. When the need will rise as the number of units expands, then the number of service providers will also increase. Many of the carriers have introduced various flavors of the

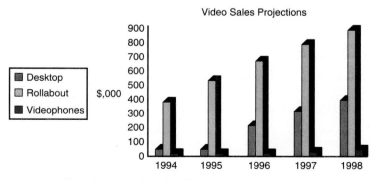

Figure 28.3 The sales projections for video systems.

multipoint conferencing systems. To do this, however, they have purchased an MCU from each of the players in the industry to maintain compatibility. Very few will discuss the fact that they try to accommodate a user's needs through a system that matches the end-user system, rather than adapting to the standards. We know that this will change over time, but the costs associated with using a public bridging system are artificially kept high because so many pieces of the equipment from various manufacturers are required.

The MCU is typically provided as a dial-up service by the major players (Figure 28.4). Here an end user can be provided the connectivity needed for a multi-location teleconference.

Connecting the Pieces

Once the decision to use a video conferencing system is made, the obvious decision to make is what system to use: the boardroom version, the roll-about unit, or the desktop model. Depending on the needs of the organization, this could involve multiple units. One of the authors has had several clients who have purchased multiple units in the three categories, at each of their sites. This is a financial and a business decision. Whatever can be accomplished with a single system is great. However, if users have to wait in line for a system to become available, they might try to avoid using this technique and revert back to the old travel arrangements. A view of each of these systems is presented in the next few pages.

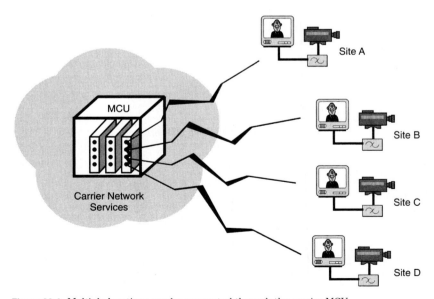

Figure 28.4 Multiple locations can be connected through the carrier MCU.

TABLE 28.2 Summary of Equipment in a Boardroom Environment

Description	Number/type
Monitors (presentation)	1 at 27" or 35", auto-scanning, multiple inputs at composite video, RGB, S-video, high resolution at 640 × 480.
Cameras	2 high quality with the following features: 8–10x zoom; auto-focus; motorized pan, tilt, and zoom; lens focal points at 6–48 mm; zoom ratio at 8 × 1; total pixels > 400,000; effective usable pixels > 380,000 (494H × 768V); horizontal resolution 450 lines; vertical resolution 350 lines, fade capable.
Codec	1 each, using H.261 plus video coding specifications, H.320 compatible, provides picture in picture service, RS-449 and G.704 standards, echo cancellation, G.728 audio output, encryption capability, public or private line service.
Controller	1 with the following features: pan, tilt, and zoom on cameras; preset camera positions; separate control buttons for camera, video, and graphics input; and variable volume controls.
Document camera	1 CCD single chip design.
Speeds supported	Variable at P × 64 from 128 Kbits/s to 1.544 Mbits/s.
Monitor	2 at 35" for large room facilities with dual RCA jack for audio and video delivery to the monitor.
Microphones	Multiple omni-directional microphones desk or table mounts.
FAX machine	For reception of documents from remote locations this should be CCITT group III standard.
VCR	Used for record or playback of video conferencing sessions.
I Mux	1 with variable inputs to draw up various channels at 64 Kbits/s, from 2 to 30 using international standards.

The boardroom version

This is the larger of the systems. It is not a mobile unit by any stretch of the imagination. To accommodate the very high-end type conferences and presentations, organizations still use the fixed room location. This is a unit that can start at $60,000 and run up into the $100,000+ range. Further, this unit will require a specially treated facility. Although the special treatment is not a system requirement, if you are dealing with the senior levels of an organization, they will want everything to be perfect. This will take some doing. First in the design of the equipment, the following pieces will likely be used (Table 28.2). These are the base pieces that will accommodate this type of a conferencing system. Added pieces can be brought into the equation, depending on the amount of functional systems needed to satisfy the business need.

A graphic representation of this type room is shown in Figure 28.5. This is the area where large amounts of money will be spent to support this environmental conditioning and conferencing equipment. The systems are shown as standalone, although there are several ways to connect the video service. This can be through the I Mux and ISDN lines from the outside world, at the BRI or PRI rates. If BRIs are used, the I Mux can support pulling the three BRIs together into a single data stream at 384 Kbits/s. If the I Mux is connected to the PRI service, it can allocate as many channels as necessary to support the full P × 64-channel capacity (Figure 28.6). Another way to connect this service is to use an I mux behind a PBX (Figure 28.7). This is a generic connection that allows the various opportunities to integrate the communications technologies into an enterprise strategy. No one design will work in all cases, but the general layout will at least give the individual a basic understanding of how it all works.

The rollabout system

The rollabout is the system that can be used in between multiple conference rooms. This is particularly attractive when the organization does not have the resources or the ability to dedicate a specific room for video conferencing systems. The users merely book the system and bring it to the most conve-

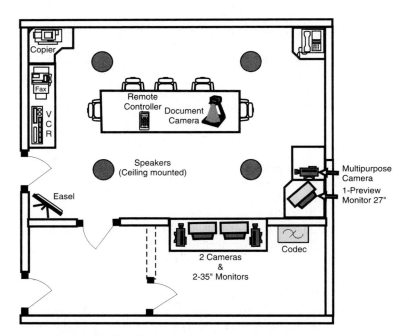

Figure 28.5 A typical boardroom layout.

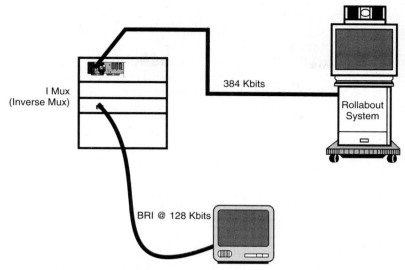

Figure 28.6 The I-Mux is used to pull channels together for video.

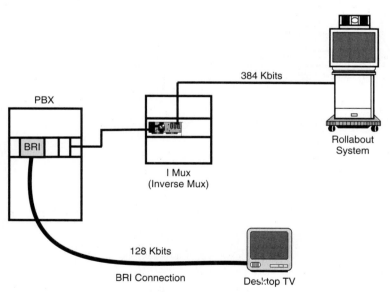

Figure 28.7 The I-Mux can sit behind a PBX, or a BRI card in the PBX can deliver desktop video.

nient conference room or individual office environment (Figure 28.8). This system costs less than the boardroom configuration, so there might be many of these purchased in an organization and rolled about from room to room. One must be careful that trained personnel know how to move this from room to room to avoid tipping and breakage. Although these units are designed to roll around, and they do, care must still be exercised in moving them. A sample checklist of equipment concerns for any video conferencing system is provided in Table 28.3. This checklist includes such pieces as would be considered for any system and whether they are standard or optional. Along with this information, the checklist also includes some pricing information so that a comparison can be made of the various systems.

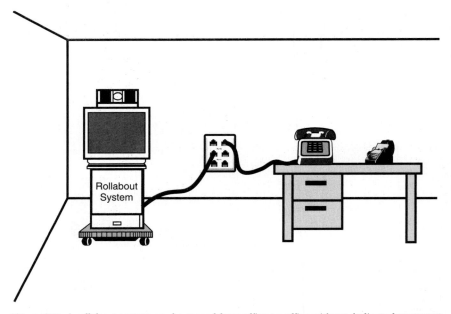

Figure 28.8 A rollabout system can be moved from office-to-office without dedicated rooms set aside.

TABLE 28.3 A Generic List of Options and Pricing for Video Systems

Description	Standard vs. optional	Pricing
1. Codec		
2. Monitors • small (12–18") • medium (20–25") • large (26–35")		
3. Cameras • auto-focus • document		
4. Multiplexer or inverse multiplexer • speeds • number ports • channels that can be banded		

TABLE 28.3 Continued

Description	Standard vs. optional	Pricing

5. Microphones
   - voice actuated
   - lapel
   - desktop
   - built-in
   - upgraded available

6. Controller
   - handheld for remote pan, zoom, and tilt
   - central desktop mounted

7. Input for other devices
   - VCR record and playback
   - slide presentation
   - white board
   - electronic blackboard
   - PC input for playback

8. Interfaces available
   - RJ45s (number available)
   - TNC/BNC
   - RS449
   - V.35 (dual) or RS366
   - ISDN (PRI or BRI) (number available)
   - Dual DSU
   - I Mux (ports, capacities, speeds)

9. Carts
   - self-contained
   - custom or standard

10. Outputs
    - NTSC
    - PAL
    - SECAM

11. Other associated pieces (vendor list of options)
    -
    -
    -

12. Installation

13. Wiring

14. Testing

15. Training (number of personnel)

16 Total system price

17. Maintenance
    - Warranty period
    - Parts and labor 5 days × 8 hours on-site
    - Parts and labor 7 days × 8 hours on-site
    - Extended periods on-site
    - Factory return

The PC-based system

The initial systems were mounted in the boardroom, then they migrated to the rollabout capability. The most recent innovation is to use a system that is desktop mounted or PC-based. Cards that have the chipsets built onto them can now be installed in various forms of PCs. The standards include boards for ISA, EISA, MAC, and S-Bus architectures. The Macintosh lines also have cards available for the NU-Bus. The convenience of placing the video conferencing right at the desktop facilitates the impromptu meetings and the ability to share documents on a computer. Some of these cards might take up a single slot in the PC, whereas others require two slots. The expansion slots on a bus are prime real estate; therefore, the movement is to find a system that will only occupy a single slot. Regardless of the number of slots, these systems are designed to work at the 56/64- to 112/128-Kbits/s rate. They will not go any higher than that today. The small camera is a fixed-focus "lipstick" camera mounted atop of the PC's monitor. Two dial-up lines are available, specifically with the BRI/ISDN service. The picture on the set can be full screen or a small corner of the screen. The software in these systems will allow document sharing and video simultaneously. This allows for the collaborative development and work group formation on the fly. If you can get used to seeing a very small image and a grainy one, then this system will fit the need for most impromptu meetings. However, experience shows that the number of participants that can easily work together on these systems is between one to two. Any more than two and the players involved in the meeting will be fighting to see what is going on. This being the case, the rollabout system will probably be a better solution The cost of the PC-based systems have been moderate, with prices ranging from $2000 to $6000. For a desktop system, this might well be the price that we'll see for the next couple of years. However, as the system becomes more mature and more applications develop to run the impromptu meeting, the prices should fall. The only significant problem with these systems is the control software needed to orchestrate the entire hardware platform simply. In numerous demos, the author has seen these software systems lock up or fail to perform as they had been advertised. This is because they are new and the bugs have not all been worked out yet. You can expect in the near future, that these problems will become nonissues. The next issue is just getting the bandwidth to the desk, the available connectivity through the building and the applications necessary to share with other workers.

The cost of this system will be a factor in the acceptance rate. With a base entry into video conferencing at this price range, more and more small branch offices, small businesses, and home-based businesses will be looking to this system. The PC-based system is shown in Figure 28.9. The number for the system is above and beyond the cost of the regular PC hardware and monitors that are required.

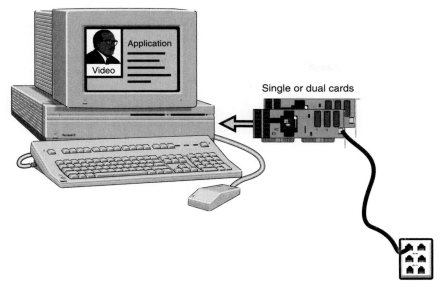

Single or dual cards

Figure 28.9 A PC-based system serves the impromptu meeting well. A single or dual card is used.

29

Finances for Telecommunications

Introduction

Up until about 1980, most communications, both data and voice, were regarded in most companies as "costs of doing business." There were many reasons for this.

1. For the first half of the century, only voice communications existed (in electronic form).

2. Voice communications capability was supplied by "the telephone company."

3. The telephone company really did not offer a great variety of options to its customers.

4. What it did offer worked well, but management could make few choices.

5. The "telecommunications manager's" main responsibility was to place and coordinate orders with the telephone company, and to track the bills to ensure that they were correct.

6. Because there was little choice in what to order, "managing telecommunications" was almost an oxymoron.

Also, people who managed communications tended not to have a management point of view. They were not oriented to providing management enough information to make business decisions. After all, presenting a telecommunications budget as a *fait accompli* was much easier than justifying each project, one by one. This orientation is still seen in many telephone bills rendered by Postal Telephone and Telegraph authorities (PTTs); the level of detail seen in a normal bill in the United States is often, outside the United States, either a separate (priced) service, or is simply unavailable.

Beginning in the late 60s, choices began to proliferate. Data communications began to grow as a percentage of the total communications budget. Still, the mentality that regarded communications as an overhead function with little potential to be managed—certainly not a possible profit center—remained at most companies.

However, as more and more United States companies felt the cold winds of international competition, and as each market niche started to fill up with more and more competitors, every area of expenditure in organizations came under scrutiny. Telecommunications (used here to include both voice and data) constitutes a significant expenditure of funds in many companies—and tends to increase steadily. In many organizations, this is the number two or three largest cost on the balance sheet. As an overhead function, it is a prime candidate for cost control and optimization, especially with the increasing number of ways to satisfy any given communications requirement.

Cost is always an issue no matter what financial position your organization is in. Management expects that the best pricing and value for its dollars are being achieved. Yet, management still sees telecommunications as an overhead function, a necessary evil. This image must be dispelled. The use of telecommunications as a strategic resource will:

- increase productivity
- reduce costs
- improve the bottom line

All you have to do is convince your management of these benefits. Moreover, companies are increasingly becoming aware that telecommunications can be used as a competitive weapon in the marketplace—a method of improving their attractiveness to customers, and in some cases a way of directly generating additional revenue. Of course, turning a new set of technologies into a "competitive weapon" can be much more difficult than it sounds. Nonetheless, it has been done enough times that the very idea no longer sounds as implausible as it once did.

In fact, notwithstanding the brouhaha about the "information highway," modern communications technologies should indeed be regarded as means to improve revenues and productivity. But expenditures on these technologies could be wasted if they do not rest on a sound financial basis. Like any other major capital or expense decision, the financial effects of such choices should be considered before in addition to during and after the choices are made. In some cases, the dollars involved might drive a decision. It is important to understand when and how to cost-justify a decision, or perform a comparative financial analysis among several alternatives, in order to prove to upper management that the investment and/or monthly expenditures will appropriately benefit the organization.

Sometimes the benefits are so obvious or overwhelming, or the costs of *not* making the investment so horrendous, that no analysis need be done. This chapter will not be helpful in such situations. This chapter focuses on the financial aspects of the decision to invest money, when the decision whether or not to make the investment has yet to be made.

- How do you approach this tender subject without alienating management?
- How do you gain acceptance that the communications expenditures will generate greater returns than many other forms of expense?

The answer is different in every organization. First and foremost you have to begin to speak in management's terms. All too often, telecommunications and telephony people want to describe technologies, applications, and "state of the art" concepts to their management. Unfortunately, the jargon and use of technical terms turns management off. Consequently, when vying for the same corporate dollars as all the other departments, telecommunications tends to lose ground. If you learn the language of business, and apply business principles to your purchasing decisions, you will stand a better chance of getting the funds needed to support the organization.

In order to analyze or justify an expenditure, one must identify two key aspects of the transaction and technology:

- Benefits
- Costs

One must also understand the normal hurdles such decisions face in one's own organization in order to perform the appropriate cost/benefit analysis.

Benefits

When considering any expenditure, the first step is to identify the expected benefits. Such benefits are normally categorized into "hard" and "soft" benefits.

Hard benefits

Hard benefits are those which can easily have assigned to them objective and predictable dollar figures. Examples of hard benefits might include some of the following ideas:

- Reduce dial-out voice communications costs by $3000 per month
- Decrease PBX maintenance costs by $500 per month
- Avoid a one-time PBX acquisition cost of $150,000 in the next fiscal year
- Allow direct charge-back of appropriate calls to clients (vs. absorbing them as overhead) in the amount of $3500 per month
- Avoid the requirement to acquire a new T1 (costing $4000 per month)
- Reduce travel costs (airline, hotel, meals) by $8000 per month
- Control overnight package shipment costs by avoiding $2500 in expected monthly bills
- Reduce the cost of dial-in data communications by $5000 per month

Every one of these examples shares a few key elements. It states a particular type of expenditure, a specific dollar amount to be saved, and a (usually recurring) time period (typically per month). Most communications and maintenance bills are stated in terms of months, so your savings will usually also be stated that way, initially. Ultimately, as you work with the figures, they will be translated into yearly figures, as management normally thinks in terms of yearly budgets. This translation also benefits you, because yearly savings are normally much more impressive, at least for savings that recur monthly.

Soft benefits

Soft benefits are results that you expect to be good for the company, but are more difficult to quantify, and sometimes even to identify. The object of the exercise is to make the "soft" benefits as "hard" as possible. This will make more sense later on.

Identifying soft benefits is typically a two-pass activity. On the first pass, one simply makes a general identification of areas that are expected to benefit from the planned expenditure. Examples might include:

1. Improve productivity of call-handlers in the customer service area.
2. Reduce the time required to add items into inventory.
3. Increase the productivity of our consultants by giving them quicker access to reference information.
4. Reduce the cycle time required to develop a new product (or troubleshoot a problem in an existing one).
5. Diminish the number of errors experienced in information retrieved by our customers when they dial in to our network.
6. Optimize the routing of our delivery vehicles.
7. Minimize the time to respond to trouble calls.

Any organization would jump at the chance to achieve any of these goals, right? Well, perhaps, but a few questions will probably arise before management signs on the dotted line. Those questions are, in summary:

1. "How much money will this save or generate, and

2. How much do I have to pay *now* to experience this saving *later*?

3. And when is *later*?"

We will address how much must be paid below. But it is your job to calculate probable savings to be derived from the expenditure you propose. How do you do it?

After all, it is quite possible that the real reason you are proposing an expenditure is to achieve a soft rather than hard saving. That is, you really do not *know* exactly how much will be saved—but you believe (or your internal client believes) that the change will be a beneficial one. But you must quantify this belief; "wishing doesn't make it so"—or convince management to loosen the purse strings.

An additional pass is required to refine the benefit statements. Soft benefits often refer, in one way or another, to improvements in the efficiency of human beings. In these examples, items 1, 2, 3, and 6 express the hope that such a benefit will occur. Quantifying such benefits is fairly straightforward, although a number of assumptions must be made. *Always state your assumptions.* You will most likely use a spreadsheet for your analysis. If someone questions—or corrects—an assumption, you can easily change it and redo the analysis. If your assumptions are not clearly stated, management cannot evaluate how realistic your predictions are.

To quantify the benefits expected from an improvement in productivity, you need to determine or assume the following:

- Exactly what activity is to be shortened? Examples include length of time on a phone call, the average time to travel from one customer location to the next, or the length of time to check *one* item into inventory.

- How many times does that activity occur per year?

- How much time is to be cut off each occurrence of the activity? For example, if an average new order call takes five minutes, the average length of such a call might be reduced by 30 seconds.

- What is the cost (average salary, hourly rate, etc.) of a person that typically performs this activity? This should be a "fully loaded" figure; that is, include medical benefits, retirement fund, vacation, etc., if applicable. The place to start to get this information is the supervisor; however, a trip to human resources might be required.

Once you have established this, it is time to hit the spreadsheet. Figure 29.1 illustrates a sample calculation based on a proposal to add a computer-integrated telephony function to a PBX. It is assumed that if many of the customers' records can be brought up on screen automatically for customer representatives, then an average of 30 seconds will be saved during each order by speeding the recording of address, telephone, and other customer information that is usually static.

It would be nice to be able to point to row seven and say, "See, this will save us over $40,000! Let's do it!" But this is a raw number. Much remains to consider before the analysis is ready for management's review. Does management expect to hire additional customer reps? If so, then the number of calls is probably expected to increase. Should you do a five-year analysis, incorporating a volume growth factor? Would the use of this improvement forestall the additional hires? Would the organization be able to make do with the same number of agents, if a thirty-second reduction per call is achieved? (This is possible, if you are handling tens of thousands of calls per day and you save 30 seconds each, this can equate to significant hours of savings.) Refer to Chapter 9 on traffic engineering, which had a similar example of the savings of time. This includes the ability to save more than just lines, trunks, and other ancillary equipment. It can also account for substantial people savings.

How sensitive is this analysis to changes in the assumptions? You should be able to get accurate figures on average salary. But how precise is your 30-second estimate? Be prepared with either studies done in your industry that bear out your calculations, or perhaps the results of a study done with a stopwatch showing how much time is used by taking down the information to be avoided. Do not forget that not all calls fit your paradigm; new customers will not be in your database, and many will call from locations other than their own telephone (thus not allowing a match-up of their telephone number with your customer records in your database).

1	Expected time saving per call (in seconds)	30
2	Annual telephone orders	250,000
3	Expected seconds saved (row 1 times row 2)	7,500,000
4	Expected hours saved (row 3 / 3600)	2,083
5	Customer Rep. salary (fully loaded)	$ 40,000
6	Customer Rep. cost per hour (row 5 / 2000)	$ 20
7	Potential saving (row 4 times row 6)	$ 41,667

Figure 29.1 A sample calculation on a proposed CTI solution.

Costs

The benefit figures, whether soft or hard, are "gross"; that is, they identify changes in or avoidance of expenditures, but they do not identify what must be done or spent to realize those savings.

Cost is a relative term. We think of cost as an outlay of dollars. Yet, true costs include both the long-term investments and support, coupled with the returns on investment. This is important to selling concepts to management. Systems in raw dollars cost approximately $700 to $1000 per station when dealing with PBXs. The long-term dollars over a five-year period can be three times that. However, if the telephone set is used to produce revenue of $1,000,000, what is the cost?

Cost/benefit calculations

As should be obvious from the name, a cost-benefit analysis balances the two key aspects of an opportunity (also known as an *investment* or, more commonly, an *expenditure*). As discussed, there are "hard" costs and "soft" benefits. If your goal is to justify an expenditure, you should do your best to turn soft benefits into hard benefits. But once you have reduced everything to dollars, your work has just begun.

Life cycles

Contrary to popular opinion, a life cycle is not (only) a piece of equipment found at your local health club. Rather, it is a key concept in the analysis of acquisitions of both communications and data processing equipment. It is not applicable to items paid for on a monthly basis, but rather only to equipment with an expected useful lifetime of several years.

What is an item's life cycle? The answer is not so obvious. It is not the length of time the equipment could successfully operate (if properly maintained). It is not even necessarily the length of time it could provide service more economically than other alternatives. Rather, it is the length of time that an organization can realistically expect to use the item in its planned role before discarding it or replacing it. But, you might say, coming up with such an expectation is only slightly more accurate than fortune telling with a crystal ball! Indeed. The process of determining the life cycle of a piece, or category, of equipment is one that involves forecasting technology trends, the expected health of the business, possibly interest rates, and other factors. In short, it is largely a matter of opinion.

Nonetheless, you must try to nail down a figure—for financial, operational, and career reasons.

Financial

Assume that someone is developing an annual budget for desktop personal computer purchases. For simplicity, we'll assume a ten-user group needs to be equipped. A reasonable approximation of the cost of one desktop unit, with software, is about $5000. If you further assume that each machine will serve productively for an average of four years, what should the annual budget be? Ten people times $5000 is $50,000 in the first year. After that, we'll spend $50,000 every four years, or $12,500 per year on average.

But wait! It has just been determined that all of the computers will have to be replaced on an average of three years rather than four. The startup cost remains the same, but now $50,000 will be spent every three years—$16,667 per year.

PBXs typically serve for a long time, often as long as a decade or more. But technology is changing more rapidly than in the past. Assuming that a PBX bought now will still be in service 10 years from now could be a big mistake.

Payback periods

One of the simpler methods of evaluating investments is the use of *payback analysis*. This isn't tripping someone that hit you in the schoolyard 30 years ago. Rather, it is an approach to taking into account the time value of money without actually doing the requisite calculations. However, it does give an initial "rough cut" feel for an investment's viability as compared to alternatives.

Notice that we are using the term *investment* in this section, rather than *expenditure*. To have a payback period at all, an investment must bring in, over time, revenue greater than that spent to get the revenue. Otherwise, there is a negative payback, suggesting that this would be a poor investment indeed!

Payback analysis, also sometimes described as break-even analysis, is straightforward. A revenue stream or periodic savings figure associated with an investment is estimated (for example, $10,000 per year in reduced calling costs by installing a virtual voice network). The costs associated with that stream of revenue are totalled (for example, $20,000 in one-time costs, plus a marginal increase in the cost of access lines to the network of $5000 per year). A little work with a spreadsheet, and *voila*! The break-even point is three years from the date of investment. Table 29.1 shows the numbers used to perform the calculations on the spreadsheet. From this

TABLE 29.1 Summary of Calculations in a Payback Analysis

Year	Savings	Cumulative savings	Cost	Cumulative costs
0	$10000	$10000	$25000	$25000
1	10000	20000	5000	30000
2	10000	30000	5000	35000
3	10000	40000	5000	40000
4	10000	50000	5000	45000
5	10000	60000	5000	50000

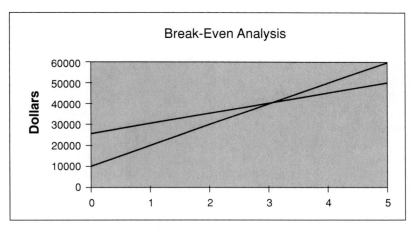

Figure 29.2 The savings vs costs graphed to show B/E.

table we then graphed the information shown in Figure 29.2. Note that what is graphed are the *cumulative* figures.

In today's business climate, a three-year break-even will most likely *not* generate intense management excitement.

Notice that with break-even or payback analysis, the "time value of money" is typically disregarded. When comparing alternative projects with similar time horizons, this is not unreasonable. But when management begins to focus more carefully on your numbers, you most likely will need to create a more rigorous analysis. One way to do that is with return on investment calculations.

ROI calculations

Return on investment (ROI) is where the rubber meets the road in investment analysis. One way to approach ROI is to use an Internal Rate of Return (IRR) analysis. To quote the help available with Microsoft's Excel spreadsheet, the IRR:

> *Returns the internal rate of return for a series of cash flows represented by the numbers in values. These cash flows do not have to be even, as they would be for an annuity. The internal rate of return is the interest rate received for an investment consisting of payments (negative values) and income (positive values) that occur at regular periods.*

In English, the IRR is comparable to the interest rate your company might get on the money it is considering investing if it put it in a bank instead of into your tender clutches. The IRR *does* take into account the time value of money, so if you have included all of the factors it does provide a good decision tool when selecting among alternative investments.

Any modern spreadsheet should be able to do this for you. Using the previous example, the net initial cost in the first year is $15,000. Each year thereafter there is a net savings of $5000. The IRR for this stream of outgo and income, during the first five years, would be about 13%. Better than passbook savings—but of course, considerably more risky.

When this analysis of the life cycle concept is mixed in, the results become much more interesting. That is because ROI is intimately tied to the total length of the expected revenue stream. It is a statement as to the comparable *average* interest rate when compared to the expected cash flow over the duration of the analysis. Table 29.2 shows how much better the investment looks if a six-year period is assumed: 24% internal rate of return! On the other hand, if the technology is supplanted in two years, we lost money at the annual rate of 23%.

TABLE 29.2 Internal Rate of Return Analysis

Year	In/out	IRR
0	–15000	
1	5000	–66.7%
2	5000	–23.2%
3	5000	0.0%
4	5000	12.6%
5	5000	19.9%
6	5000	24.3%

Pricing Considerations

If you had to plan for a new system today, the world would be at your beck and call. Vendors and carriers alike are in a slump. The sales of equipment, peripherals, and services are all slowing down. Increased pressure is being applied on the sales forces to increase revenue. Thus, it would imply that a buyers' market exists. This is true, if you know what your needs are and how to negotiate with the vendors.

Unfortunately, many telecommunications and telephony personnel might be good technically, but lack the skill sets to deal with other issues. Thus, you have to understand the pricing schemes in order to get your best deal, and be prepared to wait them out, if necessary. For equipment purchases, many of the vendors use a one-third rule. One third of the quoted price is for the equipment, one third for the installation, and one third for profit margin. The final third is where you have room to negotiate.

For example, a digital PBX system would cost approximately $750 to $1000 per line. If you use a 1000-line system, you can assume that you will pay between $750,000 to $1,000,000. Request a budgetary price from your vendors and see what they quote. You might be pleasantly surprised, or you might get "sticker shock" when you read the budgetary price. First, the price might be quoted on a rudimentary scheme of 1000 lines, 50% single line sets and 50% digital multiline sets. So, let's use $750,000.

However, the vendor might also assume that you'll want features and peripheral equipment above and beyond a rudimentary system. Therefore, you might see duplicate CPUs, battery back-up, enhanced features, station message detail recording (SMDR), enhanced networking, least cost routing, etc. The cost might come in at $1,500,000. How did it get so expensive? What happened? If they start in a position of strength (at $1500 per line) they have plenty of room to negotiate. They will offer you a 50% reduction on initial station equipment and features, right up front. The price only dropped 15% to approximately $1,275,000, and you believe you did your job. However, all that was reduced is the station equipment, at the inflated price! Remember, the system should have started at $1,000,000. How can the vendor come back and reduce the price even further? They haven't even touched the price or the profit! That's how!

The decision-making process

Once you have determined that a true need exists for the purchase of any communications peripheral or system, the next step is to evaluate everything on both a financial long-term basis, and a functional basis. Do not let yourself get trapped into the "love that technology" routine. Too many systems are bought on the basis of emotional appeal, rather than on company need. This can be the kiss of death.

Some suggestions to follow are:

- *Buy only when necessary, but allow plenty of time to start the process* This gives you the edge on the buying cycle. If you wait too long, then your leverage of time is gone, and you'll have to buy before you get the right deal.

- *Remember the financial picture* Recruit a team of others to assist in the decision-making process from finance, purchasing, facilities, and legal departments. Bring them up to speed immediately. Other sets of eyes helps to maintain focus and objectivity.

- *Don't get too close to the salesperson* This is business not friendship. If things are not going their way, the vendor (salesperson and manager) will go right around you to your management and keep you out of the loop. Control your project and your vendors.

- *Accept no verbal agreements* Everything must be in writing! A short pencil is worth a long memory. You'd be surprised how many agreements are forgotten after the contract is signed.

- *Keep everything above board* Don't accept gifts, or favors. This can ruin your career. Even if there is nothing going on, the perception is what will get you.

Justifying costs

There is no one rule for the justification process. You must learn what the criteria is within your organization. Many organizations will use the return on investment (ROI) method, or the internal rate of return (IRR) method. Still others use a discounted cash flow analysis (DCF) to authorize expenditures. Many smaller companies might well use an emotional decision and justification process. It is really up to the individual management in the organization. So, how do you know how to go about this?

- *You have to get away from the technical role and become more of a business manager.* You need to speak the language of the organization. This can be in financial terms (which is a universal language), or in production or cost containment/cost reduction terms. Whatever the language of the management, you have to get on their platform.

- *Ask for help* Find a department manager or peer who has been successful in getting approval for new projects. Go with a winning style.

- *Don't reinvent the wheel* When you are trying to justify a system, ask others to review your information. They might well give you pointers that will help. If a similar size project has been recently approved for any department, find out what that project manager did. Use a format that is already approved, something that management is obviously comfortable with.

- *If you are not comfortable with the justification process, get a champion or mentor (Senior Manager, Financial Manager) who will run with the project for you* This might be a risk because the mentor or champion might try to take full control after the approval is gained. The projects are usually highly visible, so a lot of credit could be up for grabs.

Maintenance issues

Maintenance issues should be discussed with the vendor in advance of a purchase decision. You might be replacing a system that has become obsolete or is continually breaking. The last thing you want to do is jump into a decision that might have the same problems. You certainly do not want to buy a system that is about to be discontinued or that will not be supported before the end of its financial and useful life cycle.

Determine:

- What the long-term maintenance costs of the system will be.
- How much the maintenance will cost in years two through five, and six through ten. Can you prepay maintenance at a reduced rate?
- How long the warrantee period is. It should be at least one year.
- When does the warrantee start, after cutover or after acceptance? There can be significant time differences between the two dates. Is this a moving window?
- What are the normal maintenance periods (9 to 5 Mon–Fri) and what will the costs be if maintenance is required during other periods?
- What will the costs be to add one more set, 10 more, 100 more, etc.?
- When or where are the big bumps in price?
- What will the remedies be for nonperformance?
- How will the vendor handle major and minor outages?
- Does the vendor have spares on site? If not where are they and how available are they?
- How will the vendor keep your system current? At what cost?
- Can you train your own people? If so for how much? If not, why not?
- What about initial user training, refresher training, new employee training? How often and how much?
- Can you trade back excess equipment? Can the trade be applied toward upgrades and new releases?

Reduced rates vs. improved service

What are your objectives for the purchase of a new system, reduced rates, or improved service? Many users really never address the issue of their buying motives. If the objective is to reduce rates, then the issues might vary from the improvements in service. Of course, the best choice is both, but you might have to decide between these two issues.

If reduced rates are the primary objective, you could give up some of the features and niceties of the systems. Keep in mind the long-term benefits of a system. Often, reducing rates are short-term goals that might well prove to be more expensive in the long run. If, for example, you choose the least possible expense today, by selecting a PBX with all single-line sets, you might experience a dissatisfaction from the users. Ultimately, they will begin demanding multiline sets, with added features and capabilities. Now you'll have a group of single-line sets in a closet, paid for but not used. Further, the add-on costs for the multiline sets might now be more expensive after the installation.

Improved service will point you in a direction of serving the end user. These are your customers. Although the initial costs might be higher for the services they need or ask for, the long-term capital outlay should be less. The acceptance of the system, customized to meet their needs will also go a long way in the overall payback of the system. Increases in user satisfaction, productivity, and revenue are soft-dollar benefits for telecommunications systems. However, they are real values to the organization. Don't lose sight of the total picture.

Sample calculations

Before you can present your case to management or approve the purchase of the system/service, you have to understand the total cost both in initial outlay and ongoing cost. To do this, you have to gather all the facts. This will start with the:

- Initial costs, purchases, installation, training, preparation, etc.
- Add-on costs over the period of years, based on system and peripherals.
- Maintenance costs.
- Human resources needed to operate the system.
- The obsolescence factor: how long the system will be good for.

Initial costs

Assume that you are planning to buy a new PBX that will serve multiple users (1500 to 2000). To purchase such a system, you've done your homework and estimate that the system can be purchased for $1000 per line. Your initial assessment will be approximate between $1,500,000 to $2,000,000 to purchase the system, installed with one-year warrantee. This system will be needed in one year because the existing system is too small, and continually breaks down.

A further analysis should reveal additional cost to include that might not be as obvious, but must be included in the calculations to be fair:

Construction costs of the PBX room (30×50)	$ 50,000
Raised flooring	10,000
UPS System	20,000
Fire detection/suppression system	15,000
HVAC add-on tonnage	20,000
Security system (card-key or other)	15,000
New conduits for new wiring	25,000
Electrical work	7,000
Removal of old system	5,000
Handling of old batteries (*hazardous*)	5,000
Training costs (*1500 users @ 1 Hr. @$20*)	30,000
Project management	40,000
Total	$242,000

The figures are obviously based on some industry averages. They reflect approximately 15% additional costs over and above the initial purchase price. These numbers can/should be included in the overall cost analysis for the system because they represent real costs.

Some added long-term costs which should be considered over the life cycle are:

1) Space (*rental*) 4.00 sq. ft./ann	$ 6,000/ann.
2) Insurance @ 0.50 per $1,000±	8,500/ann.
3) Spares (*parts/sets etc.*)	10,000/ann.
4) Test equipment	20,000/one time
5) Heat/light/utility costs $1.00 sq. ft.	1,500/ann.
6) Inspection costs for HVAC/fire suppression	500/ann.
7) Contingencies (*5-10% total ann. exp.*)	2,650/ann.
Total other	$29,150/Annually
	$20,000/one time

Be sure to work with various departments to understand the operating costs. These might include:

- Purchasing
- Finance
- Facilities
- Insurance
- Safety/Security

Add-on costs

Assuming that the base price for the system is as outlined above, the add-on prices might include such items as:

Automatic call distributor (100 LINE)	$ 50,000 (1)
Voice Mail/Automatic Attendant	80,000 (1)
Modem Pooling (simple)	10,000 (2)
LAN interface (56 Kbits/s Bridge)	15,000 (2)
T1 Direct Interface (5 cards @$2800)	14,000 (1)
Station Message Detail Recording	20,000 (1)
Four Wire E&M T/L Cards for V-Mail (4 @$500)	2,000 (1)
Paging Access (excludes speakers)	12,000 (2)
Power Failure Transfer (48 line @$6000)	6,000 (2)
Total add-ons	$209,000 *

*Spread over 2 years (166,000 yr. 1/43,000 year 2).

By now, you might be getting the feel that the cost of a telecommunications system is not just something that you pick off the shelf. Nor are the prices simple to calculate because so many other components are involved. You have to look at all of the added costs for construction, maintenance, human resources, insurance, taxes, etc. This is the only way to truly justify to management. They have to perform these calculations if you do not. Admittedly, they will use the financial department to perform these calculations, but if you have no input in that process, they might use different assumptions and disprove the benefit that you anticipated. Consequently, the onus is on the individual responsible for the technology to prove or disprove. How can they use different assumptions? Look at a simple point here, which is summarized on Table 29.3. This is how we compare the differences in perceptions.

TABLE 29.3 The Differences of Years Used for a Single Decision

Person responsible for considering numbers	Number of years expected
Telecommunications or information management	3 years payback or sooner
IRS	5 year depreciation at accelerated rate
Finance Department	7 year straight line depreciation
Senior Management	10 year life, don't want to do this very often

30

Facsimile

Facsimile Transmission

Just what is facsimile? Facsimile is the transmission of documents, preprinted text, and graphics that are already in paper form, converted into a graphical form and sent across a telephone line. The process is straightforward, in that a prepared preprinted document is scanned into the facsimile machine via a scanning process that reads every character as a series of dots. The scanning and transmission is across the page, left to right, using a dot- (or pixel) by-dot basis, as shown in Figure 30.1. Facsimile has been around for over 120 years. Yet, the technology was very slow in developing because there were some complexities in the total process.

The use of facsimile has always been regarded in the past as a tedious job. Newer machines have made the process far less painful, but it still takes a lot of patience. Part of the reason is that as facsimile evolved, the machine used this dot-by-dot scanning process to transmit a series of ever-changing dots across the older analog dial-up telephone network. Using this concept, and knowing that the original analog network was built to carry analog voice communications, we have learned that this network was unreliable. The unreliable nature of this network left a lot to be desired. There were always problems in the transmission process. Overcoming the problems in a voice mode was easily accomplished, but for the transmission of data, the problems were magnified. As the network degraded over time, the transmission process also degraded. This left a lot of information on the network that was not properly received.

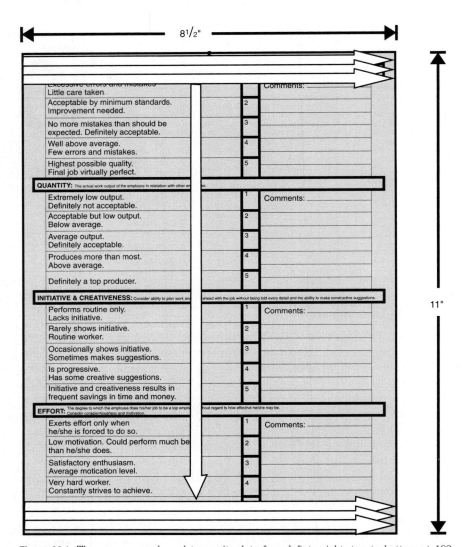

Figure 30.1 The scanner reads and transmits dots from left to right; top to bottom at 192 columns and 200 lines of resolution.

Because the transmission is on a dot-by-dot basis, rather than a character-by-character basis, transmission was both slow and very prone to errors because of line noise. A true ASCII or EBCDIC code set, as was covered in the data communications chapter in this book, was not used. Actually, the preferred alphabet in the scanning process used a modified Huffman and Read code set. This again, was based on dots rather than a true alphabet. The result was a fuzzy code set prone to line noise in the transmission process.

Types of Facsimile (Fax)

Facsimile machines are broken down into types or groups based on the standards set by the CCITT. These groups will define the scanning process, the line speed, and the modulation technique that is used. There have been four different groups established over the years. These have all met with different results and acceptance from end users. However, the older groups (CCITT Group I and II) are just about extinct. The predominant service (Group III has been widely accepted) is robust and proliferating everywhere because of the price drops in purchasing these services. Newer techniques (Group IV) are slow in evolving and replacing some of the older systems.

Group I fax

The first evolution of the fax machine was adopted as a standard under the CCITT. This was called the *CCITT Group I machine*. The fax machine was set up to use a rotating drum where a scanner reading a dot would transmit an electrical signal to a distant machine that used an electrical stylus. For those of us old enough to remember the TV series "The FBI," the rotating drum fax machine was used extensively in this series. A picture of the top 10 most wanted criminals was transmitted across the dial-up telephone network from FBI office to any other law enforcement office. If it was necessary to transmit this picture to a multitude of locations, the process was serial and handled one site at a time. The techniques of storing and forwarding the page for multiple transmissions did not exist. This again was a very expensive and time-consuming process. Back then, the cost per minute for a long-distance call was still in the $0.65 per minute rate.

The stylus can be compared to an electric pen of sorts. As the receiver received the data, it generated an electrical pulse and sent it to the stylus. When the stylus received the electrical current from the receiver, it would move back and forth on the blank page of pre-treated paper. This back-and-forth motion would scratch off an aluminum coating on the paper in the receiving machine. This aluminum oxide, when scratched off would produce a black/gray mark on the page in the form of a circular dot. The dots were spaced closely together to form the pixels of a letter, number, or graphical image. This created the alphanumeric codes that we use in our normal business correspondence. The dots were not perfectly rounded, so a result of this transmission process left a lot to be desired. The actual output could be very fuzzy (Figure 30.2).

A serious by-product of the process was that it also produced an odor that made people want to scatter. The odor caused by the chemical reaction was so pungent that the machine might be relegated to a special room. Then, only those brave souls that needed to send a page of information across the network would venture into this hidden area.

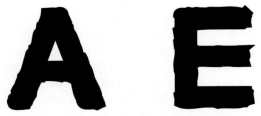

Figure 30.2 The actual output from the fax produced a fuzzy appearance due to the scribing process.

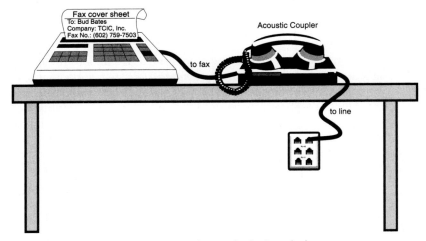

Figure 30.3 An acoustic coupler was used to send a fax in early days.

Both the receiver and the transmitter portions of these machines were slow. Using an amplitude modulation technique, it was very common for a transmission to take six minutes to send or receive a normal typed page of text. The Group I machines used an acoustical coupler to provide the input and output on the line (Figure 30.3). The acoustic coupler is a descendant of the Hush-A-Phone from the mid-1940s. By using the acoustic coupler, the operator would have a protracted procedure to establish a connection. These systems were not automatic answer, and the process was again labor-intensive. The process of setting up a call goes as follows:

To establish a facsimile call using the older Group I facsimile machines:

1. First, the operator would take the first sheet of the document and wrap the sheet around the rotating drum. This would clamp the document in place.

2. Once the first sheet was on the drum, the operator would then pick up the handset of the telephone set next to the facsimile machine, and

dial the receiving end's number. The transmitting operator would place the machine in the transmit mode, which meant that the machine was set to scan the document, not use the stylus to scribe but to read.

3. When the receiving end got the ringing tone from the voice call, an operator would pick up the phone and answer. At this point, the transmitting operator would negotiate the connection. The transmitting end would state that a fax needed to be sent to the receiver. The receiver would ask how many pages there would be, so that they could be sure that enough of the pre-treated sheets were available for this reception. It would not be uncommon at this point for the receiving end to "beg off" from receiving the document.

A number of reasons existed for this reaction:

- The operator might not have the time to stand there and receive the document.

- Sufficient paper might not be available.

- The time of day when sending to different time zones could become a problem.

Personally, I can remember trying to send a fax to a distant location, only to be told by the receiving operator that he/she was going to lunch and I would have to call back later. When trying to get back later the phone on the other end would never get answered. Why? Because they knew that I wanted to send a multipage document that would take a long time, and they did not want to receive the information. Also, the normal operator might not be on hand, and some designated alternate was at the other end, who did not feel comfortable with the setup or did not like to operate the machine.

4. Assuming that all was going according to plan, the receiving end agreed to accept the fax and made sure that enough paper was on hand. The operator at the receiving end had to wrap a piece of the pretreated paper around the drum of the receiver. Once the page was locked in place, the operator placed the machine in the receive mode, this meant that the receiver would use the stylus to scribe on the page. The telephone is still in the talk mode between both parties and they can converse while the process is being set up.

5. Upon mutual agreement, the two operators would then agree to go into the data mode. The telephone set was used to click things into place, where the operators would then place the telephone handset into the acoustic coupler and start the machines at both ends simultaneously. This process started the drums on both ends to revolve and started a signaling arrangement between the two ends.

6. As the two machines came up to speed, the stylus on one end would read the data on the sheet and convert this to an electrical pulse. A "1" represented a change in the amplitude of the carrier tone generated between the two machines, and a "0" meant nothing. The zero, after all, represented that there was no dot (pixel), while the one represented the presence of a dot. This was the amplitude modulation of the carrier tone (covered earlier in this book).

7. Once the page was completely transmitted, the operators on both ends would reach over and grab the telephone handset from the coupler. The drums would stop revolving and come to a stop so that the pages could be disengaged from the drum and removed from the machine. The receiver would verify that the information was received properly (a process that was not guaranteed). This paper was difficult at best to read because the aluminum oxide coating was a silver color and the dots scratched on the page were a gray tone. However, the agreement would be handled and the next step would begin, for each page that had to be received/transmitted, steps 1 through 6 were repeated.

At any step along the way, the process could fail and the call would have to be reestablished. A page could be illegible, the information could be trashed from glitches on the line or any of a number of problems, rendering a transmitted page unusable. Talk about frustration. A facsimile transmission of four pages could take as much as 45 minutes in this scenario, with all the manual labor and the delays in setup on a page-by-page basis.

Pictures and graphical representations could take much longer because of the amount of dots needed. Because the machine was a single page wrapped around a spinning drum, the process was manually intensive. Not a welcome chore for anyone using the machine. You can imagine the frustrations of having to send a multi-paged fax using this machine. At six minutes per page, plus the extra overhead for handling and reading time, and the telephone transmission costs being $0.65+ per minute, coupled with the need for an operator to stand by and wrap a page at a time around the drum, the cost was both expensive and undesirable. The operators were not pleased when a densely populated graphic needed to be transmitted. With all the disadvantages of this system and the reluctance of the users to even try the technique, coupled with the associated costs in transmission and labor, it is a wonder that facsimile ever really survived the technology wars.

Another problem that kept creeping up in this process included incompatibilities between different manufacturers' equipment. Even though the CCITT developed standards on how these machines would talk to each other, there was no guarantee that two machines manufactured by two different manufacturers would communicate with each other. Talk about frustration!

Group II fax

To improve on the process and the acceptance of facsimile, The CCITT developed a new standard and the manufacturers created a newer technology, using a much faster process. The older analog transmissions were limited to the amplitude modulation process, but modem communications had evolved to a faster transport using frequency modulation. Remember from the data communications section, the FM procedure doubled the rate of data transfer on the analog telephone network. In the transmission technique for Group II facsimile, a frequency modulation technique was used, reducing the time per page from six minutes to four—and under special conditions, two. You might look at this and wonder why the transmission didn't drop to a standard three minutes (half of the AM method). Obviously, with the bits-per-second rate, this would be true. But we are looking at transmitting dots, rather than characters comprised of bits. It takes a lot more dots to paint a character than it does in an ASCII code set. Further, the transmission of a page in the MA technique was six minutes, but when you added all of the overhead, the effective rate was much longer. In facsimile transmission, a page of data is transmitted, then the machines go through a whole new handshake to make sure that everything is still functioning properly. This handshake also makes sure that the page was received, the end of page was recognized and that the machine is ready to receive the next page. This all takes time in the communications process; therefore, the average time per page is better than one half the AM rate. At four minutes a page, the time was reduced for transmission by 33%. This depended on the machine to machine communications. Many of the manufacturers of facsimile machines had their own proprietary code, used between two machines of the same manufacturer.

Around that time period, the vendors were introducing a new feature: white space skip. If the machine scanned a line and saw no pixels, it would speed up until it recognized the next sequence of dots. This meant that all the spacing on a typical business document could be bypassed. Look at how much white space is on this page as you read it. If you have a line of spaces, then there is no sense in slowly scanning nothing.

The white space skip was one of the proprietary methods used by the vendors. Other techniques were also used, but they were dependent on compression techniques, skips, inter-page handshakes, and so on. All of these were fine if the machines on both ends were the same. If not, then the system would revert back to the standard mode issued under the CCITT guidelines.

Unfortunately, no laws required that a vendor adopt the standards, so if one end was CCITT compatible and the other was not, the communications might not go through. This still left confusion in the industry and a bitter thought in many people's minds. They were reluctant to use a fax because of these constant complications. We can recall some organizations that had various fax machines in a room; depending on who the document had to be transmitted to, a machine of the same manufacture was used. This ensured that at least they could communicate. However, this is an expensive way to conduct business.

The printing process changed also on a Group II fax. The introduction of a roll of paper using the aluminum oxide coating and the electrical scribing from a stylus was preserved. The roll of paper allowed the receiving end to get multiple pages of facsimile without requiring an operator to manually change on a sheet-by-sheet basis. This roll of paper also created the ability to transmit a portion of a page of information. The rolls came in two different sizes: the 164-foot and the 328-foot roll. This meant that approximately 179 to 358 pages of 8×11 pages could be received without any form of paper handling. This never worked out that nicely because cover sheets are used, transmissions might still fail and trash a sheet of paper, jamming occurs . . . the list of problems is endless. Each problem detracts from the actual vs. expected results.

A derivative of this printing process using the aluminum oxide-treated paper was a wet chemical process, which introduced an ammonium process. This paper reeked from the smell of the chemicals, and the paper was prone to turning black after exposure to light for extended periods of time. It was not uncommon to have a fax receipt and pull the page out—only to find it dripping wet from the ammonium liquid. This was a problem, because users were still uncomfortable with the smell and now they had to contend with the use of wet chemicals. Loading the liquid chemicals and the drying agents left a lot to be desired. This paper was equally sensitive to light as the aluminum oxide paper. In order to file a fax, the user had to reproduce the fax via a copier. Where a roll of paper was used, multiple receptions could be handled in sequence, which was a continuous sheet of paper (Figure 30.4). The receiving end had to cut or tear the pages off at the respective page breaks. This meant that confusion could exist when the pages were separated from each other. If a user cut the pages off at the wrong location, the result could also yield lost or misplaced pages of fax.

Another improvement in the facsimile was the connection. Many of the Group I machines were acoustically coupled to the line, the Group II introduced a jack to connect to the line. The machine might (or might not) have a phone built in, or attached to the fax machine with two jacks (one for line, one for the telset). If a call comes in, the process might even be an automatic receive mode that did not require the user to place the fax into the data mode. As a matter of fact, this was one of the more significant developments on the fax machines. The ability to send or receive a fax without a human on both ends did much to improve its acceptability. Because the older machines required a user to be there, fax transmissions were often limited to business day calls only. With the auto answer, on-line capability, fax transmissions could be done off hours to take advantage of reduced long-distance rates. Because the human did not have to be there, this could be accomplished easily. However, many a case has been recorded where

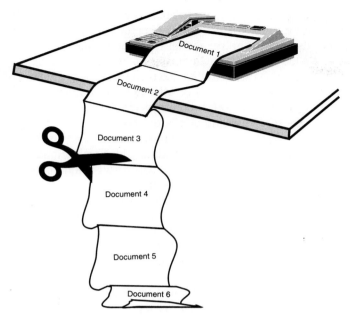

Figure 30.4 The roll would produce a continuous stream of paper. Users had to cut the pages apart.

line problems occurred during the transmission and the fax cut off the call. This meant that the fax would require a retransmission the next day. Further, the use of the roll-fed paper eliminated the need for a human to stick around after hours just to receive a fax.

These strides helped, but did not solve all of the problems. Although this was better, it was still not universally accepted as a viable transmission medium and copy reception. Only those transmissions that had to be there by the next day (or in some cases right now) were sent to the fax. More users still felt the comfort of sending documents out via an alternate means such as: mail, telex, TWX, or data. However, back in the early days of fax transmission, the telex, TWX, and data transmission services were not very comfortable to the end user, and many of the documents were already in typed form. Therefore, the time to retype a document was all wasted. Graphics and pictures were relegated to the mail as the only alternative. This delayed the flow of information in an organization well beyond tolerable levels. Something still had to be done!

Group III fax

The next development in facsimile was the CCITT Group III standard. Although this had significant improvements over Group II, the machine was expensive in the beginning. This technique still was not readily accepted as a means of reliable transmission. Only the larger organizations could migrate with the standards and the newer model equipment. Machine costs were in the range of $6000+. However, the speed of transmission was increased to 9.6 Kbits/s using a QAM technique, transmitting a page of information in one minute or less. Clearly, this was moving along with the evolution of the modem communications techniques. The CCITT standards (CCITT T.30) set the stage for the ability of a fax machine to act more like a modem. The machines could be used to transmit much finer and dense information. Additionally, if they had to transmit to the older equipment, the equipment could sense that the receiver was a Group II and could fall back in its speed to match the receiver speed. This meant that the system was backward compatible. How wonderful when standards are adhered to!

The printing techniques also changed, although the Huffman and Read codes were still used. The graphical nature of the page was still maintained as a dot pattern. Yet, the introduction of a thermal printing process on the paper improved the users' acceptance of the methodology. No longer did we have to worry about the stench of the stylus-based systems with aluminum oxide or ammonium oxide. The thermal paper was easier to handle and use. To increase the throughput on the machines a "white space skip" technique was used as part of the standard. No longer was this strictly a proprietary development and implementation. If no dots of information were represented in a block of space, the machine immediately skipped ahead, instead of transmitting nulls (zeros). This helped, especially because the transmission costs were still relatively high. Transmission costs were dropping, but the average cost per page was $0.35 to $0.40. The paper was roll-fed in the early installations of Group III fax using a 328' roll of thermal paper.

Group III was slow in starting off. Mainly, the cost of the machines was the impairment to acceptance. Even though the past ills had been fixed, the organizations considering fax transmissions were stymied by the $6000+ price tags. Ultimately, in the early 1980s, fax began to pick up the pace and acceptance was starting to roll out. Another factor was the drop in prices for the machines because of mass production and enhancements in the scanner and chipsets in the machine. An automatic cutter was added, making an easier received page, so the user received the pre-cut page, rather than having to tear off each message. The transmitting end could use a document feeder to stack 10 to 30 pages into the machine for transmitting without someone standing by the machine all the time. The manufacturers began to use a microprocessor to allow a delay-dial feature for later transmission, retry a busy number and confirm delivery. A newer modular jack and dial pad on the machine assisted the sender/receiver by directly attaching to the line. Built-in modems were used. Further developments were being introduced by the makers of fax, but user acceptance was still slow. They had been through the trials and tribulations of the evolution, and did not like to deal with this technique. Users had written off the fax machine, so they would not consider it in their normal decision methodology to send information out the door.

The thermal paper, although a great improvement over the oxide and wet processes, was still an area of contention with the user. Once the thermal paper was exposed to the heat treatment in the fax machine, the user still had to deal with curled paper; a result of running a roll-fed and heated paper. Secondly, the paper would fade, so any reception had to be copied on a photocopier for long-term storage or retention. The fax copy could not be integrated into a document, so the information had to be retyped to include it into a letter, report, etc.

Printing Options

More recent options have surfaced in the ability to print with the facsimile equipment. The introduction of the microprocessor and the addition of hard disk drives in the machine allows for the reception into the hard drive for later printing. Further, laser quality on plain bond paper has enhanced the printing of text and graphics in facsimile uses. The laser-quality output on bond paper allows for more acceptable quality and later storage of the paper. The user does not have to go through an added step, the paper is acceptable for inclusion into a document. For storage, the bond paper can be filed away without the added handling and cost of copying. This enhancement has clearly added to the acceptance of fax in the business environment. Not that this technique is inexpensive, but you get what you pay for. Using a plain-paper laser-quality printed output can be justified in terms of the reduced copier and personnel costs associated with the fax.

Another enhancement is the storage of outbound faxes in the hard drive for later transmission. 20, 40 Mbyte, and larger hard drives can be used on these machines. Of course, the hard drives, laser printers, and other enhancements all drove the costs of these facsimile machines up to the $14,000+ range, which only makes this type of machine acceptable with the larger organizations, but not as an attractive option for the smaller companies. Who can afford it?

But this has also helped to drive the cost of the lower-end machines down through the floor. The Group III thermal roll-fed fax machine of just a few years ago cost in the $2000 to $3000 range, can now be bought for $400 to $600. Now the smaller organization can afford Group III fax with thermal printing and still communicate with the larger organizations at subminute speeds. Transmission improvements, data compression, and error-correction techniques allow us to send a page of information in as little as 20 seconds. What an improvement over the original machines, which took six minutes!

These improvements have led to the introduction of departmental fax machines and have really proliferated. The larger machines in the high-end dollar range are fine for the centralized facsimile departments with dedicated operators. However, the casual user in a specific department might not need all the power of the high-end machines. Therefore, the $400- to $600-machine provides a suitable replacement or a complementary capability in any organization. With these enhancements and price differences, fax machines began to proliferate and acceptance began to increase. The smaller organizations were using more and more of these services. Home-based businesses are now in a mode that they equip their offices with the telephone, a PC with modem, and a fax machine as standard equipment.

Fax Boards

Probably the single most important step in the facsimile world was the invention of the fax board. This group III device was developed as an add-in board for a personal computer (PC). As a PC-based device, the fax card imports a true ASCII code set and transmits it to a Group II or III fax machine anywhere in the world. The same features available in the higher end machines can be used in the PC-based fax card, such as:

Store and forward

Stores the data in the PC's file structure for later retrieval and transmission. This can be set up automatically without a human being there, so long as the PC and fax modem are powered on.

Reception right into the PC

What used to be brought into a fax machine then rekeyed or scanned into a PC can now be sent directly into the PC for access later. This needs some special software because the fax received will be saved as a fax file (bit-mapped graphics).

Later printing

The document can be sent into the PC and stored. Paper has not been created initially. Only if a printed copy is needed later, the user can output the file to a printer.

Laser or dot matrix printing

The printer attached to the PC is available for the output of a fax file. Therefore, whatever printing capability is available can be used. This eliminates the need for a separate fax and PC printer.

Files kept as graphic or ASCII

Depending on the software used, the files can be saved as a graphic file, or can be imported or exported as an ASCII file.

The original fax boards (1985) started out as special cards costing around $2995, but have quickly dropped to today's price of $100 to $200, making it both economical and compatible with other fax capabilities. Newer twists in the use of the fax board are to transmit from a fax to a PC equipped with a fax board, and import text and graphics (such as pictures, logos, etc.) right into the PC. This compensates for the lack of a desktop scanner at the desktop. If a user needed a logo or picture scanned and did not have this resource available, the document could be inexpensively sent via a true fax machine (this has a scanner) into the PC fax card. The scanned image can then be imported into any other file, such as word processing, desktop publishing, etc., again depending on the software available. The faxed image might be grainier; therefore, a software that allows for a graphics touch-up service might be required. The major drawbacks to all of this on a PC are the size of the graphics files because they can be substantial. Secondly, the use of the fax board on a PC means that if the PC should fail, you also lose your facsimile capability. However, this doesn't appear to concern many who have readily endorsed this added technology with open arms.

Common Uses of Fax

Now that fax has worked its way into the work place as an acceptable technology, the users are looking for more. The use of fax has become the easiest form of electronic mail (e-mail) on the market. If you can transmit image or text from a single device, why would you need a separate e-mail system?

However, another technology has also been rapidly proliferating in the work place at an equal or faster pace. The use of local area and wide area networks (LANs and WANs) has driven a newer approach to facsimile. Because users all are getting either terminals or PCs on every desktop, the organization cannot afford to place a fax on every desk with the existing LAN-attached PC. And because the LAN is designed around the ability to share resources on the single cable system, a new demand went out to the industry. Give them a fax server on the LAN! Manufacturers were ecstatic, newer needs and demands for the use of fax on the LAN introduced a new market for the existing technology.

However, the fax servers are still in their infancy, and users will have to suffer through various growing pains, or wait and see what evolves. Foremost in this development is the ability for a user to send a fax from an attached device on a LAN, but how do you get messages in from the LAN?

If you can dial into the LAN (which brings its own security risks), how do you get to the sub-address of the actual PC you want to reach? The present way is to have manual or human intervention. The faxes are received on the server, and an administrator would have to review and forward the fax to the intended recipient. This also makes the fax a common document which might violate some confidentiality of the information. We'll see more in this arena over the next few years.

Future Machines

Obviously, the development and acceptance of facsimile will continue to proliferate. Newer capabilities will be introduced. Some of the future systems that are in various stages of development and introduction in the industry are:

Group IV fax, which really isn't that new. However it has lagged in the implementation stages. Group IV operates digital at 56/64 Kbps, transmitting a page of information in approximately 3 to 4 seconds. The cost of the machines will continue to be high as the pioneers use this technology. The backward compatibility to a Group III or II machine has to be considered, and does not yet exist. The cost and availability of dial-up 56/64 Kbits/s services has to evolve further. Federal Express tried to implement this technology years ago as a pioneer in the industry. However, being ahead of their time, the pilot known as *Zap Mail* failed.

Cellular fax is also coming. Today cellular facsimile modems operating at 9.6 Kbits/s are available, but not as well accepted. Until a reliable package of cellular modems becomes available at a reasonable price, this technology will be used by the folks on the leading (or bleeding) edge of technological implementation.

Laptop manufacturers are now developing facsimile modems to be used in conjunction with their machines. The use will depend on the costs. Because cellular communications is not inexpensive, the costs associated with the equipment will possibly hamper the widespread use.

Chapter
31

Cabling Systems

Introduction

Before the development of local area networks, organizations that needed within-building data communications cabling systems had a limited number of choices. Terminals used RS232C, unshielded twisted pair (UTP), or in some cases (especially block mode IBM terminals) coaxial cable. High-speed connections in an IBM environment were handled with special in-room cables (maximum distance, about 200 feet) called *Bus and Tag* cables. (Voice always used unshielded twisted pair wires—no choice at all.)

A key characteristic of these cabling systems is that in general, a need for a particular cable was met by pulling just enough cable to satisfy that need. The cost of installing such a cable run (to connect one device) typically ranged from about $200 to over $1000, depending on limitations of and available routes through the existing physical plant. Because in a business environment "moves and changes" occur frequently, the cabling cost alone for a normal data communications environment would be considerable—even after all of the initial installation was completed. This frequently occurred as soon as the initial installation was completed, because move and change activity was "frozen," or put off, until the cables were installed. Frustration existed because users might have gone through one or two moves from the beginning of the installation, but the wires were installed at the original location. There was no easy way to functionally change in mid-stream.

When local area networks were first introduced on a widespread basis, a new concept became popular, that of a *data utility*. Fundamental to this approach was the idea that, given the opportunity, it would be smart to pre-wire every plausible work location in a building. This pre-wiring would allow moves and changes to consist of just movement of the user hardware—and perhaps some software changes, although some LANs allowed moves with no software changes.

The cabling of the day was either coaxial cable (in ARCnet's case) or heavy coaxial cable (in Ethernet's case). Nonetheless, the idea of placing a jack in the wall for as-needed connections ("just like the electric company!" was the claim) rapidly took hold. But still, this facility was just for data communications—and only LAN data at that.

About this time, another idea began to gain in popularity: the idea that voice and data communications should be in some way integrated. Just how they would become, and (in many cases) avoid becoming integrated is a story for another day. However, one area in which apparent integration could be accomplished relatively easily was in wall jacks. That is, one could design jacks with built-in outlets for multiple types of communications. Jacks have been designed with some or all of the following, all at the same time:

- Voice
- Terminal
- LAN (sometimes more than one)
- Video
- Generic analog (in cases where the voice connection is for a digital PBX, analog connections are sometimes also provided to allow analog devices—for example, modems—to also gain external access)

Figure 31.1 shows a typical multipurpose outlet layout. The multiple outlets were designed to be utilitarian so that any need could be met with a single interface at the wall outlet. This was a good and noble goal, but many cases are documented proving that early implementations were not that successful. Obviously, these jacks can become quite crowded. Not so obviously, the jacks only provide the appearance of integration. After all, what is behind the jacks? Typically, each of the cabling types goes off on its own to wherever it needs to be. The office walls are neater, though.

A more recent concept is that of *structured cabling systems*. These systems typically are offered by some of the major systems vendors, as well as some other organizations. Some of the more successful vendors of such systems include:

- AT&T
- IBM
- DEC
- Others (such as AMP and KRONE)

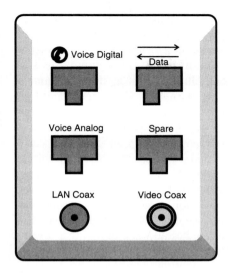

Figure 31.1 The generic wall plate emerged as the utility function for voice, data, LAN and video. These could have four or six outlets.

The idea of a structured cabling system is to predefine the type (or types) of cable to be used for each specific purpose. Then, one pre-installs wires of all identified types, coordinating the installation in order to save money and provide a consistent and neat physical plant for both current for foreseeable communications (both voice and data) needs. Such an approach is particularly beneficial when a new building is to be constructed. But buildings with older cabling that did not measure up to modern requirements have been retrofitted on occasion, at a substantial cost.

Each vendor of a structured cabling system offers essentially all media. Company management still must decide what requirements might materialize in the future. However, once a company commits to a particular system, the requirements directly drive cable selection, configuration, and installation approaches. Fewer decisions must be made. Vendor installers can work more efficiently because they are familiar with the components of the particular system.

Of course, any given vendor's system ensures that at least the products of that vendor are well-supported on that vendor's system. For example, when it was introduced, IBM's system had no provision for Ethernet cabling; but DEC's did offer such an option. In an environment where a company typically buys most of its data processing equipment from one vendor, selection of that vendor's cabling system might make sense. However, if a company wishes to sole-source the entire environment—and thereby have one company at which to point when failures occur—then using a vendor's structured cabling system would be an excellent idea. But essentially, the same system can be built by independent cabling suppliers at significantly lower cost. Secondly, if a multivendor environment exists in the building, the vendor-specific plans limit the opportunities to tie everything together neatly.

There are vendor-independent standards for building cabling, known as *EIA/TIA*,[1] *568 A/B*, and *EIA/TIA 569*. The intent of these standards is to provide in effect a highest common denominator. That is, if the standards are adhered to, they should be able to support anything—presuming that the EIA/TIA has thought of it. If you are planning to comply with the EIA/TIA guidelines, you must make your own choices as to media.

All structured cabling standards address two basic types of wiring: vertical and horizontal. *Vertical wiring* is that cabling that connects wiring closets, each of which serves multiple users. Usually, the wiring closets to be connected are each on a different floor. With luck, they will be in the same position on each floor. See Figure 31.2 for a sample of the wiring closets stacked above each other in a high-rise office building. Thus, vertical wiring usually runs up and down; hence, the name. Vertical wiring is always of at least the same bandwidth as the horizontal wiring. More often than not, higher-speed media are used for vertical wiring because connecting up groups of users (with all of their traffic appearing on the vertical wiring) can require aggregate bandwidths far exceeding that required to satisfy individual user requirements.

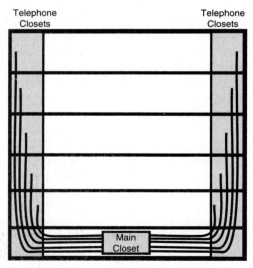

Telephone Closets   Telephone Closets

Main Closet

Figure 31.2 The telephone closets are typically stacked on top of each other in high-rise buildings. The vertical wiring feeds to each floor.

---

[1] EIA is the Electronics Industries Association; TIA is the Telecommunications Industry Association. These bodies set the standards for the structured wiring systems.

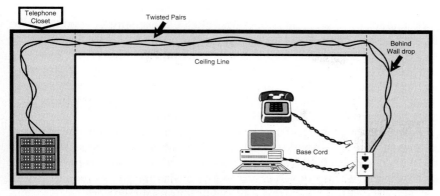

Figure 31.3 The components of the horizontal distribution.

Horizontal wiring is the wiring that runs from the wiring closet above the ceiling to the wall jack in each of the office locations. Figure 31.3 is a representation of the horizontal wiring system running to the wall jack. This includes such things as the cable, the drop inside or behind the wall, and the wall plate (jack). In many cases, the horizontal wiring will include the base cord that runs from the wall outlet and plugs into the PC or telephone set at the desk.

Interbuilding wiring is also often of concern to those wiring a campus or multibuilding environment. Like vertical wiring, it must carry higher bandwidths (only more so). It also drives special requirements, such as resistance to gophers or rodents and other environmental hazards, and presents special problems, such as appropriate grounding, when connecting between two different buildings' electrical systems.

The rest of this chapter presents some of the more likely choices currently available for these types of wiring.

Twisted Pair

Twisted pair refers to telephone wiring. This type of paired wiring usually comes as two solid copper wires, each covered with a plastic insulator, and the two wires are twisted together. It is not the "silver satin" cable often sold in hardware stores for use with telephones. *Silver satin* is a flat-wire system where all the wires are laid flat side-by-side, rather than twisted around each other in pairs. Twisted pair is often described as STP or UTP: shielded or unshielded twisted pair. UTP (Figure 31.4) has been primarily used for voice communications. It is twisted to reduce induction effects in adjacent wires, otherwise known as *crosstalk*. The more twists per foot, the better the reduction in the crosstalk. The twisted pair wiring is usually sold in various flavors, such as shown in Table 31.1. In these pairings, the number of pairs is always important because there are basic and immediate needs, then there are future needs. The best approach anyone can take when installing wiring to the desk or in the vertical distribution is to always add more.

These are typically color coded so that the pairing can be kept together with visual contact, rather than trying to figure out which wire goes with any other. This pairing color code is designed around a striped wire and a solid color wire creating a pair, a striped wire and a reverse striped color of the same combination, or in older wiring systems such as the old telephone wiring in households, it is based on a solid-color wire convention. Table 31.2 shows the more common ones. Where the convention uses a striped color code, this is represented by the two colors on the wire insulation (i.e., white/blue means a white wire with a blue stripe, blue/white means a blue wire with a white stripe). In the reverse wiring, the color code uses the two colors in the reverse; however, the normal wire is a striped and a solid wire (i.e., white/blue is a white wire with blue stripes and blue is a solid blue color wire).

If installing a telephone set (for example) that requires one pair, it would be imprudent to run the wires to the set with only one pair enclosed. Otherwise, the first thing that happens if you choose to change the device hanging off these wires is to rewire with additional pairs. Telephone systems typically can be used that require from one to four pairs at the desk. Thus, the minimum number of pairs to the desk is four. We have always adopted a standard that a minimum of two each of the four pair wiring be installed. After all, the cost of adding a few hundred feet of wire at the time of initial installation is only a few cents per foot. If, however, rewiring is necessary, the labor to go back and pull more wires will cost a significant amount (Table 31.3).

Figure 31.4 The unshielded twisted pair wiring.

TABLE 31.1 Typical Number of Pairs Used for Installation

Description	Number of pairs	Used
Analog telephone wiring	Two pairs	One pair
Digital telephone wiring	One to four pairs	One to four pairs
LAN wiring Ethernet 10BASE-T	Four pairs	Two pairs
LAN wiring Token Ring 4 or 16 Mbits/s	Four pairs	Two pairs
Horizontal telephone wires	25 pairs	One to four for individual sets

TABLE 31.2 The Color Coding for Normal Pairings

Pair number	Color coded pairing
One	White wire/blue stripe and blue wire, or white wire/blue stripe and blue wire with white stripe
Two	White wire/orange stripe and orange wire, or white/orange and orange/white
Three	White green/green or white/green and green/white
Four	White/brown and brown or white/brown and brown/white
	Older analog wiring found still in some buildings use a four-wire convention that used these color coded conventions
One	Red/green Red was the transmit and green is the receive wire
Two	Black/yellow Black is the battery and yellow is the ground for some sets, and if a two line set is used then black is transmit 2 and yellow is the receive 2

TABLE 31.3 Comparing the Difference Between Pulling a Single Cable Twice vs. Two Cables Once Shows the Real Dollar Value of Preplanning. The Main Cost Difference Is in the Labor.

Item	Cost of pulling one cable to each desk (each)	Cost of pulling two cables to each desk (each × 2)
Materials (4 pair cable, block, RJ45, cross-connect wire, base cord)	$ 9.35	$ 18.70
Labor to pull wires, punch down and test	$ 300.00	$ 320.00
Total cost per run	$ 309.35	$ 338.70
@ 50 pulls	$15,467.50	$16,935.00
Cost for additional 50 installs	$15,467.50	$0
Total cost	$30,935.00	$16,935.00
Difference		($14,000.00)

UTP is a "balanced" wiring system. This means that both wires carry similar types of electrical signals, and have similar characteristics. No special grounding is required. With the appropriate adapter, UTP can be used to support nonremote terminal traffic that would otherwise require RS232 cabling. Shielded twisted pair (STP) (Figure 31.5) was endorsed by IBM at about the same time that the company released the 4-Mbits/s Token Ring. STP was better for data communications than was early UTP—but only if

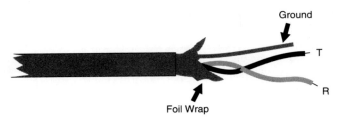

Ground

T

R

Foil Wrap

Figure 31.5 Shielded twisted pair include the foil wrap that is connected to a good earth ground.

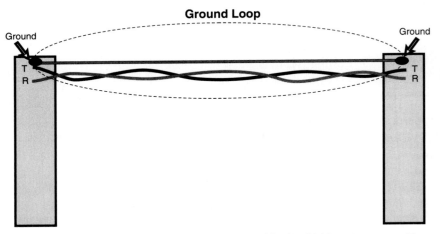

Ground Loop

Ground

Ground

T
R

T
R

Figure 31.6 If grounded on both ends at different potentials, the shield creates a ground loop. This is worse than no ground at all.

installed correctly. The shield—a foil sheath covering the sheathed copper pairs—must be properly grounded. If not, the shield can create a *ground loop* (Figure 31.6). This situation introduces the possibility for a wiring system to become an antenna, and is worse than if no shielding were used. STP is also more costly than UTP, both because of the extra foil media shielding and because the jacks and plugs cost significantly more than those used with UTP. STP was generally only used for Token Ring installations, before appropriate UTP grades and components became available. The shield, if properly installed, will provide better cable performance, such as reduced crosstalk, protection from grounds and faults, greater signal distances and protection from EMI and RFI interferences.[2]

---

[2] EMI refers to electromechanical interference, RFI refers to radio frequency interference. These are the conditions that are introduced into the copper on the twisted pair wiring.

AT&T and others experienced many queries from customers as to transmission characteristics of various UTP offerings. The company's specification sheets provided the desired information, but in a format that only an electrical engineering graduate could understand. To make buying products easier, vendors developed a set of simplified data grade wire specifications based on the underlying detailed electrical characteristics of the wire. Initially supported by the Underwriters Laboratories (UL), there were three levels, one through three. As more wiring products became available and needed, the company extended the grading scheme. Now there are five defined levels or data grades for UTP wiring, each capable of handling its own requirements as well as those addressed by all lower levels:

- Level 1, sometimes described as "barbed wire" will handle the requirements of voice communications, called plain old telephone service (POTS) and limited data rates that meet EIA 232 specific data rates (such as 19.2 Kbits/s) and that's it.
- Level 2 is certified to handle the requirements of a 4-Mbits/s Token Ring.
- Level 3 addresses Ethernet's requirements or 10 Mbits/s on twisted pairs. The specifications from various vendors suggest that this specification will support higher rates, such as 16 and 20 Mbits/s, but this is not an EIA standard.
- Level 4 is for 16-Mbits/s Token Ring speeds and the 20-Mbits/s ARCNet service. IBM is planning to introduce a 25-Mbits/s Token-Passing Ring concept, at duplex arrangements that will be called *ATM*. The specification is to use Level 3 or 4 wiring to support these speeds.
- Level 5 is certified to function properly up to 100 Mbits/s, protocol unspecified, so long as no cable run exceeds 100 meters (328 feet combined runs from the closet to the workstation). The industry is now touting that Level 5 will run up to 155 Mbits/s speeds that are supported by ATM standards.

The term "level" applies only to the wire itself. The entire system is described as "Category n" where n matches the wiring level. The entire system includes jacks, plugs, and patch panels in addition to the wiring itself. These components also need to be certified as meeting Category "whatever" requirements. Because Category 5 components cost somewhat more to install, people sometimes install Category 5 wiring, but only Category 3 components. Because replacing the components is much easier than rewiring a building, these people expect to replace the components later, if required.

Some LAN technologies require 100 Mbits/s. These include two varieties of fast Ethernet, as well as a standard for FDDI3 that runs on Category 5

UTP, now called *TPDDI* and *CDDI*.[3] No matter what LAN topology is used, UTP is almost always configured with a radial wiring plan, rarely as a physical bus. Radial (or star—radiating out from a central point) fits well into building cabling troughs, a fact that in large part accounts for the popularity of UTP data wiring.

UTP offers the lowest medium cost per foot, installed. Of available technologies, it is the most limited in distance because of the nature of the medium. For every medium, there is an inverse relationship between distance and potential speed. The specific ratio varies with media. For any given speed, UTP has the shortest distance limitation of the media discussed in this chapter.

However, there are many suppliers of UTP, and installers are familiar with it. Because of its popularity, vendors focus on getting their products to function on this type of physical plant. Unless there is a possibility that a company will need speeds greater than 100 to 155 Mbits/s to the desktop— or that cable runs will have to exceed 100 meters—a Category 5 installation is likely to handle all requirements for at least the next five years or so. If you are planning a Category 5 implementation, however, be sure to get installers experienced with Category 5, ones with more than just experience with basic telephone wiring. Installing Category 5 is not particularly difficult—but it is different than the other wiring levels. Specific considerations must be adhered to when installing this type of wiring or all efforts will be for naught. Improperly installed, the Category 5 cabling can be negated and functionally only provide speeds based on the weakest link in the system, such as Category 1. A lot of money can be spent on the wiring system only to be left with an infrastructure that doesn't meet the expectations. Or, retrofits that are expensive might be required to bring the wiring up to standard. This can be a devastating financial burden to the organization.

Twisted pair wire is typically cabled using standardized jacks and plugs. Originally standardized by AT&T, the designations of these jacks are easily understood symbols (such as RJ-11, RJ-45, etc.). We assume that someone found them easy to understand, anyway. In any case, the key differentiating factors among these jacks and plugs are:

- How many pairs of wire can terminate on the jack (two, three, or four)?
- How many pairs of wire do terminate on the jack?
- How are the pairs assigned?

Every electrical signal path on UTP requires one pair of wires. If two signal paths are required (for example, digital sending and receiving), then two

---

[3] FDDI stands for Fiber Distributed Data Interface, a specification for 100 Mbits/s on multimode fiberoptics.

pairs are required. But those pairs most likely will not be assigned in sequence (for example, left to right or right to left); that would be too easy. As an example, voice wiring on an RJ-11 jack (that can hold two pairs of wire, and thus handle two analog voice lines) typically uses the center pair of pins or wires on the jack for the first line, and the outer pair for the second line if present. If only one line is to be installed, sometimes the outer pair will not even be present on the jacks and plugs—even though the RJ-11 jacks are still used.

Every defined use of UTP (for example, 10BASE-T for Ethernet) includes a very precise specification of which positions on what jack will be used for which transmission path. 10BASE-T is different from Token Ring, and both differ from the standard for FDDI[4] over UTP, even though all of these are defined on RJ-45 jacks and plugs. Refer to Figure 31.7 for a representation of the various ways that the wires can be used in the jacks based on differing standards.

Coax

Coaxial cable is decreasing in popularity. Before fiber was developed, it was the only practical medium to achieve high bandwidths over distances exceeding a few hundred feet. Whereas a high-speed signal (up to about 100

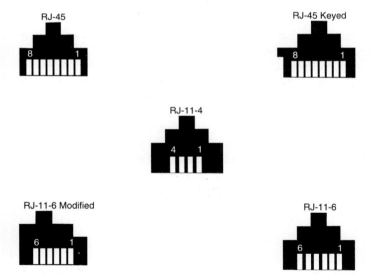

RJ-45

RJ-45 Keyed

RJ-11-4

RJ-11-6 Modified

RJ-11-6

Figure 31.7 The specific number of pairs and the pinouts differ, based on its use.

---

[4] TPDDI is the specification for the 100 Mbits/s data standard on Twisted Pair Distributed Data Interfaces, whereas the CDDI or Copper Distributed Data Interface is a vendor-specific standard to support the 100-Mbits/s data rate on twisted pair wiring.

Mbits/s) can go via UTP up to 100 meters or so, some forms of coaxial cable can carry signals requiring several hundred Mbits/s a few thousand meters. The two basic forms of coaxial cable are baseband and broadband. Baseband is used for an Ethernet connection and broadband is the typical cable used in CATV systems. The use of baseband versus broadband systems was covered in an earlier chapter.

Coaxial cable is an "unbalanced" medium; the signal goes through one conductor (the inner core), while the other conductor (the outer braid) functions as ground. Signals can be converted from a balanced to an unbalanced medium (and back) via use of a special passive device called a *balun* (*bal*anced and *un*balanced)—so long as the number of required conductors matches. This technique is used to supplement coaxial cable with twisted pair wires as an economic measure. Coaxial ends might be used on a cable run, but UTP wires are used in the middle for the vertical or horizontal run to save money in the wiring of a building. See Figure 31.8 for a typical installation using the balun.

All coaxial cables have the same basic construction. Each has an inner conductive core, usually of copper (usually solid but sometimes stranded). Around the core is the dielectric, an insulating material. Around the dielectric is a second conductive path or sheath, usually in the form of braided copper. This basic construction is illustrated in Figure 31.9 for baseband coaxial cable, and Figure 31.10 for the broadband coaxial cable structure. As with other available media, the underlying construction is varied in many ways to allow customization for specific purposes.

Variations in the thickness of the three key components, as well as in the type or types of covering used, result in a bewildering array of possible cables. Each specific type has a designation (for example, RG58A) that completely specifies its characteristics. Each type of LAN or other communications system standard typically includes a specification of on what types of cable that system is supported.

Coaxial cable now is rarely used for horizontal wiring, although Ethernet can use thin wire coax in this way. The reason is that the cost per port for thin wire is much higher than for UTP. On the other hand, coax is the second-best choice (after fiber) for vertical wiring. Its high bandwidth and its ability to handle longer distances fit the requirements of vertical wiring well. Also, it is easier to splice and repair than fiber.

Figure 31.8 The balun is used to run less expensive twisted pairs to a coax interface.

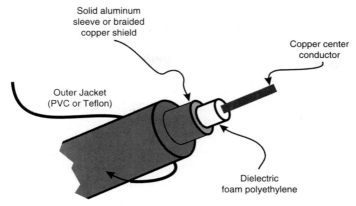

Figure 31.9 The baseband coax construction.

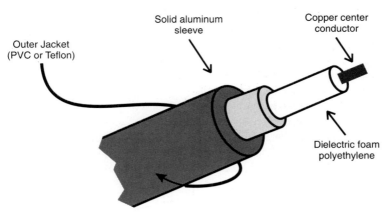

Figure 31.10 The broadband coax construction.

Coax has in the past been the first choice for inter-building wiring—that is, before fiber became generally available. But, with the exception of ease of repair and termination, fiber has now supplanted coax as the medium of choice for external wiring.

Fiberoptics

Fiberoptics is the medium of the 21st century, available now. It is a technology everyone seems to be welcoming with open arms. The available bandwidth, the potential flexibility, the range of applications available, and the ring topology all wide-area carriers appear to be installing, lend credibility to the widespread use of fiber. Considering that the technology is relatively new both as regards deployment into private telecommunications networks, as well as by the local exchange and interexchange carriers, the acceptance of the technology is positive.

Fiber can be used both as a wide-area and as a local-area cabling medium. As a WAN medium it has no peer. Essentially all new long-distance cable runs are now built using fiber technologies. Most of the interexchange carriers have adopted a ring topology in their networks. Because the bandwidth is so readily available to them, they can provide automatic reroute capabilities along the reverse route. This is also the way many of the local exchange carriers are going. See Figure 31.11 for a sample layout of a carrier network with a fiber ring. This picture shows the resiliency of a ring in the event of a cable or equipment failure.

The ring topology is only one of the available possibilities. Organizations considering installing private fiber, whether LANs or WANs, must address the issue of which topology should be deployed. The three basic topologies are the star, ring, and dual counter-rotating ring network to deliver bandwidth to the user.

The star network is fine as long as a diverse route exists as protection against failure. However, if the star is single threaded, the user and the carrier are at greatest risk of failure. The use of a diverse route tends to drive the costs up for the use of this topology. Therefore, many users might be concerned with providing only a single feed as their primary route for local service, dedicated services, and access to the POPs.

Rings automatically provide the diverse route through a single access point. Vendor and user alike gain a lot of advantages by using a ring. In the

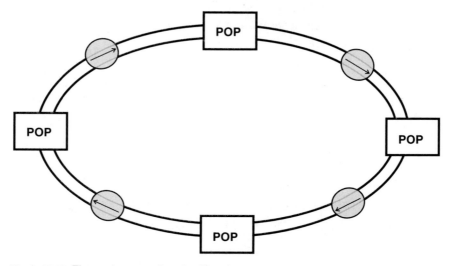

Figure 31.11 The carrier networks using fiber rings.

event of a failure of some component along the way, the ring can be automatically reversed to carry traffic in the opposite direction. Many a private user has adopted this technique. The cost is higher, but the risks are minimized. The critical point of failure might well be the entrance to the building, the common cable vault where all the services come together. Thus, before choosing to deploy fiber rings, a critical look has to be cast on the total route, and choke points eliminated.

The *dual counter-rotating ring (DCRR)* is the latest and greatest idea in the protection of bandwidth business. If the access into the building can be diverse, and a ring is established with two diverse fiber rings, running in two diverse routes to the same location, you have the DCRR. The first ring is installed normally, as any ring might be; the second is run through diverse entrances and along different conduit, pole, or direct buried routes. The most expensive solution to the network configuration, this technique is also the most versatile. Information (voice, data, graphics, images, etc.) is modulated onto both fibers and transmitted to the receiving end simultaneously. The receiver, getting both transmissions, evaluates the information and selects the best signal for use. In the event that a problem occurs along the way, causing disruptions to the signal, the receiver will only receive from one source. Therefore, the decision to select is null; the receiver uses the only signal it received.

Fiber comes in both glass and plastic varieties. Glass goes faster, but for reasonably short distances (up to about three kilometers) plastic suffices nicely. The upper bandwidth limitation on the best of the glass offerings has yet to be determined, but it is in excess of 13 gigabits (13 billion bits per second). See Figure 31.12 for a representation of the composition of the fiber. Interestingly, the gating factor limiting the speed is how quickly the electronics at the sending and receiving end can code and decode the signals. It is less difficult to generate signals much faster than can currently be recognized and processed.

Fiberoptic systems transmit digital input signals over modulated light beams (lights switched on and off work nicely for ones and zeros) that pass through the fibers. Except at cable ends, where electrical signals are converted into light for transmission, no electricity is involved. Thus, signals on fiber can neither interfere with, transmit, nor be affected by, any surrounding electromagnetic or electrical signals. This implies that:

- Fiber signals are immune to interference
- Fiber signals cannot generate any electrical interference
- Fiber signals are (almost) immune to security violations, at least while on the fiber

Glass fiber comes in two basic types: single mode and multimode. (All plastic fiber is multimode.)

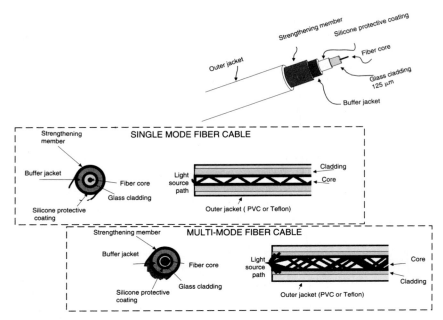

Figure 31.12 The composition of optical fiber.

Single mode is better. It is the type that has no known upper bandwidth limitation.[5] However, it costs more and is generally overkill for non-WAN applications. Telcos and long-distance carriers use it for runs of hundreds or even thousands of miles without the need of using repeaters. The most recent single mode Trans-Atlantic cable was installed over three thousand miles without any repeaters. This is obviously important because the repeaters require the same electronic componentry as the ends, which is quite expensive. Normally end users do not include single-mode fiber in their environments, but price decreases in the electronics and the fiber are making this far more attractive. Clearly, this is happening in major campus environments such as:

- Hospitals
- Military installations
- Corporate parks
- High-rise buildings
- Others

---

[5] The theoretical limit of single-mode fiber is 100 terabits per second (Tbps), although no means exist to test this speed.

As the price of the glass drops and the cost of the electronics moves closer to that of multimode fiber, one will expect to see far more of this installed. The carriers have moved to high end electronics today that are specified to carry speed on a standard called *SONET (synchronous optical networks)*. SONET specifies the speeds in which the equipment can multiplex signals from various sources into high-speed carrier services (Table 31.4). Today, we have commercially available products that will multiplex up to 2.488 Gbits/s onto the single-mode fiber.

The upper speed limit on multimode cable is currently under one gigabit, fast enough for most applications. See Table 31.5 for a summary of the multimode speeds based on the thickness of the glass. This is something that has been around for some time now, and evolutions on the thickness of the glass are continually moving up. The various thicknesses listed in the table are based on just that evolution. The current multimode fiber predominantly used is on 62.5 micron glass and uses a 125-micron outer cladding. Other standards exist, but the specifications for FDDI and other associated protocols call for the 62.5 by 125. The cable costs are getting to the point of being equal these days, but the electronics for the multimode fiber are significantly less expensive. How long that will continue is only a mystery. However, we do know from the chapters that you have already gone through thus far, the ATM, SMDS, and other higher-speed transport systems are all proliferating the use of the higher speed transport systems. As this continues and other technologies roll out, you can expect that the fiber will become a commodity-priced item.

TABLE 31.4 Typical Speeds that Are Currently Specified on Single Mode Fiber

Single-mode fiber 10 MC using SONET standards (the range of thickness is between 8.3–10 microns)									
OC1	OC3	OC9	OC12	OC24	OC36	OC48	OC96	OC192	OC255
51.84 Mbits/s	155.52 Mbits/s	466.56 Mbits/s	622.08 Mbits/s	1.244 Gbits/s	1.866 Gbits/s	2.488 Gbits/s	4.976 Gbits/s	9.953 Gbits/s	13.92 Gbits/s

TABLE 31.5 Typical Speeds for the Various Thickness of Multimode Fiber

Multimode fiber	
Microns/thickness	Typical speeds
100–140	40–100 Mbits/s
62.5–125	100–565 Mbits/s
50–125	466–622 Mbits/s

Applications for the Use of Fiber

Table 31.6 is a summary of the driving applications that will cause the proliferation of more and more fiber in the carrier and the end-user networks. Each of these applications can be satisfied by the current technologies and transport systems. However, as the needs of the integration of various pieces of these applications continues, the current speeds will not be enough. Thus, the computers we know today will demand more speed and larger file transfers. The video to the desktop or onto a LAN will get closer and closer to reality for the masses. The wiring infrastructure must, therefore, be prepared to support raw bandwidth to the desktop, rather than a slower shared medium among multiple users.

TABLE 31.6 Application Demands for the Speed of Fiber

Uses of fiber systems
Voice: Although voice does not require the speeds of fiber in itself, it is the multiplexing of thousands of voice calls onto a single fiber that will aid in the deployment.
Data: Data speeds will continue to increase beyond what we currently use. At dial today the user can get 28.8–56 Kbits/s depending on the network used. Future speeds in the multimegabits per second will not be uncommon.
Video: Real time video requires 243 Mbits/s which at today's pricing is unrealistic and prohibitively priced. However, video to the desk for simultaneous conversations at dozens to hundreds of desks will require much higher aggregated bandwidth than only fiber can deliver.
Computer conferencing: Host-to-host and client server computing systems are going to demand much more from our communications infrastructure.
CAD/CAM: CAD files are getting larger every day and more complex. Newer CAD systems produce three-dimensional, solid graphics. These files that were once a few hundred kilobytes are now in the hundreds of megabytes. Futures may be in the gigabytes or terabytes.
LAN to LAN: Transparency across LAN boundaries at speeds approximating 155 Mbits/s to every desk will obviate all other technological advances and demand the fiber to the desk.
LAN to WAN: Just as LAN boundaries need to be crossed, the wider area will be stressed to keep up with the demands of the future bandwidth.
HOST-HOST already mentioned but large data transfers and real time journaling and backup between host computers will keep up the pace on the bandwidth needs.
Multimedia: Linking voice, data, LAN, and video traffic all together simultaneously will be a big hit in the future. The fiber in the backbone and to the desk will make it happen.
Medical imaging: Teleradiology, tele-medicine and other imaging systems (CAT, MRI, etc.) will all demand the capacities of the fiber to be placed to any location and at any desk for on-demand dial-up services.

Fiber Differences

There are actually two types of materials in any fiber, both glass (or both plastic) but with different indexes of refraction. The inner core is actually where the signal is carried; the surrounding material, called the *cladding*, has a reflective inner surface (by virtue of the different refraction index) that keeps most of the light within the inner core.

Fiber comes in various sizes, all very small. One measures the diameters of both types of fibers in microns (a micron is one millionth of a meter). The diameters are always specified with two numbers, separated by a slash or dash. The first number is the diameter of the inner core; the other measures the outer diameter of the cladding. The most popular and most used multimode fiber's specification is 62.5/125; the generally used monomode fiber has an eight micron center but a similar outer diameter, so its specification is 8/125.

Fiber is the medium of choice for interbuilding wiring because of its non-electrical nature (eliminating interbuilding grounding issues, as well as, if buried, being entirely immune to the effects of lightning strikes) and the great distances it can cover without intermediate repeaters. It is also often used for vertical wiring, both for its bandwidth and its electrical immunity. Its small size also recommends it for this application—especially in buildings whose risers are already mostly full.

Why would one not use fiber wherever possible?

- The material still costs somewhat more than UTP.
- Fewer people know how to work with it.
- Splicing and terminating remains difficult, although this is improving with new automated systems developed for the purpose.
- Components using it are not generally available for all types of LANs—although they are for Ethernet.

LANs implemented on fiber do not necessarily operate any more quickly than those based on UTP. For example, Ethernet runs at 10 Mbits/s, period. To gain the speed benefits of fiber, a different type of LAN must be used, implying the need for more expensive active components throughout the network. Use of fiber effectively eliminates the upper speed limit for future uses of the LAN cabling; only the active devices would require upgrading to increase operational speeds. Category 5 systems can operate LANs up to 100 Mbits/s (but only up to about 100 meters away from a hub), faster than most communications managers think they will have to go in the near future.

Fiber Futures and Risks

A serious study is underway among many of the research houses, the long-haul carriers, and many universities to determine the impacts of an all-fiber network, using the various topologies outlined above. Expect more information and future developments in this arena in the near future. However, a problem with some of the older fiberoptic cables is starting to rear its ugly head. The fiber installed in Florida several years ago has since started to cloud, which will impede the light source from getting through the cable. This could be a forerunner of problems to come. If all the fiber being installed today has short life cycles, then we might have to look for a new technology for the future.

Additional work is also being done on the automatic recovery of networks. The bandwidth on fiber allows the carriers the luxury of completely protecting their networks. You might remember the disaster we wrote about earlier, regarding the cable cuts in October 1990, in Illinois. Had the dual-feed ring been in place, the problem would have only amounted to an operational hiccup. Instead, because the fiber ring was not completed, the outages were significant.

Research laboratories everywhere are currently looking at a process whereby they will create a solution to our "backhoe fade" problem on fiber. Free Space Optics is a technology being developed that will give us the bandwidth of the fiber (1.2 to 13.9G) over an airborne media. We can expect this to really take off if and when it becomes commercially available.

Another system, Aerial Fiber, is now being used in the industry. Fiber cables wrapped in steel strands are replacing the ground wire (or guide wire) for the electrical utilities along their rights of way. The fiber is strong enough, when wrapped in the steel stranded outer jacket, to sustain without damage a lightning-hit to a pole. Extra cable is left along the route to allow enough slack for a pole to fall to the ground without breaking the cable.

Index

Illustrations are indicated in **boldface**.

About the Authors

REGIS J. (BUD) BATES, JR., has more than 27 years of experience in the telecommunications and management information systems, with specific experience in end-user management, systems integration, disaster avoidance and recovery planning, strategic planning, and cost containment or reduction. Mr. Bates is the president of TC International Consulting, Inc., in Phoenix, Arizona—a management consulting firm specializing in voice and data, LAN communications services and implementation, technical training development and instruction, and consulting services at high levels of domestic and international organizations. He is an active lecturer (having presented to nearly 15,000 users) and author of numerous articles and books, including *Disaster Recovery Planning: Networks, Telecommunications and Data Communications* (McGraw-Hill, 1990), *Disaster Recovery for LANs: A Planning and Action Guide* (McGraw-Hill, 1993); and *Introduction to T1/T3 Networking* (Artech House, 1992).

DONALD GREGORY has over 18 years of information processing experience and is presently senior consultant with CSC Partners, a consulting subsidiary of Computer Sciences Corporation.

About the Series

McGraw-Hill Series on Computer Communications is McGraw-Hill's primary vehicle for providing communications professionals with timely concepts, solutions, and applications. Jay Ranade, series advisor and editor in chief of J. Ranade IBM, DEC, and Workstation Series has more than 125 published titles in various series.